SOLVING TRANSCENDENTAL EQUATIONS

SOLVING TRANSCENDENTAL EQUATIONS

The Chebyshev Polynomial Proxy and Other Numerical Rootfinders, Perturbation Series, and Oracles

John P. Boyd
University of Michigan
Ann Arbor, Michigan

Society for Industrial and Applied Mathematics
Philadelphia

10 9 8 7 6 5 4 3 2 1

Library of Congress Cataloging-in-Publication Data

Boyd, John P. (John Philip), 1951- author.
Solving transcendental equations : the Chebyshev polynomial proxy and other numerical rootfinders, perturbation series, and oracles / John P. Boyd, University of Michigan, Ann Arbor, Michigan.
pages cm. -- (Other titles in applied mathematics)
Includes bibliographical references and index.
ISBN 978-1-611973-51-8
1. Transcendental functions. 2. Chebyshev polynomials. 3. Transcendental numbers. I. Title.
QA353.T7B69 2014
515'.22--dc23 2014017078

Contents

Preface

"Rootfinding was much more important than we expected ... Zeros, maxima, minima—all depend on rootfinding."

—Lloyd N. "Nick" Trefethen, FRS, U.S. National Academy of Engineering, Past President of SIAM, at the Workshop of the *Chebfun* Project, Oxford, September, 2012.

The goal of this book is to teach the art of finding the root of a single algebraic equation or a pair of such equations. We shall quote theorems and theory as necessary, but this is not primarily a book of theorems and lemmas. No rootfinder was ever bailed out by a Beppo-Levi space. It is rather a narrative cookbook. By "cookbook," we mean that it is filled with simple and practical recipes for solving various classes of nonlinear equations. By "narrative," we mean that the recipes are not presented in isolation from one another but rather are woven into a connected whole. At the expense of a little redundancy, the book is structured so that one can dip into a single chapter and find a recipe, but collectively the concatenation of algorithms and how and why they work tell deeper stories about approximation, inversion, asymptotics, and iteration, as in a traditional textbook.

Rootfinding is fundamental to junior high algebra, but its apparent simplicity is deceiving. Mathematicians know that a subset of univariate zero-hunting—polynomial equations—spawned the immense power of group theory through Evariste Galois and his successors. Engineers and applied mathematicians know that the classroom exercise of generating a single number by bisection or the secant iteration is as far removed from engineering reality as a military ball is from trench warfare. Scientists do not want just numbers, mere isolated roots, but rather need to trace complete solution branches, including multiple root curves and the bifurcation points that connect them. Solutions to $f(x)=0$ are not isolated but are plane curves when there is a single parameter λ, the zero-valued contours of the bivariate function $f(x;\lambda)$. When the root depends upon two parameters (μ,λ), the zeros form a solution surface in the three-dimensional space spanned by (x,λ,μ), a manifold whose folds and foldings are the topological raw material of catastrophe theory. Zeros are easily missed in a thicket of intersecting solution branches and surfaces. The author's interest in rootfinding began with such a fiasco.

As a graduate student nearly 40 years ago, I solved an ODE eigenproblem with an interior singularity by marching by a Runge–Kutta scheme from each boundary to near the singularity, filling the gap with a power series. "Shooting," as this ODE-solving strategy is called, requires solving a single transcendental equation in a single unknown (the eigenvalue) to smoothly join the two half-solutions. Easy, seemingly, even though the roots were complex valued and many. Later, I found a better way, which was to apply a Chebyshev polynomial spectral method on a path that detoured off the real axis around the singularity—and discovered I had missed roots and therefore entire eigenmodes. Oops!

The Chebyshev-proxy method, which is described here in the form of a greatly expanded version of my *SIAM Review* article [96], has, I am happy to say, removed most of the uncertainty in the computation of all the zeros of an analytic function $f(x)$ on a real interval $x \in [a,b]$. The book gives an extensive discussion of similar algorithms for searching for zeros in disks or other areas of the complex plane. It also describes adaptive Chebyshev polynomial interpolation, which generates the proxies, generalizations to an unbounded interval, and symmetry-exploiting variants.

But what if the solution branch is singular at one or both endpoints, or $f(x)$ itself has pathologies? There is a sort of arithmurgical Murphy's law that solution branches end in singularities quite often in applications—see the humble arcsine, arccosine, and arctangent, which combine singularities with an infinite number of branches. Fortunately, modern Chebyshev polynomial and Fourier spectral methods are up to almost anything, as explained throughout the book with a profusion of examples.

An infinity of roots is best handled by perturbation theory. My student-self of 40 years ago is intensely jealous of my ready access to modern computer algebra systems like Maple, Mathematica, and Reduce, which can calculate a perturbation series to 200 terms in a few seconds and then extend the domain of convergence by forming Padé approximants and Hermite–Padé ("Shafer") approximants from the series.

Consequently, this book has a chapter entirely devoted to regular perturbation series for roots. It is unusual to combine numerical algorithms with perturbation series and computer algebra in a single volume, but it is highly appropriate here. The roots of J-type Bessel functions, for example, are best calculated by a mathematical partnership: the Chebyshev-proxy-companion matrix method for small zeros and asymptotic series for the infinite number of larger roots.

Murphy's law applies to analytical methods, too, of course, so there is also a chapter on *singular* perturbation series. These may be loosely defined as expansions that combine powers of ϵ, the small parameter, with powers of $\log(\epsilon)$ or perhaps $\exp(-1/\epsilon)$, or even $\log(\log(\epsilon))$. It turns out that by using Newton's iteration to discover the form of the series and by employing a bivariate expansion in (ϵ, δ) as though ϵ and $\delta \equiv \exp(-1/\epsilon)$ are two completely independent parameters, singular perturbation series can be very effective, too.

Computer algebra also greatly increases the range of explicit methods; Maple and Mathematica can often find solutions in an instant that would baffle the mathematician. But an explicit solution in terms of the Lambert W-function is not very useful if you know nothing about $W(\lambda)$, and human intuition can still give semi-intelligent software essential succor. The book therefore has a chapter on explicit solutions: polynomials, trigonometric polynomials and more exotic transcendental functions all fall to analytic strategies at least sometimes.

The chapters on the cubic, quartic, and quadratic equations were also the result of authorial epiphanies. For 40 years, I've investigated a class of planetary-scale oceanic and atmosphere and traveling waves called "equatorial waves." Each of the three wave species travels at a speed that is a real root of a cubic equation. I had studied Rossby and gravity waves for over a decade, published a dozen papers, and attended multiple Equatorial Theoretical Panel workshops before I discovered, in an article by Arthur Loesch, that there was a lovely arccosine formula that gives, without error, all three real roots of a cubic equation. It is not merely a personal failure, but a failure of the whole community, that equatorial dynamicists calculated phase speeds by crude approximations or Newton's iteration when something much better was known in the nineteenth century. Yet it is not even mentioned in the *Handbook of Mathematical Functions* and its ilk.

My second epiphany came from reading early chapters of Werner Rheinboldt's *Numerical Analysis of Nonlinear Parameterized Equations*. He explained that by dividing out the coefficient of the highest power and then substituting $x \rightarrow y+a$, where a is chosen to eliminate the coefficient of y^2, the general cubic equation can be trivially condensed to the "reduced cubic," $y^3 - \tau y + \xi = 0$. I had learned roots only as disembodied numbers; even the zeros of the same polynomial had no interrelationships. But Rheinboldt showed that all real roots in the entire τ-ξ parameter space lay on a single smooth surface in the three-dimensional space (τ, ξ, x). When the surface is folded, a vertical line through a point in the τ-ξ plane will intersect the solution surface at three points. In parts of the parameter plane where there is but one real root, the surface lacks folds. Double roots occur on the curves that are the edges of the folds where the slope of the surface is infinite. The origin $\tau = \xi = 0$ is the triple point where there are three equal roots, all zero.

In the 1960s, topologists had their own epiphany: folds and isolated triple points exhaust the possibilities for smooth manifolds with two parameters. Solution surfaces and equilibrium states of general transcendental functions can be smoothly mapped into the solution surface of the cubic equation. This topological approach to physics became known as "catastrophe theory"; the cubic polynomial equation is the prototype for the "cusp catastrophe." Similarly, the three-parameter surface for the zeros of the reduced quartic polynomial equation is the master pattern for the "swallowtail catastrophe". Hence our very thorough treatment of polynomial equations, especially those of low degree.

There are two reasons why this book, despite its title, has a long chapter on solving *pairs* of equations in *two* unknowns. First, many of the key ideas that have been so fruitful for univariate zero-hunting extend, at least haltingly, to bivariate systems, and it would be a pity to omit the extensions. The second reason is that the bifurcation points of univariate solution branches $x(\lambda)$ are solutions to a *pair* of equations, $f(x, \lambda) = 0, \partial f / \partial x(x, \lambda) = 0$, in the two unknowns (x, λ). Thus we can only complete our story of one equation in one unknown by discussing at least the basic strategies for zeros in two unknowns.

Oracles, expansions, iterations, polynomialization—Rootfinding is sometimes easy and sometimes soul-despairingly hard, especially if the goal is to find zeros not number by number but rather branch by branch. Even so, there is nothing in this book that requires more background than an engineer's standard undergraduate class. Previous completion of a numerical analysis class will allow some skipping and provide a useful perspective, but even that is not absolutely necessary.

The themes developed here are good background for assaulting the profusion of books that solve N equations in N unknowns. However, the special cases treated here—$N = 1$, with one chapter on $N = 2$—allow special tricks like the Chebyshev-proxy algorithm that have no counterparts for general N. Perhaps half the book is about general-N theory illustrated in the simplest form, and half is orthogonal to all the vast general-N literature. But this book is mostly a catalogue of what works. It includes many detailed case studies in this spirit.

The mathematician will find some deep themes and beautiful theorems here. Some interesting research topics and open questions are described along the way as well. However, the combination of the Chebyshev-proxy rootfinder with endgame iterations is very reliable.

In my other life in geophysical fluid dynamics and nonlinear waves, I solve problems and teach engineers how to solve problems. In that spirit, this work serves as a handbook of solving one important class of problems with as little style and finesse and as much reliability and conceptual simplicity as possible.

Good luck!

Acknowledgments. I was supported by the National Science Foundation for 35 years. I face a funding gap as this book is published but am still appreciative of years of past support. I also appreciate the several years of support from NASA through the kindness of my late colleague and department chair, Tom Donahue, who wrote me into his own contract until I could develop funding of my own. I was also supported for three years by the Applied Mathematics program of the U.S. Department of Energy.

I am appreciative of Robert Corless and three anonymous reviewers for their detailed comments, which were a big help in the war against typos and in broadening the usefulness of the book. I thank Nick Trefethen and Alex Townsend for their reviews and helpful remarks and encouragement. I am also grateful for friendship and intellectual stimulation of colleagues and students too numerous to name here, both at the University of Michigan and elsewhere, but individually appreciated and remembered.

Lastly, I thank the acquisitions editor, Elizabeth Greenspan, and the editorial and production staff at SIAM.

I dedicate this book to the memory of my brother, James C. Boyd, (1940–2011), my third parent, who taught me nothing about science and mathematics but lots about everything else. He was an elementary science teacher, assistant superintendent, and elementary school principal. His school honored him by naming its library after him. Although this book will never be in an elementary school library, I think this dedication would please him.

Notation

xEy	x a decimal, y a signed integer: scientific notation $x \times 10^y$
$f, \vec{f}$	Nonlinear equation whose roots are sought; arrow denotes a vector of equations
f_x, f_y	a subscripted coordinate denotes partial differentiation; $f_x = \partial f / \partial x$
n	number of unknowns in a multivariate system of equations
n_r	number of roots
$x, \vec{x}$	unknown; arrow denotes a vector of unknowns
$x_*, \vec{x}_*$	an exact root of $f(x)$ or $\vec{f}(\vec{x})$
$x_c, \vec{x}_c$	current approximation to a root of $f(x)$ or $\vec{f}(\vec{x})$
$x_s, \vec{x}_s$	point where the Jacobian matrix is singular
$\check{x}$	solution to easily solved $g(x)$ in artificial homotopy
Disc(f)	discriminant of the polynomial $f(x)$
$\mathcal{G}_K$	grid level K for the Clenshaw–Curtis nested Chebyshev–Lobatto grids
$\mathfrak{G}(x;N)$	Wilkinson polynomial of the second kind, $\prod_{k=1}^{N}\left(x - \frac{1}{2^{k-1}}\right)$
$\mathcal{H}$	artificial homotopy
$\vec{\vec{I}}$	Identity matrix
K	grid level in Clenshaw–Curtis adaptation ($N = 2^K + 1$ points)
$\vec{\vec{J}}$	Jacobian matrix for $\vec{f}(\vec{x})$
M	number of interpolations points on a subdomain
N	number of unknowns or polynomial degree
N_s	number of subdivisions
$\mathcal{N}$	$-\vec{\vec{J}}^{-1} f(x)$ "Newton vector"
P	total number of interpolation points on $x \in [-1, 1]$
$\mathcal{S}$	$-\vec{\vec{J}}^{T} f(u)$ "steepest descent vector"
Result($f, g; y$)	y-resultant of the polynomials (f, g)
T	pseudotime coordinate for Newton flow or other iterations-as-time-marching
$\mathfrak{W}(x;N)$	Wilkinson polynomial of the first kind, $\mathfrak{W} = \prod_{k=1}^{N}(x-k)$
$\mathfrak{W}_j$	jth window function of a partition of units
$\delta^{(m)}, \vec{\delta}^{(m)}$	correction to an approximate root $x^{(m)}$ or $\vec{u}^{(m)}$ at the mth iteration
γ	underrelaxation parameter
ϵ	small parameter that is also a perturbation parameter
ϵ_{mach}	"machine epsilon"; in MATLAB and also IEEE 754 double precision; $\epsilon_{mach} = 2.2 \times 10^{-16} = 2^{-52}$
λ	parameter
Λ	artificial homotopy parameter

μ	asymptotic rate of geometric convergence
μ_p	moments of f_x/f
ν_k	multiplicity of the kth roots
ρ	residual norm function (square of L_2 norm of $\vec{f}(\vec{u})$)
ξ	M/N_s
τ	pseudotimestep for integration of Newton flow and other flows
$\aleph$	small, user-choosable parameter that turns Newton's iteration into the Levenberg–Marquardt: $x \rightarrow x - f(df/dx)/\left\{[df/dx]^2 + \aleph\right\}$
$\beth$	user-chosen accuracy tolerance for adaptive interpolation
$\daleth$	domain in parameter space of a never-failing initialization

Part I

Introduction and Overview

Chapter 1

Introduction: Key Themes in Rootfinding

"If there exists any one reliable algorithm for finding the roots of transcendental equations, it is yet to be found. We have a variety of medicines that work with varying degrees of potency ..., but the state of the art still precludes the confident writing of computational prescriptions without having looked over the patient rather closely."
—Forman S. Acton (1970) [3]

1.1 ▪ Statement of the Problem

This book will focus on a simple problem: to find the zeros of a single function $f(x)$ in a single unknown:

$$f(x) = 0. \tag{1.1}$$

("Root" will be used interchangeably with "zero.") Often, the goal is to find all the real roots on a target interval, $x \in [a, b]$. This "real-on-an-interval" problem will receive special attention. The Chebyshev-proxy rootfinder (CPR) is a very robust black box for solving this problem and has never been previously explained in a book. However, we shall also thoroughly describe schemes for computing all roots, even if these are infinite in number. We shall also describe methods for calculating complex-valued zeros.

If $f(x)$ is zero at a point, but its first derivative is not zero at that point, then the root is said to be "simple" or a "first-order zero." If both the function and its first derivative are zero at a point, then the point is a "double zero" or a "second-order zero." If the function and its first and second derivatives all vanish simultaneously, then the point is a "triple root" or a "third-order zero." Collectively, double, triple, and higher-order roots are referred to as "multiple roots." The "multiplicity" of the root is one more than the number of derivatives that vanish simultaneously with $f(x)$.

In practice, finding one root is rather easy. Finding *all* the roots on an interval or region is much harder. The first aphorism of zero-hunting is

One root easy; many roots hard (1.2)

In applications, the unknown usually depends on one or more parameters. We shall usually display only a single parameter and denote it by λ. (When performing perturbation theory, we shall replace λ by the usual symbol for a perturbation variable, ϵ.)

The need to compute the *root curve* $x(\lambda)$ as the parameter varies is very important. First, it enables powerful new *continuation* algorithms in which the ability to find the zeros for a single, special value of λ can be extended to trace entire branches of roots. Second, changes in the number of roots at *limit points* and *bifurcation points* are not mere mathematical curiosities but usually have profound engineering and physical implications, too. The Chebyshev rootfinder is superb at finding all roots on an interval for a particular value of λ, but it approximates a solution branch only as what computer scientists call a *point cloud*: a set of discrete points (x_k, λ_k) devoid of information about connectivity and the topology of solution branches. Thus, bifurcation theory and continuation algorithms and theory remain important.

1.2 ▪ An Elementary Example: Polynomializing Transcendental Functions

Reliable black box solvers for univariate polynomials, and univariate polynomials only, have been known since the Jenkins–Traub algorithm [216] was incorporated into software libraries around 1970. But what is one to do with a transcendental function like

$$f(x) \equiv \sqrt{\pi/2}\exp(\lambda x) + \frac{x}{\lambda^2 + x^2}, \qquad \lambda = 0.8475, \tag{1.3}$$

even if only real roots are desired?

V. L. Zaguskin observed more than half a century ago that a *transcendental* function can be *polynomialized* by expanding $f(x)$ in a power series [382]. However, power series have a finite radius of convergence and yield very nonuniform in x errors. For this example, the Taylor series about the origin has a radius of convergence of 0.8475, which is very unfortunate since the root is $x_* = -0.8904994770969395$.

A better strategy is to expand $f(x)$ as a Chebyshev polynomial series. It is guaranteed to converge on the whole interval $x \in [a, b]$ provided only that $f(x)$ is singularity free on this interval and the error is nearly uniform over the whole targeted interval [68]. Table 1.1 shows that by approximating $f(x)$ on the interval $x \in [-1, 1]$, the zeros of a quintic polynomial approximate $x_* \approx -0.89$ within an absolute error of only 0.014; the Chebyshev series up to and including $T_{30}(x)$ yields the root to about 15 decimal places. Showing the root on the approximation interval in braces, the lowest few approximations are

$$\begin{aligned} f &\equiv \sqrt{\pi/2}\exp(0.8475x) + \frac{x}{0.71825625 + x^2} \qquad (1.4)\\ &\approx 1.4922 + 1.7760T_1(x) + 0.239T_2(x) \qquad \{-0.947\}\\ &\approx 1.4887 + 1.8930T_1(x) + 0.242T_2(x) - 0.117T_3(x) \qquad \{-0.802\}\\ &\approx 1.4887 + 1.8619T_1(x) + 0.239T_2(x) - 0.0858T_3(x) + 0.0035T_4(x) \qquad \{-0.85417\}\\ &\approx 1.4887 + 1.8685T_1(x) + 0.239T_2(x) - 0.125T_3(x) + 0.0035T_4(x) + 0.0329T_5(x) \qquad \{-0.90493\}. \end{aligned}$$

Later, we shall thoroughly explain the magic of Chebyshev expansions and the mechanics of extracting zeros directly from the Chebyshev series coefficients. However,

Table 1.1. *Absolute errors in the real root of Zaguskin's transcendental function as approximated by the zero on $x \in [-1, 1]$ of the Nth-degree Chebyshev interpolant of Zaguskin's function on this interval.*

N	2	5	10	15	20	30	40
Error	0.056	0.014	0.00038	6.1×10^{-6}	3.8×10^{-9}	7.9×10^{-11}	1.6×10^{-15}

classical iterative methods such as the secant iteration and Newton's iteration are very valuable as an "endgame" to refine zeros to an accuracy limited only by machine precision, thus purging all of the Chebyshev approximation error. Other ideas, such as continuation in a parameter, are very useful for connecting "point clouds" of discrete roots, computed for discrete values of a parameter λ, into *continuous* solution branches.

1.3 ▪ The Shakespeare–Weierstrass Principle: A Transcendental Function Is a Polynomial in Disguise

The theoretical foundation for polynomializing a transcendental equation is the following.

Theorem 1.1 (Weierstrass polynomial approximation theorem). *Every continuous function on an interval can be approximated, within an error ϵ as small as we please, by a polynomial of sufficiently high degree N with appropriately chosen coefficients* [353].

This proposition has been generalized to multivariate functions on irregular domains under the label "Stone–Weierstrass" theorem.

Weierstrass's own proof was nonconstructive. Bernstein proved the convergence of his approximation using what are now called Bernstein polynomials. Bernstein's approximation converges at a very slow rate, with errors falling proportional to $1/N$. When $f(x)$ is analytic, Chebyshev interpolation yields an approximation on an interval with an error that decreases exponentially fast with N as asserted by the Chebyshev convergence theorem (Theorem 2.1 in the next chapter). It is then easy to generate a polynomial $f_N(x)$ that is indistinguishable from the transcendental $f(x)$ to within machine precision, or very close to it.

The Weierstrass and Chebyshev interpolation theorems collectively imply a Big Idea, which is that to analyze and compute with a transcendental function, it is almost always sufficient to replace $f(x)$ by $f_N(x)$. CPR computes the zeros of $f(x)$ on an interval by solving the much easier problem of computing the roots of the polynomial *proxy* $f_N(x)$. However, the applications of $f(x) \rightarrow f_N(x)$ are sweeping and go far beyond rootfinding. The concept of polynomial for transcendental substitution is so important that we shall state this idea in multiple forms.

Nevertheless, the central concept, which we shall dub the "Shakespeare–Weierstrass principle," is the following:

Every transcendental function is a polynomial in disguise.

Like Michelangelo mastering marble-chipping to reveal the sculpture hidden inside the stone, we shall learn to peel away the disguise to expose the polynomial within.

The same concept can be stated more poetically by paraphrasing the English playwright and poet William Shakespeare: "All the world's a stage, and all the men and women merely players" (from *As You Like It*, Act II, Scene 7). The Weierstrassian paraphrase is as follows:

**All the world's a polynomial,
and most of numerical analysis is polynomial manipulation.**

Later, we shall restate this Shakespeare–Weierstrass concept in a more nuanced form as the Shamrock Principle. This observes that f_N is computed from a set of $(N+1)$ evaluations of $f(x)$ on a Chebyshev grid on the target interval $x \in [a,b]$, its *grid point values* or *samples*. The approximation is usually written as a truncated Chebyshev polynomial series with the $(N+1)$ nonzero coefficients a_n. Through the magic of interpolation and summation by the fast Fourier transform (FFT), the duality $f(x) \Leftrightarrow f_N(x)$ becomes the triune equivalence $f(x) \Leftrightarrow \{f(x_j)\} \Leftrightarrow \{a_n\}$.

This Big Idea of polynomial for transcendental substitution is realized most fully in the MATLAB extension *Chebfun*, an open software project initiated by L. N. Trefethen, that adds the capability to manipulate functions as if they were vectors, and linear differential and integral operators as if they were matrices [354, 37]. The reason this is possible is that *functions are vectors* in "shamrock thinking," which is also "*Chebfun* thinking." *Chebfun* converts transcendental functions to vectors of Chebyshev coefficients that define the polynomial that approximates $f(x)$ so closely that the difference is zero except in the twelfth decimal place. In the realm of Chebyshev spectral methods, differential and integral operators are (after discretization) merely matrices [353, 76]. The *Chebfun* system is now the work of many hands, a powerful tool for education and for almost any computation in one or two dimensions. But though the *Chebfun* website asserts this only implicitly, *Chebfun* is merely Shakespeare–Weierstrass/shamrock thinking carried to its logical conclusion. *Chebfun* is the polynomialization of *everything*.

1.4 ▪ Applications

Transcendental equations in one unknown arise in every field of science and engineering. A small but illuminating set of examples follows:

1. The Kepler equation for determining the eccentric anomaly in celestial mechanics is

$$E - \epsilon \sin(E) - M = 0, \tag{1.5}$$

   where ϵ is the eccentricity of an elliptical orbit, M is the mean anomaly, and the unknown E is the eccentric anomaly, sometimes denoted by ψ. Habash al-Hasid derived the same equation in the parallax problem in the ninth century. It also arises in hydrodynamics, as noted in the next item.

   Many luminaries have attacked this equation, including Newton (who invented Newton's iteration to solve it), Lagrange (who developed a power series in ϵ), Cauchy (who proved Lagrange's series converged only for $\epsilon < 0.66$), and Bessel (who invented Bessel functions to obtain the infinite series solution):

$$E = M + \sum_{m=1}^{\infty} \frac{2}{m} J_m(m\,\epsilon) \sin(mM), \qquad \epsilon \leq 1, \;\; M \in [-\pi, \pi]. \tag{1.6}$$

   Indeed, Colwell's history of the Kepler equation [130] contains 465 references, and it seems likely, allowing for more recent publications and obscure works, that at least 500 articles have been devoted to Kepler's equation.

2. The method of characteristics reduces a hyperbolic differential equation to an algebraic problem. With the initial condition $u(x,0) = Q(x)$, the inviscid Burger's equation, $u_t + u u_x = 0$, for example, has the exact but implicit solution

$$u(x,t) = Q(\xi(x,t)), \qquad \text{where}\, \xi(x,t) \text{ solves } x = \xi + Q(\xi)t. \tag{1.7}$$

It is necessary to solve the transcendental equation for ξ throughout the two-dimensional parameter space, the space-time plane. When $Q(x)$ is a sine function, the equation for ξ is the Kepler equation, where the eccentricity is replaced by the time coordinate, and the parabolic orbit case ($\epsilon = 1$) is the instant of wave-breaking [296, 50, 69].

3. Wave dispersion relations are a special sort of eigenproblem in that it is often possible to derive an explicit function whose root gives the phase speed of the wave. It is usually important to trace the dependence of the wave on a parameter, thus raising the issuing of parameterizing solution branches. The dispersion relation for modons in atmospheric and oceanic fluid mechanics is

$$f(x) \equiv \frac{J_2(x)}{xJ_1(x)} - \frac{K_2(\lambda)}{\lambda J_1(\lambda)}, \qquad \lambda \in [0, \infty], \tag{1.8}$$

where $J_n(x)$ and K_n denote Bessel functions [284], and it is known that there is an infinite number of real roots for any λ [53].

4. The contours or "isolines" of a bivariate function $f(x,y)$ are the solutions to $f(x,y) = c$, where c is a constant (the label of the isoline) and the other coordinate, y, plays the role of a parameter.

5. In computer graphics, curves may be defined either parametrically, as in $x = X(t)$ and $y = Y(t)$, or implicitly, as the zero isolines of a function $\Phi(x,y) = 0$. Both representations are useful, so a basic computer graphics operation is to derive an implicit representation from the parametric form. But tracing the contours of an implicit representation is again an exercise in rootfinding. Recovering the parametric form of a curve from the implicit function $\Phi(x,y)$ is just a particular subclass of the more general problem of parameterizing the branch $x(\lambda)$ of the solution to a function $f(x,\lambda)$ that depends on a parameter.

6. The maxima, minima, and saddle points of a differentiable function are, by first-year calculus, points where df/dx is zero. Thus, the first step in computing maxima and minima is to find all the zeros of df/dx, the second step is to evaluate $f(x)$ at these derivative zeros, and the final step is merely to compare these local maxima and minima to find the globally largest or smallest.

7. The method of steepest descent is a powerful method for asymptotic approximation of integrals and of functions with integral representations. The author alone has employed steepest descents in [51, 55, 58, 62, 64, 65, 70, 75, 74, 80, 89]. Every steepest descent approximation requires the zeros of the *phase function*. It is often possible to solve for the *stationary points perturbatively*, which yields an analytical steepest descent approximation that covers the whole parameter space, modulo the requirement that the large parameter is sufficiently large.

8. Many otherwise intractable nonlinear ODEs can be solved by separation of variables. For example, with P as a polynomial and $Q = \int dy/P(y)$,

$$\frac{du}{dx} = P(u) \qquad \longleftrightarrow \qquad dx = \int_0^x \frac{1}{P(y)} dy, \tag{1.9}$$

which gives the implicit solution

$$x = Q(u) \qquad \leftrightarrow \qquad u = Q^{-1}(x). \tag{1.10}$$

The implicit solution is useless unless one can readily invert the function $Q(u)$.

9. Amore and Fernandez derived several power series approximations to compute $r(t)$, the radius of a collapsing spherical bubble. This solves $r\,r_{tt}+(3/2)(r_t)^2+\xi^2 = 0$, $r(0)=1$, $r_t(0)=0$, where ξ is a parameter [11]. Their most rapidly converging series is not for $r(t)$, but for the *inverse* function, $t(r)$. Their best series is useless unless one can easily revert the power series to compute its *inverse*, which is an exercise in univariate rootfinding.

10. Differential equation and integral equation eigenproblems can be discretized by spectral or finite difference methods into matrix eigenproblems; the eigenvalue is then a root of the characteristic polynomial of the matrix. It is inefficient to actually form the characteristic polynomial; indeed, it turns out that a powerful way to compute the roots of the polynomial is to convert it back to a matrix eigenvalue problem via a so-called companion matrix. For some problems, such as the quantum square well, it is possible to bypass the matrix eigenproblem and derive an explicit transcendental equation for $x(\lambda)$. The energy levels in the quantum finite square well potential, for example, require the solution of

$$f(x) = \begin{cases} x\tan(x) - \sqrt{\lambda^2 - x^2} & \text{[Symmetric Modes]}, \\ x\cot(x) + \sqrt{\lambda^2 - x^2} & \text{[Antisymmetric Modes]}. \end{cases} \tag{1.11}$$

11. In a variety of applications, the eigenmatrix is a *nonlinear* function of the eigenparameter x. We shall later solve an example taken from Ruhe's highly readable review [300].

12. Gil, Segura, and Temme [195] write the following in their classic book *Numerical Methods for Special Functions*: "The zeros of special functions appear in a great number of applications in mathematics, physics, and engineering, from the computation of Gauss quadrature rules in the case of orthogonal polynomials to many applications in which boundary value problems for second order ordinary differential equations arise" (p. 191). It is therefore not surprising that they devote an entire chapter to special algorithms for finding zeros of special functions.

13. An ODE *boundary value* problem can be converted into an easier *initial value* problem by defining x to be the missing boundary condition $u_y(0)$ and solving the transcendental equation $f(x)=0$, which imposes the far boundary condition. For example, with the mixed boundary condition $u_y(b)+2u(b)=0$ and $u_{yy} + \exp(u[y]) = 0$ with $u(a)=1$, the rootfinding problem is to solve

$$f(x) \equiv u_y(b;x) + 2u(b;x) = 0, \tag{1.12}$$

where $u(y;x)$ is obtained by integrating the initial value problem using fourth-order Runge–Kutta with the initial conditions $u(a)=1$ and $u_y(a)=x$.

I was motivated to invent the CPR by just such a problem—a nasty one where the differential equation is nearly singular and the eigenvalues are complex valued. Years later, I discovered that I had missed roots and therefore eigenmodes because I had then used only local methods, as already recounted in the preface [54, 60].

14. Catastrophe theory is really just a catalog of solution surfaces that can be smoothly mapped to the root manifolds of quadratic, cubic, and quartic polynomial equations. We therefore devote whole chapters to solving such humble, seemingly high-schoolish, low-degree polynomial equations because these exemplary cases contain the topology of bifurcation, branching, and catastrophes.

15. A system of N polynomial equations in N unknowns can be solved by an algorithm that begins with computation of the Groebner basis in lexicographic order for the polynomial ideal defined by the system. The Groebner basis *triangularizes* the system in the sense that one element of the Groebner basis is univariate, another is a polynomial in just two variables, and so on. One can find all possible x-values by solving the univariate polynomial. Substituting these into the bivariate polynomial yields a univariate polynomial that can be solved for all possible y-values. Plugging all pairs of (x,y) into the third Groebner basis function yields a polynomial in z alone. And so it goes.

    The irony is that this method for solving systems of *many* equations in *many* unknowns depends on repeatedly solving a *single* equation in a *single* unknown.

16. Brown's method [117, 115, 116] and bootstrapping [46] are but two of a variety of algorithms that solve *systems* of equations *one* at a time. A system $f(x,y)=0$ and $g(x,y)=0$, for example, can be attacked by the iteration

$$x^{(m+1)} \equiv \text{solution to} \, f(x^{(m+1)}, y^{(m)}) = 0, \tag{1.13}$$

$$y^{(m+1)} \equiv \text{solution (for fixed} \, x^{(m+1)}\text{) to} \; g(x^{(m+1)}, y^{(m+1)}) = 0, \tag{1.14}$$

    where each step is the solution to a nonlinear univariate equation. The generalization to N equations in N unknowns is obvious.

17. The Lyapunov–Schmidt reduction projects a singular or nearly singular problem of N equations in N unknowns into a large but well-behaved system of $N-m$ unknowns, which can be solved for the nonsingular modes with the m singular variables as parameters; substitution of this partial solution into the projection onto the m singular degrees of freedom yields a tiny system in m unknowns, where $m \ll N$. Often, there is but a single dangerous degree of freedom, and then the system of many unknowns is Lyapunov–Schmidt reduced to finding all relevant zeros of a single nonlinear equation in a single unknown.

1.5 ▪ Key Themes

Several key ideas that will be developed in the rest of the book are briefly summarized:

- proxy-solving
- Chebyshev interpolation proxy
- iteration
- initialization of iterations
- bifurcation
- continuation

1.5.1 ▪ Two Approaches to Proxy-Solving

The first Big Idea in rootfinding is to replace $f(x)$ by a simpler function that is easier to solve. The simpler function is often called a *model* for $f(x)$ in the literature, but this is a little confusing because $f(x)$ is usually itself the result of engineering or scientific modeling simplifications. To avoid confusion, we shall refer to the "model of a model" as a "proxy."

One proxy strategy is to repeatedly apply *low-order* approximations with iteration, initialization, and continuation—themselves Big Ideas that we shall discuss in a moment. This classical approach will never go out of fashion because it is a superb way to refine the output of other algorithms and it is also simple to program, simple to understand, and an excellent option for computing inverse functions.

The other approach to proxies is to compute a proxy that is such an accurate approximation to $f(x)$ that it need only be computed once. Adaptive Chebyshev approximation can generate superb polynomial approximations to a smooth function on a finite interval. The roots of the polynomial approximation can then be found as the eigenvalues of the Chebyshev companion matrix, whose elements are trivial functions of the coefficients of the truncated Chebyshev series. This new approach is both powerful and robust and can be generalized in a variety of directions. Much of this book will describe this CPR and its generalizations, never previously described in a numerical analysis book. We shall turn first to the central ideas of proxy zero-hunting and then return later to discuss the essential ingredients of the Chebyshev polynomial algorithm.

1.5.2 ▪ Low-Order and High-Order Proxies

The simplest proxy is to replace $f(x)$ by a two-term Taylor approximation:

$$f(x) \approx f^{two\text{-}term\text{-}Taylor}(x; x^{(0)}) \equiv f(x^{(0)}) + f_x(x^{(0)})(x - x^{(0)}), \tag{1.15}$$

where f_x is used as shorthand for df/dx, a notational artifice we shall employ throughout this book. (The choice of $x^{(0)}$ is the *first guess* or *initialization*, as discussed below). Because the proxy is linear, it is easy to compute its root. This zero, which is an approximate solution of $f(x) = 0$, will be denoted by $x^{(1)}$:

$$x^{(1)} = x^{(0)} - \frac{1}{f_x(x^{(0)})} f(x^{(0)}) \qquad \text{[Newton's Iteration]}. \tag{1.16}$$

This replacement by a proxy that is linear in the unknown can be repeated, linearizing about $x^{(n)}$ to compute $x^{(n+1)}$. Repeated linearization through first-order Taylor's series is Newton's iteration, also known as the Newton–Raphson algorithm.

The *secant* method is similar except that the proxy is obtained by *linear interpolation* through two initial guesses or previous iterates. The Cauchy and Muller methods extend the Taylor series and interpolation strategies, respectively, to polynomials that are second degree in the unknown, requiring solution of a quadratic equation to determine the next iterate. A linear polynomial over linear polynomial rational approximation ([1/1] Padé approximant) gives the Halley iteration, and so it goes.

In contrast, the Chebyshev rootfinder employs a truncated Chebyshev series for $f(x)$ as the proxy. Because Chebyshev series converge *exponentially* fast with polynomial degree, the Chebyshev-proxy is a very high order proxy, graphically indistinguishable from the real $f(x)$. Recomputation of the proxy as part of an iteration is unnecessary.

1.5.3 ▪ Iteration

The second key theme is iteration: repeatedly applying an algorithm to move closer and closer to a root. Iteration is essential when the proxy is low order. The iteration stops when either a user-specified maximum number of iterations is exceeded, or when successive corrections are less than some user-specified tolerance.

The Chebyshev rootfinder employs such a high-order approximation that iteration is unnecessary. However, the zeros of the truncated Chebyshev series are found by computing the eigenvalues of a matrix, and matrix eigensolvers are always iterative, albeit hidden from the user. It is also often desirable to *Newton-polish* the zeros of the proxy by applying the Newton–Raphson algorithm to iterate away the Chebyshev approximation errors and thus refine the roots to near machine precision.[1]

1.5.4 ▪ Initialization

Iteration requires a first guess or initialization. This in turn requires systematic (or sometimes unsystematic) strategies to generate initial approximations that, although they may be crude, lie within the domain of convergence of the iteration. When the parameter λ is very small or very large, perturbation theory is often effective. Sometimes, observations, laboratory experiments, or physical and mathematical intuition can yield a good guess, too.

Initialization is important because iteration schemes are usually *local* and will converge if and only if the *first guess is sufficiently close to the answer*. The good news is that there are many general strategies for generating first approximations, some of which will be described below. The bad news is that the first approximation is necessarily problem dependent; other than the Chebyshev rootfinder, there are no general strategies, only a toolbox of stuff to try.

Because a good initialization is often difficult, a major research area is to improve iteration schemes so as to make them more robust, that is, more tolerant of a bad first approximation. Some very effective strategies, such as underrelaxation, will be explained below. Even so, it is highly desirable to generate a very good first guess, making such convergence-interval-expanding stratagems unnecessary, but in practice this is often impossible. The best way to obtain reliable success from an iteration is to build in backtracking, self-monitoring, and other strategies that can often converge to a good root from a terrible first approximation.

1.5.5 ▪ Continuation

Often, one needs to solve $f(x,\lambda) = 0$ for the root x for many different values of the parameter λ. Tracking a branch of solutions as a parameter varies is called *continuation*.

Continuation is a challenge because roots may merge and migrate off the real axis, or additional complex-valued roots may become real as the parameter varies. However, continuation also creates new algorithmic strategies.

Suppose that one can solve the equation for one particular value of λ, which will be denoted by λ_0. For example, perturbation theory may give a good approximation for $\lambda = \lambda_0$. Suppose further that $x(\lambda)$ is a *continuous* function of the parameter λ. It follows

[1] *Machine epsilon* is the precision of floating point arithmetic; a formal definition is given in the glossary. MATLAB/IEEE precision will be used to indicate a machine epsilon of 2.2×10^{-16}, which is the default in MATLAB and is also double precision in the widely adopted IEEE 754 standard.

that if $\lambda_1 \equiv \lambda_0 + \tau$ where τ is very small, then

$$x(\lambda_1) = x(\lambda_0) + O(\tau). \tag{1.17}$$

This implies that once Newton's iteration (or whatever) has refined the first guess of $x(\lambda_0)$ into an accurate root, one can use this as a first guess for $x(\lambda_1)$. Once this approximation has been refined, the converged value for $x(\lambda_1)$ supplies a first guess for $x(\lambda_2)$, where $\lambda_2 = \lambda_1 + \tau$. Thus, one can trace the entire solution branch by marching in small steps in λ, using the solution at the previous point in λ as the initialization for the next value of the parameter.

This initialization can be refined in numerous ways. For example, once two solutions have been computed, one can use a linear extrapolation to estimate the next point on the solution branch:

$$x(\lambda + \tau) \approx 2x(\lambda) - x(\lambda - \tau), \qquad \tau \ll 1. \tag{1.18}$$

An even more sophisticated extrapolation is to derive a differential equation in the parameter, the so-called Davidenko equation, and solve it by a Runge–Kutta method or other initial value method to estimate $x(\lambda + \tau)$.

The combination of initialization by extrapolating continuation with Newton's iteration is very similar to *predictor-corrector* algorithms for solving initial value problems except that here, the march is to larger or smaller values of the parameter λ rather than advancing in time. The predictor is either polynomial extrapolation or the solution of the Davidenko equation on a short interval in the parameter; the corrector is Newton's method or a similar iteration that refines the root to a precision limited only by roundoff error.

"Continuation" has thus become a semantically overloaded word, denoting both the process (branch-tracing) and also an algorithmic strategy (a predictor-corrector rootfinder).

1.5.6 ▪ Underrelaxation and Optimization

The *residual norm* function is defined by

$$\rho(x) \equiv \frac{1}{2} f(x)^2. \tag{1.19}$$

(This generalizes to higher dimensions by replacing $f(x)^2$ by the sum of the squares of the functions $f_j(x_1, x_2, \ldots, c_N)$ that define the system of equations whose roots are sought.) Note that the residual norm is always a scalar even when its argument is a vector. The problem of finding the roots of f can always be equivalently formulated as that of finding minima of the residual function ρ of minimum value zero.

When they work, *direct* rootfinding methods such as Newton's iteration are preferable because they are usually much faster than optimization methods. Indeed, an early edition of a well-known collection of numerical algorithms rather scornfully criticized rootfinding by minimization. This criticism was deleted in later editions, however.

The reason is that when iterative methods fail, minimization methods and concepts borrowed from optimization theory often succeed (slowly!). For example, Newton's method will diverge or wander endlessly when the first guess is too far from a root. Later, we will review a well-known theorem that shows that when the Newton correction is multiplied by a sufficiently small factor γ, the *underrelaxed* correction is guaranteed to reduce the residual norm. It follows that by choosing sufficiently small γ—and taking many

Table 1.2. *Extensions of univariate rootfinders to higher dimensions.*

Tactic	Comments
Extend easily to systems of N equations	
Newton's iteration	Must solve $N \times N$ matrix problem at each iteration
Generalizable to systems of two or three equations	
MAG-Intersecting Contours [section 20.2]	Intersections of superimposed contour plots of f and g
MAG-Residual-Norm [section 20.2]	mesh or contour plot of $-\log_{10}(\rho)$
Cauchy	Requires subroutine to compute conic intersections ($N=2$) to compute quadric surface intersections ($N=3$)
Extend awkwardly or not at all to systems with $N \geq 3$	
graphical	Useless for more than three unknowns
Chebyshev-proxy	Cost grows rapidly with N Lack of good polynomial solvers for large N

iterative steps—one can make the residual norm grow ever smaller until the iteration is sufficiently close to the root that a standard Newton's iteration will succeed. There are some technical snags to be explained later; nevertheless, underrelaxation greatly expands the domain of convergence for Newton's method.

1.6 ▪ Extensions and Nonextensions to Multivariate Systems

Newton's iteration extends easily to multidimensional systems. Again, the underlying methodology is to expand the components in a linear Taylor series and iterate. The main practical complication when $N > 1$ is that one must compute and factor an $N \times N$ matrix, the *Jacobian matrix*, to calculate the next iterate. This can be expensive when N is huge, inspiring a large family of *quasi-Newton* methods that fudge the Jacobian so as to trade slower convergence for a much lower cost per iterate.

When computed as the usual tensor product grid with tensor product basis, the Chebyshev-proxy is completely unworkable when N is large because of the high cost of tensor product approximation in high dimensions and, worse, the lack of reliable black-box rootfinders for multidimensional Chebyshev series. Smolyak sparse grids and basis functions have been very fruitful in stochastic modeling and uncertainty quantification in high dimensions (up to $N = 50$) [321]. There is no experience in applying sparse spectral technology to rootfinding.

Graphical methods are even more hopeless when the number of unknowns exceeds three, and they are applied only rarely for $N = 3$ where isolated roots are the intersection points of three isosurfaces. In Table 1.2, we briefly catalog a variety of algorithms and concepts according to their applicability when the number of unknowns N is one, a small integer, or a large number.

Part II

The Chebyshev-Proxy Rootfinder and Its Generalizations

Chapter 2

The Chebyshev-Proxy/Companion Matrix Rootfinder

"After so many centuries, it is gratifying that a reliable method is finally available for finding the real zeros of a smooth transcendental function on an interval. "
—J. P. Boyd [96, p. 392]

2.1 ▪ A Quick Overview

When a function $f(x)$ is holomorphic on an interval $x \in [a,b]$, its roots on the interval can be computed by the following three-step procedure. First, approximate $f(x)$ on $[a,b]$ by a polynomial $f_N(x)$ using adaptive Chebyshev interpolation. Second, form the Chebyshev–Frobenius companion matrix whose elements are trivial functions of the Chebyshev coefficients of the interpolant $f_N(x)$. Third, compute all the eigenvalues of the companion matrix. The eigenvalues λ that lie on the real interval $\lambda \in [a,b]$ are very accurate approximations to the zeros of $f(x)$ on the target interval. To minimize cost, the adaptive phase can automatically subdivide the interval and apply the Chebyshev rootfinder separately on each subinterval, either to keep N bounded or to solve *dynamic range* complications. We also discuss generalizations to compute roots on an infinite interval, zeros of functions singular on the interval $[a,b]$, and slightly complex roots.

This and the next chapter are an expansion and update of the author's review [96].

2.2 ▪ Introduction

2.2.1 ▪ Polynomial Roots and Radicals

The Chebyshev-proxy rootfinder (CPR) is more complicated than Newton's iteration or the secant method. CPR is limited to computing the zeros of a smooth function on a real interval (though generalizations that relax these restrictions are described later). Even so, the CPR algorithm offers much better reliability than Newton's iteration and other local, accurate only near the roots iterations—isn't it embarrassing that undergraduates are taught to worship proof and rigor and then taught rootfinding iterations that will work *most of the time*? *Maybe*? *Frequently*?

To be sure, Newton's iteration is valuable as the endgame of CPR, and occasionally as a stand-alone rootfinder, whenever a sufficiently good initialization is available, as explained in Chapter 6. Like an infant, however, its charms are inextricably paired with a need for a lot of caregiving.

2.2.2 ▪ Three Key Ideas: Rootfinding-by-Proxy, Adaptive Chebyshev Polynomial Interpolation, and the Chebyshev Companion Matrix

The first key idea is the replacement of $f(x)$, the function whose root is sought, by a *proxy* $f_N(x)$, a sort of mathematical stunt double for $f(x)$, such that the roots of the proxy are easily found. Many ancient (but effective) rootfinders employ a proxy. In Newton's iteration, the proxy is a two-term Taylor series. The secant iteration also uses a linear polynomial as the proxy for $f(x)$, but one computed by interpolation. Cauchy's method employs the quadratic Taylor series. Halley's iteration approximates $f(x)$ by a Padé approximant that is the ratio of two linear polynomials. And so it goes, as cataloged much later in this book in Table 16.1.

The second key idea in CPR is that an adaptive Chebyshev polynomial interpolant is the optimal proxy. The Shakespeare–Weierstrass principle that "All the world's a polynomial" is nonconstructive. Interpolation at the roots of a Chebyshev polynomial, however, *never* fails for a function $f(x)$ that is analytic on the interpolation interval. This discovery makes it possible to "polynomialize" transcendental equations by replacing functions by their polynomial approximations.

In CPR, the proxy $f_N(x)$ is defined to be that polynomial of degree N which exactly matches ("interpolates") $f(x)$ at the endpoints plus the extrema of $T_N(x)$, the Chebyshev polynomial of degree N, i.e., the Chebyshev–Lobatto grid [76]. When $f(x)$ is transcendental, it is much easier to find the zeros of the proxy $f_N(x)$ than the roots of $f(x)$ directly.

The third essential idea is to apply a reliable rootsolver for a polynomial in Chebyshev form, that is, a polynomial written as a series of Chebyshev polynomials. Remarkably, the desired polynomial *zeros* are simply the *eigenvalues* of a matrix, the so-called Chebyshev–Frobenius companion matrix, whose elements are trivial functions of the Chebyshev coefficients of the polynomial. The Chebyshev companion matrix was independently discovered at least six times in the past half-century, but it was only recently that the community of spectral modelers became aware of this clever and useful numerical technology.

Fig. 2.1 summarizes the algorithm.

2.3 ▪ Solve-by-Graph, Local Iterations, and Failure-to-Find: A Taxonomy of Rootfinders

Rootfinding algorithms are as numerous as the stars in a dwarf galaxy. McNamee's two-volume tome catalogs a bewildering variety of rootfinders just for polynomials. Comparing every published method with every other would surely exceed the energies of even Didymus of Alexandria, the most prolific author of the pre-Gutenberg centuries.[2]

Our goal, rather, is to stress ideas and show how the common clay of interpolation and matrix algebra can yield a very powerful algorithm that scores high on all four dimensions of the merit-of-rootfinders scheme discussed next.

Algorithms and software for computing zeros can be judged by a five-dimensional classification scheme. The axis of reliability and the axis of cost/efficiency are common to all library subroutines and the algorithms that animate them. The other two dimensions are peculiar to rootfinders. One is the axis distinguishing local, find-one-root-at-a-time

[2]Didymus, who lived from the time of Julius Caesar to the reign of Augustus (63 BC–10 AD), wrote or compiled more than 3000 books without the benefit of word processing or a typewriter. His awed contemporaries nicknamed him "Chalkenteros," "Bronze-Guts."

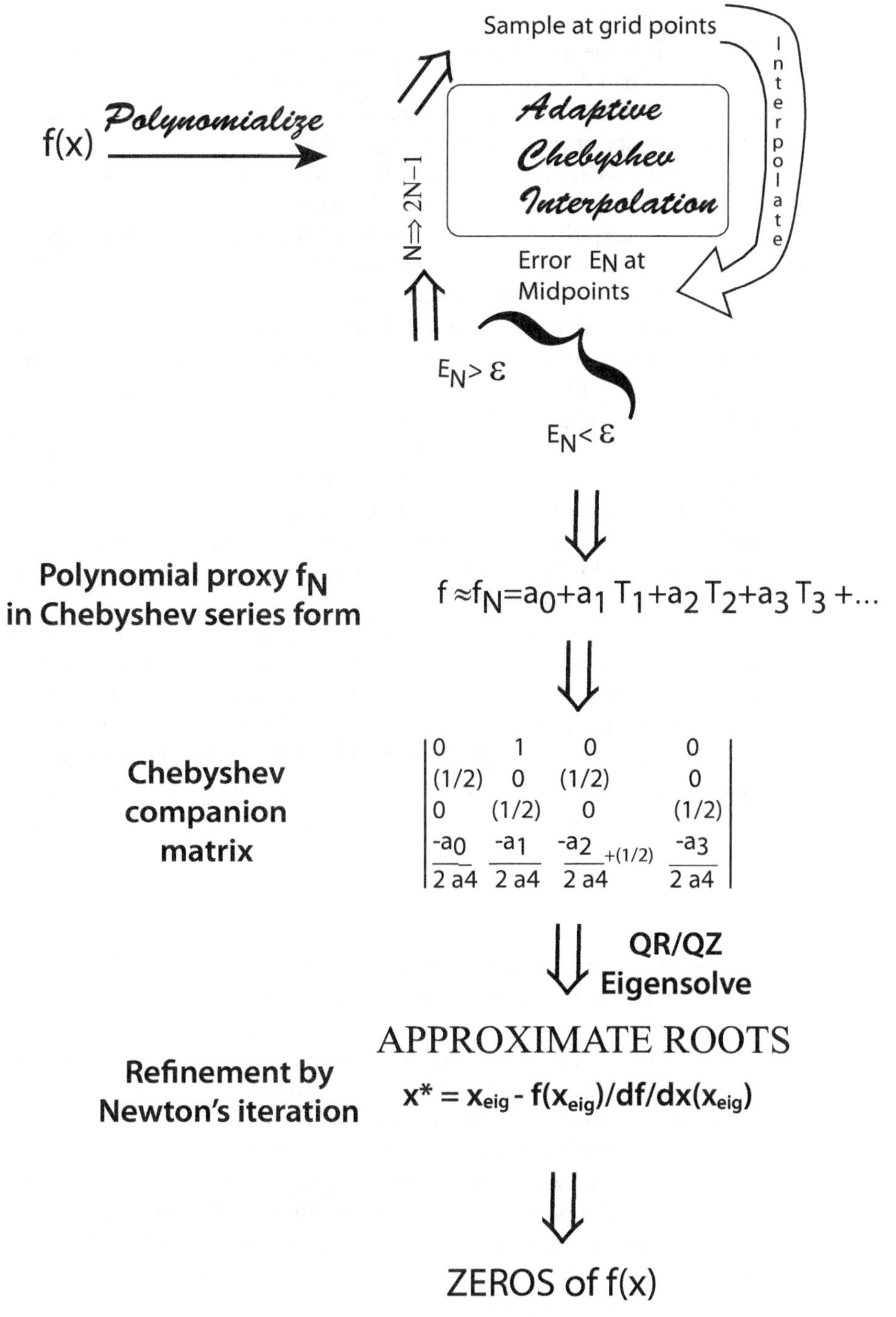

Figure 2.1. *A schematic of CPR. The user must specify an interval $x \in [a, b]$ and an error tolerance ϵ [96]. The degree of the interpolating polynomial N is adaptively determined so that $f_N(x)$ approximates $f(x)$ on the interval within the user-chosen tolerance ϵ. The elements of the Chebyshev companion matrix, given in Appendix A, are simple functions of the Chebyshev coefficients a_n. Those eigenvalues of the matrix that lie on $[a, b]$ are highly accurate approximations to roots of $f(x)$ on the interval. (The other eigenvalues are not accurate zeros of $f(x)$, barring some exceptions analyzed in the text, and should generally be discarded.)*

algorithms from global methods that compute all zeros in a region simultaneously. The fourth axis is the quantity of mandatory user-supplied information—one might call this the "axis of handholding." The fifth dimension, simplicity of programming, is applicable primarily for user-written software, since library rootfinders shield the consumer from the complexity of the underlying algorithms.

The CPR method is good by all five criteria. *Reliability:* The algorithm has a very low failure rate. *Cost/efficiency:* Automatic adaptation, the Fast Fourier Transform (FFT), and subdivision keep the cost small even when the search interval is huge. *Global-all-roots-on-a-domain* versus *local-one-root-at-a-time:* CPR is global, calculating all zeros on the given real search interval. *Handholding/mandatory user-supplied information:* CPR is a black box in the sense that the solver needs no inputs except a subroutine to compute $f(x)$, the function whose roots are sought, and the endpoints (a, b) of the search interval. *Ease of programming:* CPR requires only a few lines of MATLAB (Tables 2.1 and 2.2) or a single statement in *Chebfun* [354]).

An engineer's crude but effective first line of attack on a transcendental equation is the MAG algorithm: make-a-graph. A good visualization is more practical than a bushel of theorems in identifying the number and approximate location of well-separated roots. Unfortunately, MAG is not entirely reliable. A computer screen is only a thousand pixels wide. Pairs of closely spaced roots—what an atomic spectroscopist would call a *doublet*—can easily hide between the pixels.

Householder's monograph on solving a single equation in a single unknown [213] is a rich catalog of iteration schemes for finding roots of a transcendental equation, one at a time. The trouble is that local, one root at a time iterations can be fooled.

Bisection, for example, can be initialized by evaluating $f(x)$ on a very fine grid with points x_j. Every interval $[x_j, x_{j+1}]$ that is such that $f(x_j)$ is opposite in sign to $f(x_{j+1})$ is guaranteed to contain at least one root—but the interval might contain *more.* A pair of closely spaced zeros—a *root doublet*—is very challenging. For example,

$$f(x;\delta) \equiv x^2 - \delta^2 \tag{2.1}$$

has two roots at $x = \pm\delta$. Since δ can be arbitrarily small, one needs an arbitrarily fine sampling of $f(x)$ to guarantee the detection of the tiny interval $x \in [-\delta, \delta]$ where $f(x;\delta) < 0$.

Newton's iteration will diverge, converge to a distant zero, or orbit endlessly in a limit cycle if the first guess is insufficiently close to the root, as discussed in Chapter 6, but that is not the big issue. A more serious flaw is that one root at a time methods like bisection and Newton's iteration can easily miss roots. (A published embarrassment of the author's—omitted zeros—is corrected in [60].) A local iteration is like an antisubmarine homing torpedo: only good after one has already identified a target.

In contrast, companion matrix methods are numerically well-conditioned procedures that find all the roots of a polynomial. And Chebyshev interpolation can replace almost any function by a polynomial.

2.4 ▪ Chebyshev Interpolation: The Shamrock Principle

Polynomial interpolation is neither a new idea nor a deep one. Every small child who has drawn a connect-the-dots picture in a coloring book has performed piecewise linear interpolation, approximating a teddy bear by line segments scrawled in crayon.

The Science Museum in London has a lovely nineteenth-century conglomeration of brass pointers and gears called a parabolagraph, a sort of conjoined pair of compasses: move the three awl-like arms to three points on a curve, and the fourth arm with pencil will draw the interpolating parabola, that is, the unique quadratic polynomial through the three points.

Lagrange showed in the eighteenth century that the unique polynomial interpolant of arbitrary degree N through an arbitrary set of $(N+1)$ points $\{x_k\}$ is

$$f(x) \approx f_N(x) \equiv \sum_{j=0}^{N} f(x_j) \frac{\prod_{k \neq j}^{N+1} (x - x_k)}{\prod_{k \neq j}^{N+1} (x_j - x_k)}. \tag{2.2}$$

There is a very powerful idea embodied in interpolation. To the extent that the interpolant is an accurate approximation, the *continuous* function $f(x)$ can be replaced by the *discrete* set of $(N+1)$ samples, $\{f(x_j)\}$, its *grid point values*, in the sense that the *continuous* function can be reconstructed solely from this discrete set of numbers.

Unfortunately, interpolation at arbitrary points, or even at uniformly spaced points, does *not* always yield a good approximation even for very large N. When the grid points are evenly spaced, Runge showed at the turn of the twentieth century that polynomial interpolation of $f(x) = 1/(1+x^2)$ on $x \in [-5,5]$ *diverges* as $N \to \infty$, the *Runge phenomenon*, illustrated by Fig. 4.3 on p. 84 of [76].

However, the interpolation points do not need to be evenly spaced. Cornelius Lanczos showed that the Runge phenomenon could be utterly defeated by interpolating at the roots of the Nth Chebyshev polynomial instead [235].

Although he established no school and won neither a Nobel Prize nor a Fields Medal, Lanczos made great contributions to many areas of science and mathematics, as cataloged in his six-volume collected papers [145]. The huge symposium in honor of his centenary [114] sprawls across a vast array of arenas—general relativity, quantum mechanics, matrix methods, interpolation, Fourier theory, etc.—because *he* did.

For our purposes, though, his masterpiece was the Lanczos Thunderbolt: the 1938 paper, more than a hundred pages long, that is the origin for the vast and widely used family of numerical algorithms known as pseudospectral methods, and much more besides: in T. J. Rivlin's words, "a mighty sequoia, with branches penetrating areas of digital computation from solutions of differential equations to numerical linear algebra to approximation theory and filter design" [145, p. 3-310].

Lanczos showed that not only could the Runge phenomenon be defeated by Chebyshev interpolation, but also that the interpolant could be written in *Chebyshev form* as a truncated Chebyshev series:

$$f_N(x) = \sum_{j=0}^{N} a_j\, T_j(x), \tag{2.3}$$

where the Chebyshev coefficients $\{a_j\}$ can be obtained by multiplying the grid point values $f(x_k)$ by a square matrix of size $(N+1)$. (The mechanics and matrix elements are given in Appendix B.) The interpolation points x_j and the matrix elements are cosine functions [76], as follows from Chebyshev's nineteenth-century identity:

$$\boxed{T_n(\cos(t)) \equiv \cos(nt)} \qquad \forall x, t, \qquad n = 0, 1, 2, \ldots, \tag{2.4}$$

where T_n is the Chebyshev polynomial of degree n. The Lanczos Thunderbolt might be alternatively titled, "The Shakespeare–Weierstrass Principle Made Practical."

Later, it was discovered that the transform $\{f(x_j)\} \rightarrow \{a_j\}$ can be done even faster, in only $O(N \log_2 N)$ operations, by using the discrete cosine transform, a variant of the FFT. Lanczos and Danielson published the central idea of the FFT in 1942, more than two decades before its independent rediscovery by Cooley and Tukey [132], but even Lanczos did not appreciate how useful the FFT would be when digital computers made it possible to use really large N.

The Russian mathematician Sergei Bernstein (1880–1968) had already proved that the Chebyshev series of a function analytic everywhere on an interval $x \in [a, b]$ converges *geometrically* fast [40]. That is, the error in truncating the series after the Nth term falls as $O(\exp(-N\mu))$ for some positive constant μ. (We shall be more precise about μ in a later section, but a full discussion is given in [89], Chapter 2 of [76], and [352].) The difference between the explicit form of a transcendental function and its polynomial approximation is comparable to machine epsilon even for rather modest N.

Chebyshev thinking reached an apotheosis of sorts in the *Chebfun* system [354, 37, 285, 158]. MATLAB was originally created by Cleve Moler as a system for manipulating vectors and matrices, morphing over the years into a full-featured programming language and visualization system. The *Chebfun* system [37] is a MATLAB extension that handles piecewise analytic *functions* and differential and integral *operators* as if they were vectors and matrices, respectively. That's because in Lanczos's world, the realm of Chebyshev interpolation, functions *are* ordinary finite-dimensional vectors, and differential and integral operators are merely matrices operating on those vectors.

Chebfun approximates an arbitrary $f(x)$ on a given interval by a Chebyshev interpolant, adaptively increasing the degree until the polynomial is indistinguishable from the real thing to within machine precision. The function has thus been reduced to a vector of Chebyshev coefficients $\{a_j, j = 0, 1, \ldots, N\}$ or, equivalently, to a vector of $(N+1)$ grid point values $\{f(x_k), k = 0, 1, \ldots, N\}$. These vectors can then be manipulated by the usual vector and matrix routines that have been part of MATLAB from its beginning.

The underlying idea is that a continuous analytic function is a vector of Chebyshev coefficients in disguise. Indeed, a function wears three interchangeable masks:

- its explicit definition as a formula or the solution to a differential equation,
- its Chebyshev coefficients $\{a_j\}$, and
- its grid point values (*samples*) $\{f(x_j)\}$.

In spectral methods, this three-fold unity is the Shamrock Principle [76]: these three representations are one function, just as the three leaves of a shamrock are a single clover, as shown schematically in Fig. 2.2.[3]

To a pure mathematician, of course, this equating of the discrete with the continuous is anathema; to continue the religious metaphor, most mathematicians are Unitarians: there is only the exact analytical form, and approximations are not divine, but merely fuzzy engineering. However, the error in approximating $f(x)$ by a Chebyshev interpolant falls as $\exp(-\mu N)$ for some $f(x)$-dependent constant $\mu > 0$, as long as $f(x)$ is analytic on the interval $x \in [-1, 1]$. In the quantized world of the microchip where reality is measured not continuously but in units of machine precision, roughly 2×10^{-16} in MATLAB and IEEE standard double precision, the world view is decidedly Trinitarian.

[3]In folklore, St. Patrick, the great missionary to Ireland in the late fifth century, used the shamrock—three leaves, one stem—to explain the concept of the Trinity—three persons but one God—to the Druids.

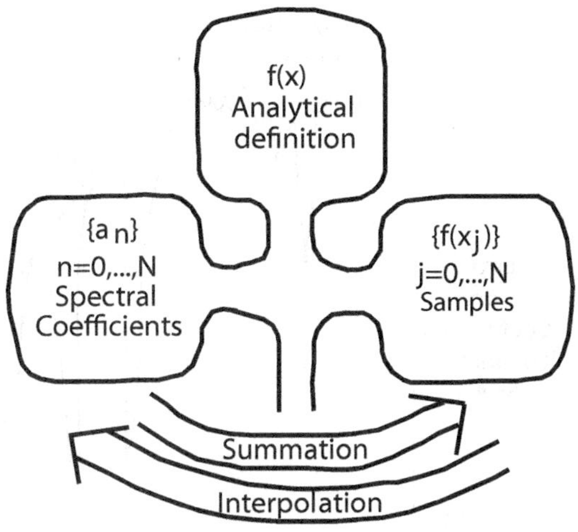

Figure 2.2. *A visualization of the Shamrock Principle, or the trinity of spectral methods. The first N Chebyshev coefficients, $\{a_n\}$, and the set of N grid point values of $f(x)$ on the interpolating grid, the samples $\{f(x_j)\}$, are* equivalent *to the analytical definition of $f(x)$, represented by the top leaf, in the sense that either of these two sets allows one to compute $f(x)$ to arbitrarily high precision for sufficiently large N* [96].

2.5 ▪ Chebyshev Convergence Theory

The output of Chebyshev interpolation is a truncated Chebyshev series

$$f_N = \sum_{j=0}^{N} a_n T_n(x). \tag{2.5}$$

The error norm is the maximum pointwise error, that is, the error in the L_∞ norm,

$$E_N \equiv \max_{x \in [-\infty,\infty]} |f(x) - f_N|. \tag{2.6}$$

Note that because $T_n(\cos(t)) = \cos(nt)$, $|T_n(x)| \leq 1$ for all $x \in [-1,1]$. It follows that the magnitude of the *term* is bounded by the magnitude of the *coefficient*:

$$|a_n T_n(x)| \leq |a_n| \qquad \forall x \in [-1,1]. \tag{2.7}$$

Fig. 2.3 illustrates typical behavior. The generic case is that $f(x)$ is analytic on the approximation interval $x \in [-1,1]$ but has singularities (in the sense of complex variable theory—poles, branch points, logarithms, etc.) at points in the complex x-plane. The Chebyshev coefficients (and maximum error on the interval) both have *geometric convergence*, that is, fall proportionally to $\exp(-n\mu)$ for some $\mu > 0$. If $f(x)$ has singularities only at infinity, then f is an *entire* function and its Chebyshev coefficients converge *supergeometrically*, that is, proportional to $\exp(-n\log(n))$ or faster. On a log-linear plot, the envelope falls more and more steeply, as illustrated in Figs. 2.5 and 2.6 on pp. 28–29 of Chapter 2 of [76]. If f is not analytic on the interval but has an infinite number of bounded derivatives (C^∞ on $x \in [-1,1]$ but not C^Ω in math jargon), the convergence is *subgeometric*, that is, proportional to $\exp(-qn^r)$ for some positive constants q and r with $r < 1$, typically $r = 1/2$. On a log-linear plot, the envelope does not asymptote to a straight line but instead slowly flattens with a smaller and smaller slope (not shown).

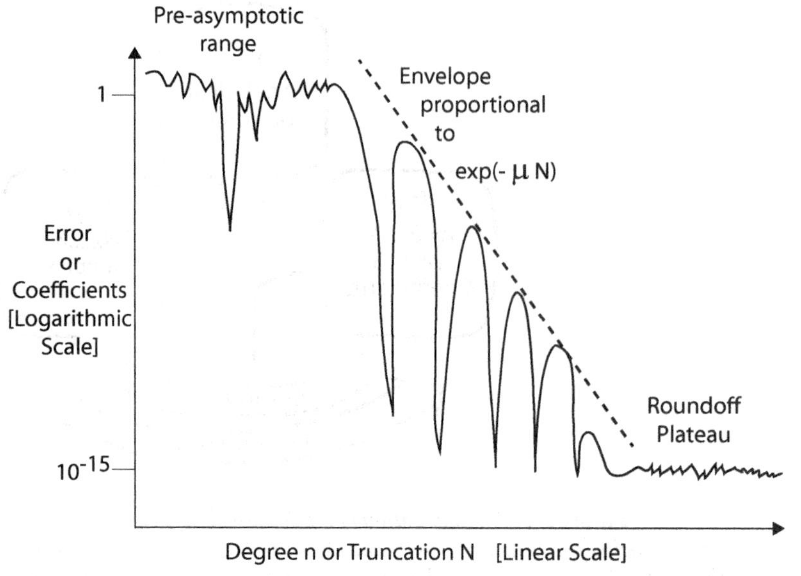

Figure 2.3. *Generic convergence of Chebyshev polynomial series. The coefficients* $|a_n|$ *and the error norm* E_N *have similar curves, and indeed* a_N *is usually the same order of magnitude as* E_N*, so this schematic applies to both. The coefficients and error may exhibit three regimes. For small degree n, the coefficients may decrease very little, or decay at a rate different from the asymptotic rate that sets in for large n. This* pre-asymptotic *regime may be absent or may be a fast decay that is replaced for larger degree by a slower decay. Because Chebyshev series converge exponentially fast, high-degree coefficients often "plateau," that is, cease to decrease but instead bounce around randomly. This plateau is due to roundoff error; it is typically at 10 to 1000 times the product of machine epsilon, with the magnitude of the largest Chebyshev coefficient* $\max_n |a_n|$*. Generically, Chebyshev coefficients and errors asymptote to an exponential rate of decay, proportional to* $\exp(-\mu n)$*, where* μ *is a positive constant, the* asymptotic rate of geometric convergence. *The exponential is often multiplied by nonexponential, slower than exponential functions. Sometimes the asymptotic decay of coefficients and errors is monotonic, but oscillations in degree n or truncation N are common. When the coefficients/error are oscillatory, it is always possible to bound the curve rather tightly by an* envelope *(dashed line) that does decay monotonically.*

Almost a century ago, Serge Bernstein proved a theorem that connects the asymptotic rate of geometric convergence to the poles and branch points of $f(x)$ in the complex x-plane.

Theorem 2.1 (Chebyshev Convergence Theorem). *Map the interval* $[a,b]$ *in* x *to* $z \in [-1,1]$ *by* $x \equiv z(b-a)/2+(b+a)/2$*. Define elliptical coordinates in the complex z-plane by*

$$\begin{aligned} \Re(z) &= \cosh(\mu)\cos(\eta), \qquad \mu \in [0,\infty], \\ \Im(z) &= -\sinh(\mu)\sin(\eta), \qquad \eta \in [0,2\pi]. \end{aligned} \tag{2.8}$$

The curve of constant elliptical coordinate μ *is an ellipse with foci at* $z = \pm 1$ *(*Bernstein ellipse*) with semimajor axis* $\cosh(\mu)$ *and semiminor axis* $\sinh(\mu)$*. If* $f(z)$ *is free of singularities everywhere inside the ellipse* $\mu = \tilde{\mu}$ *but is singular somewhere on this ellipse, then the error in the Chebyshev series, truncated after the Nth term, will fall proportionally to* $\exp(-N\tilde{\mu})$ *on the real interval* $z \in [-1,1]$ [76, 89, 352]. *The same is true for N-point Chebyshev interpolation.*

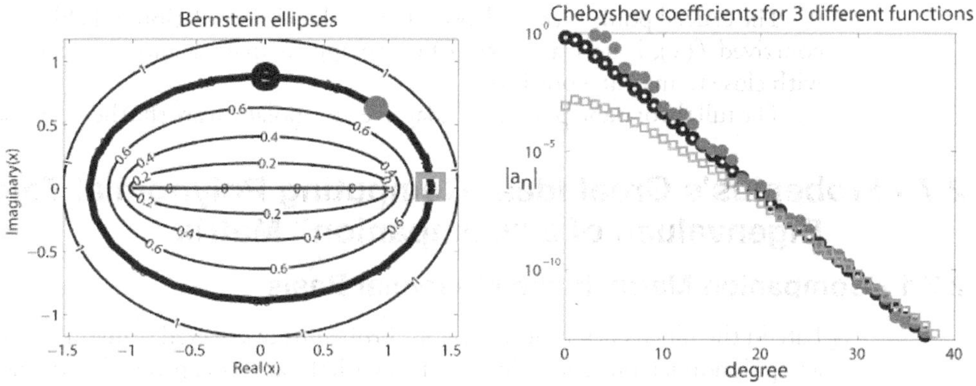

Figure 2.4. *Left: Bernstein ellipses (foci at $x=\pm 1$). Right: absolute value of the Chebyshev coefficients for three functions that are singular on the contour elliptical coordinate $\mu = 0.8$ (thick ellipse) at the points indicated by the markers. Open circles: $f_0 = (1+x)w^2/(w^2+x^2)$, where $w = \sinh(0.8)$, a simple pole on the imaginary axis. $f_1 = 20(1+x)v^2 \log\{1/(v^2+(x-v)^2)\}$, where $v = \sinh(0.8)/\sqrt{2}$, a logarithmic branch point at $x = v + iv$. $f_0 = 10^{-5}(1+x)/((x-q)^2)$, where $q = \cosh(0.8)$, a double pole on the real axis at $x = q$.*

Fig. 2.4 compares the Chebyshev coefficients of three different functions to illustrate the theorem. Although the singularities of these functions are at widely different locations (real axis, imaginary axis, and halfway between) and of three different types (logarithm, simple pole, double pole), the coefficients for all three functions have asymptotically the same slope. The reason is that all three have convergence-limiting singularities on the same Bernstein ellipse, $\mu = 0.8$, so all have Chebyshev coefficients a_n falling proportionally to $\exp(-n\mu)$.

A plethora of rules of thumb, formal convergence theorems, and so on are given in Chapter 2 of [76].

2.6 ▪ Adaptive Chebyshev Interpolation

Lanczos's 1938 paper does lack a systematic procedure for increasing the degree N until the error $|f(x) - f_N(x)|$ is everywhere less than some desired tolerance ϵ. The adaptive algorithm increases N systematically until the difference between $f_N(x)$ and $f_{2N}(x)$ is sufficiently tiny. Because a geometric rate of convergence is generic, this difference diminishes exponentially fast with N.

For efficiency, CPR exploits an observation of Clenshaw and Curtis [129]: all points on the $(2M+1)$-point Chebyshev–Lobatto grid are also points on the $(M+1)$-point grid. Thus, if the number of grid points is increased with $N = M+1, 2M+1, 4M+1, \ldots$ for some positive integer M, all samples of $f(x)$ on coarser grids can be *reused* on the finer grid. This saves a lot of execution time if $f(x)$ is very expensive to evaluate, such as the determinant of a large matrix. The Chebyshev interpolants can be calculated efficiently by a matrix–vector multiplication for small N or the FFT for large N. (The FFT is especially fast when N is a power of two.) Collectively, the efficiencies of sample recycling and the FFT make it sensible to double N at each adaptive level.

In practice, a variety of simpler N-is-enough criteria have also been applied successfully [37, 215, 129].

To reduce the cost of later steps in CPR, a final refinement is to impose an upper limit N_{max} on N such that the search interval is automatically split in two whenever adaptation requests $N > N_{max}$. This is easily programmed and is standard in *Chebfun*.

The crucial point is that adaptive Chebyshev interpolation, reliable for all but weird, contrived $f(x)$, is now a reality. One can approximate a smooth $f(x)$ by a polynomial with close to machine precision.

The full theory and practice of adaptive interpolation are the theme of the next chapter.

2.7 ▪ Frobenius's Great Idea: Computing Polynomial Zeros as the Eigenvalues of a "Companion" Matrix

2.7.1 ▪ Companion Matrix in the Monomial Basis

Late in the nineteenth century, Georg Frobenius discovered something remarkable. For any polynomial, one could find a matrix with the amazing property that the *eigenvalues* of the *matrix* are also the *zeros* of the polynomial. The elements of this *Frobenius companion matrix* are mostly zeros or ones, with a final row composed of the ratios of the lower-degree coefficients to that of the highest power. A problem of nonlinear rootfinding is really a problem in linear algebra.

However, adaptive Chebyshev interpolation yields a proxy that is a polynomial in the Chebyshev basis. Frobenius's formula for the elements of the companion matrix requires that the polynomial be expressed in the usual powers of x or monomial basis. How should one resolve this mismatch?

A polynomial in the form of a truncated Chebyshev series can always be converted into the usual monomial or power form by a matrix multiplication acting on the vector of Chebyshev coefficients [71]. Unfortunately, Gautschi [185] showed that the error in this conversion grows as $(1+\sqrt{2})^N$, where N is the degree of the polynomial [71]. This means that the conversion is acceptable for N less than 20, say, but hopelessly inaccurate in floating point arithmetic for $N > 60$. (The most recent analysis of the sins and virtues of the monomial basis is [100].)

This ill-conditioning is not a peculiarity of the conversion alone but rather reflects an unpleasant reality: the familiar power or monomial form, beloved of high school algebra teachers and pure mathematicians,

$$f_N = \sum_{j=0}^{N} b_j x^j, \tag{2.9}$$

is a very ill-conditioned way of representing a polynomial. In rational arithmetic in a high school textbook, or a computer symbolic manipulation system like Maple, algebra is, well, *exact*, and thus the student is shielded from the distasteful truth that in floating point arithmetic, roundoff errors for the power form grow exponentially with N.

In contrast, the Chebyshev form is always well conditioned, and the MATLAB extension *Chebfun* (discussed earlier) routinely and accurately manipulates truncated Chebyshev series whose degree is in the thousands. But how can one find the roots of a polynomial in Chebyshev form?

The answer is to find the eigenvalues of a matrix, the Chebyshev–Frobenius matrix, whose elements are simple functions of the Chebyshev coefficients, a_j, as given explicitly in Appendix A. The Chebyshev–Frobenius matrix was independently discovered multiple times over several decades by Good [200] (who called it the "colleague matrix"), Specht [338, 339], Barnett [22], Gol'berg and Malozemov [198], Day and Romero [146] and Stetter [344]. I. J. "Jack" Good was the first, and his clear and elementary presentation is still educational today. Barnett, who acknowledges Good's article, generalized the companion matrix to general orthogonal polynomials as part of a larger program to show that all the standard operations of polynomial algebra can be easily expressed in an orthogonal polynomial basis [24, 26, 25, 27, 28, 22, 253, 251, 252, 255]. (Grant and his collaborators gave an alternative for computing roots of a truncated Chebyshev series [201, 202].) There is no need to accept the ill-conditioning of the monomial basis for any task.

As in Frobenius's original idea, the companion matrix is a sparse matrix whose few nonzero elements are problem-independent ratios of Chebyshev coefficients. Barnett and also Day and Romero used recurrence relation ideas from the general theory of orthogonal polynomials; Stetter employed the algebra of a quotient ring in algebraic geometry to derive the same matrix. A brief undergraduate-level derivation is provided in Section A.2.

2.7.2 ▪ The Chebyshev Companion Matrix

The companion matrix for the Chebyshev interpolant for arbitrary degree is given in Appendix A. The special case of a quintic polynomial, for example, is

$$\begin{bmatrix} 0 & 1 & 0 & 0 & 0 \\ (1/2) & 0 & (1/2) & 0 & 0 \\ 0 & (1/2) & 0 & (1/2) & 0 \\ 0 & 0 & (1/2) & 0 & (1/2) \\ (-1)\frac{a_0}{2a_5} & (-1)\frac{a_1}{2a_5} & (-1)\frac{a_2}{2a_5} & (-1)\frac{a_3}{2a_5}+(1/2) & (-1)\frac{a_4}{2a_5} \end{bmatrix}. \qquad (2.10)$$

2.7.3 ▪ Costs and Reliability of the Companion Matrix Eigenproblem

The Chebyshev–Frobenius matrix method is a very accurate and well-conditioned rootfinder [146, 102, 83]. Day and Romero have shown that floating point error is well controlled if the matrix is *balanced* before the actual QR or QZ eigenprocedure is displayed, but such balancing is automatic in MATLAB eigensolving routines and can be manually invoked in all linear algebra libraries where balancing is not automatic. (In Maple, which does not automatically balance, companion matrix eigensolving failures are common, in the author's experience.)

Unfortunately, the cost is about $10N^3$ floating point operations, and this is expensive for large N. Boyd [71, 81] suggested subdividing the interval into subintervals and approximating $f(x)$ by separate but lower-degree Chebyshev interpolants on each subdomain, as mentioned previously. Boyd [81, 82] provided theorems and experiments that put subdivision on a solid footing. We revisit the theory and practice of subdivision in section 2.9.

We note some special cases where cost can be greatly reduced without subdivision in sections 2.16 and 3.9.

2.8 ▪ Dynamic Range

Subdivision can solve another difficulty. Denote machine precision by ϵ_{mach}. Define the maximum of $f(x)$ on the interval as $f_{max} \equiv \max_{x\in[a,b]}(|f(x)|)$. If a function $f(x)$ has a magnitude as small as $|f(x)| \leq O(\epsilon_{mach} f_{max})$ on a subinterval within $[a,b]$ (excluding tiny subintervals around a zero), then $f(x)$ has, to borrow a term from photography, a large *dynamic range*. An example on $x \in [-1,1]$ is

$$f(x) = \exp(-25(x-1)^2)\sin(10\pi x), \qquad (2.11)$$

whose magnitude varies wildly from the neighborhood of $x = 1$, where $f(x)$ is $O(1)$, to near $x = -1$, where $f(x)$ is $O(\exp(-100))$. A useful pragmatic criterion for a large dynamic range is that two or more adjacent samples of $f(x)$, $f(x_j)$, and $f(x_{j+1})$ are $O(k\,\epsilon_{mach} f_{max})$, where k depends on the user's tolerance for risk but satisfies $1 \ll k \ll 1/\epsilon_{mach}$.

Photographers must battle a similar dynamic range problem in the sense that it is difficult for a single photograph to capture a scene that contains portions in bright sunlight and also areas of deep shadow. If the shadows are captured by a long exposure, the bright areas will be solid white without detail, but a short exposure renders the shadows as solid black. The only completely satisfactory strategy is subdivision.

One form of photographic subdivision is to portray the scene using *multiple* photographs with *different exposures*. Often, however, this is not a good option. Another strategy is to exploit the fact that the brightness or darkness of a developed image is dependent not only on exposure and the raw intensity of light falling on the film but also on the length of time the undeveloped film is bathed in the chemicals in the darkroom. By exposing different parts of a single frame to the bath of developer fluids for different durations, shadows can be brightened and bright areas darkened. Photographers call such darkroom tricks "dodging" and "burning." Modern image-processing software mimics such darkroom craftsmanship with the "dodge" and "burn" tools in Adobe Photoshop.

Similarly, subdividing the interpolation interval into subintervals such that $f(x)$ varies by only a few orders of magnitude on each subinterval eliminates the dynamic range difficulty. Subdivision of one interval into two or more is a mathematician's way to dodge and burn. Because machine precision is 2.2×10^{-16} in MATLAB, for example, the dynamic range problem arises for Chebyshev interpolants only occasionally, but there are nontrivial examples. The first and second Wilkinson polynomials, defined as $\mathfrak{W} = \prod_{k=1}^{20}(x-k)$ and $\mathfrak{G} = \prod_{k=0}^{19}(x-1/2^k)$, are Chebyshev-analyzed in [71] and [166], respectively. A fuller discussion of dynamic range is given in the next chapter.

2.9 ▪ Subdivision

Cost and dynamic range are two reasons why the adaptive Chebyshev interpolator should sometimes split a target interval into two or more subintervals.

The QZ algorithm for computing the eigenvalues of a matrix, the default in MATLAB, has a cost proportional to the *cube* of the dimension of the matrix. *Chebfun* routinely employs $N > 1000$, but its rootfinding is relatively inexpensive. This is because *Chebfun* automatically replaces a single expansion on $[a, b]$ by subdividing the interval into several subintervals, computing lower-degree Chebyshev interpolants on each subinterval, and finding the roots on the original interval as the *union* of the zeros on the collection of subintervals. This greatly reduces cost because with M subdomains, the number of floating point operations is $O(M(N/M)^3)$, roughly M^2 cheaper than using a single Chebyshev interpolant of large degree N [105]. A more careful and rigorous analysis of the accuracy of interpolation with subdivision is given in section 3.8.4.

It is easy to embed automatic subdivision within degree-adaptive interpolation by simply adding a conditional (if statement) to split the interval in two and then calling the single-interval interpolation subroutine twice, whenever the Clenshaw–Curtis strategy calls for N larger than some user-specified limit N_{max}.

Recognition of unfavorable dynamic range is more subtle. We shall return to this question, too, in the next chapter. In broad terms, Chebyshev interpolation generates errors that are *uniformly small* in an *absolute* sense over the interval, but the *relative* errors may be large if there are subintervals where the function is oscillating with peaks and

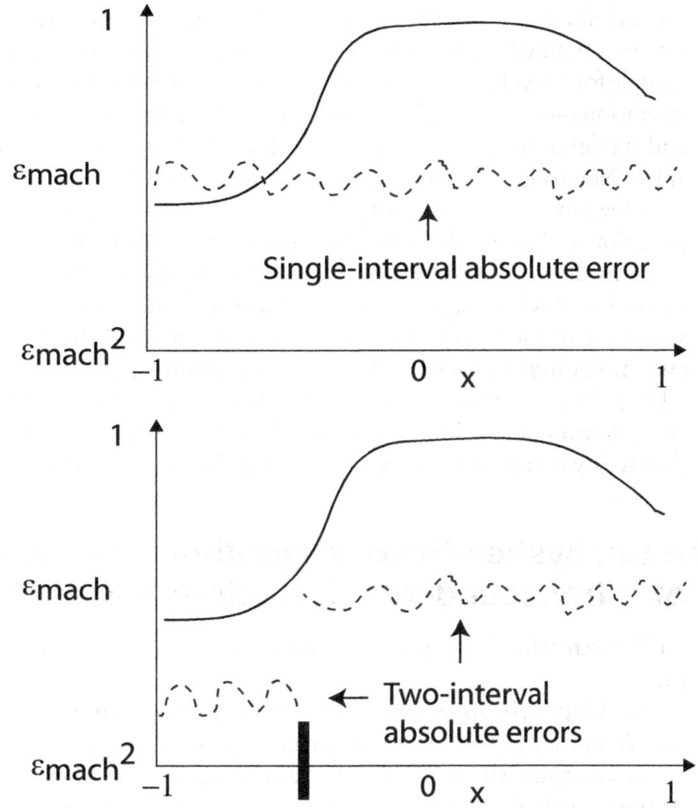

Figure 2.5. *Top: the solid curve is $f(x)$. The dashed curve, oscillating almost uniformly, is the Chebyshev interpolation error; its magnitude is comparable to machine epsilon, ϵ_{mach}. When $f(x)$ is $O(1)$ on part of the interval and smaller than machine epsilon on another part, the Chebyshev error may be larger than $|f(x)|$ where $f(x)$ is small—a huge relative error. Splitting $x \in [-1,1]$ into two smaller intervals at the point marked by the thick, black vertical line yields a much smaller absolute error on the left subinterval where f is tiny.*

valleys that are many orders of magnitude smaller than on other parts of the interval, as illustrated schematically in Fig. 2.5. The remedy is subdivision: let each subinterval of different magnitude or range of oscillation have its own Chebyshev approximation, and the difficulty of a huge range of $f(x)$ on the parent interval disappears.

2.10 ▪ Newton-Polishing

Newton-polishing is our term for the application of Newton's iteration to the output of the adaptive CPR. Newton's iteration is a poor global method, but a great local one. That is to say, without a good initial approximation, Newton's iteration can be the Flying Dutchman of Arithmurgy, wandering endlessly like the accursed never able to make landfall seaman of mythology, missing zeros entirely, never converging to anything but a useless limit cycle or converging to roots spurious for the underlying undiscretized problem. When a good approximation to the root is known, however, Newton's method will double the number of correct digits with each iteration.

Newton's iteration requires the evaluation of both $f(x)$ and df/dx at arbitrary points.

As explained in the next chapter, the Chebyshev coefficients of the polynomial df_N/dx can be obtained from those of $f_N(x)$ itself by a simple recurrence: $O(N)$ operations, a single for/next loop to program. A truncated Chebyshev series can be evaluated in $O(N)$ operations—again a single for/next loop—by either Clenshaw's recurrence (old, but fast and stable) or barycentric recurrence (new and even better). Thus, it is extremely easy to apply Newton's iteration to the Chebyshev-proxy to refine the output of the eigensolver.

One can then iterate with $f(x)$ itself to compute the roots to the full limits of machine precision and purge all Chebyshev approximation error.

Why do we recommend Newton-polishing after extravagantly praising the Chebyshev-proxy strategy? First, worrying that a root is inaccurate is an "own goal," to borrow a soccer term; a few milliseconds on a laptop can refine the zero to near machine precision and save hours of brooding about whether a three digit answer is "good enough." Second, a large dynamic range can sometimes lead to large relative errors for zeros—or even the computation of spurious zeros—in subintervals where $f(x)$ is oscillating through a range that is tiny compared to $\max_{x\in[a,b]}|f(x)|$. Newton-polishing is thus also a final safeguard.

2.11 ▪ The Chebyshev-Proxy Algorithm for Computing the Zeros of a Transcendental Equation: Mechanics

CPR is remarkably simple. To compute the real-valued roots of $f(x)$ on an interval $x \in [a,b]$:

1. Adaptively approximate $f(x)$ by Chebyshev interpolants of increasingly high degree N until $|f(x)-f_N(x)|$ is less than a preset tolerance ϵ everywhere on $x \in [a,b]$.

2. Compute the roots of $f_N(x)$ as the eigenvalues of the Chebyshev–Frobenius companion matrix (if N is small or moderate); if N is large, subdivide the interval and apply the Chebyshev-proxy scheme with moderate N on each subinterval.

3. Optionally (but highly recommended), the roots of the proxy f_N can be refined by one or two Newton iterations of $f(x)$, Newton-polishing.

The only restriction is that $f(x)$ must be a nice function on $x \in [a,b]$ in the sense that it possesses a rapidly convergent Chebyshev series on the interval, as quantified in Section 2.13. If $f(x)$ is analytic on the entire interval, then the error in truncating the Chebyshev series after the coefficient of $T_N(x)$ falls *geometrically* with N. And then a polynomial approximation $f_N(x)$ of modest degree N will be a very good proxy for the function.

A good idea is usually only half a good paper. The Chebyshev-proxy algorithm is a combination of at least three good ideas: rootfinding by proxy, Chebyshev interpolation, and a method for finding the roots of a polynomial in Chebyshev form. But even these three concepts are not enough. Ideas must be understood deeply and tested, like a textbook chapter that is not really understood until the exercises have been completed. Fortunately, the articles [71, 81, 82, 146, 102, 83] have filled in the necessary details and demonstrations. In the next few sections, we analyze difficulties, cures, and generalizations.

2.12 ▪ MATLAB and Maple Codes

The code consists of a main program, which includes a test case, as Table 2.1, and the function it calls to do the actual rootsolving is presented in Table 2.2. The codes do not implement automatic subdivision or adaptation; MATLAB implementations of these are included in *Chebfun*. The codes here do illustrate the basic components of CPR.

The function includes four user-choosable parameters. It is better to include roots slightly off the canonical interval, the real interval $x \in [-1,1]$, rather than reject a zero that has been computed as -1.0000001 or $-0.4 + i0.0000001$. The function returns as good roots all that satisfy both the inequalities

$$|\Re(x_j)| \leq 1 + \sigma, \qquad |\Im(x_j)| \leq \tau, \tag{2.12}$$

where σ and τ are tiny, nonnegative constants.

The Chebyshev companion matrix is surprisingly robust to very tiny trailing coefficients even though most nonzero elements are divided by a_N. Even so, it is silly to solve an eigenproblem of dimension 100 when the coefficients of $f(x)$ reach the roundoff plateau at $N = 40$. The function therefore approximates $f(x)$ by a polynomial f_N, where N is the smallest N such that at least one Chebyshev coefficient a_n for $n \leq N$ satisfies

$$|a_n| \geq \epsilon_{cutoff} \max_n |a_n|. \tag{2.13}$$

The default is $\epsilon_{cutoff} = 10^{-13}$.

Lastly, if there is a possibility of dynamic range problems, the program issues a warning. The criterion is

$$\min_j |f(x_j| \leq dyn\_range\_tol \max |f(x_j|, \tag{2.14}$$

where a good choice for the fourth parameter is roughly the square root of machine epsilon, or the default $dyn\_range\_tol = 10^{-8}$.

The function returns not only the roots as "froots," but also the interstitial error and the root residuals. The interstitial error is discussed in depth in section 3.3. The Chebyshev–Lobatto interpolation points are the images of a uniformly spaced set of points $t_j = \pi j / N$ under the mapping $x_j = [(b-a)/2]\cos(t_j) + (b+a)/2$. The interstitial error is the difference between $f(x)$ and its interpolant $f_N(x)$ at points that are the images of the points halfway between each of the t_j. For reasons explained in section 3.6, the L_∞ norm of this, which is actually what is returned by the function, is usually a very good estimate of the interpolation error. More precisely,

$$E_N^{interstitial} \equiv \max_{x_k} |f(x_k) - f_N(x_k)|, \; k = 1, \ldots, N; \; x_k = \frac{b-a}{2}\cos\left(\pi \frac{2k-1}{2N}\right) + \frac{b+a}{2}. \tag{2.15}$$

The vector "root_residuals" returns as its elements

$$f(r_j), \tag{2.16}$$

where the r_j are the computed roots, the elements of the vector "froots."

A nicer and more automatic version of this rootfinder has been implemented in the *Chebfun* system. Details and MATLAB software can be found at

http://www2.maths.ox.ac.uk/chebfun/.

2.13 ▪ When *f*(*x*) Isn't Smooth

In engineering reality, often $f(x)$ isn't analytic everywhere on the desired target interval. What then?

Table 2.1. *MATLAB code for the main script for computing roots using CPR.*

```
% This finds real roots of f(x) on the interval x ∈ [a, b]
% by approximating f(x) by a degree-N Chebyshev polynomial interpolant
% on [a, b]
  clear, clear all, global acoeff xiall xall faall fpoly Nt;
% specify limits of search interval, x ∈ [a, b],
  a=0, b=40
% specify Chebyshev degree "N",
  N=50
% Define the string with the name of the function.
  fnamestring='J0',
% Bessel function J_0
  [froots,Einterstitial,root_residuals ]=CPR(fnamestring,a,b,N);
% To do other functions, replace 'J0' by the name of a function file that
% computes f(x). The function must take only a single argument f.
  disp([' norm of interp. error at interstitial points=', num2str(Einterstitial)])
  disp([' number of roots=',num2str(length(froots) )] )
  disp(' roots computed by Chebyshev Proxy Rootfinder, vector "froots"'), froots
  max_residual_at_roots=max(abs(root_residuals))
  display([' maximum of |f(froots)| =',num2str(max_residual_at_roots)])
  BesselJ0_exactroots=[ 2.404825557695773 5.520078110286311 8.653727912911012 ...
  11.79153443901428 14.93091770848779 18.07106396791092 21.21163662987926 ...
  24.35247153074930 27.49347913204025 30.63460646843198 33.77582021357357 ...
  36.91709835366404]
  disp(['errors in Bessel roots']),
  errBessel=froots-BesselJ0_exactroots
  maxerror=max(abs(errBessel));
  disp([' maximum error in a Bessel zero=',num2str(maxerror)])
```

2.13.1 ▪ Piecewise Analytic Functions

A *piecewise* analytic function is one that is analytic everywhere on each subinterval $[a, b]$ *including the endpoints*. An example is

$$f(x) = \begin{cases} \cos(\pi(x-1/3)), & x \geq 0, \\ \cos(\pi(x+1/3)), & x < 0, \end{cases} \tag{2.17}$$

which is graphed as Fig. 2.6. If $f(x)$ is piecewise analytic, the remedy is piecewise Chebyshev approximation with intervals matching each piece. Accuracy is poor if any of the intervals of Chebyshev approximation include any of the discontinuities on the *interior* of the interval. Spectral accuracy will be the norm if the discontinuities between one piece and another are only at the *endpoints* of the Chebyshev search intervals.

For the example illustrated, a search interval from the origin (where the singularity is located) to any positive limit $x = b$ generates a Chebyshev series with at least geometric convergence. (Because $f(-x) = f(x)$, it is only necessary to compute positive roots.) Expanding $f(x)$ on $x \in [0, 1]$ yields

$$\begin{aligned} \cos(\pi[x-1/3]) = {} & 0.41 - 0.57T_1(2x-1) - 0.43T_2(2x-1) + 0.069T_3(2x-1) \\ & + 0.024T_4(2x-1) - 0.0022T_5(2x-1) - 0.00052T_6(2x-1) + 0.34\times 10^{-4}T_7(2x-1) \\ & + 0.58\times 10^{-5}T_8(2x-1) - 0.29e-6T_9(2x-1) - 0.40\times 10^{-7}T_{10}(2x-1) + \cdots . \end{aligned} \tag{2.18}$$

Table 2.2. *MATLAB code for the CPR function.*

```
function [froots,Einter,root_residuals]= CPR(f,a,b,N);
% Goal: compute the roots of f(x) on the real interval x ∈ [a, b].
% Input: a, b are the endpoints of the search interval
% f must be a string variable, i.e., 'f', in the calling program.
% N is degree of Chebyshev expansion [(N + 1) points]
% Output: froots is a scalar or vector containing the roots of f(x) on
% the interval x ∈ [a, b], sorted so the smallest is first.
  global acoeff xiall xall faall fpoly Nt;
% User-choosable numerical parameters. The defaults are usually fine.
  epscutoff= 1.E−13;
% trailing coefficients discarded if %abs(acoeff(j)) < trunparam*max(abs(acoeff))
  tau= 1.0E−8; % discard roots with abs(Im(root)) > tau
  sigma= 1.0E−6; % discard roots such that abs(Re(root)) > (1 + sigma)
  dyn_range_tol= 1.0E−8; % trigger for warning of small dynamic range.
  N1=N+ 1; t= (0:1:N) *pi/ N;
% trigonometric argument xi = cos(t);
% Chebyshev polynomial argument "xi"; range of xi ∈ [−1, 1]
  x= 0.5*(b − a)*xi + (b + a)*0.5;
% x ∈ [a, b] for ii= 1:N1, fa(ii)= feval(f,x(ii)); end,
% samples of f(x)
% Compute "acoeff", the Chebyshev coefficients of f(x) on x ∈ [a, b]
  phia= zeros(N1,N1); pj= ones(1,N1); pj(1)= 2; pj(N1)= 2;
  for ii= 1:N1, phia(ii,:)= ( (2/N) /pj(ii) ) * cos((ii− 1)*t) ./ pj;, end
  acoeff= phia*fa';
% Truncate the tail of the series, if very small.
  acoeffmax= max(abs(acoeff)); Nt= N; tailnorm= 0;
  for k= N: − 1:2, tailnorm= tailnorm + abs(acoeff(k + 1));
  if tailnorm <epscutoff*acoeffmax, Nt= Nt − 1; end, end,
% Part two: create Chebyshev companion matrix.
  A= zeros(Nt,Nt); A(1,2)= 1; for j= 2:(Nt − 1), A(j,j − 1)= 0.5; A(j,j + 1)= 0.5; end
  A(Nt,1:Nt)= − acoeff(1:Nt)/ (2* acoeff(Nt + 1));
  A(Nt,Nt − 1)= A(Nt,Nt − 1) + 1/2; all_roots_in_xi= eig(A);
% ← Compute roots in Chebyshev coordinate xi.
% Accept roots only if within "tau" of real axis & real(|root |) <(1 + sigma)
  icounter= 0; for ii= 1:Nt;
  if abs(imag(all_roots_in_xi(ii))) <tau * abs(real(all_roots_in_xi(ii)))
  if abs(real(all_roots_in_xi(ii) )) <= (1 + sigma), icounter= icounter + 1;
% convert roots on & near the standard interval in ξ to x ∈ [a, b].
  froots(icounter)= 0.5*(b − a)*real(all_roots_in_xi(ii)) + (b + a)/2; end, end, end
  froots= sort(froots);
% CPR ends. Checks below.
% ************ residual at roots ***************
  for j= 1:length(froots), root_residuals(j)= feval(f,froots(j)); end
% ******* Evaluate interstitial interpolation residual ********
  tinter= t(1:N) + 0.5/N; xiint= cos(tinter); xint= 0.5*(b − a)*xiint + (b + a)*0.5;
  for ii= 1:N, fainter(ii)= feval(f,xint(ii)); end
  taall= [t tinter]; faall= [fa fainter]; xiall= [xi xiint]; xall= [x xint];
  fpoly= zeros(1,2*N + 1); for j= 1:(N + 1),
  fpoly= fpoly + acoeff(j) * cos((j − 1) * taall); end,
  Einter= max( abs( faall( (N + 2):(2*N + 1))-fpoly( (N + 2):(2*N + 1) ) ));
% !!!!!!!!!!!!! Dynamic range calculation. !!!!!!!!!!!
  fmax= max(abs(faall)); dyn_range= min(abs(faall))/fmax;
  if dyn_range <dyn_range_tol,
  disp(['Warning!!: dyn_range= ',num2str(dyn_range)]), end
```

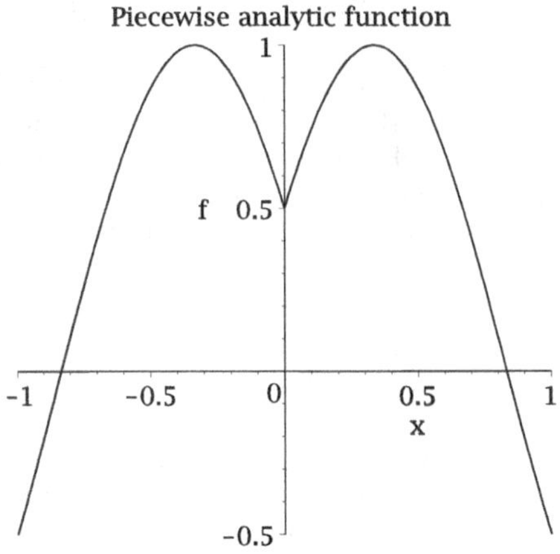

Figure 2.6. *The function $f(x)$ illustrated is piecewise discontinuous with a slope discontinuity at the origin; any Chebyshev expansion on the interval $x \in [a, b]$ will converge with only a second-order rate of convergence if $a < 0$. However, a Chebyshev series $x \in [0, 1]$ converges supergeometrically.*

The convergence is indeed rapid; recall that $|a_n|$ is a tight upper bound on $|a_n T_n(x)|$. The roots of the cubic, quintic, and $N = 10$ truncations of the Chebyshev series approximate the root $x = 5/6$ with errors of only 0.0073, 0.000042, and 5.7×10^{-10}, respectively.

2.13.2 ▪ Weak Singularities with Several Bounded Derivatives

If the singularities on $x \in [a, b]$ are *weak* in the sense that the first K derivatives exist and are bounded, then the $(N+1)$-point Chebyshev interpolation error falls proportionally to $1/N^K$ if the singularity is on the interior of the interval, or as $1/N^{2K-1}$ if at an endpoint $x = a$ or $x = b$ (Chapter 2 of [76]). If K is not too small, we can apply the Chebyshev algorithm without change except that the degree of the proxy will be larger than would be needed if $f(x)$ were analytic [90].

2.13.3 ▪ Bounded Endpoint Singularities

If $f(x)$ has a branch point at either $x = a$ or $x = b$, but only its *derivatives* are unbounded at the singularity, an exponential change of coordinate near each endpoint that is a branch point can then recover an exponential rate of convergence [61, 76], a strategy popularized by Frank Stenger [340, 341]. An example is

$$f(x) = (1+x)\sqrt{1-x^2}. \tag{2.19}$$

The transformation $x = \tanh(y)$ changes the function into

$$g(y) = (1/2 + \tanh(y))\text{sech}(y)^2. \tag{2.20}$$

The function $g(y)$ is analytic for all finite real y and decays exponentially as $|y| \to \infty$. Its expansion as a series of rational Chebyshev functions, Hermite functions, or sinc functions—any good infinite interval basis—decays exponentially fast. Spectral methods on an unbounded interval are discussed at length in section 2.14.

2.14 ▪ Roots on an Infinite Real Interval

2.14.1 ▪ Rational Chebyshev Functions

When the spatial interval is *unbounded*, it is straightforward to generalize the Chebyshev-proxy strategy merely by replacing Chebyshev polynomials by a spectral basis appropriate for an unbounded domain, such as rational Chebyshev functions or Hermite or Laguerre functions. (Endpoint singularities on a *finite* interval also generate problems on an *unbounded* interval via a change of coordinate, as explained in section 2.13.3.)

There are a couple of potential traps. First, all the basis functions discussed here yield an *exponential* rate of convergence if $f(x)$ is analytic on the interval and decays *exponentially fast* as $|x| \to \infty$. If the function decays only as an inverse power of $|x|$, rational Chebyshev functions may still yield an exponential rate of convergence, but this is not automatic. The complications of power law decay of $f(x)$ are left to [76, 62, 64, 98].

The second trap is that if there is an *infinite* number of roots, the rational Chebyshev-proxy method will yield only a finite number of zeros—those nearest the origin. The *relative* error of the spectral approximations is large where $f(x)$ has decayed to tiny values, so the output of companion matrix eigenvalues must be checked by Newton's iteration to confirm genuine zeros and exclude spurious roots.

Zeros far from the origin are best captured by asymptotic approximations, which is one strong reason why this book includes chapters on perturbation and asymptotic methods. Perturbation series often become more and more accurate as the zero index (and the absolute value of the root) increase. The chapters on regular and singular perturbation series (Chapters 14 and 15) offer details and examples.

If $f(y)$ is nonoscillatory for large y and thus has only a *finite* number of real zeros, then the rational Chebyshev-proxy will usually have no trouble finding them all.

The rational Chebyshev functions $\mathrm{TB}(y;L)$ are a good basis for $y \in [-\infty,\infty]$. These functions are the images of a Fourier cosine basis under a mapping, and also images of Chebyshev polynomials under a different change of coordinate.

These basis functions are defined by

$$TB_n(y) \equiv T_n(x) \equiv \cos(nt), \qquad y \in [-\infty,\infty], \tag{2.21}$$

where the coordinates are related via

$$y = \frac{Lx}{\sqrt{1-x^2}}, \qquad x = \frac{y}{\sqrt{L^2+y^2}}, \tag{2.22a}$$

$$y = L\cot(t), \qquad t = \operatorname{arccot}(y/L). \tag{2.22b}$$

It follows that the mechanics of applying these basis functions to find the roots of $g(y)$ on $y \in [-\infty, \infty]$ are the following:

1. Apply adaptive Chebyshev approximation to

$$f(x) \equiv g\left(\frac{Lx}{\sqrt{1-x^2}}\right). \tag{2.23}$$

2. Compute the n_r roots $x_{k,*}$ as eigenvalues of the Chebyshev companion matrix.

3. Convert the roots back to the original coordinate y:

$$y_{k,*} = \left(\frac{L\,x_{k,*}}{\sqrt{1-x_{k,*}^2}}\right) \qquad k = 1,2,\ldots,n_r. \tag{2.24}$$

For the rational Chebyshev functions on a semi-infinite domain, the procedure is the same; only the change of coordinate is different:

$$TL_n(y) \equiv T_n(x) \equiv \cos(nt), \qquad y \in [y_0, \infty], \tag{2.25}$$

$$y = y_0 + \frac{L(1+x)}{1-x} \qquad \leftrightarrow \qquad x = \frac{y - L - y_0}{y + L - y_0}; \tag{2.26}$$

$$y = y_0 + L\cot^2\left(\frac{t}{2}\right) \qquad \leftrightarrow \qquad t = 2\,\text{arccot}\left(\sqrt{\frac{y - y_0}{L}}\right). \tag{2.27}$$

Boyd, Rangan, and Bucksbaum suggested the alternative rational basis [106]

$$TM_j(r; L) \equiv \cos(jx(r)), \qquad r = L\cot(x/2)\cos(x/2), \qquad x \in [0, \pi][\text{"TM map"}]. \tag{2.28}$$

The older TL is quite satisfactory, however, so we shall not elaborate on their differences.

Hermite functions are a good basis for the unbounded domain and Laguerre functions for $y \in [0, \infty]$. The Hermite companion matrix is given in Appendix A; its Laguerre counterpart is easily derived from the formulas that apply to orthogonal polynomial companion matrices in general in [83]. The mechanics of Hermite and Laguerre interpolation are reviewed in [76].

2.14.2 ▪ Choice of the Map Parameter or Scaling Parameter *L*

One remaining complication is *universal* for all basis sets on an unbounded domain. The spectral basis functions always contain, either implicitly or explicitly, a scaling parameter that must be chosen to match the width of the basis functions to the width of the fluctuations in $f(x)$. The scale parameter difficulty arises because on $x \in [-\infty, \infty]$, it is always true that if the set $\{\psi_n(x)\}$ is a good spectral basis for the unbounded interval, so is the set $\{\psi_n(Lx)\}$ for any positive constant L. For the Hermite and Laguerre functions, no scaling parameter is to be found in the standard references, but is nevertheless *implicit* in the freedom to change the argument from x to Lx.

The rational Chebyshev functions are obtained by a change of coordinate that stretches $x \in [-1, 1]$ into either $y \in [-\infty, \infty]$ or $y \in [0, \infty]$. The basis functions $\text{TB}_n(y; L)$ (for the interval with no finite endpoints) and $\text{TL}(y; L)$ (for the semi-infinite domain) are the images of the usual Chebyshev polynomials under the mapping. The scale parameter is *explicit* in the mapping, but its role is the same as for Hermite and Laguerre functions.

The map parameter L is usually chosen experimentally (but see [62, 64]); the user should pick the value of L that gives the fastest rate of convergence for the spectral series. Here, $L = 4$ was a good choice for the numerical example later in this section. It is not necessary to carefully optimize L; usually $L = 2L_{optimum}$ and $L = (1/2)L_{optimum}$ are nearly as good as the optimal L.

2.14.3 ▪ Subdivision for an Unbounded Interval

If adaptive rational Chebyshev interpolation returns a series of very large degree N, the infinite interval can be subdivided. However, the *interior* subintervals will be of *finite width*, requiring Chebyshev *polynomials* and not rational functions.

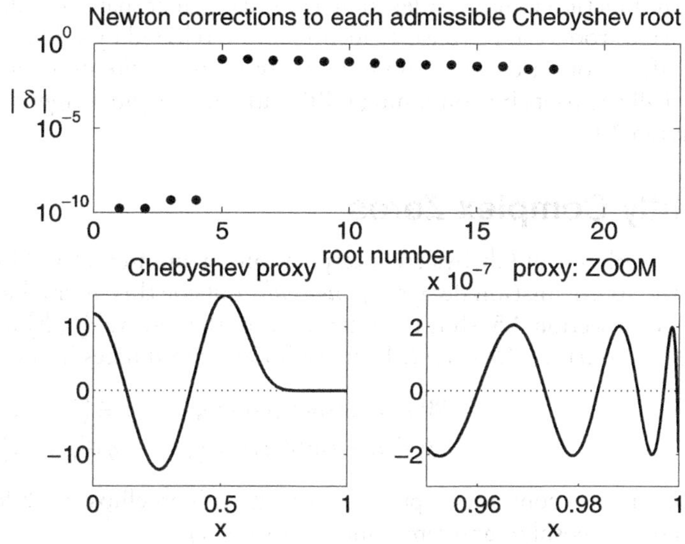

Figure 2.7. *Lower left: $f(y[x])$. Only the right half of $x \in [-1,1]$ is shown because the function is symmetric about the origin. Lower right: same as lower left but a zoom plot showing just $x \in [0.95,1]$ and graphing the Chebyshev approximation. Note that the maximum of $f(y[x])$ on the interval shown in the zoom plot is only 2×10^{-7}, although the global maximum (lower-left panel) is about 15. Upper plot: the Newton corrections $|\delta|$ for each root easily distinguish the many spurious zeros (large $|\delta|$) from the 4 genuine roots ($|\delta| \leq 10^{-10}$)* [96].

2.14.4 ▪ A Numerical Example on $y \in [-\infty,\infty]$

To illustrate numerical computation of a few roots near the origin, consider

$$f(y) = \exp(-0.5y^2)(12-48y^2+16y^4) \to \frac{df}{dy} = -4y\exp(-0.5y^2)(27-28y^2+4y^4). \quad (2.29)$$

After mapping $y \in [-\infty,\infty]$ to the finite interval, we expanded $f(y[x])$ as a Chebyshev series of degree 60. (This is equivalent to a rational Chebyshev TB basis in y.)

As illustrated in Fig. 2.7, the complication ensues because $f(y)$ decays exponentially fast as $|y| \to \infty$, which implies that $f(y[x])$ is *exponentially tiny* near the endpoints $x = \pm 1$, as evident in the zoom plot in the lower right of the figure. Although the Chebyshev expansion has tiny *absolute* errors over the whole interval, the *relative* errors are large near the endpoints. The result is that although $f(y)$ has only 4 real-valued zeros and no complex roots at all, its Chebyshev-proxy has 22 real-valued zeros on $x \in [-1,1]$.

The 18 spurious solutions can be readily identified by Newton-polishing. The upper panel in the figure shows the absolute values of the corrections δ in the first Newton's iteration to each of the proxy's roots. Four corrections do not appear because they are MATLAB "NaN"—not-a-number. The other spurious roots have Newton corrections that are $O(1)$. In contrast, the two smaller genuine roots have Newton corrections of only 1.7×10^{-10}, which are also the errors in the Chebyshev approximation of these roots; the two larger genuine zeros have errors and corrections of only 5.5×10^{-10}.

These spurious roots are a manifestation of the dynamic range difficulty; a highly accurate Chebyshev expansion is likely to have spurious zeros in regions where the amplitude of the function is many orders of magnitude smaller than its peak. It is fortunate that such spurious roots can be so easily identified and rejected by using Newton's iteration.

If, in contrast to this example, there is an *infinite* number of roots, one can often find all the roots by combining CPR with an *asymptotic approximation*, as illustrated in section 2.17.

2.15 ▪ Slightly Complex Zeros

The weakness of Chebyshev interpolation on an interval $x \in [a,b]$ is that the accuracy of the approximation decays *exponentially* outside this interval for either real or complex x. As in section 2.5, shift the interval from the interval $[a,b]$ in x to $z \in [-1,1]$, where $x \equiv z(b-a)/2+(b+a)/2$. Define elliptical coordinates in the complex z-plane by

$$\Re(z) = \cosh(\mu)\cos(\eta), \qquad \mu \in [0,\infty],$$
$$\Im(z) = -\sinh(\mu)\sin(\eta), \qquad \eta \in [0,2\pi]. \tag{2.30}$$

The curve of constant elliptical coordinate μ is an ellipse with foci at $z = \pm 1$ with semimajor axis $\cosh(\mu)$ and semiminor axis $\sinh(\mu)$.

Without approximation,

$$T_n(z) = \cos(n[\eta - i\mu]) = \cosh(n\mu)\cos(n\eta) - i\sinh(n\mu)\sin(n\eta), \tag{2.31}$$

which implies

$$|T_n(z)| = \sqrt{\cosh^2(n\mu)\cos^2(\eta) + \sinh^2(n\mu)\sin^2(n\eta)} \tag{2.32}$$
$$= \frac{1}{2}\sqrt{\exp(2n\mu) + \exp(-2n\mu) + \cos(2n\eta)} \tag{2.33}$$
$$\approx (1/2)\exp(n\mu), \qquad n\mu \gg 1. \tag{2.34}$$

This implies that along the ellipse with elliptical coordinate μ, the nth term of the Chebyshev series is no longer bounded by a_n but rather by $|a_n|\exp(n\mu)/2$. As we move farther in any direction from the canonical approximation interval, $z \in [-1,1]$, the error of the truncated Chebyshev series will increase exponentially and will diverge for $\mu > \overline{\mu}$, where $\overline{\mu}$ is the elliptical coordinate of the smallest Bernstein ellipse that contains a singularity of $f(x)$.

The implication is that in theory (and in *Chebfun*), Chebyshev interpolants can be used to find *slightly* complex roots, by which we mean roots whose imaginary parts are small.

In any event, to exclude the roots that are unreliable because they are too far from the Chebyshev target interval, $x \in [-1,1]$, the companion matrix eigenvalues should be accepted only if they pass the dual tests

$$|\Re(x_j)| \leq 1 + \sigma, \qquad |\Im(x_j)| \leq \tau, \tag{2.35}$$

where σ and τ are tiny, user-choosable nonnegative constants. Even if the goal is only real roots, a small positive value for τ such as 10^{-6} is helpful to detect a real-valued root doublet that has been perturbed into a complex conjugate pair with tiny imaginary parts. Newton's iteration can refine genuine roots to high precision or tag poor approximations as spurious, so there is some flexibility in excluding matrix eigenvalues that are outside the real interval $[-1,1]$.

The rootfinding by polynomial proxy idea can be extended to search for roots within a target region in the complex plane. Geddes and Mason showed that for approximation within a *disk*, interpolation on points evenly spaced around the boundary circle of the disk has the same optimum properties as Chebyshev interpolation on a real interval [189]. The disk or disks must be sufficiently small to be free of singularities (poles, branch points, etc.) of $f(x)$. One flaw is that overlapping disks do not tessellate the complex plane efficiently. Disks can be replaced by ellipses by interpolating at points *on* the ellipse [187] (not on a line segment inside the ellipse). More general regions can be treated by using Faber polynomials [385, 138].

A much more comprehensive discussion of complex zeros is given in Chapter 5, which is entirely devoted to zero-hunting in the complex plane.

2.16 ▪ Parity

2.16.1 ▪ Symmetries and Choice of Companion Matrix

In applications, functions that are symmetric or antisymmetric with respect to the origin are not unusual. Exploiting parity can *halve* N and reduce the cost of eigensolving by a factor of *eight*! First, we give a formal definition and then explain how to exploit such symmetry for rootfinding.

Definition 2.2 (Parity). *A function $f(x)$ is said to be* symmetric about the origin *or to possess* even parity *if for all x,*

$$f(x) = f(-x). \tag{2.36}$$

A function is said to be antisymmetric with respect to the origin *or to possess* odd parity *if for all x,*

$$f(x) = -f(-x). \tag{2.37}$$

A function that possesses one or the other of these properties is said to be of definite parity.

Theorem 2.3 (Zeros of Functions with Definite Parity). *If $f(x)$ is either symmetric or antisymmetric with respect to the origin and x_* is any positive zero, i.e., $f(x_*) = 0$ and $x_* \geq 0$, then $-x_*$ is also a root:*

$$x = -x_*. \tag{2.38}$$

If $f(x) = -f(x)$, that is, if $f(x)$ is antisymmetric with respect to the origin, $x = 0$ is always a root.

The obvious way to exploit parity is to search $x \in [0, b]$ instead of $x \in [-b, b]$ and then append the negatives of all positive roots to the list of zeros. The degree of the Chebyshev-proxy is roughly *halved.*

Other approaches are possible, such as deriving companion matrices and recurrence relations directly for basis sets composed only of even-degree Chebyshev polynomials [84]. Table 2.3 can guide the choice of companion matrix.

2.17 ▪ Additional Numerical Examples

2.17.1 ▪ Radial Basis Function Equivalence Equation

The equation

$$f(x, \lambda) \equiv \frac{1}{[2/x]K_0(\pi/x)/K_1(\pi/x) + 4/\pi} - \frac{1}{\pi}\lambda, \tag{2.39}$$

Table 2.3. *The choice of companion matrices: Symmetry, periodicity, and finite or unbounded interval.*

Situation	Basis	Strategy
$x \in [-\infty, \infty]$	Rational Cheb. TB(x)	Cheb. companion matrix, section 2.14
$x \in [-\infty, \infty]$	Hermite functions	Hermite companion matrix [83], Appendix A
$x \in [x_0, \infty]$	Rational Cheb. TL(x)	Cheb. companion matrix, section 2.14, Appendix A
$x \in [x_0, \infty]$	Rational Cheb. TL(x)	Cheb. companion matrix, section 2.14, Appendix A
$x \in [0, \infty]$	Laguerre functions	Laguerre companion matrix, Appendix A.9
Periodic: $x \in [-\pi, \pi]$	Fourier sines and cosines	Appendix A
Periodic: $f(-x) = f(x)$	Fourier cosine	Chebyshev companion matrix followed by $x_* = \arccos(\text{eigenvalue})$
Periodic: $f(-x) = -f(x)$	Fourier sine	Chebyshev 2nd-kind companion matrix; then $x_* = \arccos(\text{eigenvalue})$
Periodic: $f(-x) = f(x)$ and $f(x-\pi) = f(x)$	Fourier even cosines	$t = T/2$; apply cosine method in T section 4.6.3
Periodic: $f(-x) = -f(x)$ and $f(x-\pi) = -f(x)$	Fourier even sines	$t = T/2$; apply sine method in T section 4.6.3
Periodic: $f(-x) = f(x)$ and $f(x-\pi) = -f(x)$	Fourier odd cosines	divide by $\cos(t)$ via recurrence; then even cosines method, section 4.6.4
Periodic: $f(-x) = -f(x)$ and $f(x-\pi) = f(x)$	Fourier odd sines	divide by $\sin(t)$ via recurrence; then even cosines method, section 4.6.4

where K_0 and K_1 are the usual modified Bessel functions [284], is equivalent to computing the inverse of

$$g(x) \equiv \frac{1}{[2/x]K_0(\pi/x)/K_1(\pi/x) + 4/\pi} \tag{2.40}$$

via $x = g^{-1}(\lambda/\pi)$. Inverse functions are particularly favorable for CPR and higher-order proxies in general because an approximation for $g(x)$ on an interval $x \in [a, b]$ applies to the entire interval $\lambda/\pi \in [g(a), g(b)]$.

The interesting interval here is $\lambda \in [0, 1]$ with the corresponding range $x \in [0, 0.938]$; this equation must be solved to compute a shape parameter α for multiquadrics that makes these radial basis functions (RBFs) almost identical to Gaussian RBFs, whose shape parameter (squared) is λ [91]. (Because the upper limit of the interesting x-interval is close to one, we shall graph and expand on $x \in [0, 1]$ instead of $[0, 0.938]$, purely for convenience.) Selected Chebyshev coefficients for a single expansion on the entire target interval are listed in Table 2.4 and display the usual exponential rate of convergence. (The only technical complication is that because the arguments of the Bessel functions are π/x, $g(x)$ was interpolated on the Gauss–Chebyshev grid, which does not include the endpoints as interpolation points (see Appendix B).) Thus, the Bessel functions disappear and the original equation is replaced by the polynomial equation

$$\begin{aligned} f(x, \lambda) \approx F_{Cheb}(x, \lambda) \equiv\ & 0.183 + 0.164 T_1(2x-1) - 0.0173 T_2(2x-1) \\ & + 0.00158 T_3(2x-1) + \cdots - \frac{1}{\pi}\lambda. \end{aligned} \tag{2.41}$$

A smooth transcendental equation is just a polynomial in disguise.

Curve-fitting shows that asymptotically

$$a_n \sim 0.02 \exp(-5.55 n^{2/3}), \qquad n \to \infty. \tag{2.42}$$

Table 2.4. *Selected Chebyshev coefficients of* $g(x) = \{[2/x]K_0(\pi/x)/K_1(\pi/x) + 4/\pi\}^{-1}$ *on* $x \in [0, 1]$.

n	Coefficient of $T_n(2x-1)$
0	0.183
1	0.164
2	–0.0173
3	0.00158
4	–0.128e–3
5	0.891e–5
6	–0.449e–6
7	–0.76e–9
8	0.461e–8
9	–0.910e–9
10	0.138e–9
20	0.424e–17
40	0.434e–28
80	0.428e–45
160	0.920e–72
320	0.820583e–114
640	0.267432e–180

If $g(x)$ were analytic everywhere on the expansion domain, then the argument of the exponential would be *linear* in Chebyshev degree n. The singularity at an endpoint, a weak singularity in the sense that $g(x)$ has bounded derivatives of all orders at the point, is the reason that $n \rightarrow n^{2/3}$. In the standard jargon [76], a "geometric" rate of convergence is lowered to a "subgeometric" rate with index 2/3.

There are other options. A graph of $g(x)$ shows that it is nearly linear over the whole interval $x \in [0, 1]$. Perturbation theory is discussed at length in Chapters 14 and 15. Here, merely note that the nine Maple statements in Table 2.5 suffice to substitute a power series in λ with symbolic coefficients into $f(x, \lambda)$, collect powers of λ, impose the condition that the $O(\lambda^n)$ residual should vanish, and thus determine the series

$$x(\lambda) \sim \frac{2}{\pi}\lambda + \frac{6}{\pi^3}\lambda^2 + \frac{21}{\pi^5}\lambda^3 + \frac{75}{\pi^7}\lambda^4 + \frac{1107}{4\pi^9}\lambda^5 + \frac{4113}{4\pi^{11}}\lambda^6 + \frac{31113}{8\pi^{13}}\lambda^7 + \frac{117423}{8\pi^{15}}\lambda^8 + \cdots.$$

However, a plot of the $|p_n|$ (right panel of Fig. 2.8) shows that the coefficients first decrease exponentially, reach a minimum at $p_{14} = 6.5 \times 10^{-7}$, and grow exponentially.

A sound precept for rootfinders is

Graph early, and graph often.

However, graphs can mislead as well as lead. Because the argument of the Bessel functions is π/x, the small-x behavior of $g(x)$ is controlled by the large-x behavior of the Bessel functions. Individually, $K_0(\pi/x)$ and $K_0(\pi/x)$ have asymptotic approximations in the form of *divergent* power series in x multiplied by $\exp(-\pi/x)\sqrt{x}$, which cancel in the ratio. The series in both the numerator and denominator are divergent because $\exp(\pi/x)x^{-1/2}K_0(\pi/x)$ and $\exp(\pi/x)x^{-1/2}K_1(\pi/x)$ are infinitely differentiable but not analytic at the origin. The power series of each is asymptotic but with a zero radius of convergence.

Table 2.5. *Maple code: Perturbation series for the RBF equivalence equation.*

```
f:= (x, lambda) -> 1/((2/x)*BesselK(0,Pi/x)/BesselK(1,Pi/x) + 4/Pi) - lambda/Pi;
npert := 5; # pert. series is computed through lambda**npert;
for iorder from 1 to npert do
    x := 0;
    for n from 1 to (iorder + 3) do
        x := x + p[n]*lambda**n;
    od:
    fs := series(f(x,lambda),lambda,n + 3);
    fs := simplify(subs(csgn(lambda) = 1,fs));
    p[iorder] := solve(coeff(fs,lambda,iorder),p[iorder]);
    print ('x(lambda) =',x);
od:
```

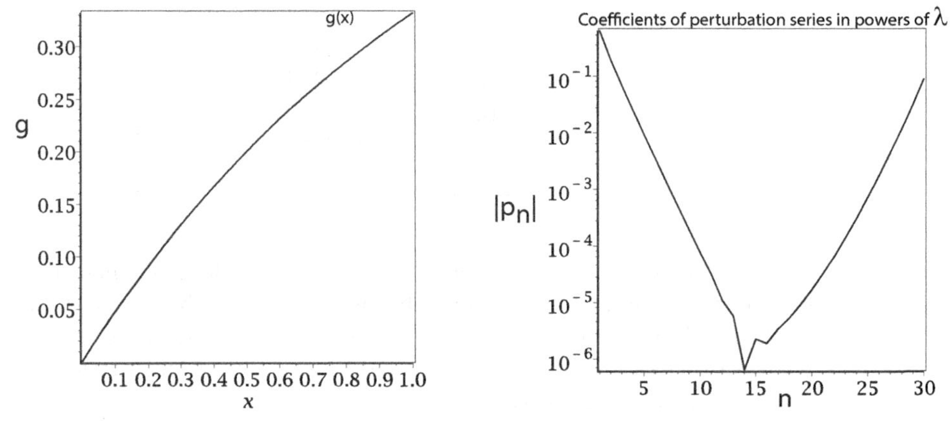

Figure 2.8. *Left:* $g(x)$ *in the RBF equivalence equation,* $f(x,\lambda) = g(x) - \lambda/\pi$. *Right: absolute value of the coefficients in the asymptotic but divergent series* $x(\lambda) \sim \sum_{n=0} p_n \lambda^n$.

Fortunately, it is still possible to use the first few terms of the series to approximate $x(\lambda)$ for $\lambda < 1$. The usefulness of asymptotic but divergent series is explained in [39, 273, 75, 79, 74] and in Chapters 14 and 15.

Indeed, this series provides a never-failing Newton's initialization (NFNI). Fig. 2.9 shows that a perturbative initialization, in this case $x^{(0)}(\lambda) \sim (2/\pi)\lambda + (6/\pi^3)\lambda^2 + (21/\pi^5)\lambda^3 + (75/\pi^7)\lambda^4$, yields monotonic convergence for all x, with the slowest convergence at $\lambda = 1$. For this worst case, Table 2.6 lists the errors in the perturbative approximation followed by the errors in the first four Newton iterates. Even the linear approximation (first line of the table), $x^{(0)}(\lambda) \sim (2/\pi)\lambda$, is an NFNI, too. The four-term initialization converges to errors less than MATLAB precision (also IEEE double precision), 10^{-16}, with just three iterations on the entire domain. The simplicity of Newton's iteration is illustrated in Table 2.7.

This function is a good example of the power of Chebyshev expansions. Even though $f(x,\lambda)$ is singular (albeit infinitely differentiable) at one endpoint, a single, exponentially convergent series completely polynomializes the problem. The Chebyshev series converges exponentially fast (proportional to $\exp(-0.55n^{2/3})$) even though a power series in the same parameter is asymptotic but divergent for all λ. Bessel functions have disappeared, and what remains is only the inversion of a polynomial.

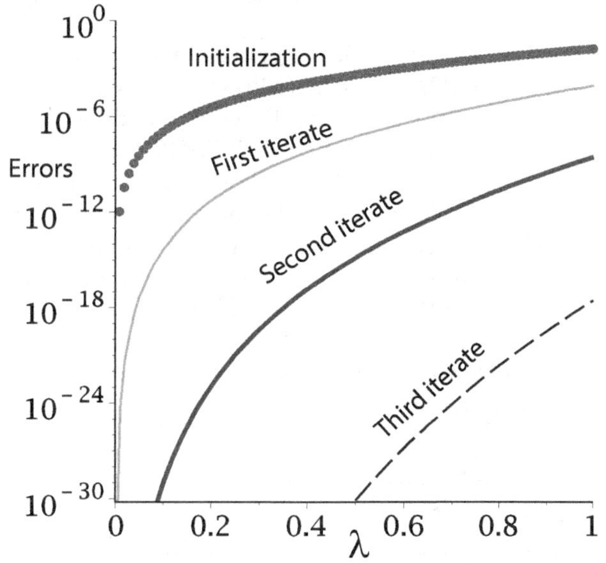

Figure 2.9. *Errors in the initialization (sum of the first four perturbative terms) and the first three Newton iterates for the entire parameter range, $\lambda \in [0,1]$, for the RBF equivalence equation.*

Table 2.6. *Errors in iterates for the RBF equivalence equation at the point of slowest convergence, $\lambda = 1$, for perturbative initializations of various orders.*

Initialization order	$E^{(0)}$ Initialization	$E^{(1)}$ 1st iterate	$E^{(2)}$ 2nd iterate	$E^{(3)}$ 3rd iterate	$E^{(4)}$ 4th iterate
1	0.3019	0.3335E−1	0.4167E−3	0.6522E−7	0.1598E−14
2	0.1084	0.4375E−2	0.7186E−5	0.1940E−10	0.1414E−21
3	0.3979E−1	0.5927E−3	0.1320E−6	0.6543E−14	0.1605E−28
4	0.1496E−1	0.8393E−4	0.2646E−8	0.2631E−17	0.6E−31

This function also illustrates that there are other good options. A few lines of Maple or Mathematica suffice to calculate the perturbation series to high order, but its asymptotic nature limits its attainable accuracy to about seven decimal places. (Padé approximants derived from divergent series may converge, as discussed in sections 14.6–14.8, but this has not been investigated for this example.) The NFNI is simple and robust and probably the best option for most users.

Once a good rootfinder has been developed, it can be rendered obsolete by using a few discrete values of $x(\lambda)$ to compute a Chebyshev or Chebyshev–Padé approximation to the root on $\lambda \in [0,1]$. Just 16 terms of the Chebyshev series are accurate to 10^{-16} for this example (not illustrated).

2.17.2 ▪ Multiple Roots and Doublets

Denoting the usual zeroth-order Bessel function by J_0,

$$f(x) \equiv (x - 10^{-4})(x + 10^{-5}) J_0(x), \tag{2.43}$$

Table 2.7. *Maple code: NFNI for the RBF equivalence equation.*

```
# Preprocessing step: Define g(x) and its derivative for use in Newton's iteration;
  g:= x → 1/((2/x)*BesselK(0,Pi/x)/BesselK(1,Pi/x) + 4/Pi);
  gx := x → − (Pi*BesselK(1, Pi/x)**2 − 2*BesselK(0, Pi/x)*x*BesselK(1, Pi/x)
          − BesselK(0, Pi/x)**2*Pi)*Pi**2/
                    (2* ( x*(BesselK(0, Pi/x)*Pi + 2*x*BesselK(1, Pi/x))**2));
  x_RBFequivalence := proc(lambda)      local x,iter;      x := (2/Pi)*lambda
  + (6/Pi**3)* lambda**2 + (21/Pi**5) * lambda**3 + (75/Pi**7) * lambda**4;
  for iter from 1 to 3 do      # Refine the initialization by 3 Newton iterations;
    x := x − evalf(( g(x) − lambda/Pi) / gx(x));
  od:
# Only 3 Newton's iterations are needed for MATLAB (16 decimal digit) precision;
  return x; end proc;
```

Table 2.8. *Roots of* $(x - 1/10000)(x + 1/100000)J_0(x)$ *as approximated by Chebyshev interpolants of various degrees.*

n	Exact	10th degree error	20th degree error	30th degree error
1	−5.520078110	1.4136e−03	−2.8873e−10	1.1546e−14
2	−2.404825558	8.9391e−03	−5.5956e−10	3.5527e−15
3	−0.00001	2.0223e−07	7.8818e−11	−7.2297e−11
4	0.0001	−2.0223e−07	−7.8830e−11	7.2288e−11
5	2.404825558	−8.9391e−03	5.5957e−10	4.4409e−16
6	5.520078110	−1.4136e−03	2.8874e−10	−6.2172e−15

is challenging because it has a doublet, which is a pair of closely spaced roots. Doublets and multiple roots are inherently very sensitive to perturbation, whether by a small physical change in $f(x)$ or a numerical perturbation due to the inherent errors of numerical approximation and floating point arithmetic. For example, perturbing

$$x^4 \Rightarrow x^4 - \epsilon \tag{2.44}$$

changes a single quadruple root to 4 simple roots at $x = \epsilon^{1/4}\exp(j i \pi/2), j = 0,1,2,3$. Note that if the perturbation $\epsilon = 10^{-16}$, for example, the roots shift by 10^{-4}, which is 12 orders of magnitude larger than the perturbation!

No rootfinding algorithm can annul these harsh mathematical realities. However, the doublet-Bessel example (Table 2.8) shows that CPR does as well as one might reasonably hope. The errors for the roots near the origin (the doublet) are larger than the errors of the other roots when the polynomial degree is 30, and increasing the degree from 20 to 30 yields no improvement for the doublet. Still, an accuracy better than 10^{-10} is quite satisfactory for most engineering purposes!

2.18 ▪ Rational Proxies

Rational approximants have become a very powerful tool in numerical analysis. A power series up to and including degree $(M + N)$ can always be formed into a rational function that is the ratio of a polynomial in degree M divided by a polynomial of degree N, a *Padé approximant*, written $f_{[M/N]}$. In Chapters 14 and 15, we show how such approximants can greatly extend the usefulness of perturbation series. Chebyshev series have been similarly generalized to Chebyshev–Padé approximants.

Table 2.9. *Maple code for Padé–Chebyshev-proxy rootfinding.*

```
restart; with(orthopoly); with(numapprox); Digits := 16;
  f := x → BesselJ(0,x);
# replace with user's choice a := 0; b := 20;
# search for roots on x ∈ [a, b]
  M := 8; N := 8;
# degrees of numerator & denominator polynomials
  fMN := chebpade(f(x),x = a..b,[M,N]);
  fnum := evalf(simplify(numer(fMN)));
  rootapprox := fsolve(fnum,x);
```

An obvious question is: How useful are rational approximations for computing zeros? Folklore asserts that diagonal approximants ($N = M$ or $N = M + 1$) are the most accurate; in comparison to the truncated Chebyshev series of degree $2N$, the rootfinding problem has been reduced to computing the zeros of the *numerator* of the rational approximation. If the QR/QZ algorithm is applied to find the zeros, halving the degree reduces the eigenvalue-solving cost by a factor of eight!

If one has a priori information that there is but a single simple zero on a given interval, one can in principle form approximations of numerator degree $M = 1$ and denominator degree as large as one pleases. The rootfinding problem is completely trivialized!

The answer to the usefulness of rational proxies is: *Definitely Maybe*.[4] First, we show some successful examples and then an analysis of their difficulties.

Example 1: Find the 6 real roots on $x \in [0, 20]$ of the Bessel function $f(x) \equiv J_0(x)$. The maximum absolute error in any of the 6 errors (before final improvement by Newton's iteration applied to the Bessel function and not its proxy) was 0.0013 for the $[8/8]$ Padé–Chebyshev approximation versus 0.00042 for the Chebyshev polynomial approximation of degree 16. The $[12/12]$ and $[24/0]$ (i.e., Chebyshev polynomial series of degree 24) reduced the errors to 1.0×10^{-8} and 6.1×10^{-9}. All computations were done in Maple in 16-digit precision (Table 2.9).

Example 2: Table 2.10 compares the diagonal Padé–Chebyshev approximation with the $[1/N]$ approximant for the function $f(x) \equiv \cos(\pi x)/(1 + x/2)$, which has a single root at $x = 1/2$ on the interval $x \in [-1, 1]$. Thirty-two-digit precision was used to extend the computations to the 20th degree. The extreme $[1/N]$ is successful, but the diagonal approximant is significantly more accurate.

These examples are encouraging, but there are difficulties. First, an approximation $[M/N]$ may have numerator zero and denominator zero at the same location [188]. The approximants of various degrees may always be arrayed in a table with rows indexed by M, the numerator degree, and columns by N, the denominator degree. The jargon used for a co-located pole and zero is the "Padé table has block structure," that is, a rectangular block of entries in the table are the same. Rational functions are the extreme of blockiness: the Padé table for any rational function always has an infinite block of identical elements all equal to the rational function.

The second difficulty is that the numerator and denominator zeros may *almost*, but not exactly, coincide. Such *Froissart doublets* are very common in applications and often appear at orders $[M/N]$ and in neighborhoods in the complex plane where lower-order approximations have apparently converged very close to the true $f(x)$.

[4]To borrow the title of one of the many fine science fiction novels by Boris and Arkady Strugatsky.

Table 2.10. *Errors in zeros from the Padé–Chebyshev diagonal and* $[1/N]$ *proxies for* $f(x) \equiv \cos(\pi x)/(1+x/2)$ *(multiple precision arithmetic).*

N	Error from $f_{[1/N]}$	Error from $f_{[(N/2)/(N/2)]}$
2	0.0060	0.0060
4	0.00048	0.00042
6	0.000036	0.0000026
8	2.6E–6	2.4E–8
10	1.8E–7	1.1E–10
12	1.3E–8	9.7E–15
14	9.5E–10	1.0E–15
16	6.8E–11	1.3E–16
18	4.9E–12	3.2E–21
20	3.5E–13	2.6E–23

Lastly, Padé and Padé–Chebyshev approximations are rather ill conditioned. This is irrelevant when the coefficients are rational numbers, manipulated without error in Maple or Mathematica, but in floating point arithmetic, very bad things can happen.

Gonnet, Güttel, and Trefethen [199] have vastly improved this dreary situation by solving the necessary linear algebra through an SVD (singular value decomposition) factorization. Their code detects rank-deficiency and lowers the order of the returned approximation when "blockstructure" is detected. And of course SVD is a powerful weapon against ill-conditioning whenever matrices need be factored or inverted.

However, the expense of the SVD and subsequent steps greatly diminishes the potential cost savings of rational proxy rootfinding relative to polynomial proxy computation of zeros. And sometimes the Gonnet, Güttel, and Trefethen algorithm returns a very low degree approximation when a high-degree rational function was requested. So, rational proxies for zero-hunting remain an interesting topic for the future—still definitely maybe.

2.19 ▪ Hermite–Padé Approximants

The Padé approximant satisfies the linear equation

$$Q_N(x) f_{[M/N]}(x) - P_M(x) = O(x^{M+N+1}). \tag{2.45}$$

An obvious generalization is to define an approximation to be a solution to a polynomial equation

$$\sum_{n=0}^{D} Q_n(x) \left\{ f_{[d_0/d_1 \ldots d_D]}(x) \right\}^n \approx 0, \tag{2.46}$$

where the coefficients $Q_n(x)$ are polynomials of degree d_n whose coefficients are chosen so that the polynomial equation is satisfied up to some power of x.

The quadratic approximant is $f[K/L/M]$, which solves [320]

$$P(f_{[K/L/M]})^2 + Q f_{[K/L/M]} + R = 0, \tag{2.47}$$

where the polynomials P, Q, and R are of degrees K, L, and M, respectively. The constants in P and Q are normalized to one without loss of generality, so there are $N = K + L + M + 1$ degrees of freedom in the three polynomials.

When $f_{[K/L/M]} = 0$, the quadratic implies that

$$R(x) = 0. \tag{2.48}$$

Table 2.11. *Approximate zeros from the quadratic Hermite–Padé proxies for* $f(x) \equiv \sin(\pi x)/(1+\pi^2 x^2)$.

[10/10/10] Digits = 16	[10/10/10] Digits = 64	[0/15/15] Digits = 16	[0/15/15] Digits = 64	[0/15/15] Digits = 16
-	−3.58	-	−3.93	-
−2.79	−3.018	2.57	−3.00021	-
−2.0032	−1.9999987	−2.0065	−1.9999965	−1.76
−0.999999936	−1.000000000000096	−0.999999932	−1.0000000000000017	−1.0011
0. and 0.0000237	0	0 and −0.58E−26	0 and −0.72D-29	0 and −0.58E−8
0.999999936	1.000000000000072	0.999999932	1.0000000000000017	1.0011
2.0032	1.9999987	2.0065	1.9999999875	1.76
2.79	2.98	2.57	3.00021	-
-	-	-	3.93	-

Table 2.11 shows the roots of R for various quadratic Hermite–Padé approximations for

$$f(x) \equiv \frac{\sin(\pi x)}{1+\pi^2 x^2}. \tag{2.49}$$

The sole ingredient for generating the Hermite–Padé proxy is the truncated Taylor series of $f(x)$, up to and including order $K+L+M$. For this $f(x)$, the radius of convergence of the power series is only $|x| \leq 1/\pi \approx 0.318$ because the function has simple poles at $x = \pm i/\pi$. Nevertheless, the high precision [0/15/15] approximation gives accurate approximations to the roots at ± 3—more than nine times the radius of convergence of the power series that was its parent! This ability to extend a power series far beyond its radius of convergence will be exploited in Chapters 14 and 15 on perturbation theory.

The table also shows that numerical roundoff error in 16-decimal precision can significantly degrade accuracy. The Hermite–Padé coefficient $R(x)$ has a double zero at the origin even though $f(x)$ has only a simple zero.

Padé and Hermite–Padé approximations in both power series and Chebyshev variants are an intriguing topic for the future. In the rest of the book, though, we shall stick to Chebyshev polynomial and other linear approximations—safe, reliable, and inexpensive.

2.20 ▪ Summary

The simple algorithm described above can be extended in a number of directions. Chapter 4 describes how to compute the roots of a general *trigonometric* polynomial by a companion matrix method [84, 95, 97]. An extension to a trigonometric polynomial augmented by a nonperiodic or *secular* term is in [107] and also Chapter 4. Polynomial-proxy methods can be extended to search for roots in regions of the complex plane, too, as described in Chapter 5.

Chapter 3

Adaptive Chebyshev Interpolation

"However it has been the authors' experience that $|a_N|$ in general provides a satisfactory error estimate even when the convergence of the corresponding Fourier-Chebyshev expansion is slow, and that it dominates the error when convergence is rapid."
—W. Fraser and M. W. Wilson [177].

3.1 ▪ Introduction: An Overview of Chebyshev Interpolation

Chebyshev interpolation is shorthand for "polynomial interpolation at the roots or the endpoints and extrema of the Nth Chebyshev polynomial." Whatever the name, this approximation converges geometrically fast with the degree N of the interpolating polynomial if the function approximated, $f(x)$, is analytic everywhere on the target interval. *Geometric* convergence means that the errors fall proportionally to $\exp(-\mu N)$, where μ is a positive constant (the *asymptotic rate of geometric convergence*), usually multiplied by slower than exponential functions of N such as a power or logarithm of N.

The central question is: How do we obtain an approximation such that

$$E_N \equiv \max_{x\in[a,b]} |f(x) - f_N(x)| \leq \daleth, \tag{3.1}$$

where $\daleth \ll 1$ is a user-chosen tolerance? For certain classes of functions, convergence theory yields tight bounds on Chebyshev accuracy. In the latter half of this chapter, we shall therefore review Chebyshev convergence theory and proffer some important illustrations.

For most $f(x)$, however, convergence theory is mostly useful for moral support. Fortunately, there is a black-box algorithm that, using only a set of evaluations of $f(x)$ at a set of discrete points (the *sample set*), generates fairly reliable answers to the following question: When is N big enough to make the error smaller than $\daleth$? This black box has many similarities to the Clenshaw–Curtis [129] quadrature scheme. It is, as we shall explain, not the cheapest strategy, but it is both inexpensive and robust.

Other criteria for deciding when "enough is enough" have been applied in the literature. We shall briefly catalog these, too, not merely as abandoned dead ends, but rather as options that illuminate the Clenshaw–Curtis adaptive scheme and its limits.

In the next section, we describe the mechanics of the first step, which is to compute the Chebyshev interpolant for a single, fixed-degree N. In the following section, we shall show that this procedure is the heart of an adaptive algorithm.

3.2 ▪ Chebyshev Interpolation of a Function *f*(*x*): Chebyshev–Lobatto Grid

Adaptive Chebyshev interpolation is completely mechanical. The key procedure is to compute a truncated Chebyshev series $\mathcal{I}_N f$, up to and including $T_N(x)$, on the user-chosen interval $x \in [a,b]$ so that $\mathcal{I}_N f$ interpolates the target $f(x)$ at $(N+1)$ interpolation points on $x \in [a,b]$. After accuracy testing, the interpolation step is repeated with $(N+1) \rightarrow (2N+1)$.

The iteration stops when either (i) N exceeds a user-chosen maximum, triggering subdivision of the interval into two, or (ii) the error is less than a user-specified tolerance $\beth$.

The Chebyshev interpolation step (for fixed N) has four parts.

The first step is to create the interpolation points by the explicit formula

$$x_j \equiv \frac{b-a}{2} \cos\left(\pi \frac{j}{N}\right) + \frac{b+a}{2}, \qquad j = 0, 1, 2, \ldots, N. \tag{3.2}$$

In much of our theoretical discussion, we assume that the domain is the canonical interval for Chebyshev polynomials, $x \in [-1, 1]$. There is no loss of generality in so assuming this because the grid formula is a linear, affine mapping from the canonical interval to $x \in [a,b]$, where a and b may be freely chosen by the user.

Next, compute the grid point values or samples of $f(x)$, the function to be approximated:

$$f_j \equiv f(x_j), \qquad j = 0, 1, \ldots, N. \tag{3.3}$$

Step three is to compute the elements of the $(N+1) \times (N+1)$ interpolation matrix. Define $p_j = 2$ if $j = 0$ or $j = N$ and $p_j = 1, j \in [1, N-1]$. Then the elements are

$$\mathcal{I}_{jk} = \frac{2}{p_j\, p_k\, N} \cos\left(j \pi \frac{k}{N}\right). \tag{3.4}$$

Lastly, the coefficients are evaluated through a vector–matrix multiply:

$$a_j = \sum_{k=0}^{N} \mathcal{I}_{jk} f_k, \qquad j = 0, 1, 2, \ldots, N. \tag{3.5}$$

(The vector–matrix multiply can be accelerated by the FFT, but since this vector–matrix multiply costs $O(2N^2)$, while the eigensolving cost is $O(10N^3)$, the FFT is scarcely worth the bother.)

The approximation is

$$f \approx \sum_{j=0}^{N} a_j T_j\left(\frac{2x-(b+a)}{b-a}\right) = \sum_{j=0}^{N} a_j \cos\left\{ j \arccos\left(\frac{2x-(b+a)}{b-a}\right)\right\}. \tag{3.6}$$

To summarize, the grid point representation, $\{f(x_j)\}$, generates through interpolation the coefficients $\{a_n\}$ of the spectral representation. Summation of the Chebyshev series is the reverse transformation $\{a_n\} \rightarrow \{f(x_j)\}$.

3.3 ▪ Adaptation without a Priori Knowledge

For some special but important classes of $f(x)$, such as trigonometric and Bessel functions, the Chebyshev coefficients are known in explicit, analytical form and rigorous bounds are possible. For all other targets, an adaptive procedure is necessary.

We exploit an observation of Clenshaw and Curtis [129]: all points on the $(M+1)$-point Chebyshev–Lobatto grid are also points on the $(2M+1)$-point grid. Thus, if the number of grid points is increased, with $N = M+1, 2M+1, 4M+1, \ldots$, for some positive integer M, all samples of $f(x)$ on coarser grids can be *reused* on the finer grid, which is very economical.

It is convenient (though not strictly necessary) to further choose N to be a power of two (plus one) and define a *grid level* K:

$$\mathcal{G}_K = \{\cos(\pi j/2^K), j = 0, 1, \ldots, 2^K\}, \tag{3.7}$$

so that level K is a grid with $N = 2^K + 1$ points (Table 3.1).

Table 3.1. *Chebyshev–Lobatto grids. Points in boldface on level K are not in grid $\mathcal{G}_{K-1}$ and are therefore interstitial points for level $(K-1)$. Each point is listed both as its exact, analytical form (upper entries in each row) and its numerical value (immediately under the explicit form).*

K	N	Points	Trig, $j=0,\ldots,N$
1	3	$0, \pm 1$	$\cos\left(\pi \frac{j}{2}\right)$
2	5	$0, \frac{\pm 1}{\sqrt{2}}, \pm 1$ $0, \pm\mathbf{0.7071}, \pm 1$	$\cos\left(\pi \frac{j}{4}\right)$
3	9	$0, \frac{\pm\sqrt{2-\sqrt{2}}}{2}, \frac{\pm 1}{\sqrt{2}}, \frac{\pm\sqrt{2+\sqrt{2}}}{2}, \pm 1$ $0, \pm\mathbf{0.3827}, \pm 0.7071, \pm\mathbf{0.9239}, \pm 1$	$\cos\left(\pi \frac{j}{8}\right)$
4	17	$0, \pm T_7(\zeta), \frac{\pm\sqrt{2-\sqrt{2}}}{2}, \pm T_5(\zeta), \frac{\pm 1}{\sqrt{2}}, \pm T_3(\zeta), \frac{\pm\sqrt{2+\sqrt{2}}}{2}, \pm\zeta, \pm 1$, where $\zeta \equiv \frac{\sqrt{2+\sqrt{2+\sqrt{2}}}}{2}$ $0, \pm\mathbf{0.1951}, \pm 0.3827, \pm\mathbf{0.5556}, \pm 0.7071, \pm\mathbf{0.8315}, \pm 0.9239, \pm\mathbf{0.9808}, \pm 1$	$\cos\left(\pi \frac{j}{16}\right)$

The second observation is that since, by construction, interpolation is exact at each of the interpolation points, the error in approximating $f(x)$ by the Nth-degree polynomial oscillates between the grid points. The local maximum of the absolute magnitude of the error is roughly halfway between each of the grid points. (In the trigonometric coordinate $t = \arccos(x)$, the grid is uniform, and the midpoints of each subinterval $t \in [t_k, t_{k+1}]$ are grid points of the Chebyshev–Lobatto grid with $2N+1$ points.) These local error maxima are near to, though not exactly coincident with, the N *interstitial midpoints*. The interstitial midpoints are by definition on the $2N+1$-point Lobatto grid, but not on the Lobatto grid with $N+1$ points, as illustrated in Figs. 3.1 and 3.2. These interstitial midpoints are

$$\mathcal{G}^{interstitial}(K) \equiv \mathcal{G}(K+1) - \mathcal{G}(K). \tag{3.8}$$

The error at these interstitial midpoints, dubbed the *interstitial error*, is

$$E_N^{interstitial} \equiv \max_{x_k \in (\mathcal{G}(K+1) - \mathcal{G}(K))} |f(x_k) - f_N(x_k)|, \quad k = 1, 2, \ldots, 2^K = N;$$

$$x_k = \frac{b-a}{2}\cos\left(\pi \frac{2k-1}{2^{K+1}}\right) + \frac{b+a}{2}. \tag{3.9}$$

As shown in Table 3.2, the interstitial error norm (the maximum error on a set of discrete points) is very close to the maximum error on the continuous interval, the error in the L_∞ norm. The progression from level to level is visualized in Fig. 3.3.

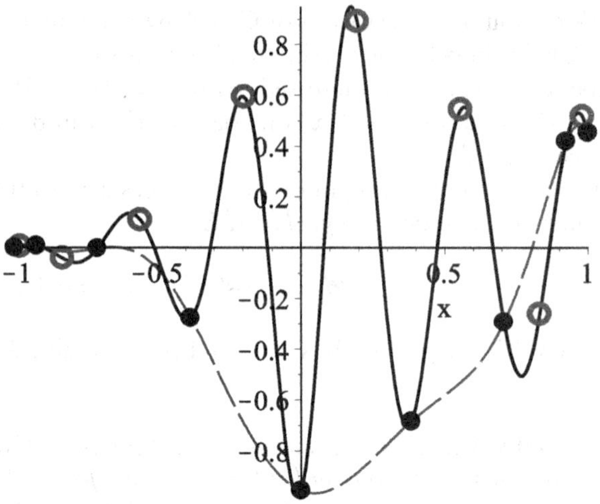

Figure 3.1. $f(x) = \cos(16(x+1/5))(1+x)\exp(x^2)/(1+9x^2)$ *is the solid curve. The 9-point interpolant to it is the dashed curve. The intersections of the two at the Lobatto grid points are marked by solid black circles. The 8 interstitial points—part of the 17-point Lobatto grid—are marked by open circles.*

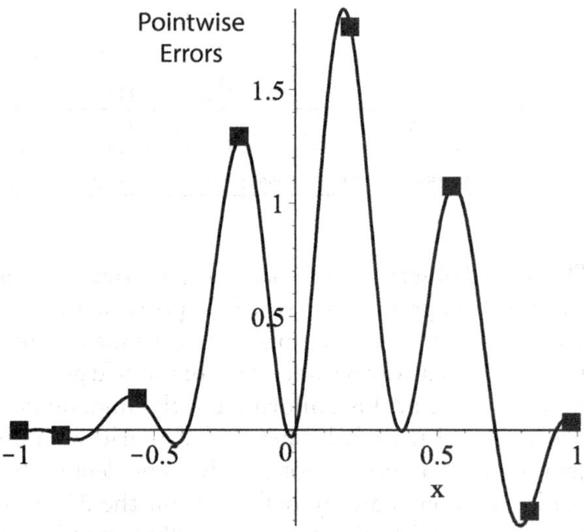

Figure 3.2. *Same as previous figure, but showing the pointwise error* $E_9(x) = |f(x) - f_9(x)|$ *as the black curve; the interstitial errors are marked by solid boxes.*

Our termination criterion is to stop once

$$|E_N^{interstitial}| \leq \daleth, \tag{3.10}$$

where $0 < \daleth \ll 1$ is a user-specified error tolerance. This termination criterion is justified in the next section.

In practice, a variety of simpler *N is enough* criteria have also been applied successfully [37, 215, 129], as cataloged in Table 3.3. We shall later offer some exemplary expansions to explain why we prefer a more conservative strategy.

Table 3.2. *Interstitial and L_∞ error norms for different grid sizes for a representative example,* $f(x) = \cos(16(x+1/5))(1+x)\exp(x^2)/(1+9x^2)$.

K	N	Max. of interstitial error	L_∞ error norm
1	3	0.5235	1.832
2	5	0.2023	1.952
3	9	1.776	1.856
4	17	0.4725	0.4725
5	33	0.002526	0.2526
6	65	7.171×10^{-8}	7.171×10^{-8}
7	129	3.96×10^{-14}	4.13×10^{-14}
8	257	7.15×10^{-14}	7.37×10^{-14}
9	513	1.257×10^{-13}	1.744×10^{-13}

To reduce the cost of later steps in CPR, a final refinement is to impose an upper limit N_{max} on N such that the search interval is automatically split in two whenever adaptation requests $N > N_{max}$. This is easily programmed and is standard in *Chebfun*.

The crucial point is that adaptive Chebyshev interpolation, reliable for all but weird, contrived $f(x)$, is now a reality. One can approximate a smooth $f(x)$ by a polynomial with almost machine precision.

3.4 ▪ Truncation and the Roundoff Plateau

When N is sufficiently large, the coefficients (as computed in finite precision arithmetic) "plateau" at some level, which is typically $N \epsilon_{mach} \max_{j\in[0,N]} |a_j|$, where $\epsilon_{mach} = 2.2 \times 10^{-16}$, as illustrated in Fig. 3.4.

The *error* is bounded in order of magnitude by a number that is typically

$$N^2 \epsilon_{mach} \max_{j\in[0,N]} |a_j|,$$

an order of magnitude larger than the fluctuations of individual high-degree coefficients.

3.5 ▪ Oscillating Coefficients

A Chebyshev series is really a Fourier cosine series in disguise, where the disguise is the change of coordinate $x = \cos(t)$. It follows that we can learn much about Chebyshev series by studying Fourier series.

By using simple trigonometric identities, one can prove the following theorem, which shows that if a function has coefficients all of the same sign, translating the function along the real axis by a distance x causes the coefficients to *oscillate*. In other words, it is *generic* rather than exceptional for Fourier coefficients to oscillate.

Theorem 3.1 (Fourier Series with a Phase Shift). *Define*

$$f(t) = \alpha_0 + \sum_{j=1}^{\infty} \alpha_j \cos(jt) + \sum_{j=1}^{\infty} \beta_j \sin(jt). \tag{3.11}$$

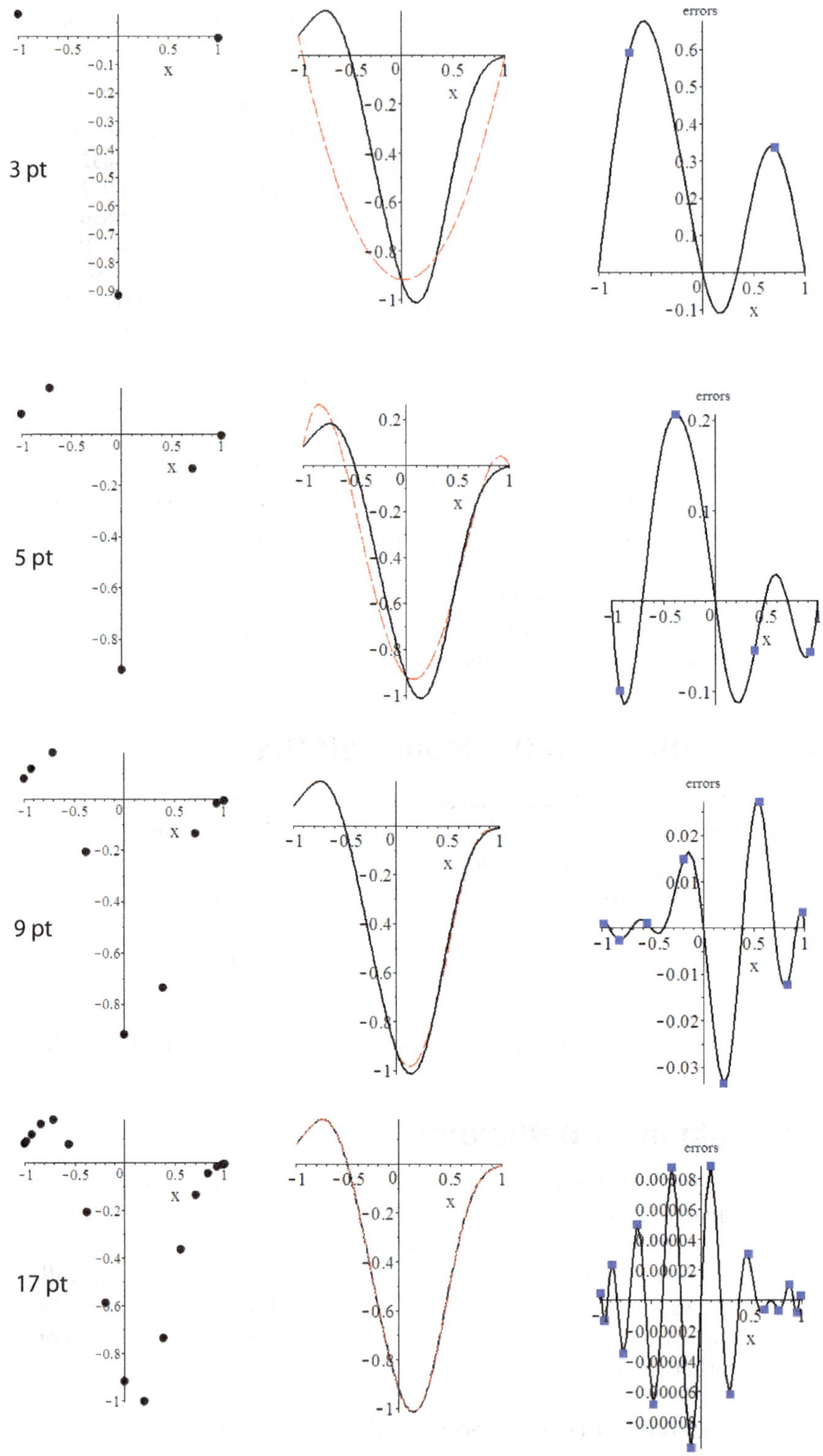

Figure 3.3. *The levels of Clenshaw–Curtis adaptive Chebyshev interpolation run vertically downward. The left column shows the samples of $f(x)$ as black disks. In the center column, the black curve is $f(x)$ and the red dashed curve is the interpolant at that level. In the right column, the curves are the interpolation error; the blue squares are the errors at the interstitial points at each level.*

Table 3.3. *Adaptation criteria for Chebyshev and Legendre series. $\beth$ is a user-chosen tolerance and N is the highest retained Chebyshev degree.*

Reference	Error test
Clenshaw and Curtis [129]	$k^2\|a_{N-2}\|, k\|a_{N-1}\|, \|a_N\| \leq \beth$ k typically 1/8
Fraser and Wilson [177]	$\|a_N\| \leq \beth$
Sloan and Smith [333]	Separate treatment of even and odd coefficients $p\ \exp(-\mu n)/(1-\exp(-\mu)) < \beth$, where p, n fitted to last 3 coefficients of each parity
Jacobs [215]	$\|a_{N-2}\|, \|a_{N-1}\|, \|a_N\| \leq \beth\ \max_n \|a_n\|$
Battles and Trefethen [37]	$\|a_{N-1}\|, \|a_N\| \leq 2$ machine epsilon
Mason and Handscomb [258], pp. 192–195	Fitting of last few coefficients by decaying exponential $p\ \exp(-\mu n)$: $p\ \exp(-\mu n)/(1-\exp(-\mu)) < \beth$
Boyd [71]	$\sum_{n=(2/3)N}^{N} \|a_n\| \leq \beth$
Boyd [78, 83]	$\max_{j\in[0,2N]} \|f(x_j) - f_N(x_j)\| \leq \beth$ x_j are the points of a $(2N+1)$-point Lobatto grid
Mavriplis [259]	Fit of last 4 coefficients to $a(n) = c\exp(-\sigma n)$; then $E_{estimate} = \sqrt{\frac{a(N)^2}{(N+1/2)} + \int_{N+1}^{\infty} \frac{(a(m))^2}{n+1/2}\, dm}$

Then

$$f(t+s) = \alpha_0 + \sum_{j=1}^{\infty} \left\{ \alpha_j \cos(js) + \beta_j \sin(js) \right\} \cos(jt) + \sum_{j=1}^{\infty} \left\{ \beta_j \cos(js) - \alpha_j \sin(js) \right\} \sin(jt).$$

The translated Fourier series is a mixture of sine and cosine coefficients and therefore cannot be converted into a Chebyshev series. However, we can easily prove the following theorem.

Theorem 3.2 (Shifted Chebyshev Series). *Define*

$$f(x) = a_0 + \sum_{j=1}^{\infty} a_j\, T_j(x). \tag{3.12}$$

Then

$$[f(\cos(\arccos(x)+s)) + f(\cos(\arccos(x)-s))]/2 = a_0 + \sum_{j=1}^{\infty} a_j \cos(js)\, T_j(x).$$

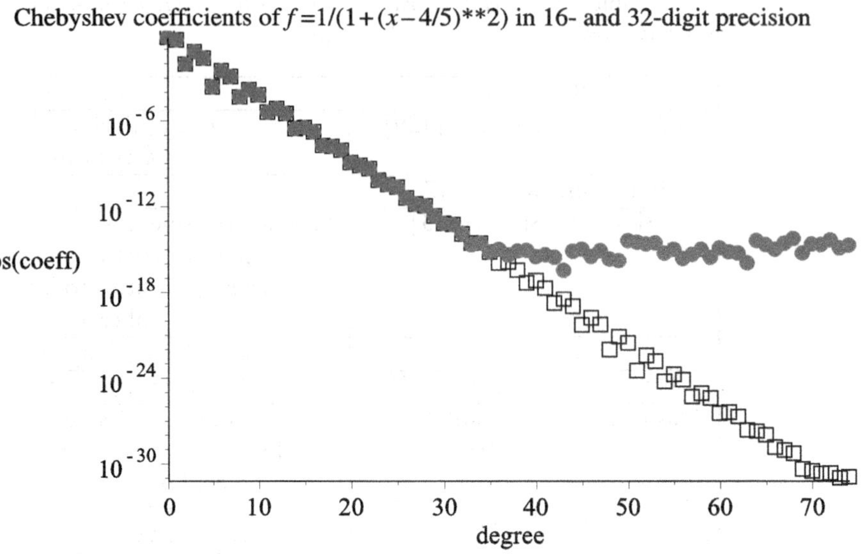

Figure 3.4. *Numerically generated spectral coefficients for a typical function as computed using 16-digit (MATLAB) precision (solid squares) and 32 decimal digit precision (hollow squares). Let a_{max} denote the maximum absolute value of the spectral coefficients a_j for all j. Then when the exact coefficients fall below $\epsilon\, a_{max}$ for some very small ϵ, spectral algorithms will compute roundoff-corrupted coefficients that will flatten out at roughly $|a_n| \sim \epsilon |a_{max}|$ for all sufficiently large n. ("Roughly" means that coefficients in the "roundoff plateau" fluctuate randomly about the indicated magnitude.) Typically, ϵ is about 10 to 1000 times as large as the roundoff error ϵ_{mach}, alias machine epsilon, where $\epsilon_{mach} \approx 2 \times 10^{-16}$ in IEEE double precision, which is also the default in MATLAB. ϵ increases with N but is $f(x)$ dependent. When the precision is increased, the roundoff plateau drops accordingly, as shown in the lower curve.*

Proof: Define $f(\cos(t)) \equiv g(t)$. Then

$$f(x) = a_0 + \sum_{j=1}^{\infty} a_j\, T_j(x), \tag{3.13}$$

$$g(t) = a_0 + \sum_{j=1}^{\infty} a_j\, \cos(jt), \tag{3.14}$$

$$g(t \pm s) = a_0 + \sum_{j=1}^{\infty} a_j \cos(js)\cos(jt) + \sum_{j=1}^{\infty} \mp a_j\, \sin(js)\sin(jt), \tag{3.15}$$

$$[g(t+s) + g(t-s)]/2 = a_0 + \sum_{j=1}^{\infty} a_j \cos(js)\cos(jt),$$

$$\{g(\arccos(x)+s) + g(\arccos(x)-s)\}/2 = a_0 + \sum_{j=1}^{\infty} a_j \cos(js)\, T_j(x),$$

$$\{f(\cos(\arccos(x)+s)) + f(\cos(\arccos(x)-s))\}/2 = a_0 + \sum_{j=1}^{\infty} a_j \cos(js)\, T_j(x).$$

∎

Typical spectral coefficients thus show *oscillatory* decay. As explained at much greater length in Chapter 2 of [76], the coefficients can be bounded from above by an *envelope* that *does* decay monotonically.

3.6 ▪ Stopping Criteria, or When Enough Is Enough

If $f(x)$ is analytic on the expansion interval, then its Chebyshev coefficients will fall geometrically (section 2.5). (The decay is usually oscillatory, but this is of little concern.) If the coefficients are of the asymptotic form

$$a_n \sim p\,\exp(-\mu\, n), \tag{3.16}$$

where p and μ are positive constants, then when the series is truncated after a_N, the truncation error E_N is bounded by [76, Chap. 2]

$$E_N \leq p\exp(-\mu N)\frac{\exp(-\mu)}{1-\exp(-\mu)}. \tag{3.17}$$

Unless $\exp(-\mu)$ is very close to one, the order of magnitude of the error is the same as $|a_N|$, which is the "last coefficient rule of thumb" of [76]. (If the coefficients oscillate, then $\exp(-\mu n)$ is the envelope, a tight upper bound on the decaying, oscillating coefficients.)

Other authors have used more elaborate criteria, as cataloged in Table 3.3. Some have tried to fit the asymptotic form (3.16). However, as explained with examples in Chapter 2 of [76] and in the previous section, it is common for Chebyshev coefficients to *oscillate* with degree. Convergence theory guarantees that the coefficients a_n can be bounded from above by $p\,\exp(-\mu n)$, the envelope of the coefficients, for some p and μ, but the decay is often nonmonotonic. Functions with parity, such as $\cos(x)$ (symmetric about the origin, $\cos(-x) = \cos(x)$) are extreme examples: all odd coefficients are zero. For this reason, Sloan and Smith fit the even-degree and odd-degree coefficients separately and then increase the number of grid points if the larger of the two errors exceeds the user-chosen tolerance [333]. Similarly, other authors, such as Jacobs [215] and Battles and Trefethen [37], examine the *last three* or *last two* coefficients.

Parenthetically, note that similar convergence criteria have been useful in adaptive spectral elements, which employ Legendre series closely related to Chebyshev expansions [259, 208].

3.7 ▪ Justification of the Stopping Criterion

Our recommended stopping criterion is motivated by the following theorem, stated formally later in this section: If $\delta_{2N}(x) \equiv f_{2N}(x) - f_N(x)$, then we can prove that the L_∞ norm of $\delta_{2N}(x)$ is less than five times the interstitial error for all $N < 500$, with a proportionality factor that grows only logarithmically above five for larger N. Thus, we can guarantee that if the interstitial error is tiny, the correction to $f_N(x)$ will be tiny also. This theorem is powerful because it does not depend on curve-fitting the coefficients or other specific assumptions about spectral behavior.

To prove the theorem, we need to briefly recall the concepts of cardinal functions and the Lebesgue constant.

For any set of $(N+1)$ interpolating functions $\phi_j(x)$ such that the matrix problem imposing the interpolation conditions is nonsingular, it is always possible to rearrange the basis into N linear combinations of the ϕ_j such that the new basis functions C_j

have the "cardinal property" of vanishing at all grid points but one, the "adelphic point" x_j associated with the jth cardinal function:

$$C_j(x_k) = \begin{cases} 1, & j = k, \\ 0, & j \neq k. \end{cases} \tag{3.18}$$

In other words, $C_j(x)$ is the solution to the interpolation problem when the vector of samples of $f(x)$ is the jth *identity vector*. The coefficients of C_j in the original basis are just the elements of the jth column of the *inverse* of the interpolation matrix. Lagrange showed how to explicitly construct such cardinal functions for polynomial interpolation on an arbitrary set of interpolation points. The Chebyshev–Lobatto cardinal functions for $N+1$ points are given by

$$C_j(x) \equiv (-1)^{j+1} \frac{(1-x^2)}{p_j N^2 (x - x_j)} \frac{d T_N(x)}{dx}, \tag{3.19}$$

where $p_j = 1$ if $0 < j < N$; $p_N = p_0 = 2$.

The cardinal basis is conceptually useful because the coefficients of the interpolant $f_N(x)$ in the cardinal basis are just the samples of $f(x)$:

$$f_N(x) = \sum_{k=0}^{N} f(x_k) C_k(x; N). \tag{3.20}$$

The Lebesgue function and Lebesgue constant Λ_N are defined by the following definition.

Definition 3.3 (Lebesgue Function). *Let $C_j(x; x_0, x_1, x_2, \ldots, x_N)$ denote the cardinal functions for polynomial interpolation associated with the grid of points $\{x_0, x_1, x_2, \ldots, x_N\}$. Then the Lebesgue function is*

$$L_N(x; x_0, x_1, x_2, \ldots, x_N) = \sum_{j=1}^{N} |C_j(x; x_0, x_1, x_2, \ldots, x_N)|. \tag{3.21}$$

Definition 3.4 (Lebesgue Constant). *The Lebesgue constant is the L_∞ norm of the Lebesgue function*

$$\Lambda_N(x_0, x_1, x_1, x_2, \ldots, x_N) \equiv \max_{x \in [a,b]} |L_N(x; x_0, x_1, x_1, x_2, \ldots, x_N)|. \tag{3.22}$$

Theorem 3.5. *Define*

$$\delta_{2N}(x) \equiv f_{2N}(x) - f_N(x) \tag{3.23}$$

to be the difference between the $(N+1)$-point and $(2N+1)$-point interpolants. Then

$$|\delta_{2N}(x)| \leq \Lambda_N \, |E_N^{interstitial}| \; \forall x \in [-1, 1], \tag{3.24}$$

where Λ_N is the Lebesgue constant for an N-point Chebyshev–Lobatto grid.

Proof: By definition, f_{2N} interpolates $f(x)$ at the $(2N+1)$ Chebyshev–Lobatto grid points, that is, $f_{2N}(x_k) = f(x_k)$ at each such point, and similarly f_N interpolates $f(x)$

on the $(N+1)$ Chebyshev–Lobatto grid. Also by definition, $\delta_{2N}(x)$ is the interpolant of $f_{2N}(x) - f_N(x)$ on the finer grid. The interpolant can be written in terms of the Chebyshev cardinal functions, where it follows that

$$\delta_{2N}(x) = \sum_{k=1}^{N} \{f(x_{2k-1}) - f_N(x_{2k-1})\} C_{2k-1}(x; 2N), \tag{3.25}$$

where the interpolation points and cardinal functions are those of the $(2N+1)$-point grid. The definition of the interstitial error, (3.9), is

$$E_N^{interstitial} \equiv \max_{x_k \in (\mathcal{G}(K+1) - \mathcal{G}(K))} |f(x_k) - f_N(x_k)| \tag{3.26}$$

$$= \max_{x_k \in (\mathcal{G}(K+1) - \mathcal{G}(K))} |f_{2N}(x_k) - f_N(x_k)|, \tag{3.27}$$

since $f_{2N}(x_k) = f(x_k)$ on the $2N+1$-point Lobatto grid; this implies

$$|f(x_{2k-1}) - f_N(x_{2k-1})| \leq E_N^{interstitial}, \tag{3.28}$$

where the bound is tight. The definition of the Lebesgue function implies that

$$\sum_{k=1}^{N} |C_{2k-1}(x; 2N)| \leq \Lambda_{2N+1} \qquad \forall x \in [-1, 1]. \tag{3.29}$$

It follows that

$$|\delta_{2N}(x)| \leq E_N^{interstitial} \Lambda_{2N+1} \qquad \forall x \in [-1, 1]. \tag{3.30}$$

■

The interstitial error stopping criterion is very conservative because it examines N pieces of information, the differences $f(x_k) - f_N(x_k)$, $x_k \in \mathcal{G}(K+1) - \mathcal{G}(K)$. In contrast, Jacobs, Sloan and Smith, and Clenshaw and Curtis examine only the last three Chebyshev coefficients of $f_N(x)$, others just examine the last coefficient, and still others fit an asymptotic form to only the last few coefficients.

Unfortunately, it is not possible to guarantee that adaptation will succeed for all $f(x)$ even with Boyd's interstitial error termination. For example, suppose $P_N(x)$ is an arbitrary polynomial of degree N and $q(x)$ is a function analytic on $x \in [a, b]$ but otherwise arbitrary, and let $\Omega_{2N+1}(x) \equiv \prod_{j=0}^{2N}(x - x_j)$, which by construction is zero at each of the $2N+1$ Chebyshev–Lobatto grid points x_j. Then any function of the form

$$f(x) = P_N(x) + \Omega_{2N+1}(x) q(x) \tag{3.31}$$

will trigger Boyd's stopping criterion at the level of $N+1$ interpolation points because by construction, $P_N(x)$ is the interpolant of $f(x)$ on both the Lobatto grid with $(N+1)$ points and the Lobatto grid with $(2N+1)$ points; the interstitial error $E_N^{interstitial}$ is zero. However, the difference between $f(x)$ and its interpolant $P_N(x)$, $\Omega_{2N}(x)q(x)$ may be large and poorly approximated by a polynomial of degree N.

All of the other stopping criteria in the table can also be fooled, and much more easily: a run of three or four tiny coefficients will trigger these other termination criteria even if higher-degree Chebyshev coefficients are not small. It is obvious that $f(x)$ that satisfy N conditions, as implicit in (3.31), are much rarer than $f(x)$ whose Chebyshev series satisfies only the three conditions that a_{N-2}, a_{N-1}, and a_N are all zero for some N, even though higher coefficients are nonzero. Like a kindly financial advisor, we shall leave each reader to find his or her own comfort point on the risk versus cost scale. The probabilities of being tripped up by the various strategies cataloged in the table are at present uncertain.

We shall instead try to offer further illumination by examining some interesting classes of functions.

3.7.1 ▪ Lacunary Chebyshev Series

Lacunary Chebyshev series can fool many stopping criteria.

Definition 3.6 (m-Lacunary Series). *A Chebyshev series of the form*

$$f(x) = \sum_{n=0}^{\infty} a_n T_{nm}(x) \tag{3.32}$$

is an m-lacunary series if each nonzero Chebyshev coefficient is followed by $(m-1)$ zero coefficients.

Warning: The term "lacunary," which is a Latin adjective that means "having gaps," is often applied in the narrow sense to denote a power series or Fourier series whose gaps of zero coefficients increase as a geometric progression or faster, a function that cannot be analytically continued beyond a "natural boundary."

Theorem 3.7 (Lacunary Chebyshev Series). *Suppose*

$$f(x) \equiv g(T_m(x)), \tag{3.33}$$

where m is an integer and $g(x)$ is analytic everywhere on $x \in [-1,1]$ with the Chebyshev series

$$g(x) \equiv \sum_{n=0}^{\infty} g_n T_n(x). \tag{3.34}$$

Then all Chebyshev coefficients of $f(x)$ whose degrees are not integer multiples of m are zero. More precisely,

$$f(x) \equiv \sum_{n=0}^{\infty} g_n T_{nm}(x). \tag{3.35}$$

Proof: Chebyshev's identity is $T_m(\cos(t)) = \cos(mt)$ for all nonnegative integers m. The trigonometric change of coordinate $x = \cos(t)$ then yields $f(\cos(t)) = g(T_m(\cos(t))) = g(\cos(mt))$. With the further change of coordinate $T \equiv mt$, this becomes $g(\cos(T))$, which, because of the required analyticity of g, has a nice Fourier cosine series in T:

$$g(\cos(T)) \equiv \sum_{n=0}^{\infty} g_n \cos(nT). \tag{3.36}$$

Converting this series back to t gives a Fourier series in which all terms except those whose degrees are integral multiples of m are missing:

$$g(\cos(mt)) \equiv \sum_{n=0}^{\infty} g_n \cos(nmT). \tag{3.37}$$

Converting from t back to x gives the theorem. ∎

(A special case of the theorem is the well-known identity $T_n(T_m(x)) = T_{mn}(x)$.)

For example, the function

$$\frac{1}{1+qT_m(x)} = \frac{1}{q\sqrt{\frac{1}{q^2}-1}} - \frac{2}{q\sqrt{\frac{1}{q^2}-1}} \sum_{n=1}^{\infty} \left(-\frac{1}{q} - \sqrt{\frac{1}{q^2}-1} \right)^n T_{mn}(x), \qquad q^2 < 1, \tag{3.38}$$

has an infinite number of gaps of $(m-1)$ consecutive zero Chebyshev coefficients, where m may be arbitrarily large.

Thus, a function like $f(x) = (3/4)/(5/4 - T_7(x)) = 1 + 2\sum_{n=1}^{\infty}(1/2)^n T_{7n}(x)$ can fool adaptive schemes that examine just the last three coefficients. However, it is unclear whether such functions with "runs" of three or more zero coefficients ever arise in applications.

3.7.2 ▪ Wall of Poles Functions

The wall of poles function is defined by

$$\mathbb{W}(x;q,M) \equiv \frac{\prod_{j=0}^{M}(-1)^j / \{(x - j(2/M)+1)^2 + q^2\}}{\prod_{j=0}^{M}(-1)^j / \{(1 - j(2/M)+1)^2 + q^2\}}. \tag{3.39}$$

The function is constructed to have $2M+2$ simple poles at the locations

$$x_p = -1 + j\frac{2}{M} \pm iq. \tag{3.40}$$

That is, there are $(M+1)$ poles in the upper half-plane and an equal number in the lower. (Also, $\mathbb{W}(1,q,M) = 1$ for all q and M.) All poles have locations whose real parts are evenly spread on the interval $x \in [-1,1]$ and whose imaginary parts are at $\pm qi$, as shown by the disks in the top panel of Fig. 3.5.

By increasing the number of poles, M, it appears that one can create arbitrarily long flat plateaus in the coefficient plot, as shown in Fig. 3.6.

We know of no applications where either large-gap lacunary Chebyshev series or wall of poles–like series have been observed.[5] Indeed, we conjecture that simple Chebyshev error tests are probably rather like Gaussian elimination [355]: the worst case is very bad, but in a probabilistic sense, the chances of trouble are actually rather small. Unfortunately, at present, no way is known to quantify "typical" Chebyshev series.

We therefore recommend a very conservative strategy, which is to apply Boyd's stopping criterion, terminating interpolation only when the interstitial error of our interpolant is smaller than some tiny, user-chosen tolerance.

3.8 ▪ Dynamic Range Revisited

3.8.1 ▪ Example: Nonlinear Matrix Eigenproblems

Ruhe [300] gives a good discussion of nonlinear eigenproblems in which the eigenvalue appears nonlinearly in the matrix. One illustrative problem is to find the zeros (eigenvalues) of the determinant of a matrix $\vec{\vec{T}}$ that is the sum of three matrices:

$$\vec{\vec{T}} = (\exp(\lambda)-1)\vec{\vec{B}}_1 + \lambda^2 + \vec{\vec{B}}_2 - b_0\vec{\vec{I}}, \tag{3.41}$$

[5] If a bivariate function is invariant under the $2m$-element dihedral group D_m, its Fourier series in the polar angle is a cosine series containing only $\{1, \cos(m\theta), \cos(2m\theta), \ldots\}$. But these are m-lacunary *cosine* series, not Chebyshev.

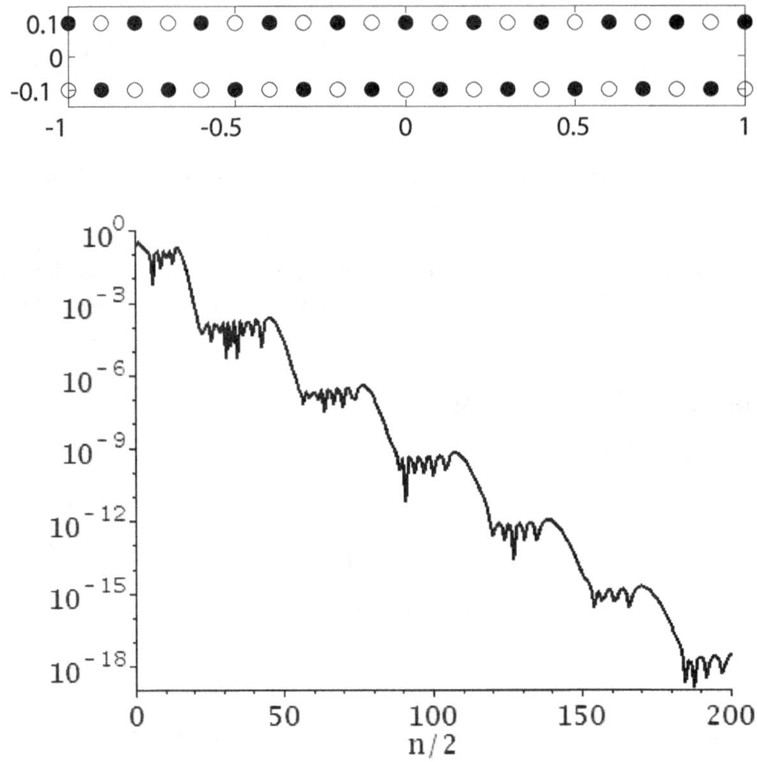

Figure 3.5. *Top: arrangement of singularities in the complex plane for the wall of poles function. The disks show the location of poles, all a distance q (here, $q = 1/10$) from the real axis; open and filled circles denote residues of opposite signs. Bottom: the absolute values of the even-degree Chebyshev coefficients up to a degree of 400 of $\mathbb{W}(x; 1/10, 20)$. The odd coefficients are all zero.*

where $\overset{\rightrightarrows}{I}$ is the identity matrix and $\overset{\rightrightarrows}{B}_1$ and $\overset{\rightrightarrows}{B}_2$ are $M \times M$ matrices with the elements

$$B_{1,jk} = (M+1-\max(j,k))jk, \qquad B_{2,jk} = 1/(j+k), \qquad j = 1,2,\ldots,M, k = 1,2,\ldots,M.$$

We will consider the particular case $M = 12$ and $b_0 = 100$. The eigenvalues, that is, roots of the determinant of the matrix $\overset{\rightrightarrows}{T}(\lambda)$, are given in Table 3.4. A 12×12 matrix is obviously not very large. Even so, if we were to use a single expansion on the smallest domain with integer endpoints that spans the roots, $\lambda \in [-13, 3]$, the determinant of $\overset{\rightrightarrows}{T}$ would vary from 6×10^{17} to 10^{38}—a ratio of about 10^{21}, as shown in Fig. 3.7. In standard MATLAB/IEEE double precision, machine epsilon is 2.2×10^{-16}. With this precision, subintervals where $f(\lambda)$ is oscillating between peaks and valleys smaller than $10^{-16} \max |f(\lambda)|$ will be invisible to the Chebyshev series. As shown in the inset plot, the Chebyshev approximation completely misses half a dozen roots in this "invisible valley."

The errors derived from a single MATLAB precision expansion on $\lambda \in [-13, 3]$ are shown just to the right of the exact eigenvalues in Table 3.4. It is important to note that in 16-digit precision, increasing the size of the Chebyshev expansion beyond the degree-66 truncation used for the table does not produce any improvement. It is a ubiquitous characteristic of Chebyshev series that the coefficients level off at a magnitude 1 or 2

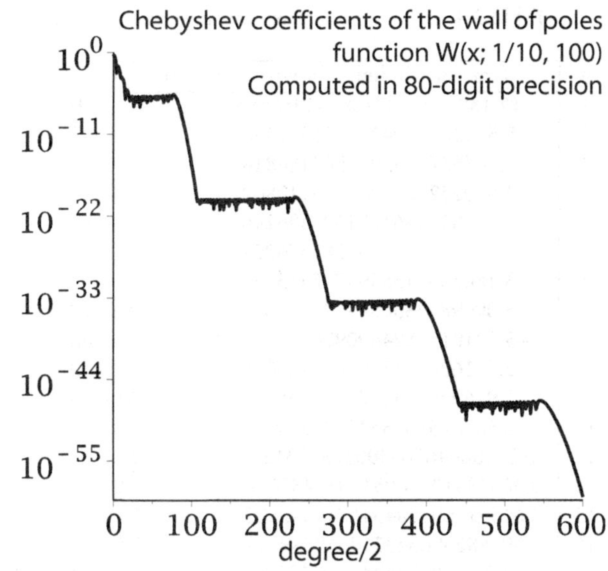

Figure 3.6. *Same as previous figure but with* 100 *poles on each side of the real axis, each at* $\Im(x) = \pm 1/10$. *The coefficients have plateaus of roughly* 100 *coefficients even though the overall rate of convergence is geometric. As in the preceding example, all odd coefficients are zero; the even coefficients up to* a_{1200} *are shown.*

or 3 powers of 10 larger than the product of machine epsilon with the absolute value of the largest Chebyshev coefficient, and fluctuate randomly around this mean level, as discussed in section 3.4. Higher Chebyshev coefficients for $f(\lambda)$ here are part of the roundoff plateau illustrated in Fig. 3.8.

The only remedies are to (i) increase the precision, as by increasing the environmental variable "Digits" in Maple, or (ii) subdivide. We employed the former strategy to compute the benchmark eigenvalues listed in the table, which are accurate to all 24 decimal places shown. We used 4 separate expansions to generate the errors shown in the rightmost column of the table; the intervals were $\lambda \in [-13, -6.5]$, $[-6.5, 0]$, $[0, 2]$, and $[2, 4]$. To avoid feeding inaccurate coefficients from the roundoff plateau into the Chebyshev companion matrix rootfinder, we computed expansions using 129 points on each subinterval and then deleted all coefficients a_n with $n \geq N_t$, where

$$N_t = \text{smallest } N \text{ such that } \frac{|a_n|}{\max_n |a_n|} \leq \epsilon_{cutoff} \, \forall n \geq N, \tag{3.42}$$

where here ϵ_{cutoff} was chosen as 10^{-14}.

3.8.2 ▪ Wilkinson's Polynomial of the Second Kind

James Wilkinson introduced two famous examples of polynomials that are very challenging for rootfinding. His second example [372, p. 44], which we shall dub the "Wilkinson polynomial of the second kind," is

$$\mathfrak{S}(x; N) = \prod_{k=1}^{N} \left(x - \frac{1}{2^{k-1}} \right). \tag{3.43}$$

Table 3.4. *Exact and approximate eigenvalues of Ruhe's nonlinear eigenproblem.*

n	Exact to all digits shown	Errors, $\lambda \in [-13, 3]$	Errors, four subintervals
1	−12.1381079287408250956585	2.6E−14	5.50E−15
2	−5.5912872894242715521456	3.5E−7	9.8E−15
3	−4.1675377832963562186818	0.37	1.3E−9
4	−3.6602327004865753110862	0.21	3.1E−8
5	−3.4226365461051308596246	0.11	9.3E−6
6	−3.2901157972756087376002	Missed	0.00012
7	−3.1863493158593717535357	Missed	0.00080
8	−3.0969852856884720250658	Missed	0.0011
9	−3.0218159524690906018849	Missed	0.00055
10	−2.9616268915690694189076	Missed	0.0027
11	−2.9169995231004815858475	Missed	0.0022
12	−2.8696758598530409889611	0.53	0.00037
13	0.054666462654905734731254	0.00011	1.7E−14
14	0.3034841065695119354870144	0.00017	3.9E−14
15	0.6159176912843467440438021	0.00001	6.2E−14
16	0.8908825965433126692890443	0.00050	1.4E−13
17	1.102617252690990070375139	0.0022	4.1E−14
18	1.271812680363519373590447	Missed	3.4E−14
19	1.444479469689578228156115	Missed	3.1E−14
20	1.643216904877135661240017	0.00003	8.6E−6
21	1.873415846541880535813628	0.00005	1.4E−6
22	2.137488145434155593118924	3.5E−5	1.9E−6
23	2.429750488271480764082818	9.9E−7	1.9E−7
24	2.712521182120578699060367	4.5E−10	2.1E−8

Gautschi [184] and Farouki and Rajan [167] also discussed the condition number of its roots when expressed in power form, Bernstein form, and Chebyshev polynomials of the first and second kinds. For the Wilkinson polynomial of the second kind, the power basis and Bernstein basis are well conditioned but the Chebyshev basis is poor. The problem is dynamic range: Although the maximum of $\mathfrak{G}(x;N)$ is 0.005, the function is less than 10^{-18} everywhere on $x \in [0, 0.13]$, an interval that contains a bunch of roots and yet is, because of the small range of $f(x)$ upon it, invisible to the Chebyshev series in floating point arithmetic.

Fig. 3.9 shows the multiple scales of the function.

3.8.3 ▪ Criteria for Recognizing Dynamical Range Problems and Subintervals of Large Relative Error

"I do it careful."

—Emma Boyd to her mother after climbing out of her crib at the age of 27 months.

"I do it *very* careful."

—Emma Boyd, one day later, after her second crib escape.[6]

It is obviously important to recognize intervals where the dynamical range is large. This is a subset of the larger issue for all applications of Chebyshev series, which is to recognize

[6]The crib was promptly retired; insufficiently careful arithmurgists will be fired!

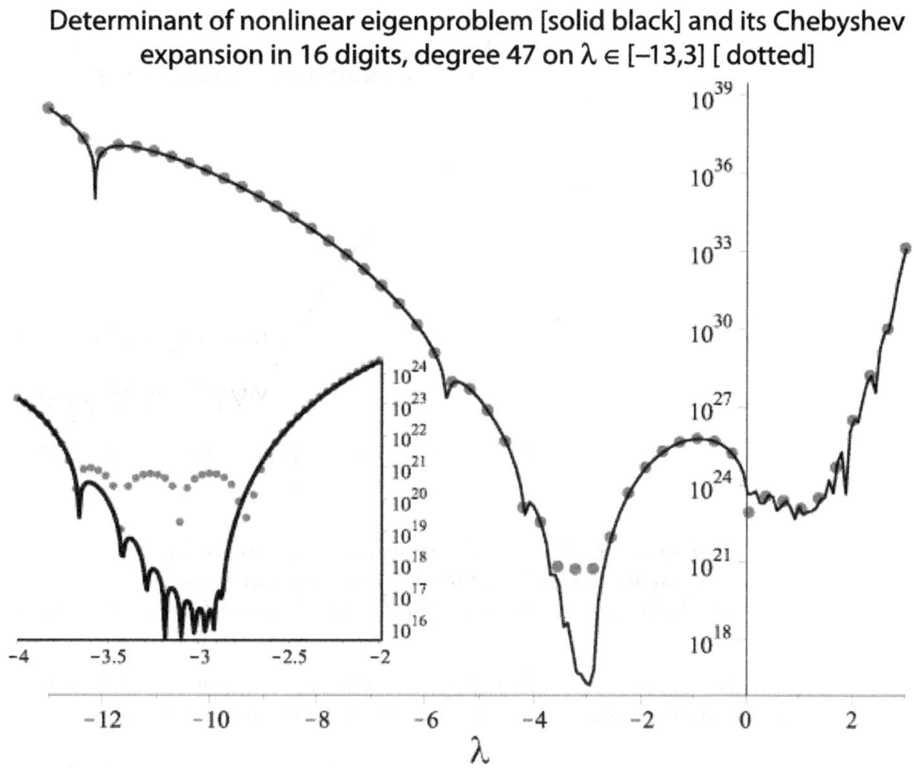

Figure 3.7. *Although the dimension of the matrix is only* 12*, the upper curve (solid black) shows that on the interval* $\lambda \in [-13, 3]$*, which barely spans the spread of the zeros, the determinant* $f(\lambda)$ *varies by a factor of almost* 10^{21}*. The inset shows that if the Chebyshev series of* $f(\lambda)$ *is computed until the coefficients plateau at somewhat larger than machine epsilon, a valley from* $\lambda \in [-3.7, -2.7]$ *that contains half a dozen roots is missed by the determinant-approximating polynomial expansion, whose values at discrete points are plotted as disks in both panels.*

portions of the approximation interval where the *relative* error is large. Strategies for doing so include the following:

1. MAG: Make a graph.
2. Inspect a vector of samples of the function, $f(x_k)$, for runs of tiny values, i.e.,
$$|f(x_j)|, |f(x_{j+1})|, \ldots, |f(x_{j+r-1})| \ll \max_{x \in [a,b]} |f(x)|, \tag{3.44}$$
where r is the run length.
3. Check slope df/dx and curvature d^2f/dx^2 at all interpolation points where $f(x)$ is tiny.

Good automatic detection of subintervals of large *relative* error is still an important research problem. The run length r is deliberately left vague for this reason. Two further precepts are useful.

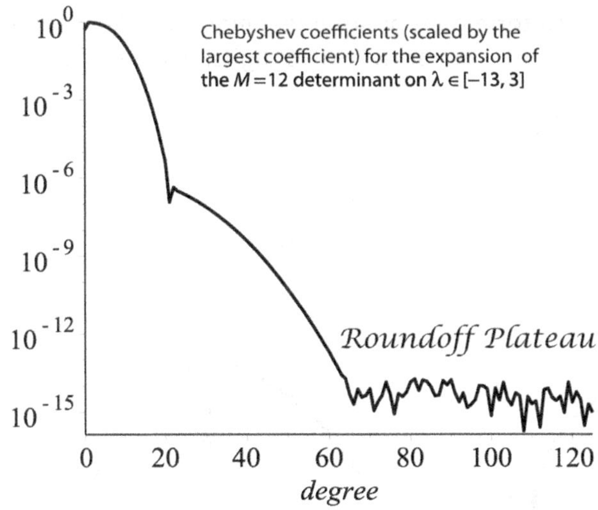

Figure 3.8. *Absolute value for the coefficients of the Chebyshev series of $f(\lambda)$, the determinant whose roots are the eigenvalues of the $M = 12$ nonlinear matrix eigenproblem, on the interval $\lambda \in [-13, 3]$. The coefficients have been divided by the absolute value of the coefficient of largest magnitude.*

The first is to be a good physicist or a good applied mathematician. John Wheeler, the eminent physicist, passed on to all his students the following:

Precept 1 (Wheeler's First Moral Principle). *Never do a calculation until you already know the answer.*

Wheeler's point in this bit of physics Zen is that unless one has thought about a problem, done back-of-the-envelope estimates of the outcome, and so on, Very Bad Things are likely to happen when you do careful, detailed computations. A good physicist or applied mathematician will usually have enough insight into $f(x)$ to anticipate dynamic range problems.

The other precept is the Emma Boyd strategy: Do it *very* careful! Make friends with the function. Let anomalies worry you, and make plots and zoom plots now and then. Reliable computations are the product of good algorithms applied by mildly paranoid users.

3.8.4 ▪ Theory of Errors in Multidomain Interpolation Using the Chebyshev Grid on Each Subdomain

"Splitting an interval into subintervals is the safest practical strategy [to cope with a large dynamic range] so far. Perhaps the reader can suggest something better!"

—J. P. Boyd [71, p. 6]

One motive for subdividing the search interval is to minimize cost. Since the cost of a QR eigenvalue calculation scales as the *cube* of the size of the matrix, it is much cheaper by a factor of N_s^2 to solve N_s problems of size M than a single problem of size $N = N_s M$. Special eigensolvers that exploit the sparsity of companion matrices can reduce cost and the advantages of subdivision, but unless some miracle delivers an eigenvalue solver that is linear in N, subdivision is cheaper, provided that N_s problems of size M yield the same accuracy as a single problem of size $N = N_s M$. But is this accuracy hypothesis true?

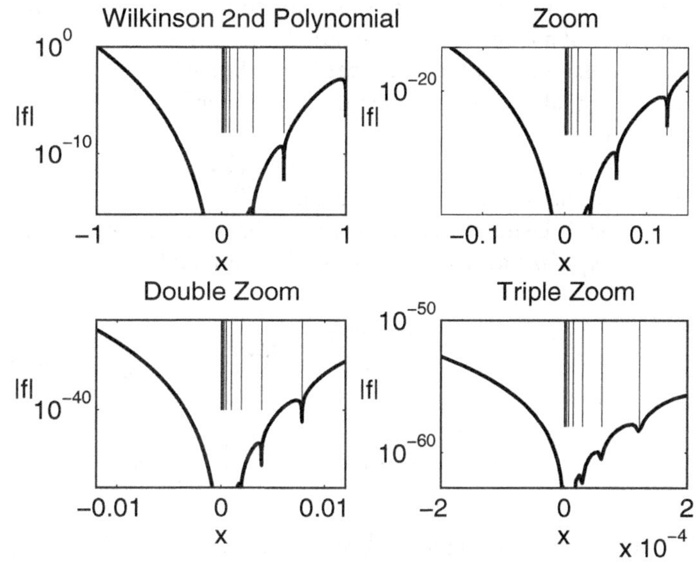

Figure 3.9. *Wilkinson's second polynomial. The vertical line segments near the top of each panel point to zeros. Each panel, ordered from left to right across the top and then left to right again below, lowers the bottom axis by another factor of machine epsilon,* 2.2×10^{-16}; *the horizontal scales also contract at a geometric rate. Nevertheless, some roots are below the axis, inaccessible to a Chebyshev series on this interval, even on the "triple zoom" plot.*

The answer is: almost. There is a penalty for subdivision but not a large one. Berzins's review [44] provides a good background for multi-interval polynomial interpolation but concentrates on uniform grids, which are irrelevant here. The most pertinent treatment of subdivision with Chebyshev grids is Boyd and Ong [105].

With N_s subdomains and polynomial interpolation of degree M on all subdomains, the total number of interpolation points on $x \in [-1, 1]$ is

$$P = 1 + N_s M. \tag{3.45}$$

If we fix the ratio

$$\xi \equiv \frac{M}{N_s}, \tag{3.46}$$

then the generalized theory of Lagrangian polynomial interpolation of [105] shows that the scheme *always converges* as M and N_s simultaneously tend to infinity.

If, instead, M is fixed, one can again easily prove convergence, but the error falls only at finite order, proportional to P^{-M-1} as $P \to \infty$. However, this is no drawback because Newton-polishing will remove the small errors in roots computed from the Chebyshev approximations on each subdomain.

When the interpolation points are the roots of a Chebyshev polynomial, it is well known that the Runge phenomenon is abolished, but we need to look deeper than this.

One happy aid to the analysis is that the Chebyshev coefficients for Runge's exemplary function are

$$\frac{p^2}{y^2 + p^2} = \frac{p^2}{\sqrt{p^4 + p^2}} + \frac{2p^2}{\sqrt{p^4 + p^2}} \sum_{n=1}^{\infty} (-1)^n \left(2p^2 + 1 - 2\sqrt{p^4 + p^2}\right)^n T_{2n}(y). \tag{3.47}$$

As explained in Chapter 2 of [76], this function is an illuminating example because the properties of a Chebyshev series are dominated by the *locations* of its singularities in the complex plane and not by the *type* of singularity (pole, logarithm, square root, etc.). Thus, a detailed analysis of the error in approximating this function is enlightening for general $f(x)$.

Let a_n denote the coefficients of a Chebyshev series. Two theorems of Chapter 2 of [76] show that the error in a truncated Chebyshev series is bounded on $x \in [-1,1]$ by $\sum_{n=N+1} |a_n|$. Using this, one can prove that the maximum pointwise error of the $(M+1)$-point *interpolant* for the series above is bounded by

$$E^{Cheb,bound} \leq 2 \sum_{n=N+1}^{\infty} |a_n| \equiv 2 \frac{2p^2}{\sqrt{p^4+p^2}} \frac{\delta^{M+1}}{1-\delta}, \tag{3.48}$$

$$\delta(p) \equiv \sqrt{2p^2+1-2\sqrt{p^4+p^2}} \approx \begin{cases} 1-p, & p \ll 1, \\ \frac{1}{2p}, & p \gg 1. \end{cases} \tag{3.49}$$

This bound is rather tight and becomes an equality in the limit $p \to \infty$. The convergence-limiting poles are on the imaginary axis at $y = \pm i p$. As p increases, the poles move farther from the real axis, the peak of the function on the real y-axis broadens, and $\delta(p)$ decreases monotonically since smoother implies faster convergence.

Suppose the goal is to approximate a function with singularities $x = \pm i c$, such as

$$f(x) = \frac{c^2}{x^2+c^2} \tag{3.50}$$

on $x \in [-1,1]$. If we employ a multidomain procedure with an ever-increasing number of subdomains N_s, using a Chebyshev grid on each subdomain, then the problem on the middle subdomain is to approximate, as the subdomain shrinks, $p^2/(y^2+p^2)$ on $y \in [-1,1]$, where $y \equiv xN_s$ with

$$p = cN_s. \tag{3.51}$$

The dependence of p on the number of subdomains N_s arises because when we add more subdomains, we must rescale the local subdomain coordinate; consequently, f varies more smoothly on each subdomain as the subdomains shrink and the poles move farther from the real axis in y. (The pole locations remain at $x = \pm ic$ in the original unscaled coordinate x.)

It is convenient to have the freedom to restate the error estimates in terms of the total number of points P on the union of all subdomains and the ratio ξ. The interrelationships between parameters are

$$P \approx N_s M \;\&\; \xi \equiv \frac{M}{N_s} \quad \to \quad M = \sqrt{P}\sqrt{\xi} \;\&\; N_s = \frac{\sqrt{P}}{\sqrt{\xi}}. \tag{3.52}$$

There are two regimes in the error. When $cN_s \ll 1$, which requires that the singularities of $f(x)$ be very near the real axis and N_s be small or moderate, $p \ll 1$ so that $\delta \approx 1-p \approx \exp(-p)$. The subdomain error is proportional to

$$\exp(-cMN_s) = \exp(-cP), \tag{3.53}$$

where P is the total number of points on the union of all the subdomains. Thus, when $cN_s \ll 1$, the error is (approximately) *independent* of the number of subdomains.

In the limit as M and N_s simultaneously tend to infinity, the error moves into the second regime where $cN_s = p \gg 1$:

$$E^{Cheb,bound} \approx \frac{1}{2}\frac{\sqrt{\xi}\,e^{-1/2\sqrt{P}\sqrt{\xi}(\ln(P)+2\ln(c)-\ln(\xi)+\ln(2))}}{c\sqrt{P}}. \tag{3.54}$$

This error bound shows that as M and N_s increase, there *is* a penalty for using subdomains. The error of a single-interval approximation falls *geometrically* as $\exp(-P|\log(\delta(c))|)$, whereas the rate of convergence of multidomain Chebyshev approximation converges only *subgeometrically*, with an error proportional to $\exp(-(1/2)\sqrt{\xi}\sqrt{P}\log(P))$ as $P \to \infty$ as illustrated in Fig. 3.10.

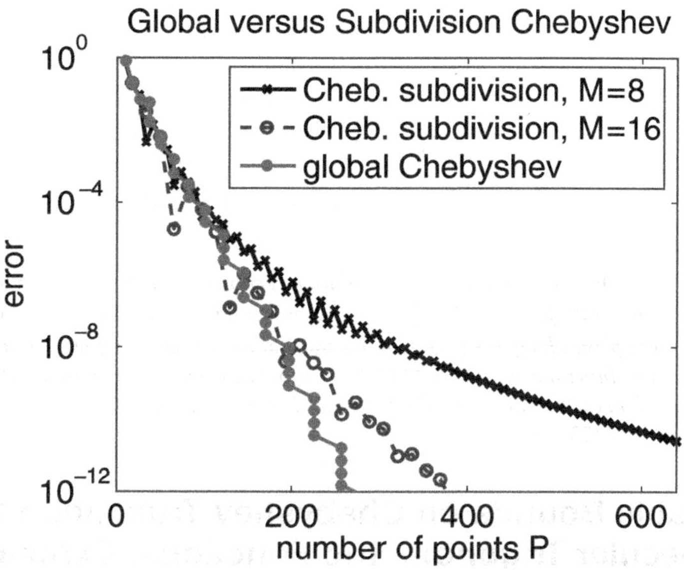

Figure 3.10. *A comparison of the maximum pointwise error in approximation by the usual single-domain Chebyshev interpolant (bottom curve), and Chebyshev multi-interval approximants with a fixed number of points on each subdomain (*17 *for the middle curve,* 9 *for the top), where the label "M" denote the degree on each subdomain. The horizontal axis* $P = N_s M + 1$ *is the* total *number of interpolation points on the union of all subdomains. In order to have near-singularities and peaks all over the interval,* $f(x)$ *was chosen as* $f = 1/(1+100[x+1]^2)+1/(1+100[x+1/2]^2)+1/(1+100x^2)+1/(1+100[x-1/2]^2)+1/(1+100[x-1]^2)$.

Fig. 3.11 gives another perspective, which is to fix the total number of points at $P \approx 100$ and plot the error for Chebyshev multidomain interpolation for four different choices of c, the nearness of the singularities of $f(x)$ to the real axis. When the singularities are close to the real axis—upper curves, narrow peaks in $f(x)$—the error is almost independent of M and N_s *separately* but depends only on their *product* P. Only when the singularities are far from the real axis (very smooth $f(x)$, bottom curves) is there a pronounced error minimum for small N_s. For Chebyshev interpolation, this dip in error is very shallow: using a lot of subdomains gives rather large error, but the minimum at $N_s = 5$ is at an error only slightly smaller than for a single global expansion. In practice, the penalty for subdivision is rather small.

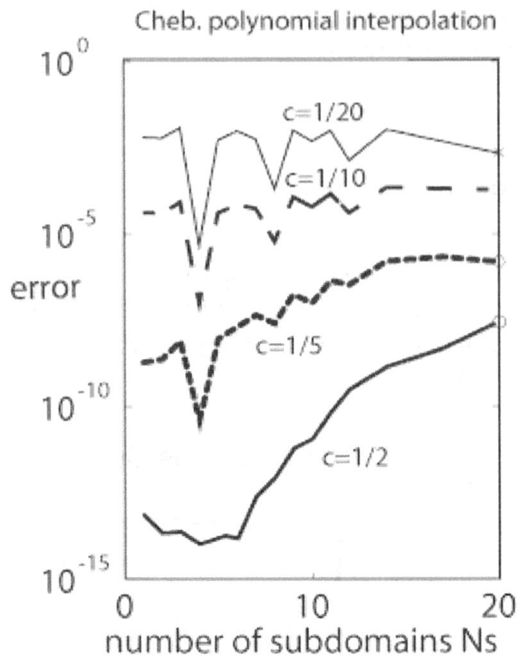

Figure 3.11. *Maximum pointwise error in multidomain polynomial approximations to four functions of the form* $f(x) = \sum_{j=0}^{4} c^2/((x+1-j/2)^2 + c^2)$, *i.e., five poles spaced evenly in the upper complex plane along* $\Im(x) = c$ *plus five more at the complex conjugate locations in the lower half-plane. The four functions have* $c = 1/20, 1/10, 1/5, 1/2$ *from top to bottom. The usual Chebyshev grid was applied on each subdomain. The total number of grid points on the union of subdomains was as close as possible to* 100.

3.9 ▪ Provable Bounds on Chebyshev Truncation Error for Secular Trigonometric Functions, Exponentials, Bessel Functions, and Other Special *f*(*x*)

For special but important classes of functions, the Chebyshev coefficients are known. The asymptotics of these explicit expressions allow rigorous, provable error bounds that make the Clenshaw–Curtis strategy unnecessary.

However, if and only if $f(x)$ is a *periodic* function, it is more efficient to approximate it by *trigonometric* interpolation, as discussed in the next chapter.

Yudell Luke and colleagues assembled the largest catalog of analytic formulas for Chebyshev coefficients in [246]. Some illustrations follow.

3.9.1 ▪ Secular Trigonometric Polynomials and the Kepler Equation

Definition 3.8 (Linear Secular Trigonometric Polynomial). *Let B denote a constant. A truncated Fourier series, augmented by a linear term of the form*

$$f_N(t) \equiv Bt + \sum_{j=0}^{N} a_j \cos(jt) + \sum_{j=1}^{N} b_j \sin(jt), \tag{3.55}$$

is a linear secular trigonometric polynomial of degree N [107].

The polynomial of degree N contains a total of $2N+2$ terms.

There is an obvious generalization, which is to replace the linear polynomial in t by an algebraic polynomial of higher degree.

The Kepler equation of celestial mechanics is $x-\epsilon\,\sin(x)-\lambda=0$. This is *not* a periodic function because of the linear term, x. When ordinary and trigonometric polynomials are thus mixed together, the best strategy is to use Chebyshev expansions.

One can do a particularly good job here because the coefficients of the Chebyshev series of $\sin(x)$ are Bessel functions. Furthermore, the error in a truncated Chebyshev series is bounded by the sum of the absolute values of all neglected coefficients [175, 76]. The combination of the general Chebyshev truncation bound together with asymptotics of Bessel functions eliminates the need for Clenshaw–Curtis style adaptive approximation. If the Kepler equation is polynomialized by replacing the sine function by its Chebyshev approximation of degree 15,

$$\begin{aligned}&-\lambda+\pi y-\epsilon\left\{3.14y-5.16771277519855y^3+2.55016394839721y^5\right.\\&-0.59926386322604y^7+0.08214347708860\,y^9-0.00736564609504y^{11}\\&\left.+0.00046097562573\,y^{13}-0.00001877013878y^{15}\right\}=0,\end{aligned}\qquad(3.56)$$

where $y\equiv x/\pi$, the single real root of this polynomial equation on $y\in[-1,1]$ approximates the root of the Kepler equation to better than 10 decimal places for all $\epsilon\in[0,1]$, $\lambda\in[0,\pi]$ [85].

For secular trigonometric polynomials in general, Boyd and Sadiq prove the following theorem [107].

Theorem 3.9. *The error in truncating the Chebyshev series up to and including the term of degree M for a linear secular trigonometric polynomial $f_N(x)$,*

$$f_N(x)\equiv Bx+\sum_{j=0}^{N}a_j\,\cos(j\pi x)+\sum_{j=1}^{N}b_j\,\sin(j\pi x),\qquad(3.57)$$

is rigorously bounded by

$$|f_N(x)-\sum_{k=0}^{M}g_k T_n(x)|\le E_{bound}(M;\{g_k\})\equiv\sum_{n=M+1}^{\infty}|g_k|,\qquad(3.58)$$

where the coefficients required in the bound are

$$g_{2k}=(-1)^k\,2\sum_{j=1}^{N}a_j J_{2k}(j\pi),\quad g_{2k+1}=(-1)^k\,2\sum_{j=1}^{N}b_j J_{2k+1}(j\pi),\;k=1,2\ldots,\qquad(3.59)$$

Proof: The Chebyshev coefficients of the sine and cosine are Bessel functions as in (3.59). ■

3.9.2 ▪ Other Provable Bounds

The Bessel functions themselves are another good family of examples because their coefficients are *products* of Bessel coefficients, such as

$$J_0(ax)=J_0^2(a/2)+2\sum_{n=1}^{\infty}(-1)^n J_n^2(a/2)T_{2n}(x),\qquad(3.60)$$

$$J_1(ax)=2\sum_{n=0}^{\infty}(-1)^n J_n(a/2)J_{n+1}(a/2)\,T_{2n+1}(x).\qquad(3.61)$$

It follows that by applying known Bessel asymptotics, we can accurately bound the error in polynomializing Bessel functions through truncated Chebyshev series.

There are two reasons why provable bounds are likely to remain a small niche. The first is that special functions often occur in rather complicated forms. Flierl, in a study of free oscillations in ocean basins, needed to solve

$$f(x)=\sum_{n=0}^{\infty}\frac{J_n^2(\alpha)J_n(\gamma)}{(1+\delta_{n0})J_n(y)}, \tag{3.62}$$

$$\alpha=\frac{\beta}{2x}, \qquad \gamma=\sqrt{\frac{\beta^2}{4x^2}-\Gamma^2}, \tag{3.63}$$

where β and Γ are constants, geophysical parameters explained in Flierl's paper, and where x is the frequency [172]. His (15a) is another dispersion relation with multiple infinite series of Bessel functions, which sprawls over four lines, too complicated to repeat here. Deriving bounds on infinite series from the known asymptotics of the Chebyshev coefficients of Bessel functions is not trivial.

The second reason why function-specific bounds are only mildly advocated is that adaptive Chebyshev interpolation is extremely reliable—especially reliable—precisely for those smooth, nonpathological functions for which the Chebyshev coefficients are known analytically.

3.9.3 ▪ Roots of Radial Basis Functions

Radial basis function (RBF) approximations are of the form

$$f(x)=\sum_{j=1}^{M}b_j\,\phi(x-x_j), \tag{3.64}$$

where $\phi(x)$ is a symmetric function called the *kernel*. It has been proved that even when the interpolation points x_j are irregular, several species of RBF kernels yield an exponential rate of convergence for functions that are analytic on the interpolation interval. Popular spectrally accurate RBFs are Gaussians (GA) and inverse quadratics (IQ):

$$\phi(x)=\exp(-\epsilon^2x^2) \qquad [\text{GA}], \tag{3.65}$$

$$=\frac{1}{1+\epsilon^2x^2} \qquad [\text{IQ}], \tag{3.66}$$

where ϵ is the *shape parameter* (an inverse width).

Unfortunately, no companion matrices are known for RBFs. The best strategy for computing the zeros of an RBF expansion is to re-expand it in Chebyshev polynomials. This is easy for Gaussians and inverse quadratics because the exact Chebyshev coefficients for individual basis functions are known:

$$\exp(-\epsilon^2x^2)=\exp(-[1/2]\epsilon^2)\left\{I_0(\epsilon^2/2)+2\sum_{n=1}^{\infty}(-1)^nI_n(\epsilon^2/2)T_{2n}(x)\right\}, \tag{3.67}$$

$$\frac{1}{1+x^2/a^2}=\frac{a}{\sqrt{a^2+1}}+\frac{2a}{\sqrt{a^2+1}}\sum_{n=1}^{\infty}(-1)^nq(a)^n\,T_{2n}(x), \qquad x\in[-1,1], \tag{3.68}$$

where $q=2a^2+1-2a\sqrt{a^2+1}$. To be sure, these expansions are only the basis functions with $x_j=0$, that is, peaked about the origin. However, for reasons explained in

Chapter 2 of [76], the series for off-centered peaks converge *faster* than when the same function is centered at the origin. Thus, if the Chebyshev expansions shown here are accurate to within a given, user-chosen tolerance with a given truncation N, the accuracy in approximating all the other $\phi(x-x_j)$ will be even better.

The error in a Chebyshev series is bounded by the sum of the absolute values of all the neglected coefficients, as proved in Chapter 2 of [76]. Therefore,

$$E_N^{GA} = \max_{x\in[-1,1]} \left| \exp(-\epsilon^2 x^2) - \exp(-[1/2]\epsilon^2)\left\{ I_0(\epsilon^2/2) + 2\sum_{n=1}^{N}(-1)^n I_n(\epsilon^2/2) T_{2n}(x)\right\}\right| \tag{3.69}$$

$$\leq 2\exp(-[1/2]\epsilon^2)\sum_{n=N+1}^{\infty} I_n(\epsilon^2/2) \tag{3.70}$$

$$\approx 2\exp(-[1/2]\epsilon^2) I_{N+1}(\epsilon^2/2),$$

$$E_N^{IQ} = \max_{x\in[-1,1]} \left| \frac{1}{1+\epsilon^2 x^2} - \frac{1}{\epsilon\sqrt{1/\epsilon^2+1}} + \frac{2}{\epsilon\sqrt{1/\epsilon^2+1}} \sum_{n=1}^{\infty} (-1)^n q(\epsilon)^n \, T_{2n}(x)\right| \tag{3.71}$$

$$\leq \frac{2}{\epsilon\sqrt{1/\epsilon^2+1}} \sum_{n=N+1}^{\infty} q(\epsilon)^n \tag{3.72}$$

$$\approx \frac{2}{\epsilon\sqrt{1/\epsilon^2+1}} q(\epsilon)^{N+1}, \tag{3.73}$$

with $q = 2/\epsilon^2 + 1 - 2(1/\epsilon)\sqrt{1/\epsilon^2+1}$, where the last line in each equation group is the last coefficient rule of thumb from Chapter 2 of [76]. Exponentially convergent series have terms decaying so rapidly that the infinite error sum is well approximated by the first term in the series.

These bounds are, strictly speaking, error bounds for *truncated* Chebyshev series. However, a theorem in Chapter 2 of [76] shows that the error bound for *interpolation* is never more than double that of truncation of the infinite series. Thus, the bounds on interpolation error are provably no worse than $2E_N^{GA}$ and $2E_N^{IQ}$.

The best bound on the absolute error of the RBF series is then $E_N = E_N^{GA} \sum_{j=1}^{M} |b_j|$, and similarly for IQ RBFs. One can then guarantee an error in the Chebyshev series smaller than δ by increasing N until

$$E_N^{GA} \sum_{j=1}^{M} |b_j| \leq \delta. \tag{3.74}$$

This can be harder than one might expect because RBF coefficients have the vice of sometimes being orders of magnitude larger than the maximum of $f(x)$.

Many will use the adaptive Clenshaw–Curtis interpolation anyway. It is nice that provable bounds on the Chebyshev approximation error are available, but unnecessary. The important point is that the Chebyshev-proxy method is the best way to find zeros of series of analytic RBFs.

3.10 ▪ Mechanics of Newton-Polishing

3.10.1 ▪ Overview

Newton-polishing is our colloquialism for applying Newton's iteration to the eigenvalues of the Chebyshev companion matrix. Another iteration can be substituted for Newton's.

The crucial point is that the accuracy of an iteration is limited only by the accuracy with which one can evaluate $f(x)$ itself.

Newton's iteration is

$$x^{(n+1)} = x^{(n)} - \frac{f(x^{(n)})}{df/dx(x^{(n)})}. \tag{3.75}$$

If df/dx is unknown or difficult to compute, a finite difference approximation may be substituted instead:

$$x^{(n+1)} = x^{(n)} - \frac{f(x^{(n)})}{(f(x^{(n)}+h) - f(x^{(n)}-h))/(2h)}, \tag{3.76}$$

where the default choice for h is the square root of machine epsilon for a one-sided difference (with combined finite difference and roundoff error also $O[\epsilon_{mach}]^{1/2}$) and is the cube root of machine epsilon for a centered difference (with combined finite difference and roundoff error $O([\epsilon_{mach}]^{2/3})$, as justified in section 6.8.

However, it is not necessary to compute df/dx with high precision because the iteration is self-correcting; the accuracy of the root depends only on the precision with which $f(x)$ can be evaluated. All errors in the approximation of df/dx and also in the Chebyshev-proxy are washed away by this final step.

If $f(x)$ is the determinant of a large matrix or otherwise expensive to evaluate, it may be desirable to iterate with the Chebyshev approximation before iterating $f(x)$ itself, staying with the cheap to evaluate functions as long as possible.

For efficiency, it is important to stop when further iterations produce no reduction in the correction: very close to the root, iteration stall is the spoor of hitting the roundoff barrier.

The remaining three subsections provide the simple formulas needed to apply Newton's algorithm.

3.10.2 ▪ Recurrence for the Coefficients of *df/dx*

Newton's iteration for a truncated Chebyshev series requires the derivative of the Chebyshev series. Fortunately, the coefficients of the derivative may be calculated from those of $f(x)$ in $O(N)$ operations by a simple recurrence. Denote the coefficients of $f_N(x)$ by a_n and those of its first derivative by b_n:

$$f_N \equiv \sum_{j=0}^{N} a_j T_j\left(\frac{2x-(b+a)}{b-a}\right), \qquad x \in [a,b], \tag{3.77}$$

$$\frac{df_N}{dx} \equiv \sum_{j=0}^{N} b_j T_j\left(\frac{2x-(b+a)}{b-a}\right). \tag{3.78}$$

Then, applying the recurrence in descending degree,

$$\tilde{b}_{N+1} = \tilde{b}_N = 0, \tag{3.79}$$

$$\tilde{b}_{k-1} = \frac{1}{c_{k-1}}\left\{2k\,a_k + \tilde{b}_{k+1}\right\}, \qquad k = N, N-1, N-2, \ldots, 1,$$

where $c_k = 2$ if $k = 0$ and $c_k = 1$ for $k > 0$. We then perform

$$b_n = \frac{2}{b-a}\tilde{b}_n. \tag{3.80}$$

3.10.3 ▪ O(N) Coefficient-Based Formula: Clenshaw's Recurrence

$$f_N \equiv \sum_{j=0}^{N} a_j T_j\left(\frac{2x-(b+a)}{b-a}\right) = \sum_{j=0}^{N} a_j \cos\left\{ j \arccos\left(\frac{2x-(b+a)}{b-a}\right)\right\}. \tag{3.81}$$

To avoid the evaluation of a lot of trigonometric functions, one may use the faster recurrence relation given here:

First, initialize by defining

$$y = \frac{2x-(b+a)}{b-a}, \qquad b_1 = 0, \qquad b_2 = 0. \tag{3.82}$$

Second, apply the loop $N+1$ times:

$$b_0 = 2y b_1 - b_2 + a_{N+1-j}, j = 1,2,\ldots,N, \\ b_3 = b_2, \qquad b_2 = b_1, \qquad b_1 = b_0. \tag{3.83}$$

The sum of the truncated series at x is then

$$f_N = (1/2)(b_0 - b_3) + (1/2)a_0. \tag{3.84}$$

3.10.4 ▪ O(N) Sample-Based Formula: Barycentric Interpolation

The barycentric interpolation formula was first published by H. E. Salzer [306]. Although the divisions *look* ill conditioned whenever $|x - x_j| \ll 1$, N. J. Higham proved they are forward stable for either of the two Chebyshev grids and more generally for any set of interpolation points with a small Lebesgue constant [209]:

$$\boxed{f_N(x) = \frac{\displaystyle\sum_{j=0}^{N} \frac{w_j}{x - x_j} f_j}{\displaystyle\sum_{j=0}^{N} \frac{w_j}{x - x_j}}}, \tag{3.85}$$

where $f_j \equiv f(x_j)$.

The weights w_j are not known for the Legendre grid but are known analytically for both Chebyshev grids:

$$x_j = \cos\left(\pi \frac{2j+1}{2N+2}\right) \to w_j = (-1)^j \sin\left(\pi \frac{2j+1}{2N+2}\right) \qquad \text{[Gauss/Roots Grid]}, \tag{3.86}$$

$$x_j = \cos\left(\pi \frac{j}{N}\right) \to w_j = (-1)^j \delta_j, \; \delta_j = \begin{cases} 1/2, j = 0 \text{ or } j = \pi \text{ [Lobatto Grid]} \\ 1, \text{ otherwise.} \end{cases} \tag{3.87}$$

Interestingly, if the interval is transformed to $[a, b]$, the weights are multiplied by a common factor that *cancels* in the barycentric formula so that the w_j are *independent* of a and b.

3.11 ▪ Summary

Adaptive Chebyshev interpolation with Clenshaw–Curtis N-doubling and the conservative interstitial error termination criteria is very robust.

However, it is possible to devise peculiar series that can fool less conservative stopping criteria. It is unclear if any termination condition can be infallible. However, the interstitial error criterion has a very low failure rate.

If $f(x)$ oscillates with tiny peaks and valleys on part of the interval while being many orders of magnitude larger on other parts of the interval, then $f(x)$ is said to have a large dynamic range. Chebyshev interpolation will yield large *relative* errors in the neighborhood of the tiny maxima and minima even though the *absolute* errors of Chebyshev approximations are uniformly small over the entire expansion interval. We've described some useful guidelines for detecting such problems. Tai Tsun Wu's advice to "make friends with the function" (lecture at Harvard, 1973) is highly pertinent. The physics or engineering background of a function will usually give warning of its large dynamic range.

There are research papers still to be written on both adaptation criteria and automatic identification of large dynamic ranges. Fortunately, if one is careful and conservative, failure of stopping criteria is almost never an issue in practice.

For special functions, one can prove rigorous bounds on the error in Chebyshev approximations. The brute power and reliability of adaptive Chebyshev interpolation has made these a luxury but not a necessity.

Chapter 4

Adaptive Fourier Interpolation and Rootfinding

"Spectral methods are the manure-detector of applied mathematics. Subtleties of ill-posedness, unbounded high derivatives, and other hidden confusion are painted over by second order methods with implicit dissipation and discretization error, but cause a spectral method to keel over and die like a canary in a coal mine, warning through its collapse, "Danger! Danger!"

—J. P. Boyd (2003, unpublished)

4.1 ▪ Background

The Fourier rootfinding problem is to compute the zeros of trigonometric polynomials.

Definition 4.1 (Trigonometric Polynomial). *A truncated Fourier series of the form*

$$f_N(t) \equiv \sum_{j=0}^{N} a_j \cos(jt) + \sum_{j=1}^{N} b_j \sin(jt) \tag{4.1}$$

is a trigonometric polynomial of degree N. This contains a total of $2N+1$ terms.

There is a theorem that gives the number of roots of trigonometric polynomials.

Theorem 4.2 (Fundamental Theorem of Trigonometric Algebra). *A trigonometric polynomial of degree N, $f_N(t)$, has exactly $2N$ roots, when the roots are counted according to their multiplicity, in the strip of the complex plane such that $-\pi < \Re(t) \leq \pi$. (Note that roots with $\Re(t) = -\pi$ are excluded.) Through periodicity, each root in the fundamental strip spawns a countable infinity of others; if t_* is any root with $|\Re(t)| \leq \pi$, then*

$$t_k = t_* + 2\pi k, \qquad k = 0, \pm 1, \pm 2, \ldots, \tag{4.2}$$

for any integer k.

Proof: Substitute $t = -i \log(z)$ to convert a trigonometric polynomial in t into an algebraic polynomial in z. Apply the fundamental theorem of algebra [83]. ∎

If $f(x)$ is a *periodic* function, analytic for real x, it is more efficient to approximate it by *trigonometric* interpolation. The asymptotics of the Chebyshev series for trigonometric functions show that in the limit of large N, an accurate Chebyshev approximation of a trigonometric polynomial with $(2N+1)$ terms requires roughly $\pi/2$ as many Chebyshev polynomials as the trigonometric polynomial that it is approximating, that is, roughly πN Chebyshev polynomials. (For finite N, it is a little more than πN.)

Nevertheless, because there is so much theory and software for algebraic polynomials, a major strategy is to transform a *trigonometric* polynomial rootfinding problem into an ordinary polynomial zero calculation, not by interpolation but rather by one of *three* different *exact transformations*:

1. complex-valued substitution $z = \exp(ix)$,
2. tan half-angle substitution, or
3. new variable: $c = \cos(x)$, $\sin(x) = \sqrt{1-c^2}$.

These are now useful primarily for theory and for explicit solutions and therefore are discussed at length in section 13.4.

Numerically, computing the eigenvalues of a Fourier companion matrix is the best way to find the zeros of a trigonometric polynomial. Adaptive Fourier interpolation, very similar to its Chebyshev counterpart, can be applied to generate accurate trigonometric polynomial approximations to smooth transcendental periodic functions.

The rest of the chapter is a potpourri of topics. After a brief discussion of the undesirability of subdividing the periodic interval, the three species of Fourier companion matrices are defined and compared in section 4.3. In applications, Fourier series composed only of cosines or sines or even cosines are common; the next three sections outline the options. The following four sections present theorems: factorization, zero-free interval and zero-containing tests, and so on. The chapter ends with a short history of Fourier rootfinding. This also reviews alternatives to the recommended strategy, which is to calculate zeros by computing the eigenvalues of the CCM or FDCM Fourier companion matrices (see section 4.3).

4.2 ▪ Subdivision

Subdividing the interval for Fourier interpolation is a step that should be resisted as long as possible because subdivision destroys the periodicity. The approximations on the smaller, nonperiodic intervals must be Chebyshev interpolants, not Fourier. We have already noted that for periodic functions, trigonometric interpolation is more efficient than Chebyshev by a factor of $\pi/2$. Thus, dividing a periodic interval into M subintervals usually requires Chebyshev interpolants of degree $(\pi/2)N/M$ on each subdomain. The rest is then all Chebyshev. As a practical matter, *special methods for periodic $f(x)$ are sensible only without subdivision.*

4.3 ▪ Fourier Companion Matrices

Three companion matrices for Fourier series are currently known. All have the property that the zeros of a trigonometric polynomial that lie within the fundamental strip ($\Re(x) \in (-\pi,\pi] \otimes \Im(x) \in [-\infty,\infty]$) can be easily computed from their eigenvalues. The three were first defined and compared in [95]:

1. complex-valued companion matrix (CCM),
2. Fourier division companion matrix (FDCM),[7]
3. elimination companion matrix (ECM).

[7]Written as FCDM (a typo) in [95].

The CCM is Frobenius's companion matrix for an *algebraic* polynomial. The CCM for general N is given in Appendix A. For $N=2$, the CCM is

$$\begin{vmatrix} 0 & 1 & 0 & 0 \\ 0 & 0 & 1 & 0 \\ 0 & 0 & 0 & 1 \\ (-1)\frac{a_2+ib_2}{a_2-ib_2} & (-1)\frac{a_1+ib_1}{a_2-ib_2} & (-1)\frac{2a_0}{a_2-ib_2} & (-1)\frac{a_1-ib_1}{a_2-ib_2} \end{vmatrix}. \tag{4.3}$$

The eigenvalues are transformed into roots by means of

$$t_{k,m} \equiv \arg(z_k)+2\pi m - i\log(|z_k|), \qquad k=1,2,\dots,2N,\ m=\text{integer}. \tag{4.4}$$

Only the zeros on the *unit circle* in z correspond to *real-valued* roots in the *original* trigonometric coordinate t [83].

The FDCM is derived by specializing the derivation of the general multiplication matrix for a general set of basis functions to the special case of a Fourier series. For a quadratic trigonometric polynomial, this companion matrix is

$$\begin{bmatrix} 0 & 2 & 0 & 0 \\ -\frac{a_0}{a_2}+1 & -\frac{a_1}{a_2} & -\frac{b_1}{a_2} & -\frac{b_2}{a_2} \\ 0 & 0 & 0 & 1 \\ M_{4,1} & M_{4,2} & M_{4,3} & M_{4,4} \end{bmatrix}, \tag{4.5}$$

where

$$M_{4,1} = \frac{-a_0a_2b_1 - b_2a_1a_2 + a_0b_2a_1 - a_2{}^2b_1}{\left(b_2{}^2+a_2{}^2\right)a_2}, \tag{4.6}$$

$$M_{4,2} = -\frac{a_1a_2b_1 + 2\,b_2a_2{}^2 + 2\,b_2a_0a_2 - a_1{}^2b_2}{\left(b_2{}^2+a_2{}^2\right)a_2}, \tag{4.7}$$

$$M_{4,3} = -\frac{2\,a_2{}^2a_0 + b_1{}^2a_2 - b_1b_2a_1 - 2\,a_2{}^3}{\left(b_2{}^2+a_2{}^2\right)a_2}, \tag{4.8}$$

$$M_{4,4} = -\frac{-b_2{}^2a_1 + 2\,b_2b_1a_2 + a_2{}^2a_1}{\left(b_2{}^2+a_2{}^2\right)a_2}. \tag{4.9}$$

For each eigenvalue, we calculate the two branches of the arccosine function in the strip $\Re(t)\in(-\pi,\pi]$:

$$t_j^{\pm} = \pm\arccos\left(\lambda_j/2\right) \tag{4.10}$$

Since there are $4N$ of these numbers and the fundamental theorem of trigonometric algebra [83] shows that a trigonometric polynomial of degree N has precisely $2N$ roots, counted according to their multiplicity, with $\Re(t)\in(-\pi,\pi]$, it follows that half of these $t_j^{\pm}$ are spurious. The simplest procedure to resolve this ambiguity is to evaluate $|f(t)|$ at each of the candidates. The genuine roots have tiny residuals, whereas $|f(t)|$ is not small for the spurious zeros; the candidates generating the $2N$ largest residuals should be rejected. Newton's iteration, begun from a candidate $t_j^{\pm}$, will resolve doubtful cases.

This FDCM companion matrix fails if

1. $a_N = 0$, or

2. one or more pairs of roots are symmetric about the origin, that is, $\pm t_*$ are *both* zeros of $f(t)$ for some t_*.

However, these difficulties can be removed by the translation (change of coordinate)

$$t \equiv \bar{t} + s \qquad \Leftrightarrow \qquad \bar{t} = t - s, \tag{4.11}$$

where s is a positive constant. After computing the eigenvalues in the shifted coordinate, it is trivial to translate them back to the original coordinate.

The ECM is derived by making the substitutions $\cos(t) = c$ and $\sin(t) = s$. It is easy to eliminate the new variable s so as to obtain a univariate algebraic polynomial $P(c)$ of degree $2N$. The ECM is the Chebyshev companion matrix for this polynomial in c. A rather complicated procedure is needed to convert eigenvalues into roots, so we shall leave the details to [95].

The cubic trigonometric polynomial case is, for example,

$$\begin{vmatrix} 0 & 1 & 0 & 0 & 0 & 0 \\ (1/2) & 0 & (1/2) & 0 & 0 & 0 \\ 0 & (1/2) & 0 & (1/2) & 0 & 0 \\ 0 & 0 & (1/2) & 0 & (1/2) & 0 \\ 0 & 0 & 0 & (1/2) & 0 & (1/2) \\ (-1)\frac{d_0}{2d_6} & (-1)\frac{d_1}{2d_6} & (-1)\frac{d_2}{2d_6} & (-1)\frac{d_3}{2d_6} & (-1)\frac{d_4}{2d_6} + (1/2) & (-1)\frac{d_5}{2d_6} \end{vmatrix}, \tag{4.12}$$

where the d_k are the Chebyshev coefficients of $P(c)$.

This substitution is very popular in robotics, where it is employed to convert systems of trigonometric polynomials in the angles of the elbow and hand movements of the robots into systems of ordinary polynomial equations [278, 337]. It is demonstrated in [95] that the same polynomial in c can be obtained by taking the resultant of the polynomial system in (c, s). It is also the result of applying another popular algebraic geometry technique, the method of Groebner bases.

In spite of all these connections, however, the ECM proved inferior in both accuracy and ease of programming to the CCM and FDCM.

4.4 ▪ Equivalence of the Trigonometric Cosine Polynomial and Chebyshev Polynomial Rootfinding Problems

Fourier cosine series (with all sine coefficients zero) are common in applications. This simplifies both theory and applications. In particular, it is easy to convert a cosine polynomial to an algebraic polynomial, as expressed by the following theorem. However, it is much easier to bound derivatives of trigonometric polynomials than ordinary polynomials, so sometimes it is convenient to transform algebraic polynomials to trigonometric. In both directions, then, the following is important.

Theorem 4.3 (Polynomial/Fourier Cosine Equivalence). 1. *Let $f_N(x)$ be an algebraic polynomial of degree N with Chebyshev coefficients a_n. Let $g_N(t)$ denote a cosine polynomial* *of degree N whose cosine coefficients are also chosen to be* a_n. *Then* $g_N(t) = f_N(\cos(t))$:

$$f_N(\cos(t)) = \sum_{n=0}^{N} a_n \cos(nt), \tag{4.13}$$

$$f_N(x) = \sum_{n=0}^{N} a_n T_n(x). \tag{4.14}$$

2. *Let* $x_j, j = 1, \ldots, N$, *denote the zeros of* $f_N(x)$, *sorted so that* x_1 *is smallest. Let* t_k *denote the roots of* $g_N(t)$. *Then*

$$t_{2N+1-k} = \arccos(x_k), \qquad k = 1, 2, \ldots, N, \tag{4.15}$$

$$t_k = -\arccos(x_k), \qquad k = 1, 2, \ldots, N. \tag{4.16}$$

If $x_1 = -1$, *then* $t = \pi = t_1 = t_{2N}$ *is a double root. If* $x_N = 1$, *then* $t = 0 = t_N = t_{N+1}$ *is a double root.*

Proof: This is an immediate consequence of the identity $T_n(\cos(t)) = \cos(nt)$. ■

4.5 ▪ Equivalence of the Trigonometric Sine Polynomial and Chebyshev Polynomial Series of the Second Kind

The Chebyshev polynomials of the second kind $U_n(x)$ are defined by the identity [258, 336]:

$$U_j \equiv \frac{\sin([j+1]t)}{\sin(t)} \qquad \forall j. \tag{4.17}$$

Theorem 4.4 (Fourier Sine/$U_n(x)$ Equivalence). 1. *Let* $g_N(t)$ *denote a sine polynomial of degree N and* $h_N(x)$ *be defined by* $h_N(\cos(t)) = g_N(t)$. *Then*

$$g_N(t) = \sum_{n=0}^{N-1} b_n \sin([n+1]t), \tag{4.18}$$

$$h_N(x) = \sqrt{1-x^2} \sum_{n=0}^{N-1} b_n U_n(x) \tag{4.19}$$

$$= \sqrt{1-x^2}\, k_{N-1}(x), \tag{4.20}$$

where $k_{N-1}(x)$ *is a polynomial of degree* $(N-1)$.

2. *Let* $x_j, j = 1, \ldots, (N-1)$ *denote the zeros of* $k_N(x)$. *Let* t_k *denote the roots of* $g_N(t)$. *Then*

$$t_k = \arccos(x_k) \qquad k = 1, 2, \ldots, (N-1), \tag{4.21}$$

$$t_N = 0, \tag{4.22}$$

$$t_{2N-k} = -\arccos(x_k) \qquad k = 1, 2, \ldots, (N-1), \tag{4.23}$$

$$t_{2N} = \pi, \tag{4.24}$$

Proof: This is an immediate consequence of the identity that defines Chebyshev polynomials of the second kind. ■

4.6 ▪ Parity

If $f(t)$ has parity, the cost can be greatly reduced (by a factor of two to a factor of eight) by exploiting the symmetry. Recall Definition 2.2 in section 2.16: a function $f(t)$ is symmetric about the origin or possesses even parity if for all t, $f(t)=f(-t)$; the condition $f(t)=-f(-t)$ is labeled antisymmetric or odd parity. A symmetric Fourier series is a *cosine* series; all the sine coefficients are zero. If $f(t)$ is antisymmetric, its expansion is a *sine* series.

Fourier basis functions also have definite parity with respect to $t=\pi/2$. A function with *double* parity symmetry can be represented by a *quarter-Fourier* series, that is, one that includes the one-fourth of a general Fourier basis whose elements match both symmetries of $f(t)$. Boyd and Xiu proved that all functions with parity at both $x=0$ and $x=\pi/2$ must be periodic with period 2π [110].

4.6.1 ▪ Algorithms for a General Cosine Series

Justification: A Fourier cosine series in t for a function $g_N(t)$ is equal, with the change of coordinate $t=\arccos(x)$, to a function $f(x)\equiv g_N(\arccos(x))$ with the same coefficients as the Chebyshev series of $f_N(x)$,

$$g_N(t)=\sum_{n=0}^{N} a_n \cos(nt), \tag{4.25}$$

$$f_N(x)=\sum_{n=0}^{N} a_n T_n. \tag{4.26}$$

Algorithm for General Cosines:

1. Compute the coefficients a_n of the cosine expansion of $f_N(t)$.
2. Form a companion matrix for Chebyshev polynomials with the a_n inserted as the coefficients of T_n.
3. Compute the eigenvalues x_j of the Chebyshev companion matrix.

The zeros in $\Re(t)\in(-\pi,\pi]$ are

$$t_k=\begin{cases} \arccos(x_k), & j=1,3,\ldots,N, \\ -\arccos(x_{k-N}), & j=N+1,3,\ldots,2N. \end{cases} \tag{4.27}$$

4.6.2 ▪ Algorithms for a General Sine Series

Justification: A Fourier sine series in t is equal, with the change of coordinate $t=\arccos(x)$, to an algebraic polynomial (multiplied by $\sqrt{1-x^2}$) whose coefficients as a Chebyshev series of the second kind are identical:

$$g_N(t)=\sum_{n=1}^{N} b_{n-1}\sin(nt) \tag{4.28}$$

$$=\sin(t)\sum_{n=1}^{N} b_{n-1}\frac{\sin(nt)}{\sin(t)}, \tag{4.29}$$

$$f_N(x)=\sqrt{1-x^2}\sum_{n=1}^{N} b_{n-1}U_{n-1}(x), \tag{4.30}$$

Algorithm for Sines:

1. Compute the coefficients b_n of the sine expansion of $g_N(t)$.
2. Form a companion matrix for Chebyshev polynomials of the second kind with the b_n inserted as the coefficients of U_n.
3. Compute the eigenvalues x_j of the Chebyshev companion matrix.

The zeros in $\Re(t) \in (-\pi, \pi]$ are

$$t_k = \begin{cases} 0, & k = 1, \\ \arccos(x_k), & j = 2, 3, \ldots, N, \\ -\arccos(x_{k-N}), & j = N+1, 3, \ldots, 2N-1, \\ \pi, & j = 2N. \end{cases} \tag{4.31}$$

4.6.3 ▪ Algorithms for Quarter-Fourier Series: Even Sines and Even Cosines

When the basis consists of only the *even* cosines, make the change of variable $T = 2t$, apply the procedure for *general* cosine polynomials, and then make the trivial reverse transformation $t_k = T_k/2$.

Exactly the same procedure applies when the basis is *even* sines, i.e., $\sin(2mt), m = 1, 2, \ldots$.

4.6.4 ▪ Algorithms for Quarter-Fourier Series: Odd Sines and Odd Cosines

The following lemmas allow one to extend the analysis for even cosine and even sine polynomials to odd cosine and odd sine polynomials.

Lemma 4.5 (Dividing Odd Cosines by cos(t)). *Define*

$$\alpha(t) = \sum_{j=0}^{N} a_j \cos([2j+1]t), \tag{4.32}$$

$$\sigma(t) = \frac{\alpha(t)}{\cos(t)} = \sum_{j=0}^{N} c_j \cos(2jt). \tag{4.33}$$

Then the coefficients of $\sigma(t)$ can be computed by

$$c_N = 2a_N, \tag{4.34}$$

followed by the recurrence

$$c_j = 2a_j - c_{j+1}, \qquad j = N-1, \ldots, 1, \tag{4.35}$$

and

$$c_0 = a_0 - (1/2)c_1. \tag{4.36}$$

Proof: Use the trigonometric identities for the product of two cosines. ■

Lemma 4.6 (Dividing Odd Sines by sin(t)).

$$\beta(t) = \sum_{j=0}^{N} b_j \sin([2j+1]t) \tag{4.37}$$

and

$$\tau(t) = \frac{\beta(t)}{\sin(t)} = \sum_{j=0}^{N} d_j \cos(2jt). \tag{4.38}$$

Then the coefficients of the symmetric polynomial $\tau(t)$ can be computed by

$$d_N = 2b_N, \tag{4.39}$$

followed by the recurrence

$$d_j = 2b_j + d_{j+1}, \qquad j = N-1, \ldots, 1, \tag{4.40}$$

$$d_0 = b_0 + (1/2)d_1. \tag{4.41}$$

Proof: Use the trigonometric identities for the product of a cosine with a sine. ∎

Algorithm for Odd Cosines: Let $\alpha(t)$ be a sum of odd cosines only,

$$\alpha(t) = \sum_{j=0}^{N} a_j \cos([2j+1]t). \tag{4.42}$$

1. Use Lemma 4.5 to compute

$$\sigma(t) = \frac{\alpha(t)}{\cos(t)} = \sum_{j=0}^{N} c_j \cos(2jt). \tag{4.43}$$

2. Compute the zeros of $\sigma(t)$ by using the procedure already outlined for zeros of a function containing even cosines. That is, form the Chebyshev companion matrix using the c_j to replace each of the a_j, the coefficients of $T_j(x)$. The roots of σ, s_j, are related to the eigenvalues x_j of the Chebyshev matrix by

$$s_j = \arccos(x_j)/2, \qquad j = 1, \ldots, N. \tag{4.44}$$

The roots of $\alpha(t)$ are the union of the roots of $\cos(t)$, $t = \pm\pi/2$ (which are zeros of all odd cosines), plus the roots of $\sigma(t)$, the set s_j.

Algorithm for Odd Sines: Let $\beta(t)$ be a sum of odd sines only,

$$\beta(t) = \sum_{j=0}^{N} b_j \sin([2j+1]t). \tag{4.45}$$

1. Use Lemma 4.6 to compute

$$\tau(t) = \frac{\beta(t)}{\sin(t)} = \sum_{j=0}^{N} d_j \cos(2jt). \tag{4.46}$$

2. Compute the zeros of $\tau(t)$ by using the procedure already outlined for zeros of a function containing even cosines. That is, form the Chebyshev companion matrix, but using the d_j to replace each of the q_j, the coefficients of $T_j(x)$ (not $T_{2j}(x)$). The roots of τ, s_j, are related to the eigenvalues x_j of the Chebyshev matrix by

$$s_j = \arccos(x_j)/2, \qquad j = 1, \ldots, N, \tag{4.47}$$

where the division by two is the only compensation needed because the roots in τ are those of an even cosine series.

The roots of $\tau(t)$ are the union of the roots of $\sin(t)$, $t = 0, \pi$, plus the roots of $\sigma(t)$, the set s_j.

4.7 ▪ Factorization Theorems of a Trigonometric Polynomial

The fundamental theorem of algebra states that any polynomial of degree N may be factored into the product of N monomials, each of the form $(x - x_j)$, where x_j is one of the roots. Here, we derive analogous factorizations of a trigonometric polynomial.

Theorem 4.7 (Factorization of a Trigonometric Polynomial). *Let the general trigonometric polynomial be*

$$f(t) \equiv \sum_{j=0}^{N} a_j \cos(jt) + \sum_{j=1}^{N} b_j \sin(jt), \tag{4.48}$$

and let t_j denote its $2N$ roots, counted according to their multiplicity, on the fundamental strip $-\pi < \Re(t) \leq \pi$. Then

1.

$$f(t) = \exp(-iNt)\frac{1}{2}(a_N - ib_N)\prod_{k=1}^{2N}(\exp(it) - \exp(it_k)). \tag{4.49}$$

2.

$$f(t) = (a_N - ib_N)(-1)^N 2^{2N-1}\left\{\prod_{k=1}^{2N}\exp\left(i\frac{t_k}{2}\right)\right\}\prod_{k=1}^{2N}\sin\left(\frac{t - t_k}{2}\right). \tag{4.50}$$

3.

$$f(t) = (a_N - ib_N)(-1)^N 2^{2N-1}\exp\left\{\frac{i}{2}\sum_{k=1}^{2N} t_k\right\}\prod_{k=1}^{2N}\sin\left(\frac{t - t_k}{2}\right). \tag{4.51}$$

4. *If the highest coefficients a_N and b_N of the trigonometric polynomial $f(t)$ are real valued, then the factorization theorem can be simplified to*

$$f(t) = \pm\sqrt{(a_N)^2 + (b_N)^2}\,(-1)^N\, 2^{2N-1}\prod_{k=1}^{2N}\sin\left(\frac{t - t_k}{2}\right). \tag{4.52}$$

Proof: As explained in equation (A.41), a general trigonometric polynomial, regardless of whether its coefficients are real or complex valued, may always be converted to an *associated polynomial* $h(z)$ through the change of coordinate $t = \exp(iz)$ and multiplication by the nonvanishing factor z^N. The fundamental theorem of algebra asserts that an algebraic polynomial of degree $2N$ has the factorization

$$h(z) = h_{2N}\prod_{k=1}^{2N}(z - z_k). \tag{4.53}$$

The first proposition (4.49) is just the transposition of this factorization from z back to t.

To obtain the second factorization, suggested by the work of Angelova and Semerdzhiev [12], note that

$$\begin{aligned}
\exp(-iNt)\prod_{k=1}^{2N}(\exp(it)-\exp(it_k)) &= \prod_{k=1}^{2N}\exp(-it/2)\,(\exp(it)-\exp(it_k)) \\
&= \prod_{k=1}^{2N}(\exp(it/2)-\exp(-it/2)\exp(it_k)) \\
&= \prod_{k=1}^{2N}\exp(i\,t_k/2)\,(\exp(it/2)\exp(-i\,t_k/2)-\exp(-it/2)\exp(it_k/2)) \\
&= \left\{\prod_{k=1}^{2N}\exp(i\,t_k/2)\right\}\prod_{k=1}^{2N}(\exp(it/2)\exp(-i\,t_k/2)-\exp(-it/2)\exp(it_k/2)) \\
&= \left\{\prod_{k=1}^{2N}\exp(i\,t_k/2)\right\}(-4)^N\prod_{k=1}^{2N}\frac{\exp(it/2)\exp(-i\,t_k/2)-\exp(-it/2)\exp(it_k/2)}{2i} \\
&= \left\{\prod_{k=1}^{2N}\exp(i\,t_k/2)\right\}(-1)^N 2^{2N}\prod_{k=1}^{2N}\sin([t-t_k]/2).
\end{aligned}$$

Substituting this into the first factorization (4.49) gives the second form of the factorization (4.51).

To prove the third proposition, note that the cosine/sine form of $f(t)$ requires—if a_N and b_N are real valued—that the coefficient of $\exp(-iNt)$ be $(a_N+ib_N)/2$, the complex conjugate of the coefficient of $\exp(iNt)$. However, in the factorization (4.49), multiplying together each term independent of t in the product gives a different, independent expression for the coefficient of $\exp(-iNt)$. Equating the two gives

$$\frac{1}{2}(a_N+ib_N)=\frac{1}{2}(a_N-ib_N)\prod_{k=1}^{2N}\exp(it_k). \tag{4.54}$$

If we define Ξ, r, and θ via

$$\prod_{k=1}^{2N}\exp(it_k/2)\equiv\Xi^2, \qquad a_N+ib_N=r\,\exp(i\theta), \tag{4.55}$$

(4.54) requires that

$$\exp(2i\theta)=\Xi^2 \qquad\rightarrow\qquad \Xi\equiv\prod_{k=1}^{2N}\exp(it_k/2)=\pm\exp(i\,\theta). \tag{4.56}$$

It follows that

$$(a_N-ib_N)\prod_{k=1}^{2N}\exp(it_k/2)=\pm|a_N-ib_N|=\pm\sqrt{a_N^2+b_N^2}. \tag{4.57}$$

Substituting this into the second factorization then proves the third. ■

A weaker half-sine factorization plays an essential role in the Durand–Kerner scheme of Angelova and Semerdzhiev [12]. However, the proportionality constant of their $Q(t)$ is a function of a point $t=y$, where y is not a root but is otherwise arbitrary, and they do not explicitly state the theorem.

Other than by explicitly evaluating the product of complex-valued factors in (4.51), no simple way to determine the overall sign in (4.52) is as yet known.

4.8 ▪ Chebyshev Series for Trigonometric Functions

Luke and colleagues [246] give

$$\cos(jt) = J_0(j\pi) + 2\sum_{k=1}^{\infty}(-1)^k J_{2k}(j\pi) T_{2k}(t/\pi), \tag{4.58}$$

$$\sin(jt) = 2\sum_{k=0}^{\infty}(-1)^k J_{2k+1}(j\pi) T_{2k+1}(t/\pi). \tag{4.59}$$

The well-known asymptotic approximations of the Bessel functions show that the coefficients of $\cos(Nt)$ and $\sin(Nt)$, the highest terms in the trigonometric polynomials, oscillate until the truncation N_{Cheb} of the Chebyshev series satisfies the inequality

$$N_{Cheb} > N\pi. \tag{4.60}$$

Thus, the approximation, even a crude one, of a trigonometric polynomial of degree N (and therefore $2N+1$ terms) requires at least πN Chebyshev terms, an increase of at least 50%.

This is why it is desirable to hunt zeros in trigonometric polynomials without conversion to a finite Chebyshev polynomial series.

4.9 ▪ Zero-Free Interval Theorems

There is a vast literature on the roots of *algebraic* polynomials [269, 271]; through the transformation $z = \exp(it)$, these theorems easily extend to trigonometric roots. However, there are some useful theorems that have no obvious algebraic counterpart. A trio of examples are stated next.

These theorems can be employed in subdivision methods for finding roots of ordinary polynomials in Chebyshev form. (See the discussion of Weyl subdivision in section 20.8.) As noted earlier, a Chebyshev series is always equivalent to a cosine polynomial with the same coefficients under the change of coordinate $x = \cos(t)$. Because the cosines oscillate uniformly in t, in contrast to the Chebyshev polynomials, which oscillate very nonuniformly in x, it is much easier to flag intervals as *zero-free* in the trigonometric form.

Theorem 4.8 (Zero-Free Sufficient Condition). *Let $f(t)$ be a trigonometric polynomial of degree N:*

$$f(t) = \sum_{j=0}^{N} a_j \cos(jt) + \sum_{j=1}^{N} b_j \sin(jt). \tag{4.61}$$

If

$$|a_0| \geq \sum_{j=1}^{N} |a_j| + \sum_{j=1}^{N} |b_j|, \tag{4.62}$$

then $f(t)$ has no real zeros on the interval $t \in [-\pi, \pi]$.

Proof: Rewrite the polynomial as

$$f(t) = a_0 + \rho(t), \qquad \rho(t) \equiv \sum_{j=1}^{N} a_j \cos(jt) + \sum_{j=1}^{N} b_j \sin(jt). \tag{4.63}$$

Now because $|\cos(jt)| \leq 1$ and $|\sin(jt)| \leq 1 \ \forall t \in [-\pi, \pi]$ and for all j, it follows that

$$|\rho(t)| \leq \sum_{j=1}^{N} |a_j| + \sum_{j=1}^{N} |b_j|. \tag{4.64}$$

If the condition of the theorem is met, then $\rho(t)$ can never be sufficiently large in magnitude for *real* t to cancel a_0 and so create a root of f. ■

Theorem 4.9 (Zero-Free Interval Using First Derivative Bound). *Let $f_N(t)$ denote the trigonometric polynomial*

$$f_N(t) \equiv \sum_{j=0}^{N} a_j \cos(jt) + \sum_{j=1}^{N} b_j \sin(jt). \tag{4.65}$$

Define

$$B_1 \equiv \sum_{j=1}^{N} |j\, a_j| + \sum_{j=1}^{N} |j\, b_j|. \tag{4.66}$$

Let $t = \tau$ denote an arbitrary real point on the interval. Then the open interval $t \in (\tau - \sigma, \tau + \sigma)$ contains no zeros of $f_N(t)$, where

$$\sigma(\tau) \equiv |f_N(\tau)| \,/\, B_1. \tag{4.67}$$

Proof: B_1 is a bound on the first derivative of the trigonometric polynomial, that is,

$$\left| \frac{df_N}{dt}(t) \right| \leq B_1 \ \forall \ \text{real}\ t. \tag{4.68}$$

This follows by explicitly differentiating f_N and then observing that $|\cos(jt)| \leq 1$ and $|\sin(jt)| \leq 1 \ \forall t \in [-\pi, \pi]$. Now if $|f_N(\tau)|$ is nonzero, the polynomial cannot reduce its magnitude as t moves away from τ at a rate faster than the bound B_1 on the first derivative. It follows that

$$|f_N(t)| \geq |f_N(\tau)| - B_1 |t - \tau| \geq 0 \tag{4.69}$$

for all t such that the right-hand inequality in (4.69) is true. ■

Corollary 4.10 (First Derivative Bound/Equispaced Sampling). *Suppose that the trigonometric polynomial is sampled at the set of M discrete, evenly spaced points for arbitrary $M > 0$:*

$$t_j \equiv -\pi + \pi(2j - 1)/M, \qquad j = 1, 2, \ldots, M. \tag{4.70}$$

Then if

$$\min\left(\left| f_N(t_j) \right| \right) \geq B_1 \frac{\pi}{M}, \tag{4.71}$$

$f_N(t)$ has no real zeros.

Proof: Apply Theorem 4.9 with $\tau = t_j$ for each j in turn. If $\sigma(t_j)$ is larger than π/M, then the interval $t \in [-\pi, \pi]$ is completely covered by a set of overlapping zero-free intervals. ■

A similar theorem can be obtained from Cauchy's theorem for the error in interpolation.

First, we must prove the following lemma.

Lemma 4.11.

$$|(t-\alpha)(t-\beta)| \leq (\beta-\alpha)^2/4 \tag{4.72}$$

for $t \in [\alpha, \beta]$.

Proof: Define

$$q \equiv (t-\alpha)(t-\beta). \tag{4.73}$$

The first derivative is $q_t = 2t - (\alpha + \beta)$. The only zero is $t_m = (\alpha + \beta)/2$. Explicit calculation shows that

$$q(t_m) \equiv (\beta-\alpha)^2/4, \tag{4.74}$$

from which the lemma follows. ∎

Theorem 4.12 (Zero-Free Interval: Second Derivative Bound). *Let $f_N(t)$ denote the trigonometric polynomial*

$$f_N(t) \equiv \sum_{j=0}^{N} a_j \cos(jt) + \sum_{j=1}^{N} b_j \sin(jt). \tag{4.75}$$

Define

$$B_2 \equiv \sum_{j=1}^{N} |j^2 a_j| + \sum_{j=1}^{N} |j^2 b_j|. \tag{4.76}$$

Let $t \in [\alpha, \beta]$ denote a small interval where the endpoints are real. If

1. $\text{sign}(f_N(\alpha)) = \text{sign}(f_N(\beta))$ *and*

2. $$B_2 \frac{(\beta-\alpha)^2}{8} < \min(|f_N(\alpha)|, |f_N(\beta)|), \tag{4.77}$$

then $f_N(t)$ does not contain a zero on the interval $t \in [\alpha, \beta]$.

Proof: The linear polynomial that interpolates $f_N(t)$ on the interval $t \in [\alpha, \beta]$ is given analytically by

$$f_{linear}(t) \equiv f_N(\alpha)\frac{t-\beta}{\alpha-\beta} + f_N(\beta)\frac{t-\alpha}{\beta-\alpha}. \tag{4.78}$$

If $\text{sign}(f_N(\alpha)) = \text{sign}(f_N(\beta))$, then

$$\max(|f_N(\alpha)|, |f_N(\beta)|) \geq |f_{linear}(t)| \geq \min(|f_N(\alpha)|, |f_N(\beta)|) > 0. \tag{4.79}$$

The linear interpolant does not have a root on $t \in [\alpha, \beta]$ and furthermore is separated from zero by the smaller of $|f_N(\alpha)|$ and $|f_N(\beta)|$. However, the function that the straight line approximates, $f_N(t)$, may still have a zero on the interval if the error in the linear interpolant is sufficiently large—larger than the smaller of $|f_N(\alpha)|$ and $|f_N(\beta)|$.

Cauchy proved a theorem on the error in polynomial interpolation [144, p. 56]. Specializing this to linear interpolation gives

$$f_N(t) - f_{linear}(t) = \frac{1}{2}\frac{d^2 f_N}{dt^2}(\xi)(t-\alpha)(t-\beta) \qquad \forall t \in [\alpha, \beta], \tag{4.80}$$

where ξ is some point on $t \in [\alpha, \beta]$. Replacing some factors by their maxima gives

$$|f_N(t) - f_{linear}(t)| \leq \max_{t\in[\alpha,\beta]} \left| \frac{d^2 f_N}{dt^2}(t) \right| \frac{(\beta-\alpha)^2}{8}, \tag{4.81}$$

where we used Lemma 4.11, proved above, that $|(t-\alpha)(t-\beta)| \leq (\beta-\alpha)^2/4$. B_2 is a bound on the maximum value of the *second* derivative of the trigonometric polynomial over the entire real axis, that is,

$$\left| \frac{d^2 f_N}{dt^2}(t) \right| \leq B_2 \, \forall \text{ real } t. \tag{4.82}$$

Apply the same reasoning as used in the proof of Theorem 4.9 to show that B_1 bounds the first derivative. Inserting this into (4.81) gives

$$|f_N(t) - f_{linear}(t)| \leq B_2 \frac{(\beta-\alpha)^2}{8}. \tag{4.83}$$

It follows that if the error bound on the right is smaller than the minimum of $f_{linear}(t)$ on the interval, which is the smaller of $|f_N(\alpha)|$ and $|f_N(\beta)|$, then it is impossible for $f_N(t)$ to vanish on the interval $t \in [\alpha, \beta]$. ■

Applying this theorem to a set of same-size intervals that span $t \in [-\pi, \pi]$ proves the following.

Corollary 4.13 (Second Derivative Bound: No Real Zeros). *Suppose that the trigonometric polynomial is sampled at the following set of $M+1$ discrete, evenly spaced points for arbitrary $M > 0$:*

$$t_j \equiv -\pi + 2\pi j/M, \qquad j = 0, 1, \ldots, M. \tag{4.84}$$

Then if

$$\min\left(\left|f_N(t_j)\right|\right) \geq B_2 \frac{\pi^2}{2M^2}, \tag{4.85}$$

$f_N(t)$ has no real zeros.

4.10 ▪ Theorems That an Interval Contains at Least One Zero

Theorem 4.14 (Trigonometric Polynomial with Zero Constant Must Have a Real Root). *Let $f(t)$ be a trigonometric polynomial of degree N with real-valued coefficients a_j and b_j. Suppose that the constant a_0 is zero but some coefficients for $j \geq 1$ are* nonzero. *Then $f(t)$ has at least one root on the interval.*

Proof:

$$a_0 = \frac{1}{\pi}\int_0^{\pi} f(t)\,dt. \tag{4.86}$$

Thus, the condition that $a_0 = 0$ is really a condition that the *average* of $f(t)$ over the interval is *zero*. This is not possible if $f(t)$ is everywhere one-signed on the interval; there must be regions where f is positive and regions where it is negative. This in turn requires that f cross the axis at least once. ■

Theorem 4.15 (Small Symmetric Part Implies at Least One Zero). *Let S and A denote the symmetric and antisymmetric parts of a trigonometric polynomial with real coefficients:*

$$S(t) \equiv \sum_{j=0}^{N} a_j \cos(jt), \qquad A(t) \equiv \sum_{j=1}^{N} b_j \sin(jt). \tag{4.87}$$

Then if

$$|S(\tau)| < |A(\tau)| \tag{4.88}$$

for some real-valued $\tau \in [-\pi, \pi]$, *there is at least one real root on the interval* $t \in [-\tau, \tau]$.

Proof: Any function may be decomposed into its symmetric part $S(\tau)$ and antisymmetric part $A(\tau)$:

$$f(\tau) = S(\tau) + A(\tau), \qquad \text{such that} \qquad f(-\tau) = S(\tau) - A(\tau). \tag{4.89}$$

If $|S(\tau)| < |A(\tau)|$, then the signs of $f(\tau)$ and $f(-\tau)$ are controlled by the sign of $A(\tau)$. If, for example, $A(\tau)$ is positive, then $f(\tau) > 0$, regardless of the sign or precise magnitude of $S(\tau)$, and $f(-\tau)$ must be negative. Bolzano's theorem (section 19.2) then says there must be at least one real root on the interval $t \in [-\tau, \tau]$. ■

Many other pertinent *oracles* can be found in Chapter 19.

4.11 ▪ History of Fourier Polynomial Zero-Hunting

A naive concept of science might be abbreviated as "Arriba, siepre arriba"—"Higher, ever higher."[8] The history of companion matrices is a refutation of steady progress. The companion matrices for orthogonal polynomials were independently discovered by [338, 339], [22], [198], [344], and [146]. The early work, however, died out like *homo africanus*. Boyd's 1995 paper on solving transcendental equations by Chebyshev interpolation and its adoption by *Chebfun* brought orthogonal polynomial companion matrices into the mainstream.

Similarly, there was much activity on finding the roots of trigonometric polynomials between 1982 and 1994, but no papers in the following 11 years (Table 4.1).

The first trio of trigonometric papers by Semerdzhiev, Makrelov, and Angelova employed a generalization of the Durand–Kerner iteration for *simultaneously* finding all roots of a polynomial [12, 247, 248]. Frommer gave a more abstract justification of the trigonometric Durand–Kerner method that simplified the proof that the algorithm converges quadratically near the roots and slightly simplified the algorithm [179]. Another Durand–Kerner article is [366], by Weidner, which is the earliest appearance of the $z = \exp(it)$

[8] The dying words of Jorge Chavez Dartnell (1887–1910), who completed the first aerial crossing of the Alps, but was fatally injured in landing.

Table 4.1. *A select bibliography on Fourier zero-hunting.*

Angelova and Semerdzhiev [12]	Trigonometric generalization of Durand–Kerner
Boyd [84]	Symmetry-exploiting companion matrices for cosine polynomials, sine polynomials, and polynomials with double parity
Boyd [83]	Comprehensive review of orthogonal polynomial and Fourier rootfinding
Boyd [95]	ECM derived from the $c = \cos(t)$, $s = \sin(t)$ substitution
Boyd [97]	FDCM
Boyd and Sadiq [107]	Zeros of Fourier series with a linear secular term
Carstensen and Petkovi'c, Carstensen and Reinders, Carstensen [122, 123, 121]	Durand–Kerner method, higher-order generalizations, and interval arithmetic
Frommer [179]	Simplified Durand–Kerner iteration
Ichim and Molnar [214]	Lin–Bairstow method
Makrelov and Semerdzhiev [247]	Durand–Kerner simultaneous rootfinding
Makrelov and Semerdzhiev [248]	Primarily polynomials of exponentials
Schweikard [311]	$x = \tan(t/2)$ transformation, square-free method
Schweikard [312]	Interval arithmetic
Weidner [366]	$z = \exp(it)$ transformation to polynomial in z

transformation for converting a trigonometric polynomial in t into an ordinary polynomial in z (section 13.4.1).

Schweikard introduces three new ideas. The first is the conversion of a trigonometric polynomial to an algebraic polynomial through a different (and less simple) means than Weidner: the half-angle tan transformation [311]

$$t = 2\arctan(x) \qquad \leftrightarrow \qquad x = \tan(t/2). \tag{4.90}$$

This implies the identities

$$\cos(t) = \frac{1-x^2}{1+x^2}, \qquad \sin(t) = \frac{2x}{1+x^2}. \tag{4.91}$$

Since $\cos(jt)$ and $\sin(jt)$ can be expanded as series of powers of $\cos(t)$ or as $\sin(t)$ times a cosine polynomial, an arbitrary trigonometric polynomial can always be transformed into a rational function of x, which becomes an ordinary polynomial after multiplying through by the appropriate power of $1 + x^2$. This transformation is more cumbersome than Weidner's, but it has the advantage that trigonometric polynomials with real coefficients are transformed to algebraic polynomials with real coefficients.

The second is the observation that the concept of a *square-free* polynomial, that is, one that has only simple roots, can be extended to trigonometric polynomials [311]. The algorithm is to convert to an algebraic polynomial, calculate the square-free algebraic polynomial, and then convert back to trigonometric form. This can be done without error in a computer algebra system if the coefficients are rational or algebraic numbers.

The third idea is interval arithmetic [312]. In a trio of papers, Carstensen and collaborators developed the interval arithmetic ideas at greater length and generalized the Durand–Kerner simultaneous rootfinding method to higher-order iterations, that is, those converging faster than quadratically [122, 123, 121]. Carstensen and Petkovi'c note that Weidner's transformation, which yields an algebraic polynomial with complex coefficients even when the original trigonometric polynomial has only real coefficients, is undesirable in interval arithmetic methods because the intervals are forced to become bounding rectangles in the complex plane.

The preoccupation with the Durand–Kerner iteration is a bit surprising. For ordinary polynomials, experience shows that if the starting points are chosen on a circle in the complex plane that encloses all the roots, the algorithm usually converges. Although convergence is guaranteed to be quadratic very close to the roots, the Durand–Kerner method often converges quite slowly for the early iterates. For multiple roots, the Durand–Kerner method converges linearly, that is, as δ^m for some $\delta < 1$, where m is the iteration number. All these remarks carry over to applications to trigonometric polynomials. In particular, there is no *global* convergence theory—only theorems about rates of convergence *near* the roots. No comparisons with the Fourier companion matrix method have yet been made.

The Lin–Bairstow method is a Newton's iteration for a system of two equations in two unknowns that computes the real coefficients of a *quadratic* factor of a polynomial, regardless of whether the two roots of the quadratic factor are real or complex valued. Ichim and Molnar devised a similar algorithm for trigonometric polynomials in 1994 [214].

The CCM is fast and reliable. Perhaps for trigonometric rootfinding, this is The End of History! But probably not.

Mediocre talents please tenure committees by scores of publications. It matters not that the new algorithms are vastly inferior to existing methods. All novelty is publishable, at least in third-tier journals, at least for the persistent.

But also probably not because genius does what it must. The mature technology of successive overrelaxation was blown up by multigrid. The old and dying technology of spherical harmonic spectral methods, 40 years ancient, was resuscitated by fast Legendre transforms. Perhaps trigonometric rootfinding will be similarly energized.

In the meantime, the CCM is highly recommended.

Chapter 5

Complex Zeros: Interpolation on a Disk, the Delves–Lyness Algorithm, and Contour Integrals

"A classical adage in numerical analysis says that one should avoid the use of ordinary moments."

—P. Kravanja et al. [233, p. 218]

5.1 ▪ An Overview: Finding All Roots in a Region Ω in the Complex Plane

The CPR philosophy can be extended to complex roots by generating a polynomial proxy for $f(z)$ through adaptive interpolation on a circle-bounded or ellipse-bounded domain in the complex plane. However, this has not been the traditional approach.

In 1967, Delves and Lyness developed an algorithm that has been widely used and extended [152, 153]. They first applied an ancient *oracle* by numerically approximating a contour integral that evaluates to n_r, the number of zeros of $f(z)$ within the region Ω bounded by the contour of integration $\partial\Omega$. Instead of finding the zeros of the interpolant, however, they constructed polynomials of *small* degree n_r whose zeros match those of $f(x)$. In spite of the well-known folklore about moments noted in the epigraph to this chapter, Delves and Lyness employed moments to compute their polynomials. Because of the "moments" worry, many variants of the Lyness–Delves algorithm have been developed (Table 5.1).

A major practical issue is to devise good ways to tessellate portions of the complex plane. Disks are best for interpolation and contour integral evaluations, but to tessellate without overlap, one must truncate disks to the hexagons inscribed within them. The ratio of the area of a disk of unit radius, π, to the area of a hexagon of unit side that just fits inside the unit disk, $\pi(2/[3\sqrt{3}])$, is 1.209. Thus, covering an area of the complex plane with disks entails about 20% duplication, a wastage that has inspired many efforts to cover an area more efficiently when chasing complex zeros.[9]

Another practical issue is that searchable areas must be free of singularities of $f(x)$. A singularity *near* the target domain Ω will dramatically decelerate the convergence of the interpolant.

[9]Delves and Lyness did not use a near-hexagonal tessellation, but instead a scheme borrowed from D. H. Lehmer, illustrated in their Fig. 1, in which a central disk is surrounded by a ring of 9 disks, all 10 overlapping so as to leave no gaps.

Table 5.1. *A select bibliography on rootfinding in the complex plane.*

Reference	Method
Delves and Lyness [152, 153]	Bisection of a square into four smaller squares or disk into nine smaller circles; roots from low-degree associated polynomial constructed from moments of $d\log(f)/dz$
McCune [263]	Cauchy integral for a single root
Botten, Craig, and McPhedran [48]	Lyness–Delves algorithm with radial subdivision followed by sectorial subdivision as needed
Kravanja and Van Barel [231, 232]	derivative-free path integral for roots
Kravanja, Van Barel, Ragos Vrahatis, and Zafiropoulos [233]	Better-conditioned eigenvalue scheme to compute roots from moments
Sakurai, Kravanja, Sugiura, and Van Barel [304]	Error comparison with original Lyness–Delves; improved [233]'s scheme to compute zeros from moments
Dellnitz, Schütze, and Zheng [151]	First determine number of roots by contour integration; apply Newton's iteration with random initialization until all zeros are found
Weyl, Pan, and McNamee [286, 287, 368, 270]	A square is subdivided (quadtree); proximity test excludes some squares; suspect squares are quartered
Austin, Kravanja, and Trefethen [16]	Comparison of polynomial proxy with Cauchy integral schemes; review

5.2 ▪ Interpolation in the Complex Plane

Lyness and Delves [153] proved that interpolation on a *circle* with a *uniformly* spaced grid displays geometric convergence. If the interpolation circle is of radius R and the function being interpolated is analytic in a circle of larger radius R^{anal}, the N-point interpolant will have an error bounded by a constant times $(R/R^{anal})^N$. Geddes and Mason showed that interpolation on a *circle* with a *uniformly* spaced grid and *monomial* basis is as optimal for an *area* in the complex plane as Chebyshev interpolation is optimal for interpolation of nonperiodic functions on an interval, or Fourier interpolation is best for periodic functions on an interval [189]. This is perhaps not surprising because in the monomial basis, using polar coordinates (r,θ) centered on an arbitrary point z_0, a polynomial can be written on a circle of radius $r=R$ as

$$f_N(z)=\sum_{n=0}^{N-1} b_n(z-z_0)^n=\sum_{n=0}^{N-1} b_n R^n \exp(in\theta), \tag{5.1}$$

which manifestly is a Fourier series.

The interpolation points are

$$\theta_j=2\pi(j-1)/N, \qquad z_j=z_0+R\exp(\theta_j), \qquad j=1,2,\ldots,N. \tag{5.2}$$

Let $\vec{f}$ denote the vector whose elements are the samples of $f(x)$ at the interpolation points, $f(z_j)$. The matrix $\vec{\vec{V}}$ of the matrix multiplication transform has the elements

$$V_{j,k}\equiv\omega^{-(k-1)(j-1)}R^{-(j-1)}/N, \tag{5.3}$$

where ω is the Nth root of unity:

$$\omega = \exp(2\pi i/N). \tag{5.4}$$

Let $\vec{b}$ denote a vector whose elements are the polynomial coefficients b_n. These coefficients can be obtained from the samples of the function being interpolated by a matrix–vector multiplication:

$$\vec{b} = \vec{\vec{V}}\vec{f}. \tag{5.5}$$

When R is considerably different from unity, the power series coefficients b_n are badly scaled in the sense that they may grow exponentially with degree even for functions with power series that converge rapidly within the region of interest. It proved helpful to make the change of coordinate

$$w \equiv (z - z_0)/R \tag{5.6}$$

so that the circle of interpolation is always the unit circle centered on the origin in the complex w-plane. The same V_{jk} formula still applies, but with $R = 1$ in the expression for the matrix elements. We shall denote the coefficients of the w-polynomial by a_n,

$$f_N(w(z)) = \sum_{n=0}^{N-1} a_n \exp(in\theta), \tag{5.7}$$

and the interpolation points are $w_j = \exp(i\theta_j) = \omega^{j-1}$.

The Vandermonde matrix $\vec{\vec{V}}$ is also the matrix of the Fourier transform. The coefficients of the interpolant can be computed with greater speed by applying the FFT ([113] and Chapter 9 of [76]).

The coefficients a_n are also the coefficients of the power series of the interpolant in w with respect to the point $z = z_0$. As $|w|$ shrinks, the power series converges faster because $|b_n||w|^n \exp(in\theta)|$ shrinks as $|w| \to 0$ so that the polynomial is most ill conditioned in the disk on the boundary circle, $|w| = 1$. On the circle, however, the power series $f_N(z)$ is just the Fourier series $f(z_0 + R\exp(i\theta)) = \sum_n b_n \exp(in\theta)$ and thus is very well conditioned. As will be illustrated by the case studies below, interpolation on the unit disk is both FFT-fast and Fourier-stable.

To perform the interpolation adaptively, proceed as for Chebyshev interpolation. Apply interpolation for

$$N = 2^{K-1} \tag{5.8}$$

points, where K is an integer, the *Clenshaw–Curtis level*. Evaluate the error $f(x) - f_N(x)$ at the midpoints between each of the interpolation points on the N-point grid, and continue to move to larger and larger K until the on-the-disk interstitial residual, $E^{K,interstitial}$, is smaller than some user-specified tolerance $\beth$, where

$$E^{K,interstitial} = \max_{k=1,2,\ldots N} |f(w_k^I) - f_N(w_k^I)|, \qquad w_k^I = \exp\left(i\pi\frac{2k-1}{2N}\right), \; k = 1,2,\ldots,N, \tag{5.9}$$

where f_N is the interpolant of $f(x)$ at the points $w_j = \exp(i\pi j/N), j = 0,1,\ldots,N$.

As long as there are *no singularities* of $f(z)$ *on* and *within* the circle containing the interpolation points, interpolation on a circle converges geometrically, that is, $|f(z) - f_N(z)| \leq \text{constant}\exp(-\mu z)$ for some $\mu > 0$ everywhere within the disk.

5.2.1 ▪ The Frobenius Companion Matrix

The zeros of the polynomial in w are the eigenvalues of the $(N-1)\times(N-1)$ Frobenius companion matrix. The matrix elements are all zeros except for

$$F_{j-1,j} = 1, \qquad j = 2,\ldots,(N-1), \tag{5.10}$$

$$F_{N-1,j} = -a_{j-1}/a_{N-1}, \qquad j = 1,2,\ldots,(N-1). \tag{5.11}$$

These eigenvalues must then be converted from w to the original coordinate z by

$$z_k^{roots} = R w_k + z_0. \tag{5.12}$$

In the same way that zeros off the interval $x \in [-1,1]$, derived from a Chebyshev series, are unreliable and must be rejected, those roots of the disk polynomial that are outside the circle of the interpolation points are also unreliable.

The simplicity of interpolation in a disk belies the complexity of identifying regions in the complex plane that should be searched by overlapping disks and of ensuring that the disks are free of singularities of the function whose zeros are being sought.

First, though, we need to describe contour integration.

5.3 ▪ Contour Integration

In applications, it is sufficient to choose the path of integration to be a single smooth curve or to be the piecewise union of a finite number of smooth pieces. Each piecewise smooth curve is parameterized in the form $z(t) = x(t) + iy(t)$, where $t \in [a,b]$ for some constants a and b and t is the parameter. If the curve is not closed, it is an *arc*. The integral on an arc or closed curve is then

$$I^{ab} = \frac{1}{2\pi i}\int_a^b g(z)\frac{dz}{dt}dt \tag{5.13}$$

$$= \frac{1}{2\pi i}\int_a^b g(x(t)+iy(t))\left\{\frac{dx}{dt} + i\frac{dy}{dt}\right\}dt. \tag{5.14}$$

The integration curves are *directed* in the sense that integration around a closed curve is ambiguous until one specifies a direction. The default is that the curve is traced in a counterclockwise direction as the parameter t increases.

Important special cases are integration along a circle or an ellipse. Let R denote the radius of the circle. The curve and the contour integral are

$$z(t) = z_0 + R\exp(it), \qquad t \in [0,2\pi], \tag{5.15}$$

$$I^{circle} = \frac{1}{2\pi i}\int_0^{2\pi} g\left(z_0 + Re^{it}\right)Re^{it}\, i\, dt. \tag{5.16}$$

An ellipse with foci at $\pm A$ and the corresponding contour integral are

$$z(t) = z_0 + \cos(\mu + it) = z_0 + A\cosh(\mu)\cos(t) + iA\sinh(\mu)\sin(t), \qquad t \in [0,2\pi],$$

$$I^{ellipse} = \frac{1}{2\pi i}\int_0^{2\pi} g(z(t))\left\{-A\cosh(\mu)\sin(t) + iA\sinh(\mu)\cos(t)\right\}dt. \tag{5.17}$$

Here, μ and t are elliptical coordinates in the complex plane.

Similarly, the integral around a rectangle with lower left corner at $z = z_0$ and sides L and M is

$$I^{rectangle} = I^{z_0,z_0+L} + I^{z_0+L,z_0+L+iM} + I^{z_0+L+iM,z_0+iM} + I^{z_0+iM,z_0}, \tag{5.18}$$

where

$$I^{z_0,z_0+L} = \frac{1}{2\pi i} \int_0^L g(z_0+t)\,dt, \tag{5.19}$$

$$I^{z_0+L,z_0+L+iM} = \frac{1}{2\pi i} \int_0^M g(z_0+L+it)\,\{i\}\,dt, \tag{5.20}$$

$$I^{z_0+L+iM,z_0+iM} = \frac{1}{2\pi i} \int_0^L g(z_0+(L-t)+iM)\,\{-1\}\,dt, \tag{5.21}$$

$$I^{z_0+iM,z_0} = \frac{1}{2\pi i} \int_0^M g(z_0+iM-it)\,\{-i\}\,dt. \tag{5.22}$$

5.4 ▪ Delves and Lyness Algorithm

Theorem 19.8 in section 19.6 asserts that if a function $f(x)$ is analytic everywhere within a region Ω of the complex plane including its boundary $\partial\Omega$, then the number n_r of roots in the region is given exactly by

$$n_r = \frac{1}{2\pi i} \oint \frac{f_z(z)}{f(z)}\,dz, \tag{5.23}$$

where roots are counted according to their multiplicity.

The trapezoidal rule of elementary numerical analysis has an accuracy with N points that decreases *exponentially* fast with N for a *circular* contour. For systematically searching multiple regions without overlap, square contours are more convenient, but the trapezoidal rule has an error that decreases only as $O(1/N^2)$ for a contour with sharp corners [153]. Exponential accuracy can be recovered by applying Gaussian quadrature separately to each smooth segment of the contour. However, because the number of roots is always an *integer*, high accuracy in evaluating the contour integral is *unnecessary*.

When the number of roots in a region is large, one can subdivide the region and retest until one has identified regions with a small number of roots (or perhaps only one) in a given region. One can then apply any standard technique for computing the roots. The crucial point is that when the number of roots is known, one can repeatedly iterate until all roots have been found. Because one *knows* how many roots are within the domain Ω, even crude rootfinders become robust, guaranteed algorithms.

Dellnitz, Schütze, and Zheng [151] employ only the oracle of the Delves–Lyness scheme. To compute the zeros, they apply Newton's iteration with an initial condition that is a *randomly chosen* point within Ω. If the iteration converges to a root within Ω, then hurrah! If it fails, or converges to a root outside Ω, they try again with a different, randomly chosen initial condition. They repeat until the number of converged roots in Ω is n_r.

One practical difficulty is that the analyticity of $f(z)$ in Ω is often unknown. It is important to note that a simple *pole* within the contour of integration will contribute the *same* as a zero. However, a singularity of $f(z)$ within the interpolation disk is equally fatal to interpolation. The search/interpolation/enclosed-by-contour region must be singularity free.

After calculating n_r, Delves and Lyness use accurate evaluation of higher-order moments of the contour integral of the logarithmic derivative of $f(u)$ to simultaneously find all roots within Ω; a FORTRAN 77 implementation with careful discussion of programming issues was published by Botten, Craig, and McPhedran [48]. The moments are defined by

$$\mu_p \equiv \frac{1}{2\pi i} \oint_{\partial\Omega} dz\, z^p \frac{f_z}{f(z)}. \tag{5.24}$$

When $\partial\Omega$ is a circle, the trapezoidal rule yields the moments

$$\mu_p \approx \frac{1}{N} \sum_{j=0}^{N-1} \omega_j^{p+1} \frac{f_z(\omega_j)}{f(\omega_j)}, \tag{5.25}$$

where $\omega_j = \exp(2\pi i j/N)$ are the Nth roots of unity. The quadrature error decreases exponentially fast with the number of quadrature points N: careful error analysis is given in [153].

The trapezoidal rule can be applied to noncircular contours with spectral accuracy as long as the contour is closed and analytic; the special case of an elliptical contour is described in section 5.3.

Delves and Lyness define an *associated polynomial* that is constructed so that its roots z_k and their multiplicity ν_k match those of the (generally nonpolynomial) function $f(z)$ inside the complex domain Ω:

$$P^{assoc} \equiv \prod_{k=1}^{n_r} (z - z_k)^{\nu_k} = z^M + \sigma_1 z^{M-1} + \cdots + \sigma_M. \tag{5.26}$$

One can show that the moments of the associated polynomial are identical to those of $f(z)$. The coefficients of the associated polynomial can be found from the moments by means of

$$\sigma_k = -\left(\mu_k + \sum_{j=1}^{k-1} \mu_{k-j} \right), \qquad k = 1, 2, \ldots. \tag{5.27}$$

One can then apply a standard *polynomial* rootfinder to this associated polynomial as the next to last step to approximately compute the roots of a *nonpolynomial* $f(z)$. (The final step is to refine the zeros by Newton's iteration, as always recommended.)

For the important special case that there is only a single zero in the domain,

$$z_* = \mu_1 \tag{5.28}$$

$$= \frac{1}{2\pi i} \oint_{\partial\Omega} dz\, z \frac{f_z}{f(z)} \tag{5.29}$$

$$\approx \frac{1}{N} \sum_{j=0}^{N-1} \omega_j^2 \frac{f_z(\omega_j)}{f(\omega_j)}. \tag{5.30}$$

The degree of the interpolant N is *large* so that $f_N(z)$ is an accurate approximation to $f(z)$. In contrast, the degree of the associated polynomial $P^{assoc}(z)$ is equal to n_r, the number of zeros in the disk, which is usually much smaller. However, the ill-conditioning of moment-based computations has motivated the pursuit of more stable algorithms.

Kravanja, Van Barel, Ragos, Vrahatis, and Zafiropoulos have published an improved algorithm and a publicly available FORTRAN 90 code [233]. As recommended by Delves and Lyness themselves, the Kravanja et al. scheme automatically subdivides a rectangle if the number of roots exceeds some user-specified number M (where $M \approx 5$) so as to avoid the ill-conditioning difficulties of computing a large number of zeros simultaneously. Another part of Kravanja et al.'s strategy is to compute the roots as the eigenvalues of the generalized matrix eigenvalue problem

$$\begin{vmatrix} \mu_0 & \mu_1 & \cdots & \mu_{n-1} \\ \mu_1 & \mu_2 & \cdots & \mu_n \\ \vdots & \vdots & \cdots & \vdots \\ \mu_{n-1} & \cdots & \cdots & \mu_{2n-2} \end{vmatrix} \begin{vmatrix} x_1 \\ x_2 \\ \vdots \\ x_n \end{vmatrix} = \lambda \begin{vmatrix} \mu_1 & \mu_2 & \cdots & \mu_n \\ \mu_2 & \mu_3 & \cdots & \mu_n \\ \vdots & \vdots & \cdots & \vdots \\ \mu_n & \cdots & \cdots & \mu_{2n-1} \end{vmatrix} \begin{vmatrix} x_1 \\ x_2 \\ \vdots \\ x_n \end{vmatrix}. \tag{5.31}$$

The multiplicities of the roots are found in a second, separate step by solving a Vandermonde system. Sakurai, Kravanja, Sugiura, and Van Barel offer an error analysis of both the original and improved algorithms and show that the latter is better conditioned [304].

Dellnitz, Schutze, and Zheng devised the randomly initialized Newton scheme, described earlier, that was also very effective in identifying a large number of roots in real-world applications [151]. The Delves–Lyness moments algorithm works well, but it is by no means necessary or optimal to follow each of their steps.

5.5 ▪ Numerical Illustrations

5.5.1 ▪ Example 1: $f(z) = (z - 1/2)\,(z - i)\,(z - [1 + i])$

The interpolant on the disk proxy should be exact for all $N > 4$. The right panel in Fig. 5.1 shows that for $N = 7$, roundoff has generated errors in the interpolant roots of a part in a trillion—absurdly small, but large compared to machine precision, which here is the IEEE double precision/MATLAB standard of $\epsilon_{mach} = 2.2 \times 10^{-16}$. We have no explanation. For larger N, the root errors for both the classic Delves–Lyness and interpolant-proxy methods descend to smaller than 10^{-14}.

One might suppose that the interpolant-proxy method would fail catastrophically for $N > 4$ because then all the power series coefficients b_n for $n > 3$ will be random numbers whose order of magnitude is machine epsilon. This is worrisome because the elements in one row of the Frobenius companion matrix are ratios with a_N as the denominator. Will not these tiny, random divisors destroy the accuracy of the matrix eigenvalues? Although excessive N is wasteful in operation count, it has only a minor effect on accuracy if, as is automatic in MATLAB, the *matrix* is *balanced* before the eigenvalue computation begins. The error in the roots is about the same whether $N = 7$ or $N = 127$. (The same indifference to tiny trailing coefficients is also observed when the proxy is a Chebyshev polynomial series, as described in earlier chapters.)

In contrast, the Delves–Lyness error does not plateau but exhibits geometric convergence with increasing N. Why does one need 127 points to compute the roots of a cubic polynomial that can be exactly interpolated using only 4 points? The answer is that in the Delves–Lyness method, the key process is contour integration, and unfortunately the integrand is not the *polynomial* $f(z)$ itself, but rather the *rational function* $(df/dz)/f$.

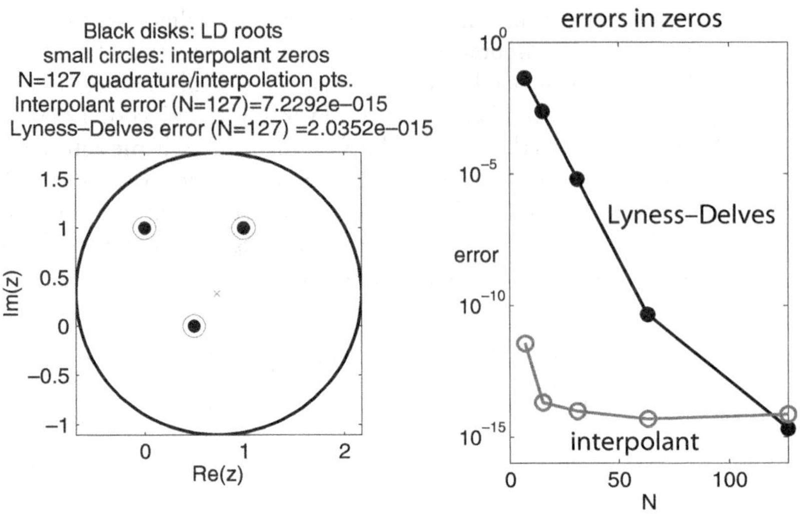

Figure 5.1. *Cubic polynomial:* $f(z)=(z-1/2)(z-i)(z-[1+i])$. *Left: the three zeros are marked by solid disks inside small circles; the samples of* $f(z)$ *used for interpolation and contour integration lie on the large circle of radius* $R=1.437$ *also illustrated; its center is at* $z_0=0.7338+(1/3)i$, *marked with a small "x." (The peculiar numbers for* R *and* z_0 *were chosen to demonstrate that the circle of integration can be chosen arbitrarily, provided there are no singularities of* $f(z)$ *inside or on it.) Right: the error norm, defined as the maximum of the absolute errors for the three zeros, plotted versus* N, *the number of sampling/quadrature/interpolation points. The upper curve is the Delves–Lyness algorithm and the lower (more accurate!) curve with circles is the error in computing the zeros as eigenvalues of the companion matrix derived from interpolation on the circle shown in the left panel.*

5.5.2 ▪ Example 2: Rational Function of Exponentials

The second example, taken from Delves and Lyness's original paper, is to find the complex root $(0.343004199843762, 1.033945790483547)$ of the derivative of a rational function of exponentials. This derivative (denoted by $f(z)$ here, but df/dz in the original), is

$$f(z) \equiv -e^{-z}\frac{43800e^{-z}-65892e^{-2z}+31021e^{-3z}-13844e^{-4z}+5961e^{-5z}-2642e^{-6z}-22700}{\left(227e^{-z}-219e^{-2z}+186e^{-3z}-38e^{-4z}\right)^2}.$$

This problem is challenging because the function also has a pole not too far from the zero.

Fig. 5.2 depicts three different choices of the center and radius of the search circle.

I. Both the pole and zero are far from the circle of integration and interpolation: $z_0 = 3/10+i$, $R=1/5$.

This is the favorable case: both the Lyness–Delves and interpolation-proxy methods converge rapidly.

II. The *pole* is close to the circle of integration and interpolation, but the root is a safe distance away: $z_0=1/5+9/10i$, $R=2/5$.

Both methods converge much more slowly, decelerated by the singularity near the contour.

III. The *zero* is close to the circle of integration and interpolation, but the pole is a safe distance away: $z_0=3/10+i$, $R=1/5$.

The interpolation-proxy method converges rapidly because it is degraded only by *singularities* of $f(z)$ near the circle where the interpolation samples lie. The Delves–Lyness

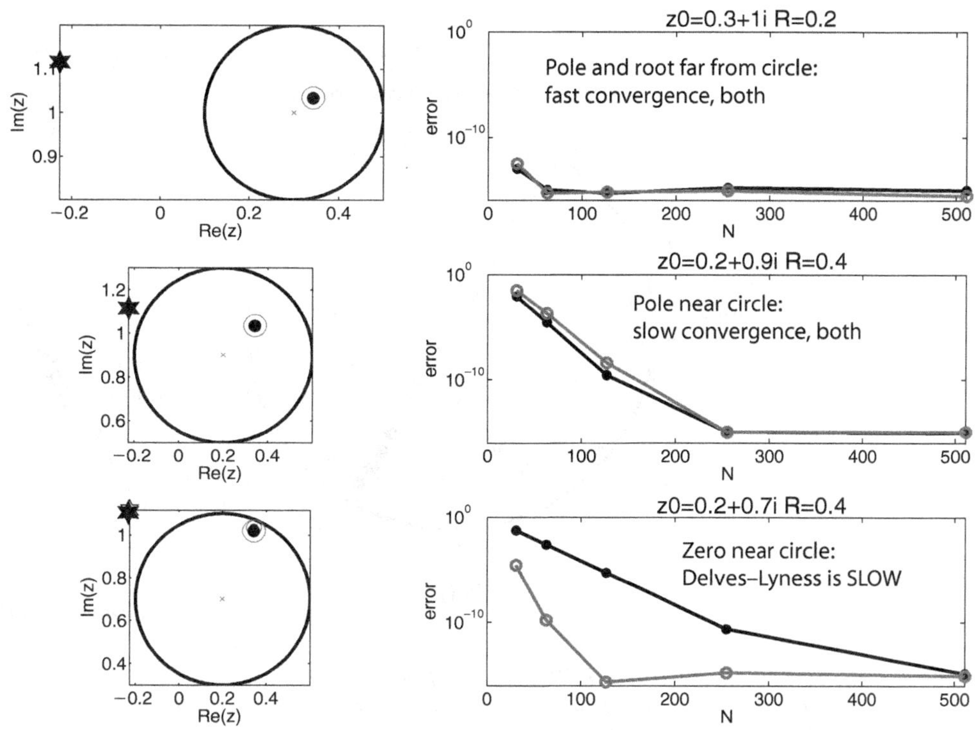

Figure 5.2. *Three different choices of a search disk for Example 2. In each diagram on the left, the single root is the disk; the pole is marked by the six-sided star. The absolute error in the root is plotted versus N in the three right graphs for both rootfinders.*

method, in contrast, is poisoned by both *roots* and *singularities* near the contour integration path.

Delves and Lyness noted in their original paper that near-the-contour zeros could be treated with antivenom, so to speak, by choosing new integration paths. They described an automated procedure for identifying poisonous roots and triggering the adjustment of integration contours. This zeros-near-the-contour fix, however, adds considerably to the complexity of their algorithm.

5.5.3 ▪ Example 3: The Product of a Sine Function with a Pole

The function

$$f(z) = \frac{\sin(4\pi \exp(i\pi/3)z)}{z-4} \tag{5.32}$$

is a sine function, made less trivial by adding a singularity and also rotating the argument of the sine. Here, this line of oscillation is rotated by an angle of $-\pi/3$ about the origin; the sine function oscillates from minus one to one along this line. In the complex plane, however,

$$\sin(x+iy) = \frac{\exp(ix)\exp(-y) - \exp(-ix)\exp(y)}{2i} \tag{5.33}$$

$$\sim (-i/2)\exp(-i\,\mathrm{sign}(y)x)\,\exp(|y|). \tag{5.34}$$

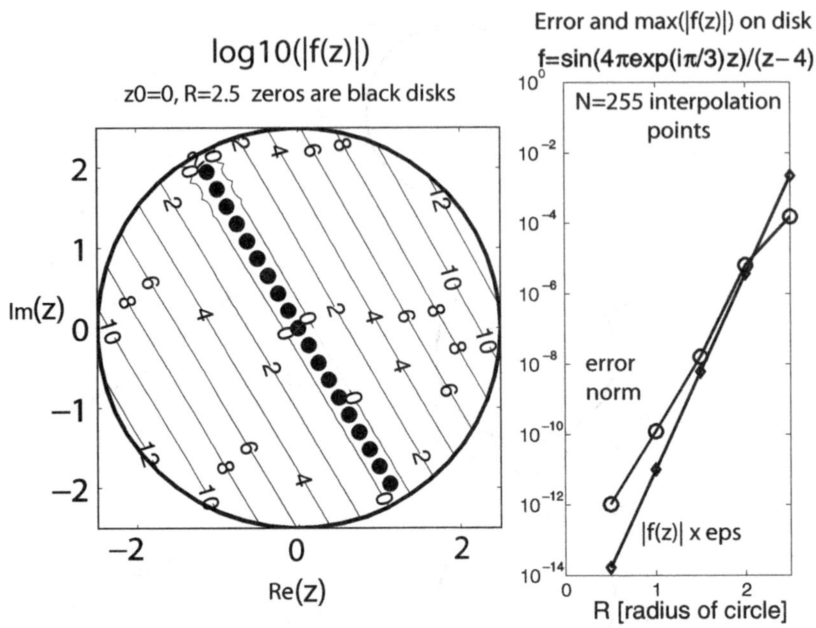

Figure 5.3. *Left: a contour plot of the isolines of the base-10 logarithm of* $|f(z)|$, *where* $f(z) = \sin(4\pi \exp(i\pi/3)z)/(z-4)$. *The zeros (black disks) all lie in the "valley" where* $|f(z)| \sim O(1)$. *Right: the curve with open circles is the largest absolute error for any of the roots within a disk generated by the interpolation-proxy with 255 points, plotted versus the radius R of the circle, centered at the origin, used for interpolation. The curve with diamonds is the product of the maximum value of* $|f(z)|$ *on the circle (and disk) with machine epsilon in MATLAB,* 2.2×10^{-16}.

As the interpolation-proxy rootfinder is applied on circles of larger and larger radius R, the largest value of $|f(z)|$ in the disk grows exponentially fast with R. No matter how many interpolation points are used, the absolute error of interpolation is limited to the product of $|f(z)|$ with machine epsilon. That is because roundoff error forces the coefficients of the spectral series to plateau at a value that is usually 10 to a 100 times machine epsilon. When R is sufficiently large that $|f(z)|$ reaches values of 10^{15}, which happens for $R = 3$ here, the absolute errors in interpolation become $O(1)$. Alas, this is the same magnitude as the oscillations of the sine function along the line containing all the roots. Interpolation is thus generating random numbers in the valley of $|f(z)|$ where the zeros lie, spawning a double chain of spurious zeros (not shown), and the rootfinder has no accuracy at all.

Fig. 5.3 shows that the errors in interpolation-proxy rise exponentially with increasing circle radius R. The power series coefficients plateau when $N = 127$. Therefore, inadequate numbers of interpolation points are not a problem for the 255-point computation shown in the right panel. The strong correlation between the errors in the zeros and the product of $|f(z)|$ with machine epsilon shows that the problem is an intrinsic one of huge dynamical range, as discussed earlier for the Chebyshev-proxy in section 2.8. The only remedy is to tessellate the target region in the complex plane by smaller disks and apply the rootfinder separately to each disk.

The Lyness–Delves algorithm is so awful for $R > 1$ that its errors are not shown in Fig. 5.3. For small, fixed R, its accuracy displays the same pattern as in Example 1: its error converges exponentially fast with increasing numbers of quadrature points along the

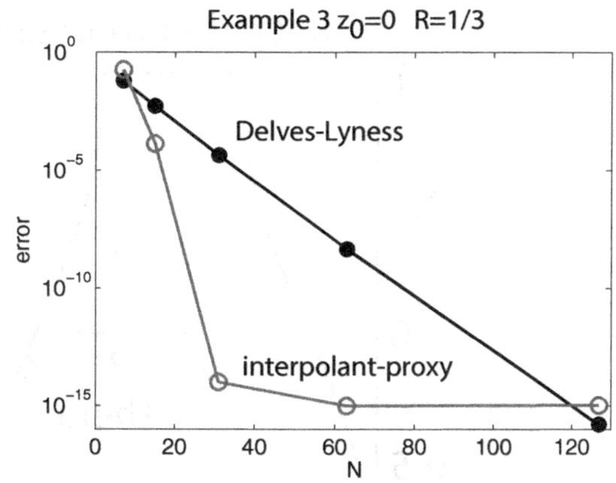

Figure 5.4. *Comparison of rootfinder error norms for Example 3. The interpolant-proxy errors are the lower curve, (with open circles); the Lyness–Delves errors are the upper curve, (with solid disks), both plotted versus N, the number of interpolation or quadrature points. $R = 1/3$, and the circle for interpolation and contour integration was centered on the origin. The error norm is defined as the maximum, for a given N, of the individual errors in those roots of $f(z)$ that lie within the disk.*

circle, N, but is orders of magnitude worse than the error computed from the companion matrix of the interpolant for small N and moderate N, as shown in Fig 5.4.

5.6 ▪ Rectangles Instead of Disks

5.6.1 ▪ Interpolation on the Boundary of a Square via Conformal Mapping of the Square to the Disk

If we map the unit square to the unit disk, we can then apply interpolation with points on the unit circle to the transformed function. If successful, such a strategy will allow us to tessellate the complex plane with nonoverlapping squares instead of overlapping disks.

Theorem 5.1 (Square-to-Disk Conformal Mapping). *Let $w = r\exp(i\theta)$ denote the complex variable of the unit disk and $z = x + iy$ the complex variable of the square. The function*

$$w = \text{coslem}\left(\frac{K}{2}\exp(-i\pi/4)(z+1+i)\right) \tag{5.35}$$

maps the unit square with vertices at $(\pm 1, \pm 1)$ in the z-plane to the interior *of the unit circle in the w-plane, where the cosine-lemniscate function is the Jacobian elliptic function*

$$\text{coslem}(\zeta) = \text{cn}(\sqrt{2}\zeta; k = 1/\sqrt{2}) \tag{5.36}$$

and where the period of $\text{cn}(u; k = 1/\sqrt{2})$ *is $4K$ in u, where*

$$K = 1/2\frac{\pi^{3/2}}{(\Gamma(3/4))^2} = 1.8540747 \tag{5.37}$$

is the complete elliptic integral of the first kind for modulus $k = 1/\sqrt{2}$. The inverse map is

$$z = \frac{2}{K}\exp(i\pi/4)\text{coslem}^{-1}(w) - (1+i). \tag{5.38}$$

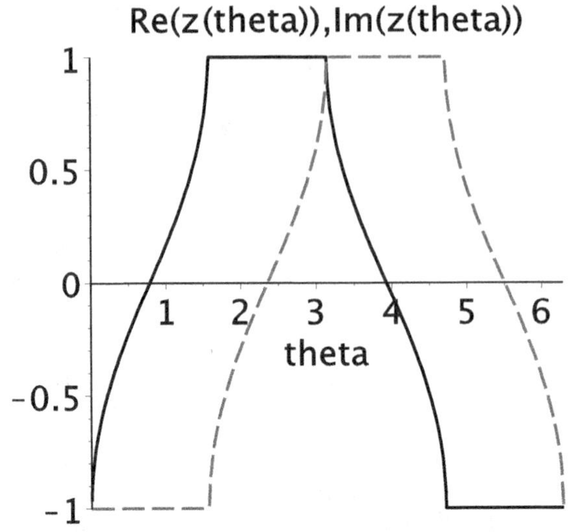

Figure 5.5. *Conformal mapping of the unit square in z to the unit disk in w. Solid line: $\Re(z(\exp(i\theta)))$, where $\theta = \text{angle}(w)$; thus, what is plotted is the (real part of the) function on the unit circle in w that is being interpolated. Dashed line: $\Im(z(\exp(i\theta))$. Both curves are continuous, but their derivatives $\Re(dz(\exp(i\theta))/d\theta)$ and $\Im(dz(\exp(i\theta))/d\theta)$ are each discontinuous at four different θ on the range $\theta \in [0, 2\pi]$—the images of the four corners of the square.*

where

$$\text{arccoslem}(w) = \int_w^1 \frac{1}{\sqrt{1-t^4}}\,dt \tag{5.39}$$

$$= K/\sqrt{2} - F(w, i), \tag{5.40}$$

where $F(w,k)$ is the incomplete elliptic integral (known in the nineteenth century [159], *but the statement here follows* [92]*; applications include* [111, 10]*).*

Unfortunately, interpolation on the square requires that $f(z[w])$ be analytic everywhere on the unit disk in w. The simplest case is $f(z) = z$, in which case the function we interpolate on the circle is $z(w)$, the inverse map itself. Alas, this map is *singular* at all four corners of the square, as manifest in the four slope discontinuities in Fig. 5.5.

There is no simple way to search the complex plane by interpolation on nonoverlapping squares.

5.6.2 ▪ Delves–Lyness Rootfinding in the Square

"Square regions have the advantage that they can be subdivided without redundancy. On the other hand, more efficient integration rules can be obtained for circles."

—L. M. Delves and J. N. Lyness, caption to Fig. 1 of [152]

The Delves–Lyness algorithm is more flexible than interpolation because contour integration around a square can be performed with an accuracy that converges geometrically fast with N. As explained earlier, it is merely necessary to treat each side separately

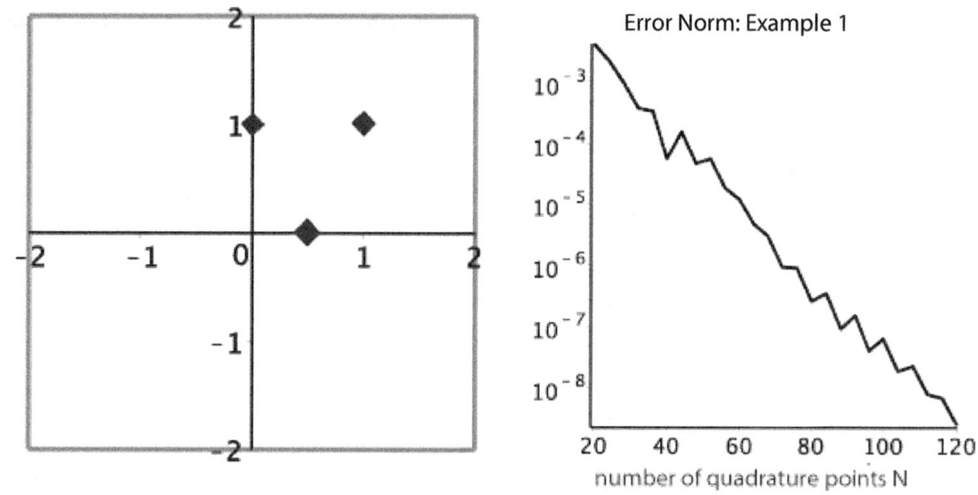

Figure 5.6. *The Delves–Lyness method was applied in a square domain with sides of length four to find the three zeros of our first example for which $f(z)$ is a cubic polynomial. The domain and the zeros (diamonds) are shown on the left. The right panel shows that the errors fall exponentially fast with N when the contour integrals along each side are computed using Clenshaw–Curtis Chebyshev quadrature with $N/4$ points.*

with Clenshaw–Curtis quadrature (Appendix B) for nonperiodic integrands and sum the results.

Delves and Lyness themselves rather obscured this point by (i) noting correctly that applying the trapezoidal quadrature rule to a square contour has only second-order accuracy and (ii) applying Romberg integration, which is repeated Richardson extrapolation, to obtain a scheme of higher but still nonexponential accuracy.

Fig. 5.6 shows that by using Clenshaw–Curtis Chebyshev quadrature instead of a finite-order method, one can in fact obtain approximate roots with an error that falls geometrically with the number of quadrature points.

Fig. 5.7 shows that exponential convergence is also achieved in the square for a transcendental $f(z)$ with a pole outside the targeted domain.

5.7 ▪ Polynomialization

In atmospheric science, a prototype Sturm–Louisville eigenproblem of the fourth kind is

$$u_{xx} + (1/x - \lambda)\,u = 0,\, u(-\infty) = 0,\, u(B) = 0,\ B > 0. \tag{5.41}$$

The solution is logarithmically singular at the origin; the branch cut must, for physical reasons, lie in the upper half-plane, which is equivalent to interpreting the pole in the coefficient of the differential equation as $1/x = \lim_{\epsilon\to 0} 1/(x - i\epsilon)$ [4]. The eigenvalues λ are *complex valued* and always lie in the upper half-plane [2]. This problem is discussed extensively by Boyd in [52, 54, 101, 61, 104].

Boyd derived accurate WKB approximations to both the eigenfunctions and eigenvalues. The *double Bessel* approximation for the eigenvalues is obtained by solving the secular trigonometric polynomial

$$f(z) \equiv z + \sin(2z)/2 - 1/2\,\pi - \frac{(n-1/4)\,\pi\,\sin(z)}{\sqrt{B}},\quad \lambda_n = \sin^2(z)/B, \tag{5.42}$$

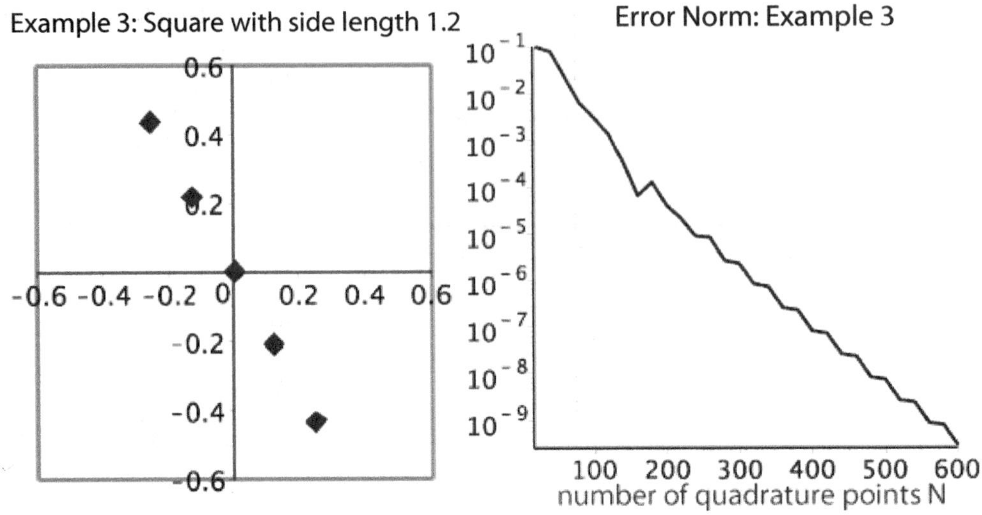

Figure 5.7. *The Delves–Lyness method was applied in a square domain with sides of length four to find five zeros of Example 3, a sine function divided by $(z-4)$. Left: the domain and the zeros (diamonds). Right: the errors fall exponentially fast with N when the contour integrals along each side are computed using Chebyshev quadrature with $N/4$ points.*

where $n = 0, 1, \ldots$ is the mode number. Because all the transcendental functions are sines, it is trivial to bypass both interpolation and the Delves–Lyness method by expanding the function $f(z)$ as a power series in z using the known expansion for the sine function:

$$
\begin{aligned}
f = & -1/2\,\pi + \left(2 - \left(n - \frac{1}{4}\right)\pi \frac{1}{\sqrt{B}}\right) z + \left(1/6\,\frac{(n-1/4)\,\pi}{\sqrt{B}} - \frac{2}{3}\right) z^3 \\
& + \left(-\frac{1}{120}\,\frac{(n-1/4)\,\pi}{\sqrt{B}} + \frac{2}{15}\right) z^5 + \left(-\frac{4}{315} + \frac{1}{5040}\,\frac{(n-1/4)\,\pi}{\sqrt{B}}\right) z^7 \\
& + \left(\frac{2}{2835} - \frac{1}{362880}\,\frac{(n-1/4)\,\pi}{\sqrt{B}}\right) z^9 \\
& + \left(-\frac{4}{155925} + \frac{1}{39916800}\,\frac{(n-1/4)\,\pi}{\sqrt{B}}\right) z^{11} + \cdots .
\end{aligned}
\tag{5.43}
$$

Table 5.2 shows that the series through z^{11} yields excellent approximations in 16-digit arithmetic to the ground-state ($n = 1$) eigenvalues over a wide range of B; the exact roots were obtained by Newton's iteration; the (relative) errors are defined as the difference between the polynomial and those of the transcendental f, scaled by dividing by the absolute value of the exact root z_*.

Boyd also gave a more accurate approximate dispersion relation labeled the "Whittaker" approximation. Although only a small perturbation of the double Bessel formula, the Whittaker formula does not have power series because it includes the term $\ln\left(\Gamma\left(1 + 1/2\frac{\sqrt{B}}{\sin(z)}\right)\right)$, which is infinite at the origin. Oops! A simple remedy is to compute the roots of the double Bessel approximation and then use these as the initialization for solving the Whittaker equation by means of Newton's iteration.

Table 5.2. *Ground-state eigenvalues in the double Bessel approximation to the Boyd Sturm–Liouville eigenproblem of the fourth kind.*

B	$\|z\|$	$B\lambda_1$ [power series (5.43)] $(B\lambda_1)$	$B\lambda[f(z),(5.42)]$ $(B\lambda_1)$	$\|\lambda_1^{poly}-\lambda_1\|/\|\lambda_1\|$
0.01	3.85	−636.90+3.8674 i	−546.45+3.1609 i	0.166
0.1	2.65	−49.588+3.3473 i	−49.090+3.2814 i	0.010
1.	1.49	−2.2702+3.3096 i	−2.2704+3.3097 i	0.000052
10.	1.30	0.97515+1.4190 i	0.97514+1.4191 i	0.000017
100.	1.33	1.1593+.54619 i	1.1593+.54614 i	0.000034

5.8 ▪ Log and Polynomials

In many applications, approximations are the sum of a polynomial plus another polynomial multiplied by a logarithm. It is not possible to expand such sums as power series about the origin because the logarithm is singular at the origin. Interpolation in a disk centered at the origin is equally impossible. The difficulty can be removed by making a transformation

$$z = \exp(g(\zeta)), \tag{5.44}$$

where $g(\zeta)$ is a smooth function. A log-polynomial becomes, taking $g(\zeta)=\zeta$,

$$\begin{aligned} f(z) &= (p_0+p_1z+p_2z^2)+(q_0+q_1z+q_2z^2)\log(z) && (5.45)\\ &= p_0+p_1\exp(\zeta)+p_2\exp(2\zeta)+(q_0+q_1\exp(\zeta)+q_2\exp(2\zeta))\,\zeta. && (5.46)\end{aligned}$$

For example, the Charney baroclinic instability problem of atmospheric and oceanic dynamics has the complex-valued phase speed c as the eigenvalue; the differential equation is solved by the Whittaker function

$$u(\eta) = W_{r,1/2}(\eta) = P(\eta)+Q(\eta)\log(\eta), \tag{5.47}$$

where $r > 0$ is a real parameter and $\eta = z-c$, where z is the vertical coordinate with $z=0$ as the ground. The phase speed c is the negative of the zero η_0 of the function

$$f(\eta) = \eta u_\eta - u(\eta) = 0. \tag{5.48}$$

The good news is that the power series $P(\eta)$ and $Q(\eta)$ have infinite radii of convergence. The bad news is that when these series are truncated at some degree N, the result is not a polynomial, but rather a log-polynomial. For example, when $r = 1/2$, close to the parameter value that gives the highest growth rate,

$$\begin{aligned} W_{1/2,1/2}(\eta) = (0.5641895835 &+ (-0.2820947918\ln(\eta)+0.2282368862)\eta\\ &+(0.07052369795\ln(\eta)-0.09232107046)\eta^2\\ &+(0.03189572988-0.01763092447\ln(\eta))\eta^3\\ &+(-0.004537932625+0.002203865560\ln(\eta))\eta^4\\ &+(-0.0002754831945\ln(\eta)+0.000636112375)\eta^5\\ &+O(\eta^6). \end{aligned} \tag{5.49}$$

Table 5.3. *Errors in solutions to the Charney baroclinic instability problem for $r = 1/2$. The exact eigenvalue is $c = 0.901016078298246361461888806 + 0.5449895075672669160349069 2\,i$. N is the degree of the polynomial and log-polynomial approximation. Only the leftmost converged value in each row is shown (in boldface); the (identical) entries to the right are replaced by dashes. All computations in 16-decimal-digit precision.*

N / N_ζ	20	40	60	80	100
2	**1.2267**	–	–	–	–
3	0.22578	**0.22575**	–	–	–
4	0.34640e–1	**0.31968e–1**	–	–	–
5	0.27258	**0.51584e–2**	–	–	–
6	0.39063	0.57731e–3	**0.57687e–3**	–	–
7	0.45407	0.67112e–4	**0.61361e–4**	–	–
8	0.40229	0.42978e–2	0.48372e–5	**0.48372e–5**	–
9	0.41609	0.67381e–1	0.36668e–6	**0.36851e–6**	–
10	0.43038	0.85613e–1	0.15450e–6	**0.22287e–7**	–
11	0.38352	0.15997	0.36635e–5	**0.13195e–8**	–
12	0.32366	0.12918	0.51474e–4	0.65060e–10	**0.64682e–10**

Our degree convention is that "degree N" means that the polynomial part of the truncated series is of degree N and this is also the degree of the polynomial multiplying $\log(\eta)$. Thus, (5.49) displays a log-polynomial of degree five.

The zeros were computed by the following steps:

1. Approximate $f(\eta)$ by truncating the η power series at degree N.
2. Convert the coordinate η to ζ where $\eta = \exp(\zeta)$. ($f(\zeta)$ is now a sum of terms of the form
$$f(\zeta) = \sum_{k=0} q_{0k} \exp(k\zeta) + q_{1k}\, \zeta \exp(k\zeta), \tag{5.50}$$
where q_{jk} are numerical coefficients.)
3. Re-expand $f(\zeta)$ as a power series in ζ up to and including degree N_ζ. (Usually, N_ζ must be considerably larger than N, as illustrated in Table 5.3.)
4. Compute the zeros of the polynomial in ζ.
5. Apply Newton's iteration to each polynomial root.
6. Accept only converged zeros.

The results for a typical value of the Charney parameter r are shown in Table 5.3. Each row show the absolute errors for a given N as approximated by polynomials of various degrees N_ζ (columns).

The exponential change of coordinate followed by polynomial rootfinding is capable of generating good initializations for Newton's iteration even for N as small as 10. However, the zeros are not accurate prior to Newton-polishing until N_ζ is considerably larger. It is better to err on the side of caution and choose $N_\zeta > 20$ so as to avoid missing zeros.

The entire algorithm, excluding the verification of roots through Newton's iteration and output, is executed by a handful of lines in Maple, as shown in Table 5.4.

Table 5.4. *Maple code: Zeros of a log-and-polynomia: Charney baroclinic instability dispersion relation.*

```
restart; Digits := 16;
N := 10; # order of eta power series;
Nzeta := 80; # order of zeta series, Nzeta/N ≫ 1 ;
r := 0.5; # Charney parameter;
ff := eta*((1/2 − r/eta)*WhittakerW(r, 1/2, eta) − WhittakerW(r + 1, 1/2, eta)/eta)
                                                    − WhittakerW(r,1/2,eta) ;
fp := evalf(convert( series(ff,eta,N + 1),polynom));
f := evalf(convert(series(subs(eta = exp(zeta),fp),zeta,Nzeta + 1),polynom));
froots := solve(f,zeta);
nroots := numelems([froots]);
for j from 1 to nroots do print(j,evalf(exp(froots[j]) ) ); od:
```

5.9 ▪ Explicit Contour Integral Ratio for a Single Root

Luck and Stevens offer a simpler exact expression for a zero of a function $f(z)$ that is analytic in a disk of radius R with center at $z = z_0$ that contains a single, simple root. Let $g(z) = 1/f(z)$ or some other function such that the simple zero has become a simple pole [245]. Define

$$w(\theta) \equiv \frac{1}{f(z_0 + R\exp(i\theta))}. \tag{5.51}$$

Then the root is without approximation the ratio of two contour integrations

$$z_* = \frac{\oint w(\theta)\exp(2i\theta)d\theta}{\oint w(\theta)\exp(i\theta)d\theta}. \tag{5.52}$$

Luck and Stevens note that if the contours of integration are chosen to be circles of radius R about the point $z = z_0$, then the quadratures can be performed by the Fast Fourier Transform (FFT). The major criticism of their method is that it depends on the presence of but a single, simple zero within the disk bounded by the path of integration; in contrast, polynomial interpolation along the same circular contour yields a polynomial whose roots are accurate even if many zeros or multiple roots lie within the disk, as reviewed earlier.

The Delves–Lyness algorithm, which can find many zeros within a domain in the complex plane, simplifies, when there is but one root, to

$$z_* = \frac{1}{2\pi i}\oint_{\partial\Omega} z\,\frac{f_z}{f(z)}. \tag{5.53}$$

When the contour is a circle of radius R, the trapezoidal rule is geometrically convergent, that is, the error falls proportionally to $\exp(-qN)$ for some $q > 0$. Then with $z_j \equiv R\exp(i\pi j/N)$ as quadrature points spaced uniformly around the unit circle,

$$z_* = \frac{1}{2\pi i}\oint_{\partial\Omega} z\,\frac{f_z}{f(z)}dz \tag{5.54}$$

$$\approx \frac{1}{N}\sum_{j=0}^{N-1} z_j^2\,\frac{f_z(z_j)}{f(z_j)} \tag{5.55}$$

Table 5.5. *Errors in applying the trapezoidal rule on the unit circle for the single zero Delves–Lyness method:* $f(z) = \cos(z - 7/10)/(4 + z^2)$.

10	0.2920
20	0.058
30	0.014
40	0.0035
50	0.00086
60	0.00022

Table 5.5 shows the exponential convergence when solving for the single root within the unit disk of $f(z) = \cos(z - 7/10)/(4 + z^2)$.

Luck and Stevens describe a couple of refinements. First, the function $1/f(z^*)$ can be replaced by any other function such that a root at $f(z)$ within the disk bounded by the circle of integration yields a pole in the integrand. Second, this explicit expression can be differentiated to obtain an expression for the derivative of the root with respect to the parameter λ.

5.10 ▪ Weyl Exclusionary Quadtree Algorithm for Squares in the Complex Plane

Weyl [368] in 1924 (!) proposed a simple algorithm for finding the zeros of a polynomial in the complex plane. Begin with a square. Subdivide into four smaller squares. In each, perform a *proximity* test to estimate the nearest zero, or otherwise flag some squares as provably zero-free. Squares that might have zeros are labeled "suspect" and each subdivided into four. Eventually, all the roots are isolated within a small number of very tiny suspect squares. Newton's iteration can then efficiently finish the job.

Pan highly praises the Weyl quadtree scheme in his reviews [270, 286, 287], but only for *polynomials*. However, the method can be applied to general $f(z)$ in conjunction with a suitable test for flagging zero-free squares of f.

We shall not offer a longer discussion here because the algorithm is extensively discussed with case studies in Chapter 20 to solve real systems of two equations in two unknowns.

5.11 ▪ MAG/Complex Plane

A graphical strategy for finding complex-valued roots of a function is to make a contour plot of just the zero isoline for $\Re[f(x + iy)]$ and a similar plot for $\Im(f)$, and superimpose the two. The roots are the intersections of these zero contours [294]. Fig. 5.8 is an illustration.

Larkin [237], Pfeiffer [294], and Scarton [308] describe two-dimensional graphical methods for chasing complex roots. Scarton calls it the "method of eigen-valleys," a name that emphasizes that every zero is a valley of $|f(x)|$ in the complex x-plane.

With little a priori knowledge of the eigenfrequencies for weakly damped quasi-geostrophic normal modes in irregularly shaped ocean basins, LaCasce and Pedlosky made heavy use of contour plots of their dispersion relation [234].

Alternatively, one can make contour plots or surface mesh plots of the absolute value of $f(x)$ in the complex x-plane. However, the roots are often rather shallow minima. It is usually better to plot contours of the *logarithm* of the absolute value, or to make a mesh

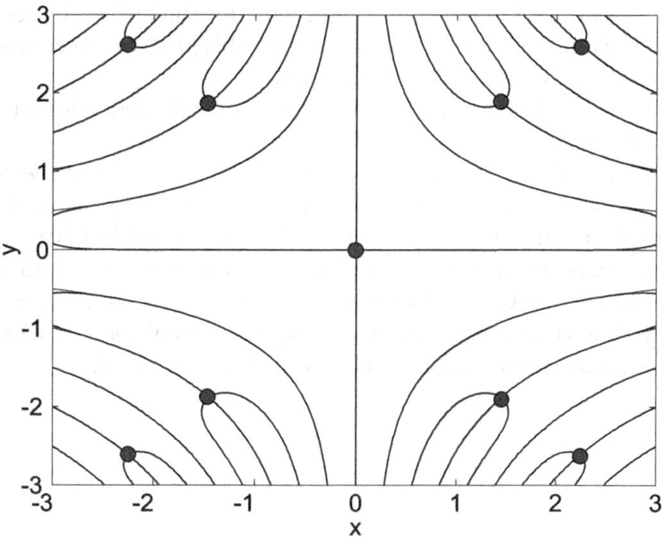

Figure 5.8. *MAG/complex plane. The zero isolines of the real part and also the imaginary part of erf*$(x + iy)$ *are superimposed. The contour intersections show nine roots of the error function in the complex plane.*

plot of the *negative* of the *logarithm* so that roots appear as tall narrow spikes. However, even with these tricks, it is difficult to distinguish between roots and near-roots, so the superimposed zero contours method is recommended.

If $f(x)$ has a finite number of roots in the entire complex plane, it may be possible to find them all by employing a change of coordinate in both the real and imaginary parts of x to map the entire u-plane to a finite domain in a new complex variable w.

5.12 ▪ Summary

The Delves–Lyness method for a disk has a much lower operation count than the interpolation-proxy scheme because the Delves–Lyness scheme does not need to do an $N \times N$ matrix eigensolution. However, the Delves–Lyness method is poisoned by zeros near the circle of integration and often converges with N much more slowly than its chief competitor. When the target domain is a disk, we recommend the interpolation-proxy strategy as preferable because of its robustness.

Domains bounded by smooth curves can be substituted for a disk. Geddes has shown that interpolation is easy and geometrically convergent on a domain bounded by an *ellipse* in the complex plane [187]. Domains bounded by contours with *corners* have singular maps to the disk, so tessellating an area with squares or hexagons or triangles is not an option for the interpolation-proxy method. However, the Delves–Lyness method can be applied to square or rectangular regions without difficulty.

Thus, reliable methods to search a single domain in the complex plane for zeros are now available. The challenge to the software engineer is to write code that will automatically subdivide a large area into many smaller domains to exhaustively search for roots. Dodging singularities of $f(z)$ is a nontrivial challenge to the applied mathematician. As with rootfinding on an interval, the code must automatically identify regions of large dynamic range and subdivide them. The program must also subdivide when the number of zeros in a region is too large, especially for the Delves–Lyness moments procedure, which is reliable only for domains with (roughly) five or fewer zeros, based on the practical experience of those who have applied it.

Nevertheless, searching big disks or small rectangles for roots is now well established and reliable.

The Riemann–Hilbert strategy (section 13.7) and the contour integrals for a single root inside the path are well established, too. Proponents claim that these algorithms are explicit solutions, so these methods are discussed in Chapter 13. We leave it to the reader as to whether a contour integral that must be evaluated numerically rates as an "explicit" solution. However, the Riemann–Hilbert method has been used effectively as part of an analysis of all the countable infinity of zeros of some functions, as in [329]. The contour integrals are thus legitimately more than numerical.

Part III

Fundamentals: Iterations, Bifurcation, and Continuation

Chapter 6

Newton Iteration and Its Kin

"Where the statue stood
of Newton with his prism and silent face,
the marble index of a mind forever
voyaging through strange seas of thought, alone."
—William Wordsworth

6.1 ▪ Newton's Iteration: The Basic Algorithm

In the latter half of this book, a wide array of analytical methods for solving nonlinear equations are described. When these fail—most of the time, alas—one must retreat to arithmurgy.[10] Almost all numerical rootfinders involve some form of iteration. In the Chebyshev-proxy rootfinder (CPR), for example, the iterations are hidden from the user, but the eigensolver that translates the companion matrix into zeros is iterative, and Newton's iteration is used to polish the roots and thus eliminate the errors of the Chebyshev interpolant. Iteration is an inescapable ingredient of zero-hunting.

The first step in iteration is to create a first guess through intuition, perturbation theory, or a random number generator. The second step is to repeatedly refine the approximation until the root is accurate to within the limits of machine roundoff error.

The fundamental rootfinding iteration is Newton's iteration. (This is often called the Newton–Raphson method, in part to distinguish the rootfinder from other algorithms labeled "Newton" and in part to credit Raphson, whose treatment of zero-computing was much closer to the modern approach, but we shall usually employ the shorter label.) The key idea is to replace $f(x)$ by a linear Taylor series approximation, centered on the current iterate $x^{(n)}$. Because the linear Taylor series is only an approximation to $f(x)$, the root $x^{(n+1)}$ of the linear series is only an approximation, too, to the root of $f(x)$. However, if the current iterate is sufficiently close to the true zero, then the Taylor series approximation will be accurate at and near the root. This in turn implies that the next iterate will be closer to the exact root of f than the current iterate so that one can approximate the root as accurately as possible (within roundoff limits) by repeating the iteration sufficiently often.

The linear Taylor series, expanded about the current approximation $x = x^{(n)}$, is

$$f(x;\lambda) \approx f(x^{(n)}) + \frac{df}{dx}(x^{(n)})(x - x^{(n)}). \tag{6.1}$$

[10]"Number-working" from the Greek $\alpha\rho\iota\theta\mu o\sigma$, "number," and -$\epsilon\rho\gamma o\sigma$, "working."

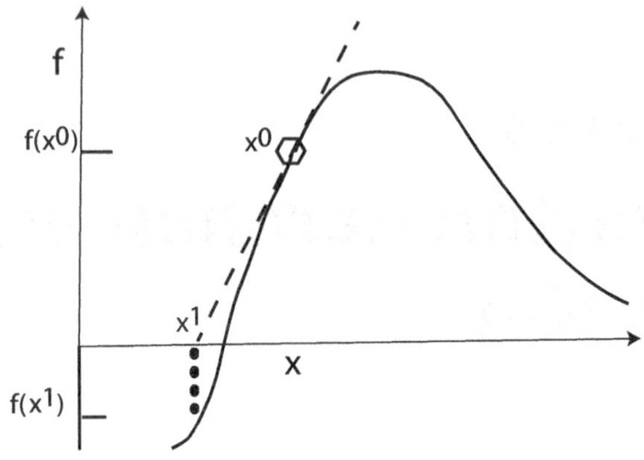

Figure 6.1. *Newton's iteration can be performed graphically by drawing at the initial point (hexagon) a line that is tangent to $f(x)$ at $x = x^{(0)}$ (dashed). The intersection of the tangent with the x-axis is $x^{(1)}$.*

The correction

$$\delta^{(n+1)} \equiv x^{(n+1)} - x^{(n)} \tag{6.2}$$

is the zero of the polynomial, linear in x, that is the first-order Taylor series:

$$\frac{df}{dx}\delta^{(n+1)} = -f(x^{(n)}) \qquad \leftrightarrow \qquad x^{(n+1)} = x^{(n)} - \frac{f(x^{(n)})}{df/dx(x^{(n)})}. \tag{6.3}$$

6.2 ▪ Geometric Interpretation: Newton's Iteration as the Tangent Line Approximation

In a single unknown, Newton's iteration is equivalent to approximating $f(x)$ by a straight line that is tangent to the graph of $f(x)$ at the nth iterate, $x = x^{(n)}$, and then finding the next iterate $x^{(n+1)}$ as the intersection of the tangent line with the x-axis.

Newton's method is often taught in precalculus mathematics courses by appealing to a diagram such as Fig. 6.1. If $f(x)$ does not curve away from the tangent line too rapidly, then the root of the line, which by definition is the intersection of the line with the x-axis, will be a good approximation to the root of $f(x)$, and the iteration will rapidly converge. The error in a first-order Taylor series is proportional to the second derivative of f, as shown by the usual error formula

$$f(x) = f(x^{(n)}) + \frac{df}{dx}(x^{(n)})\left(x - x^{(n)}\right) + \frac{1}{2}\frac{d^2f}{dx^2}(\xi)\left(x - x^{(n)}\right)^2, \tag{6.4}$$

without approximation for some ξ. However, this error formula also has a geometric interpretation. The curvature of $f(x)$, or in other words the rate of change of the slope of $f(x)$, is also proportional to the second derivative. It follows that the larger the second derivative, the more rapidly $f(x)$ will curve away from the tangent line and therefore the greater the error in Newton's iteration for a given distance $|x^{(n)} - x_*|$ from the root.

The geometric visualization also explains why Newton's iteration has only a finite domain of convergence. Fig. 6.2 shows that if the first guess is too far from the nearest root, Newton's iteration will diverge. The domain of convergence is discussed in much greater detail in section 6.10.

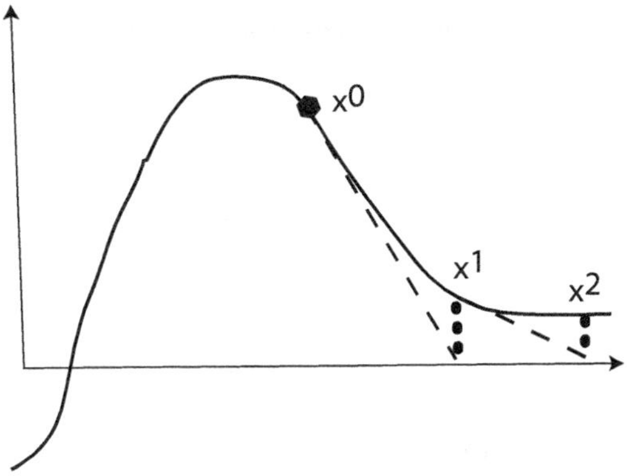

Figure 6.2. *Divergence in Newton's iteration. Tangent lines are dashed; dotted lines project from* $(x^{(n)}, 0)$ *to* $(x^{(n)}, f(x^{(n)}))$*. When the initial iterate is on the right side of the hill, the iterates diverge to* $+\infty$*, moving* rightward *even though the root is* left *of* $x^{(0)}$.

6.3 ▪ Digit-Doubling ("Quadratic") Convergence

Theorem 6.1 (Convergence Rate of Newton's Iteration). *Assume that* $f(x)$ *is an analytic function in the vicinity of a root* x*. (This stipulation can be considerably weakened without invalidating the theorem.) Denote the error after the* n*th iteration by*

$$e^{(n)} \equiv x - x^{(n)}. \tag{6.5}$$

Suppose that the root is simple, that is, $df/dx \neq 0$ *at the root. Then sufficiently close to the zero, that is, when* $|e^{(n)}|$ *is sufficiently small,*

$$|e^{(n+1)}| \sim constant\, |e^{(n)}|^2 \tag{6.6}$$

for some proportionality constant independent of the iteration number n*. In words, this means that if the error is in the third decimal place after* n *iterations, it will be in the sixth decimal place after the next iteration and* $O(10^{-12})$ *after* $(n+2)$ *iterations. This rate of error-squaring convergence is said to be* digit-doubling *or* quadratic convergence.

Proof: See [154, 224]. ∎

One unsatisfactory aspect of this theorem is the vagueness of the phrase "sufficiently close." Empirically, it is common to observe a roughly geometric rate of convergence during the early iterations, that is, the error decreases by a factor of two or three with each additional iteration, before the digit-doubling convergence finally appears.

Fig. 6.3 illustrates the digit-doubling convergence for Kepler's equation of celestial mechanics. More than 120 different methods have been proposed for this equation [130], but the graph shows that Newton's method is quite sufficient—with a good first guess. Given an adequate initial approximation, Newton's iteration converges so rapidly that the *rate* of convergence is never an issue.

Unfortunately, convergence will happen only if the first guess is sufficiently close to the root: the success of Newton's iteration is always linked with (metaphorical) chains of iron to the quality of the first guess. Therefore, *initialization* will be a major theme.

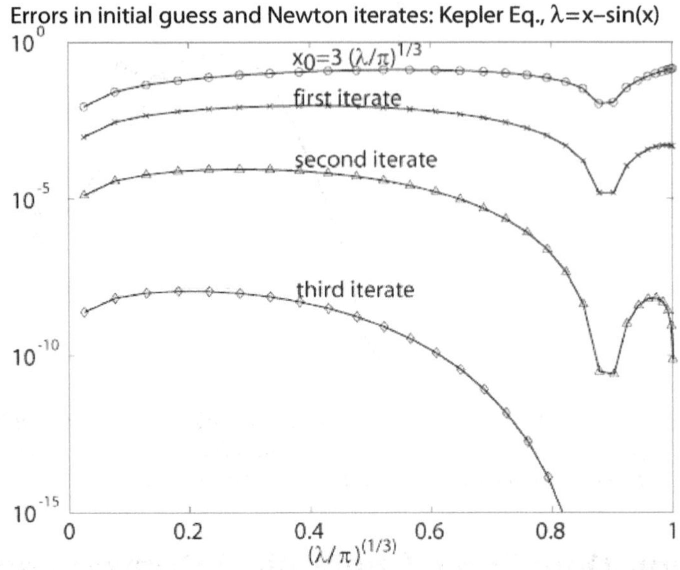

Figure 6.3. *Errors in the first guess $x_0 \equiv 3(\lambda/\pi)^{1/3}$ (top curve) and the first three Newton iterates for solving Kepler's equation in the parabolic case, $x - \sin(x) = \lambda$, plotted versus the cube root of the parameter λ. The maximum errors of the first guess and the first four iterates are $0.14, 9.4\times10^{-3}, 8.8\times10^{-5}, 1.1\times10^{-8}, 9.4\times10^{-16}$. (The fourth iterate is below the bottom axis and so is not visible.) As asserted by the convergence rate theorem, the number of correct digits—equivalently, the logarithm of the error—roughly doubles at each iteration until limited by machine precision, and this is true for the full range of the parameter λ. Only the positive range $\lambda \in [0, \pi]$ is illustrated because $x(\lambda)$ is antisymmetric with respect to $\lambda = 0$, and so are the errors.*

6.4 ▪ Flaws of Newton's Iteration

There are several things that can go wrong with the simple algorithm described above, including

1. failure to converge to a root,
2. convergence to the wrong root or to a root distant from the initialization, and
3. trapping within an endless limit cycle.

In addition, there is the perennial difficulty that an iteration always requires an initialization, a first guess for the root.

Each zero is surrounded by a finite domain of convergence. There are several remedies for a bad initialization. Underrelaxation, described in a later section, significantly expands the domain of convergence for most problems. The continuation method, which is explained in Chapter 8, is a systematic way to generate very accurate first guesses by increasing the parameter λ in small steps and using the solutions for previously computed values of λ to extrapolate a first guess for the next value of λ. A third strategy is to replace Newton's iteration by an algorithm that uses a higher-order approximation with a wider range of accuracy than the linear Taylor series of Newton's; higher-order schemes in a single unknown are described at the end of this chapter; tensor methods for bivariate and larger systems are explained in [309, 169, 18, 17, 19].

The second difficulty is that the derivative df/dx may be zero. The correction $\delta^{(n)}$ is then infinite. Generically, the derivative of a smooth function is zero on a set of discrete points. It is thus improbable that an iteration will land *exactly* on a singular point, but if an iteration is merely *close*, then the Newton correction will be huge. Unfortunately, it is quite common for the iteration to land sufficiently close to a point where $df/dx = 0$ that the iteration is thrown away to infinity like a comet ejected from the solar system by a close encounter with Jupiter. Underrelaxation helps to manage the near-singularity problem, too.

6.5 ▪ Higher-Order Roots

At a double root where both the function $f(x)$ and its first derivative are zero, the digit-doubling convergence rate of Newton's iteration is replaced by a geometric rate of convergence, with each iteration reducing the error by roughly $1/2$ *near* a double root.

A closely spaced pair of simple roots (*root doublet*) will produce a similar rate of convergence when the initial guess is much farther from the pair than the distance between the roots: the iteration converges geometrically at first, but eventually digit-doubling convergence appears. This is illustrated in Fig. 6.4, which solves the exemplary problem

$$f(x) = x^2 - \lambda \qquad \rightarrow x_* = \pm\sqrt{\lambda}, \tag{6.7}$$

with the Newton's iteration

$$x^{(n+1)} = x^{(n)} - \frac{(x^{(n)})^2 - \lambda}{2x^{(n)}}. \tag{6.8}$$

We can analytically demonstrate this geometric-to-digit-doubling transition by writing the current iterate as $x^{(n)} = \sqrt{\lambda} + \epsilon(n)$, where ϵ is thus the error in the nth iterate. The next Newton approximation is then

$$\begin{aligned} x^{(n+1)} &= \sqrt{\lambda} + \epsilon - \frac{2\epsilon\sqrt{\lambda} + \epsilon^2}{2\sqrt{\lambda} + 2\epsilon} && (6.9) \\ &\approx \sqrt{\lambda} + \epsilon/2, \qquad \epsilon \gg \sqrt{\lambda} && \text{[Geometric, Error-Halving Convergence]} \\ &= \sqrt{\lambda} + \frac{\epsilon^2}{2\sqrt{\lambda}}, \qquad \epsilon \ll \sqrt{\lambda} && \text{[Digit-Doubling, Error-Squaring Convergence].} \end{aligned}$$

The last line shows that the usual digit-doubling convergence is restored in that each iteration reduces the error from ϵ to $O(\epsilon^2)$ during the endgame. But farther from the root, the error is merely halved at each iteration. The Newton correction is too small by about a factor of two, which can be remedied by overrelaxation, as discussed in the next two sections.

Fig. 6.5 shows the same behavior for a more realistic example, whose solution is $x(\lambda) = \arccos(\lambda)$.

6.6 ▪ Underrelaxation and Overrelaxation: Newton's Iteration with Line Search

"The addition of line searches to the Newton iteration greatly increases the range of nonlinear equations ... that can be solved .... Line searches are the most usual way of extending the Newton iteration to force convergence from poor starting approximations."

—M. D. Powell

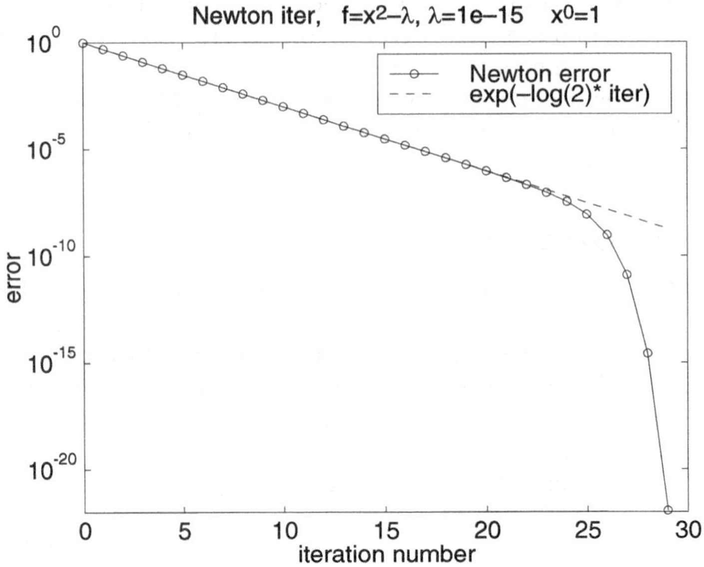

Figure 6.4. *Newton's iteration for* $x^2 - \lambda = 0$, *whose solution is* $x(\lambda) = \sqrt{\lambda}$. *The graph shows the iteration for* $\lambda = 1 \times 10^{-15}$, *which gives a pair of very closely spaced simple zeros, a root doublet. The dashed guideline is* $\exp(-\log(2)\, n)$, *where* n *is the iteration number.*

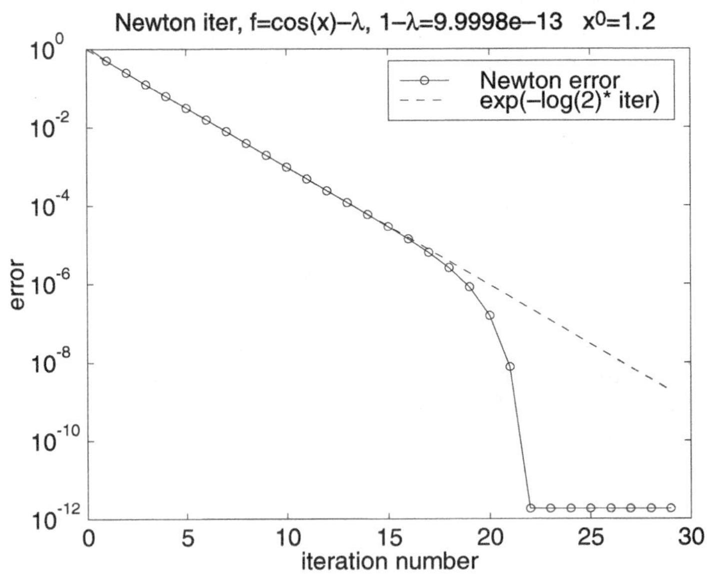

Figure 6.5. *Newton's iteration for* $f(x) = \cos(x) - \lambda = 0$, *whose solution is* $x(\lambda) = \arccos(\lambda)$. *The graph shows the iteration for* $\lambda = 0.999999999999$, *which is very close to the branch point at* $\lambda = 1$. *The dashed guideline is* $\exp(-\log(2)n)$, *where* n *is the iteration number. There is no improvement for iteration number* > 22 *because the iteration has reached a plateau of* $O(\text{machine epsilon})$. *Variable precision arithmetic systems offer the freedom to change this graph. Maple, for example, offers a parameter "Digits," which is the number of decimal digits of precision. If the calculation is repeated in a variable precision arithmetic system, the error will level off at new plateau slightly larger than* $10^{-\text{Digits}}$.

6.6.1 ▪ Underrelaxation and Overrelaxation

The rate of convergence can be accelerated, and the domain of convergence expanded, by multiplying the Newton correction by a constant γ.

Definition 6.2 (Underrelaxation/Overrelaxation). *Define the generalized Newton's iteration by*

$$x^{(n+1)} = x^{(n)} + \gamma_n \, \delta^{(n)}_{Newton}, \tag{6.10}$$

where $\delta^{(n)}_{Newton}$ *is the usual Newton correction. The choice* $\gamma_n = 1$ *is Newton's method. An* overrelaxed *method is obtained by choosing*

$$\gamma_n > 1 \qquad \textit{[Overrelaxed]}, \tag{6.11}$$

whereas underrelaxation *is*

$$\gamma_n < 1 \qquad \textit{[Underrelaxed]}. \tag{6.12}$$

Overrelaxation by a factor of two greatly accelerates convergence from a mediocre first guess near a root doublet.

6.6.2 ▪ Line Search

A good strategy to optimize γ_n is an Armijo line search. The residual $f(x)$ is evaluated for

$$\gamma_{n,j} \equiv 1/2^{(j-1)/2}, \qquad j = 0, 1, 2, \ldots, j_{max} \tag{6.13}$$

for some j_{max}. The final correction at the nth iterate sets γ_n equal to whichever $\gamma_{n,j}$ gives the smallest $|f|$.[11] The combination of Newton's method with this line search is the "Newton–Armijo" iteration.

When there are many unknowns, the cost of factoring the Jacobian is an order of magnitude greater than the cost of a single evaluation of $f(x)$, so this line search strategy increases the cost per iteration only slightly and is thus very attractive. For a single unknown, alas, the line search greatly increases the cost per iteration, but who cares? Iterations when there is but one unknown cost only a microsecond on a modern laptop. The important point is that underrelaxation significantly expands the interval of convergence.

In higher dimensions, the adjective "line" is appropriate because the search is restricted to a ray in a multidimensional space of unknowns. When there is only one unknown, the line search is a minimization over a connected portion of the entire space of unknowns.

In the endgame, where x is very close to the minimum, the search will find $\gamma_n = 1$ is best and thus reverts to the standard Newton's iteration.

To accelerate convergence near a double root, it suffices to also include overrelaxation by a factor of two in the line search.

6.7 ▪ Minimizing the Residual Norm: Steepest Descent and the Levenberg–Marquardt Flow

With N unknowns, a necessary and sufficient condition for a zero of the system $f_1, f_2, \ldots, f_N$ is a minimum of zero of the *residual norm* function

$$\rho(x_1, x_2, \ldots, x_N) = \|\vec{f}(\vec{x})\| \equiv \frac{1}{2}\left\{f_1^2 + f_2^2 + \cdots + f_N^2\right\}. \tag{6.14}$$

This implies that *rootfinding* can always be recast as a *minimization* problem.

[11] Other criteria for the goal of the search are described in [224], but minimizing the residual $|f|$ is the simplest.

When the initial guess is within the digit-doubling convergence domain, minimization methods are inferior to Newton's iteration. When the first guess is poor, however, minimization thinking can be very useful.

There are three important variations on Newton's iterations that incorporate minimization thinking:

1. Newton flow,
2. steepest descent, and
3. Levenberg–Marquardt.

When the number of unknowns N is greater than one, Newton's iteration and these three variants give four different correction vectors, that is, four vectors that differ in both direction and magnitude. When there is but a single unknown, however, an iterate can only move to the right or to the left. A celebrated theorem asserts that a correction parallel to the Newton correction always lowers the residual norm, in any number of unknowns, for sufficiently extreme underrelaxation. This implies that for $N = 1$, the underrelaxed Newton's iteration can duplicate steepest descent, Newton flow, and the Levenberg–Marquardt flow; directionally, these four are the same. Therefore, we omit a detailed description of minimization schemes and refer the reader to [154, 74].

Nevertheless, the steepest descent idea—march always downhill to eventually reach a minimum—provides a rationale for the Armijo line search. The Levenberg–Marquardt iteration modifies the Jacobian so that the modified derivative is always positive; in one direction, $-f/f_x$ is modified to $-f_x f/(f f_x + \aleph)$ for some small constant $\aleph$, where $\aleph$ is adaptively varied to minimize the residual.

Lastly, the concept of the Newton flow is useful to prove the reliability of underrelaxation and simultaneously to compute the domain of convergence of Newton's iteration in the limit of severe underrelaxation.

6.7.1 ▪ The Newton Flow Equation

To apply the powerful methods of dynamical systems theory as well as minimization ideas, it is conceptually useful to recognize that Newton's iteration is equivalent to applying a time-discretization method to the ordinary differential equation in a pseudotime T,

$$\frac{dx}{dT} = -\frac{f(x)}{df/dx} \qquad \text{[Newton Flow Equation]}. \tag{6.15}$$

Newton's iteration is equivalent to integrating this by the forward Euler time-marching method with unit step in pseudotime. Underrelaxation is equivalent to applying the same Euler forward scheme with a pseudotimestep less than one.

Very close to a root x_*, defining

$$\delta \equiv x - x_* \tag{6.16}$$

and employing

$$\left(\frac{df}{dx}\right)^{-1} f(x) \approx \left(\frac{df}{dx}\right)^{-1} \left\{ f(x_*) + \frac{df}{dx}(x - x_*) \right\} = x - x_*, \tag{6.17}$$

the Newton flow equation can be written approximately as

$$\frac{d\delta}{dT} = -\delta \qquad \text{[Newton Flow Near Root]}. \tag{6.18}$$

The quadratic, digit-doubling convergence of the classical Newton's iteration is not obvious from this form. However, it is clear that $\delta = 0$, that is, $x = x_*$, is an attractor for this ordinary differential equation in pseudotime. With sufficiently strong underrelaxation (so that the forward Euler method accurately tracks the analytical solution of the Newton flow equation), it is also clear that the error δ will decay exponentially fast toward the root. This can be formally stated as a rigorous theorem.

Theorem 6.3 (Exponential Decay of the Newton Flow). 1. *From any initial conditions and without restrictions except that $df/dx \neq 0$, the solution to the Newton flow equation*

$$\frac{dx}{dT} = -f(x)/\frac{df}{dx}(x) \tag{6.19}$$

makes $f(x(T))$ decay exponentially with pseudotime as

$$f(x(T)) = \exp(-T)\, f(x(T=0)). \tag{6.20}$$

2. *With sufficiently small underrelaxation parameter (pseudotime step) γ_n, each step of the underrelaxed Newton's iteration always decreases the residual norm function*

$$(1/2) f(x(T))^2.$$

3. *Sufficiently close to a root x_*, the difference between the Newton flow and the root decreases exponentially with pseudotime T, that is,*

$$E(T) \approx \exp(-T) E, \qquad E \equiv \vec{x} - \vec{x}_*. \tag{6.21}$$

Proof: The first and third propositions come immediately from (6.18) and the linearized analysis that follows. The only restriction is that the Newton flow equation is singular at points where $df/dx = 0$. Note that $\frac{df}{dx}\frac{dx}{dT} = \frac{df}{dT}$ for any composition of functions $f(x(T))$.

The second proposition follows from noting that the underrelaxed Newton's iteration is simply the Euler forward time-marching scheme for integrating the Newton flow equation. For sufficiently small γ where the underrelaxation parameter is the timestep in this interpretation, the Euler integration (iteration) will closely track the Newton flow and therefore have the same property of monotonically decreasing the norm of the residual function. ∎

Unfortunately, the theorem is a little deceptive. The problem is that if the iteration approaches a singularity where $f_x = 0$, then the Newton correction will become larger and larger. For the iteration to closely follow the Newton flow and also so that the residual norm is reduced at each iteration, it is necessary to use a smaller and smaller underrelaxation time step. It remains true that increasing the pseudotime by one will reduce the residual by $\exp(-T)$, but if γ has been reduced to $O(10^{-6})$ so that we need a million iterations to increase T by unity, then in a practical sense, the iteration fails. Furthermore, the Newton flow may follow the residual norm downhill to a *local* minimum where the minimum is nonzero and therefore is not a root.

The convergence domain of the Newton flow has a much simpler geometry than its counterpart for an unmodified, no-underrelaxation Newton's iteration. The latter is usually fractal and multiply connected, often with an infinite number of disjoint intervals as the convergence of a single zero, as discussed at greater length in section 6.10. In contrast, the Newton flow will march downhill in the residual norm, monotonically reducing $|f(x)|$ all the way to the root, as long as the starting value lies on the interval

$$x^{(0)} \in (x_L, x_R), \tag{6.22}$$

where x_L is the zero of df/dx that is nearest to the root, but to the left of the root, and similarly x_R is the closest zero of the slope of $f(x)$ to the right of x_*.

The range of convergent initializations can be expanded by replacing Newton's iteration by schemes based on higher-order proxies, such as the Cauchy and Shafer approximations. We therefore postpone a discussion on enlarging the interval of good starting values until we have explained these higher-order rootfinders.

6.8 ▪ Quasi-Newton: Cheating on the Derivative

When there are many unknowns, computing and factoring the Jacobian matrix at each iteration is expensive, inspiring a large variety of "quasi-Newton" algorithms that approximate the inverse of the Jacobian. In one dimension, too, sometimes it is difficult or relatively expensive to compute f_x.

One useful quasi-Newton strategy is to approximate the derivative by finite differences:

$$\frac{df}{dx} \approx \frac{f(x^{(n)}+h)-f(x^{(n)})}{h}. \tag{6.23}$$

The roundoff error E_r in either a first- or second-order difference is $O(\epsilon_{mach}/h)$—the error in the difference divided by h. The truncation errors are $E_{fd,1} \sim (h/2)f_{xx}(x^{(n)})$ for a one-sided difference and $E_{fd,2} \sim (h^2/6)f_{xxx}(x^{(n)})$ for a second-order centered approximation. If the derivatives are $O(1)$, then the sum of errors is

$$E_{total} \sim \begin{cases} O(\epsilon_{mach}/h)+O(h) & \text{[One-Sided Difference]}, \\ O(\epsilon_{mach}/h)+O(h^2) & \text{[Two-Sided Difference]}. \end{cases} \tag{6.24}$$

The optima are (Fig. 6.6)

$$\begin{aligned} h_{min} &\sim O(\sqrt{\epsilon_{mach}}) \to \min(E_{total}) \sim O(\sqrt{\epsilon_{mach}}) && \text{[One-Sided Difference]}, \\ h_{min} &\sim O((\epsilon_{mach})^{1/3}) \to \min(E_{total}) \sim O((\epsilon_{mach})^{2/3}) && \text{[Two-Sided Difference]}. \end{aligned}$$

There is no accuracy penalty because Newton's iteration, even with an inexact derivative, is self-correcting like all convergent iterations, and the final accuracy of the zero is limited only by the accuracy with which one can evaluate $f(x)$ itself. When df/dx is inexact, the negative consequences are that the digit-doubling convergence of the true Newton's iteration is lost, replaced by a geometric rate of convergence, and the domain of convergence slightly shrinks.

Another strategy, which is known variously as the "chord iteration" or "frozen Jacobian" scheme, is to compute the derivative df/dx *just once* and use this value throughout the entirety of the iteration without updating. The Shamanski algorithm splits the difference between chord and Newton by updating the derivative *occasionally*. Again, the cost per iteration is reduced but the number of iterations is slightly increased and the domain of convergence is slightly reduced.

6.9 ▪ The Kepler Equation: A Case Study of Newton's Iteration

The Kepler equation of celestial mechanics is the two-parameter equation in a single unknown,

$$\lambda = x - \epsilon \sin(x), \tag{6.25}$$

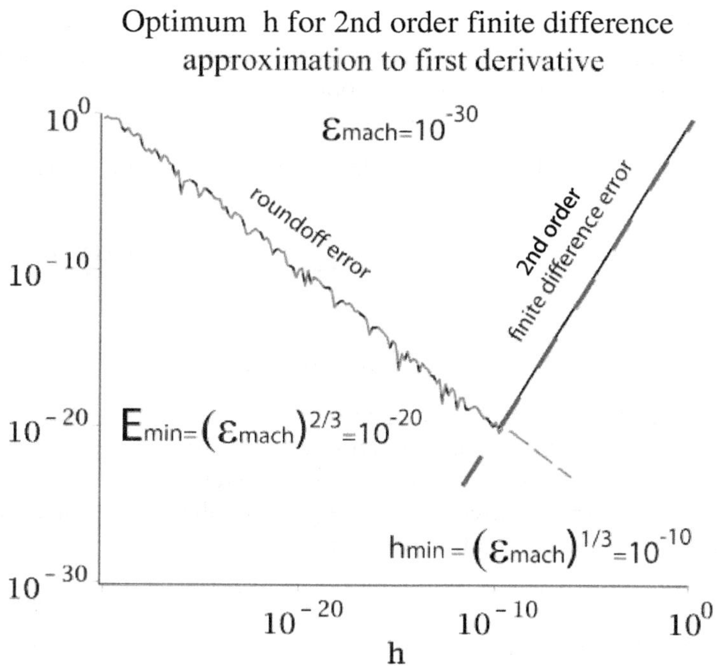

Figure 6.6. *The left jagged curve is roundoff error in a centered finite difference approximation to df/dx. The right dashed curve is the approximation error of the finite difference formula, that is, $|df/dx - (f(x+h) - f(x-h))/(2h)|$ as computed in infinite-precision arithmetic; this error is $O(h^2)$. Denoting machine epsilon by ϵ_{mach}, the best choice for a second-order accurate difference is $h \sim O(\epsilon_{mach}^{1/3})$. The total error is then $O(\epsilon_{mach}^{2/3})$.*

where λ is the so-called mean anomaly (usually M in astronomy), ϵ is the eccentricity of an elliptical orbit, and the unknown x is the eccentric anomaly (ψ in astronomy). The parameter ranges are $\lambda \in [-\pi, \pi]$ and $\epsilon \in [0,1]$; the equation has a single real root everywhere in this region except for a triple root at $\lambda = 0$, $\epsilon = 1$.

Because of its importance, at least 120 algorithms have been applied to this equation [280, 130]. Newton himself proposed two methods, the simpler of which "was, according to Whitehead [editor of Newton's collected papers], the first publication by Newton [1687, in his *Principia*] of what we know as Newton's method" [124, p. 223]. However, this plethora of algorithms is a little silly, at least in the computer age: Newton's method is sufficient over the whole parameter range:

$$x^{(n+1)} = x^{(n)} - \frac{x^{(n)} - \epsilon \sin(x^{(n)}) - \lambda}{1 - \epsilon \cos(x^{(n)})}. \tag{6.26}$$

However, the standard textbook presentation of Newton's method as a plug-in-the-numbers algorithm is rather misleading, too. Kepler's equation illuminates a couple of key issues with Newton's iteration.

First,

$$df/dx = 1 - \epsilon \cos(x). \tag{6.27}$$

When $\epsilon = 1$ and $\lambda = 0$, the derivative df/dx is zero, too, causing potential overflow errors in Newton's iteration due to the zero denominator. Embarrassingly, the singularity is at the trivial case where the root is $x = 0$.

There are a couple of remedies:

1. Apply the Levenberg–Marquardt iteration, which is a perturbed Newton's iteration such that the denominator can never vanish:
$$x^{(n+1)} = x^{(n)} - \gamma_n \frac{df/dx(x^n)f(x^{(n)})}{\left\{df/dx(x^{(n)})\right\}^2 + \aleph}, \tag{6.28}$$
where $\aleph$ is a small positive constant.

2. Include an "if" statement in the code that will replace Newton's iteration with the perturbative approximation $x \approx \lambda + \epsilon \sin(\lambda)$ when $|df/dx|$ is very, very small, such as less than 10^{-8}. This, too, is a general strategy; it is the ironic truth that Newton's method is often slow converging or failing in parameter regions where perturbative series are rapidly convergent.

3. Switch to a higher-order iteration such as Cauchy's (described in section 6.11).

Strictly speaking, the singular point $\lambda = 0, \epsilon = 1$ is a triple root, and Newton's iteration converges exponentially fast (though not at a digit-doubling rate) to higher than first order roots, as noted in section 6.5. However, the risk of overflow in floating point calculations is considerable. Replacing Newton's by the perturbative approximation is the best strategy for writing robust software.

The second difficulty is that it is an arduous endeavor to devise a first guess that is uniformly successful. For small ϵ, perturbation theory gives the lowest-order approximation $x \approx \lambda$ and the first-order initialization $x \approx \lambda + \epsilon \sin(\lambda)$. Unfortunately, as illustrated by the red regions in the upper two panels of Fig. 6.7, both perturbative approximations fail in the sense that Newton's iteration diverges in the neighborhood of $\lambda = 0$, $1 - \epsilon \approx 0$. These parametric regions are small in ϵ, which is why a logarithmic scale in $1 - \epsilon$ is used for the plots. Unfortunately, these regions are important because comet orbits are highly eccentric, with ϵ very close to one.

The difficulty around $\epsilon = 1, \lambda = 0$ is that Kepler's equation has a *triple root* at this point. Along the line $\epsilon = 1$, the initialization
$$x(\lambda, \epsilon = 1) \approx 6^{1/3}\, \lambda^{1/3} \tag{6.29}$$
yields rapid Newton convergence for all λ. It is possible to generate a uniform approximation, accurate for all ϵ and not merely $\epsilon = 1$, by approximating $\sin(x)$ by $x - x^3/6$ and solving the resulting cubic equation. (The approximation (6.29) is the special case $\epsilon = 1$ of this strategy.)

Using the analytic solution of the cubic equation described in Chapter 11, the solution to $\lambda = x - \epsilon(x - x^3/6)$ is
$$\rho = \sqrt{8(1-\epsilon)/\epsilon}, \tag{6.30}$$
$$x^{(cubic)} = \rho \sinh\left\{\frac{1}{3}\operatorname{arcsinh}\left(24\frac{\lambda}{\epsilon\rho^3}\right)\right\}. \tag{6.31}$$

When $\epsilon = 1$ precisely, the sinh-arcsinh formula should be replaced by the special case (6.29), which avoids evaluating the arcsinh function with an infinite argument.

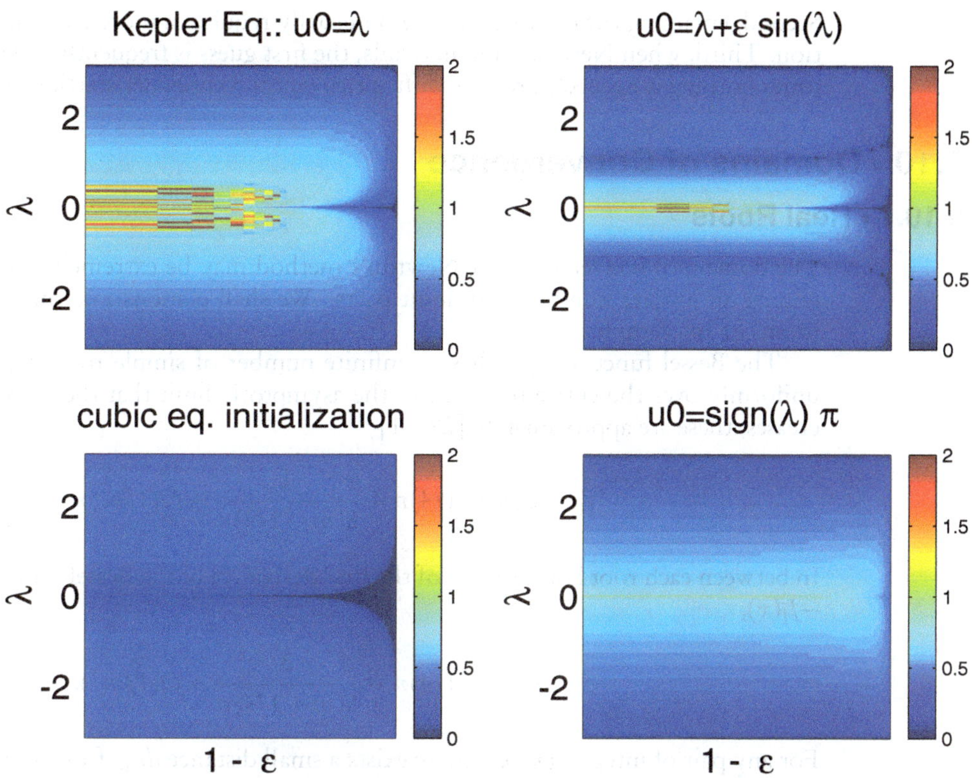

Figure 6.7. *Shaded plots showing the base-10 logarithm of the number of iterations required to reach an accuracy of* 10^{-8} *for solving the Kepler equation,* $\lambda = x - \epsilon \sin(x)$, *from 4 different initializations. For each plot, the entire parameter range is shown, but note that the horizontal axis plots* $1-\epsilon$, *not* ϵ, *on a* logarithmic *scale. Note also that the shading encodes the base-10 logarithm of the number of iterations; thus, the intermediate shade of "1" codes for 10 iterations. The number of iterations was stopped at 100, so the dark red areas, which represent "100 iterations" (labeled "2" on the colorbars) are really regions where Newton's iteration* diverges *for the given first guess. Upper left:* $x^{(0)} = \lambda$. *Upper right: the improved small-*ϵ *approximation,* $x^{(0)} = \lambda + \epsilon \sin(\lambda)$. *The lower left panel, which converges over the entire parameter plane with no more than 3 iterations, employs a first guess that solves a cubic polynomial equation, as explained in the text. The lower right panel shows that the initialization of Charles and Tatum [126] requires more iterations but is much simpler and is always successful. The smallest value of* $1-\epsilon$ *is* 8.5e−5; *the smallest* $|\lambda|$ *is* 7.0E−6. *For the extreme values of* $\epsilon = 1$ *and/or* $\lambda = 0$, *Newton's iteration may overflow or underflow unless the denominator is slightly perturbed.*

This cubic equation approximation is obviously more complicated than the two perturbative approximations. However, as shown in the bottom panel of Fig. 6.7, the cubic initialization is so good that Newton's iteration converges to an absolute error of less than 10^{-8} in just three iterations, and to machine precision in four.

Charles and Tatum [126] proposed an even simpler initialization:

$$x^{CT} = \text{sign}(\lambda)\,\pi. \tag{6.32}$$

This isn't as accurate as the cubic initialization, but they show that Newton's iteration converges from this starting guess for all $\epsilon < 1$ and all λ.

This example illustrates a couple of general principles about initialization. First, a *uniformly* good first guess is hard to find, even for an equation as simple as Kepler's.

Second, enlarging the convergence region usually requires a more complicated initialization. Third, when Newton's method fails, the first guess is frequently the culprit. Sometimes nothing is needed to resolve difficulties except a shrewder starting point.

6.10 ▪ Domains of Convergence

6.10.1 ▪ Real Roots

The domain of convergence for Newton's method may be extremely nonsmooth, containing an infinite number of disjoint parts. We shall demonstrate this proposition by means of an example.

The Bessel function $J_0(x)$ has an infinite number of simple roots $j_{0,n}$ spaced quasi-uniformly over the entire real axis; in the asymptotic limit that the zero number n increases, these are approximately [284, 1]

$$j_{0,m} \approx (m-1/4)\pi + \frac{1}{8(m-1/4)\pi} + O(1/m^3). \tag{6.33}$$

In between each root is a zero $j_{1,m}$ of the first derivative of the Bessel function, $dJ_0/dx = -J_1(x)$,

$$j_{1,m} \sim (m+1/4)\pi - \frac{3}{8(m+1/4)\pi} + O(1/m^3). \tag{6.34}$$

For any pair of integers (m,k), there exists a small distance $d_{k,m}$ from $x = j_{1,m}$ such that the first Newton's iterate, initialized with $x^{(0)} = j_{1,m} + d_{k,m}$, is exactly $x^{(1)} = j_{0,k}$. This distance $d_{k,m}$ solves

$$j_{0,k} = j_{1,m} + d_{k,m} - \frac{J_0(j_{1,m}+d_{k,m})}{\frac{dJ_0}{dx}\left(j_{1,m}+d_{k,m}\right)}. \tag{6.35}$$

The solution is, expanding the arguments for the Bessel functions, and implicitly replacing the Bessel zeros by their asymptotic approximations,

$$d_{k,m}^{asymptotic} \approx (j_{0,k} - j_{1,m}) + \frac{J_0(j_{1,m})}{\frac{dJ_0}{dx}\left(j_{1,m}\right)}, \qquad |m-k| \gg 1. \tag{6.36}$$

Table 6.1 shows that the approximation is quite good. (It is not applied to the trivial case when the initial point is the root itself, 8.65.)

This analysis implies that (i) the vicinity of the mth root of the first derivative of the Bessel function includes parts of the Newton domains of convergence of all of the infinity of zeros of the Bessel function itself and (ii) the domain of convergence to the kth root of the Bessel function includes an infinite number of disconnected pieces, one piece in the vicinity of each of the local maxima or minima of $J_0(x)$.

Again, a diagram is illuminating, as shown in Fig. 6.8. Newton's iteration proceeds by drawing a line tangent to the initial point $x^{(0)}$. A geometrical way of describing the basin of attraction for Newton's method is to say that there is an infinite number of tangent lines intersecting the kth zero of $J_0(x)$. Each tangent also intersects the graph of the Bessel function near one of its local maxima or minima, as shown in the diagram.

Table 6.1. *The first six initial values for which Newton's iteration converges to the third root of the Bessel function $J_0(x)$ in a single iteration.*

m	$x^{(0)}$	$d_{3,m}$	$d_{3,m}^{asymptotic}$
1	4.0515	0.2198	0.2074
2	8.4778	1.4621	0.6104
3	8.6597	8.6597	not applicable
4	13.1045	−0.2192	−0.2141
5	16.3418	−0.1288	−0.1279
6	19.5243	−0.0915	−0.0912

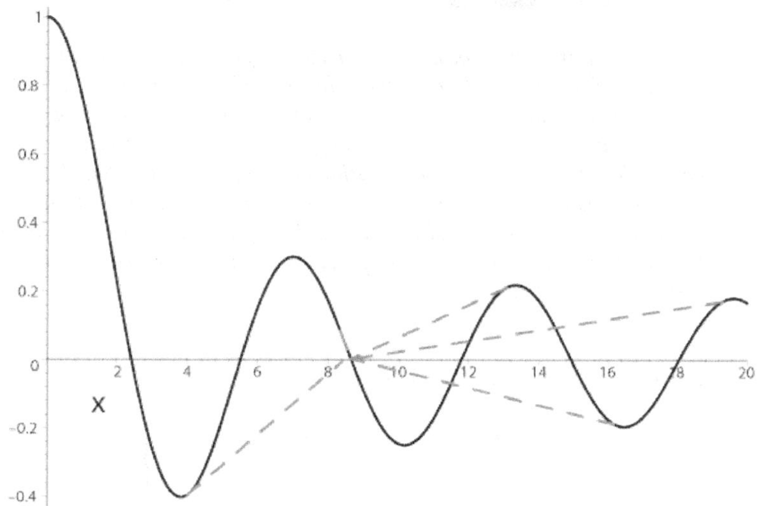

Figure 6.8. *Solid: $J_0(x)$. The dashed lines connect initial values from which Newton's iteration will move to the third root of the Bessel function in a single step. Each of these lines is tangent to the graph of the Bessel function at one end, the initial point, and at the other end intersects the axis at the third root, $x = 8.653727913$.*

6.10.2 ▪ Domain of Convergence for the Newton Flow (Underrelaxed Newton's Iteration)

In contrast, the domain of convergence of the Newton *flow* to a real root is straightforward. Since the Newton flow always moves to smaller $|f(x)|$, it is useful to borrow a term from hydrology as follows.

Definition 6.4 (Drainage Basin). *The* drainage basin *of a root x_* is the interval around the root where f/f_x is negative to the left of the zero and positive to the right of the zero so that the Newton flow, from any starting point on this interval, will monotonically converge toward the root.*

Theorem 6.5 (Newton Flow Convergence Domain). *Let x_* denote a real root of any multiplicity. The Newton flow will converge to that root if and only if the initialization $x^{(0)}$ is on the interval*

$$x^{(0)} \in [x_L, x_R], \tag{6.37}$$

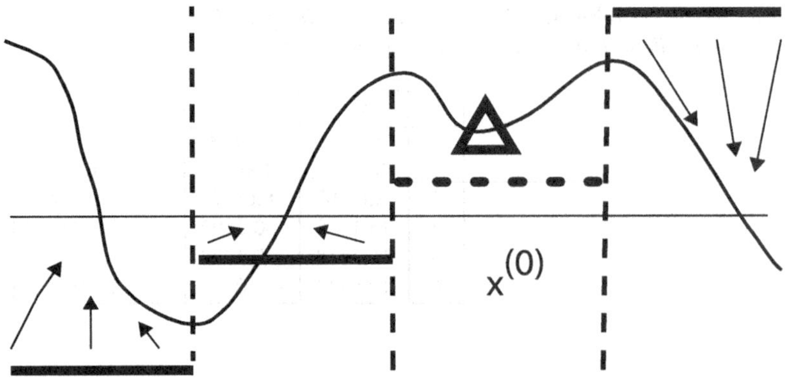

Figure 6.9. *Schematic of the convergence intervals for the Newton flow (not the Newton's iteration). The horizontal axis is the value of the initial condition $x^{(0)}$. The vertical axis is $f(x)$. The three thick horizontal bars span the intervals of convergence to a root, as indicated by the arrows. The triangle marks a local minimum of $|f(x)|$, where the second derivative at that point is the same sign as the function itself, violating the condition of the theorem. The dotted horizontal line spans the interval of initial conditions that converge not to zero but rather to a local minimum of the magnitude of the function. The dashed vertical lines are walls, located where the slope df/dx is zero and the signs of the second derivative and $f(x)$ are opposite; these are boundaries between the Newton flow intervals of convergence to different roots.*

where $x_L < x^{(0)} < x_R$ and where

$$\frac{df}{dx}(x_L) = 0, \qquad \text{sign}\left\{\frac{d^2 f}{dx^2}(x_L)\right\} = -\text{sign}\{f(x_L)\}, \tag{6.38}$$

$$\frac{df}{dx}(x_R) = 0, \qquad \text{sign}\left\{\frac{d^2 f}{dx^2}(x_R)\right\} = -\text{sign}\{f(x_R)\}. \tag{6.39}$$

Proof: The diagram in Fig. 6.9 is more illuminating than a formal proof. Recall that the Newton flow always flows downhill, that is, toward smaller values of $|f(x)|$. It is obvious that for a smooth, continuous function, the Newton flow must always descend to a zero or to a nonroot local minimum of $|f(x)|$, from which the theorem follows immediately. ■

In other words, the drainage basin of the root in the Newton flow is bounded by the *nearest zeros* of the *first derivative* of df/dx. The theorem is a more formal statement of what was already noted in the Newton flow section, section 6.7.1.

6.10.3 ▪ Complex Plane Dynamics

Saari and Urenko review Newton's iteration in the complex plane. It has been rigorously proved that chaotic behavior (in the narrow meaning of dynamical systems theory) is common [302]. The fantastically intricate convergence domains are well described in the fractal book of Peitgen and Richter [291] and the article by Peitgen, Prüfer, and Schmitt [292].

Underrelaxation is a powerful tool for expanding the domain of convergence, as discussed previously. In so doing, underrelaxation smooths away the fractal boundaries of the Newton convergence domain, replacing this domain with a much smoother and topologically much simpler domain. In the limit the underrelaxation parameter goes to zero, the underrelaxed Newton convergence domain asymptotes to the Newton flow convergence domain.

6.11 ▪ Higher-Order Proxies: Cauchy, Muller, and Shafer Iterations

When there is only one unknown, it is simple to improve upon Newton's iteration by replacing the linear polynomial that serves as a proxy for $f(x)$ by a polynomial of higher degree or by a rational function. With but one unknown, it is easy to find the roots of such a proxy. Halley's iteration employs the $[1/1]$ Padé approximant,[12] which yields the modified Newton's iteration

$$x^{(n+1)} = x^{(n)} - \frac{f}{f_x} \frac{1}{1 - 0.5 f_{xx} f / (f_x)^2}, \tag{6.40}$$

where $f(x)$ and all its derivatives are evaluated at $x = x^{(n)}$. Halley's converges faster than Newton's in the endgame, but this is of little practical value. Its domain of convergence is often larger than Newton's without underrelaxation, but Halley's will move toward the root only when f/f_x is negative to the left of the root and positive to the right, so that the simply connected part of its domain of convergence can be no larger than the drainage basin of the Newton flow, that is, no improvement upon Newton's iteration with line search and underrelaxation. Therefore, we shall discuss it no further.

The Cauchy, Muller, and Shafer iterations all require solving a quadratic equation and therefore all return *two* corrections. The usual procedure is to accept the *smaller* correction and discard the larger. We shall return to a more broad-minded strategy after first discussing the mechanics of each iteration.

Cauchy's method uses a three-term Taylor series,

$$f \approx f(x^{(n)}) + f_x(x^{(n)})\delta + (1/2) f_{xx}(x^{(n)})\delta^2,$$

retaining the quadratic term in $\delta \equiv x^{(n+1)} - x^{(n)}$. This quadratic has two roots. The usual high school formula, $(-f_x \pm \sqrt{f_x^2 - 2 f_{xx} f}) f_{xx}$, gives both branches, but unfortunately, one of the two branches is the tiny difference of small terms when x is close to a zero. To minimize roundoff errors, we therefore use an alternative form for this small-magnitude root and write the two branches of the quadratic as

$$\delta = \frac{-2f}{f_x + \text{sign}(f_x)\sqrt{f_x^2 - 2 f_{xx} f}}, \tag{6.41}$$

$$\delta^{2d} = \frac{-f_x - \text{sign}(f_x)\sqrt{f_x^2 - 2 f_{xx} f}}{f_{xx}}. \tag{6.42}$$

The first Cauchy correction δ reduces to the Newton correction, $-f/f_x$, as $f \to 0$.

Shafer's method exploits five terms of the Taylor expansion about the current iterate to compute the $[2/2]$ Padé approximant, the ratio of two quadratic polynomials:

$$f_{[2/2]}(x) = \frac{a_0 + a_1 x + a_2 x^2}{1 + b_1 x + b_2 x^2}. \tag{6.43}$$

[12]This justification of Halley's iteration is anachronistic in that Padé's thesis appeared over a century after Halley died, but the Padé connection is the most illuminating modern derivation.

The roots are the zeros of the numerator, therefore the roots of a quadratic, just as for Cauchy's iteration:

$$\delta = \frac{-2a_0}{a_1 + \mathrm{sign}(a_1)\sqrt{a_1^2 - 4a_2 a_0}}, \tag{6.44}$$

$$\delta^{2d} = \frac{-a_1 - \mathrm{sign}(a_1)\sqrt{a_1^2 - 4a_0 a_2}}{2a_2}. \tag{6.45}$$

The input is the quartic Taylor approximation

$$f^{(quadratic)} = f + f_x \delta + \frac{f_{xx}}{2}\delta^2 + \frac{f_{xxx}}{6}\delta^3 + \frac{f_{xxxx}}{24}\delta^4, \tag{6.46}$$

whence comes

$$b_1 = \frac{2f_{xxx}f_{xx} - f_x f_{xxxx}}{4f_x f_{xxx} - 6f_{xx}^2}, b_2 = \frac{3f_{xxxx}f_{xx} - 4f_{xxx}^2}{24 f_x f_{xxx} - 36 f_{xx}^2}, \tag{6.47}$$

$$a_0 = f, \qquad a_1 = f_x + b_1 f, \qquad a_2 = f_{xx}/2 + b_1 f_x + b_2 f. \tag{6.48}$$

6.11.1 ▪ Multithreaded Iterations

One interesting extra capability of the Cauchy and Shafer methods is that both can track a second root and thereby cope with the close proximity of two solutions near a bifurcation point. Fig. 6.10 shows that for the arccosine, which has an infinite number of branches confined (for real x) to $\lambda \in [-1,1]$, the Cauchy second solution is graphically indistinguishable from the lower branch for $\lambda \in [0.95,1]$, whereas the Shafer approximation has the same merit for roughly triple the interval. Neither interval of accuracy is large compared to the total range in λ. In the limit $n \to \infty$ such that $x^{(n)}$ has converged to the upper branch, the analytical forms of the second solutions, $x^{Cauchy,2d} = \arccos(\lambda) - (2/\lambda)\sqrt{1-\lambda^2}/\lambda$ and $x^{Shafer,2d} = \arccos(\lambda) - (4/3)\sqrt{1-\lambda^2}/\lambda - (2/3)\sqrt{1-\lambda^2}\lambda$, show that both are infinite at $\lambda = 0$! Nevertheless, as discussed in the next chapter, points where two branches meet are singular points (bifurcation points); Newton's iteration, and indeed every scheme that returns a unique answer, must have difficulties in the neighborhood of a point where there are *two* solutions.

One strategy with these high-order iterations is what we shall dub "multithreading": instead of discarding the larger correction, a multithreaded Cauchy iteration further iterates *both* solution branches. At the successive stages, two branches become four, and eight, and sixteen, and so on. As a matter of practicality, it is therefore necessary to impose an upper limit on the number of threads. Furthermore, one must "prune" the set of converged threads by deleting duplicate roots. Multithreading is not expensive when solving a single equation in a single unknown. The converged threads will generate much duplication of roots while greatly increasing the probability that no roots in the neighborhood of the starting point are missed.

6.11.2 ▪ Expanding the Domain of Convergence

Higher-order proxies for $f(x)$ usually expand the domain of convergence compared to Newton's method. Fig. 6.11 illustrates why: the higher-order proxies are simply more accurate approximations over a wider range in x than is the tangent line approximation that justifies Newton's iteration.

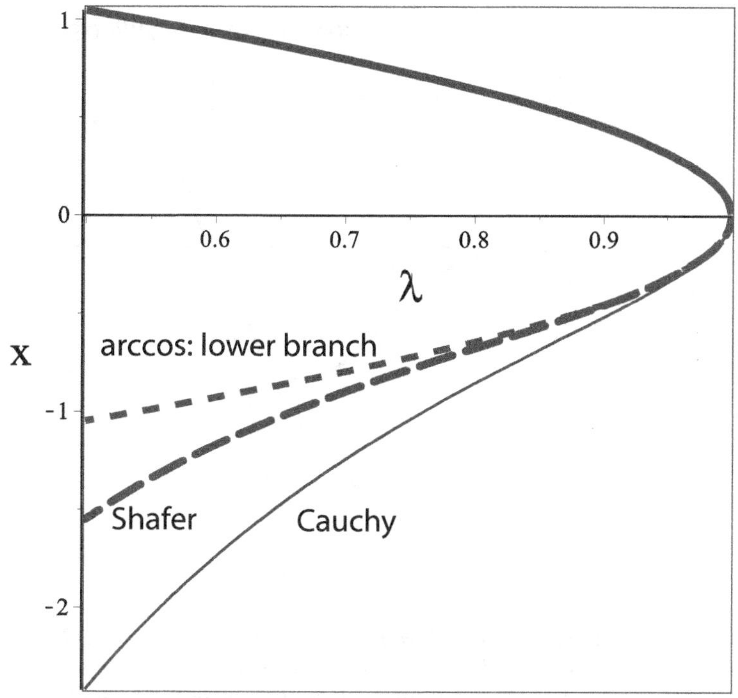

Figure 6.10. *Cauchy and Shafer approximations to branches of the arccosine ($f(x, \lambda) = \cos(x) - \lambda \rightarrow x(\lambda) = \arccos(\lambda)$). The thick topmost curve is the principal branch of the arccosine. When the Cauchy and Shafer iterations have converged to this principle branch, their second solutions approximate the lower branch (short dashes) near the limit point; the Shafer approximation (long dashes) is much more accurate than Cauchy's (thin bottom-most curve).*

Fig. 6.12 and Table 6.2 show that underrelaxation widens the range of convergent starting values from the Newton interval of convergence, $x \in [-0.707, 0.07]$, to a domain bounded by the roots of df/dx (at $x = \pm 1$) that are closest to the left and right of the desired zero, $x = 0$. The higher-order Halley and Cauchy methods do as well as the underrelaxed Newton iteration, but without underrelaxation. The Shafer iteration is an "over-the-horizon" method, capable of converging even from a starting point on the far side of the hill between the initialization and the root. A doubly threaded Shafer iteration has an even wider domain of convergence. Fig. 6.13 illustrates these same concepts with a different $f(x)$.

6.12 ▪ Newton's Iteration for Newton's Exemplary Polynomial

Isaac Newton illustrated his rootfinding methods by computing the one real root of the cubic polynomial[13]

$$f(x) \equiv x^3 - 2x - 5, \qquad x_* = 2.094551482. \tag{6.49}$$

[13]This example has been used over and over again, illustrating not only various rootfinding techniques, but also illuminating the longevity of canonical examples. Augustus de Morgan wrote to William Whewell, January 20, 1861: "The reason I call $x^3 - 2x - 5$ a celebrated equation is because it was the one on which Wallis chanced to exhibit Newton's method when he first published it, in consequence of which *every* numerical solver has felt bound in duty to make it one of his examples. Invent a numerical method, neglect to show how it works on this equation, and you are the pilgrim who does not come in at the little wicket (*vide* J. Bunyan)."

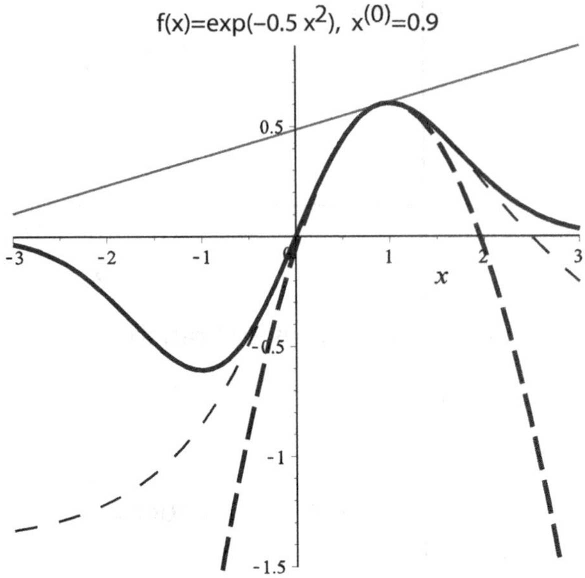

Figure 6.11. *Three proxies for $f(x) = x\exp(-(1/2)x^2)$ when $x^{(0)} = 0.9$. The thick solid curve is $f(x)$; the long dashed curve is the Cauchy approximation (a parabola), and the thin dashed curve is the Shafer approximation (the ratio of two quadratic polynomials). This starting point, $x^{(0)} = 0.9$, is outside the domain of convergence to the root at the origin for Newton's iteration, which is $x \in [-0.707, 0.707]$. The tangent line from $x^{(0)} = 0.9$ is very flat and hits the axis only for $x = -3.8$, and from there the Newton iteration diverges to $x = -\infty$. In contrast, the Cauchy and Shafer approximations have roots at $x = 0.036$ and -0.0022—remarkably good approximations to the roots from a starting point where Newton's iteration diverges.*

This polynomial is a good illustration of Newton's iteration because it is possible to analyze it very completely.

The Newton flow converges to the one real root if and only if

$$x^{(0)} \in [\sqrt{2/3}, \infty], \tag{6.50}$$

where $\sqrt{2/3} = 0.816\ldots$ is a root of df/dx. The Newton flow converges to a *nonroot local minimum* at $x = -\sqrt{2/3} = -0.816\ldots$, the other zero of df/dx, for all $x \in [-\infty, 0.816]$.

The nth iterate is a rational function of the initialization $x^{(0)}$ (as true for any $f(x)$ that is a polynomial). The numerator degree is 3^n and the denominator degree is $3^n - 1$. The two iterates are, simplifying the formulas by using y as an abbreviation for $x^{(0)}$,

$$x^{(1)} = \frac{2y^3 + 5}{3y^2 - 2}, \tag{6.51}$$

$$x^{(2)} = \frac{16y^9 + 255y^6 - 270y^4 + 300y^3 + 180y^2 + 210}{(12y^6 - 18y^4 + 60y^3 + 24y^2 + 67)(3y^2 - 2)}. \tag{6.52}$$

All of the iterates have poles at $x^{(0)} = \sqrt{2/3}$, the zero of df/dx, because an initial guess at a zero of the derivative will, unless the point is a double zero, be ejected to infinity. The iterates have additional poles to the left of $x^{(0)} = \sqrt{2/3}$.

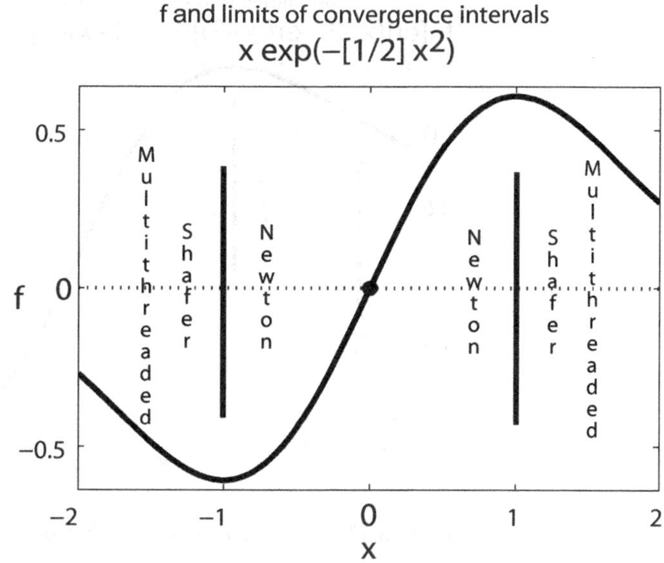

Figure 6.12. *The thick black curve is $f(x) = x\exp(-(1/2)x^2)$, which has roots at the origin and infinity. The vertical text blocks "Newton" bound the interval of starting values for which Newton's iteration converges to the zero at the origin, $x^{(0)} \in [-1/\sqrt{2}, 1/\sqrt{2}]$. Underrelaxation widens the domain of convergence to $x^{(0)} \in [-1, 1]$, as marked by the long vertical solid lines; the Halley and Cauchy methods do as well with or without underrelaxation. The outer vertical texts delineate the convergence intervals for the Shafer flow and multithreaded Shafer (outermost).*

Table 6.2. *Interval of convergence with and without underrelaxation.*

$f(x)$	$x\exp(-(1/2)x^2)$	$\tan(x)/x - 3/2$	$0.5(\text{erf}(x+2) - \text{erf}(x-2))$ $-2(1+x)\exp(-x^2) - 0.5$
Root	0	0.9674026382	−0.6428882495
Roots of df/dx	±1	0	−1.179, 0.372
Poles	-	$\pi/2$	-
Newton	[−0.707, 0.707]	[0.15π, 0.5π]	[−1.046, 0.117]
Newton Flow	[−1, 1]	[0, 0.5π]	[−1.178, 0.373]
Halley	[−1, 1]	[0, 0.5π]	[−1.178, 0.373]
Cauchy	[−1, 1]	[0, 0.5π]	[−1.178, 0.373]
Shafer	[−1, 1]		[−1.342, 0.449]
Shafer Flow	[−1.26, 1.26]		[−1.342, 0.449]
Multithreaded Shafer	[−1.53, 1.53]		[−1.91, 0.71]

When the initialization is not too far from the root, Fig. 6.14 shows that Newton's iteration converges exponentially fast.

6.13 ▪ Never-Failing Newton's Initialization

6.13.1 ▪ Definitions and an Example: The Square Root

Newton's iteration is powerful in three situations. First, it is very useful for refining approximate zeros calculated by other algorithms (Newton-polishing). Second, Newton's is an excellent tool to trace a solution branch as a function of the parameter λ when the

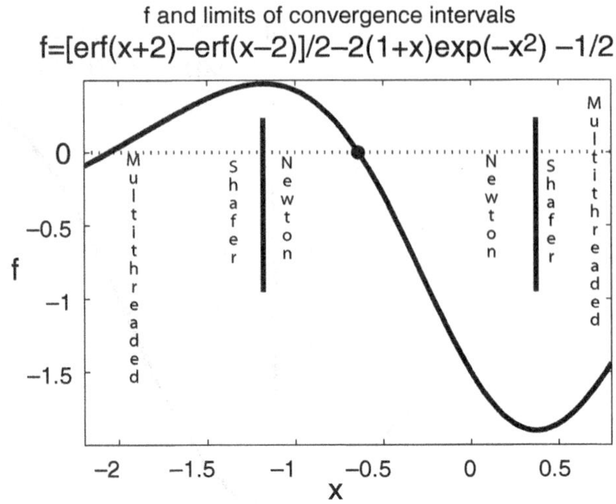

Figure 6.13. *Same as previous figure, but for a different $f(x)$.*

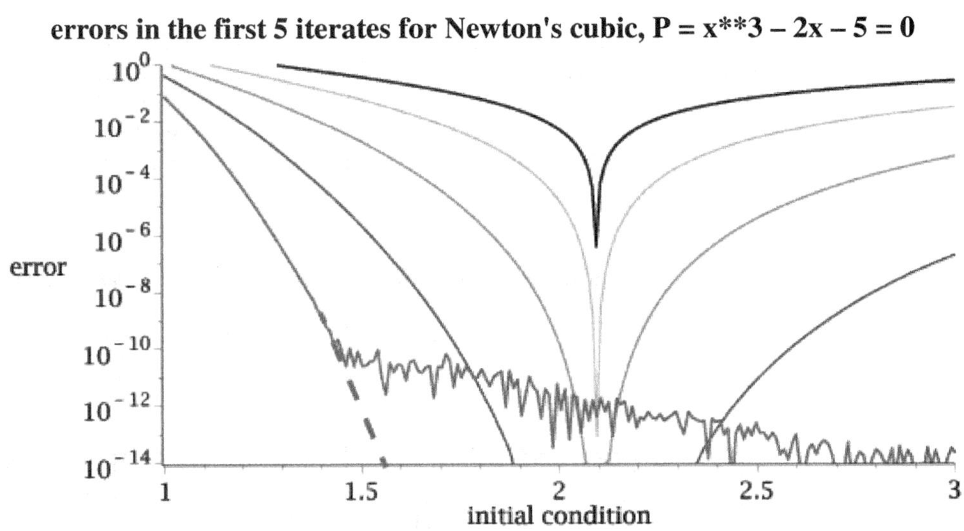

Figure 6.14. *The errors in the first* 5 *Newton iterates for Newton's cubic polynomial. The uppermost curve is the first iterate; the bottom curve (with noisy oscillations) is the fifth iterate as computed in* 16*-digit precision. In higher precision, the fifth iterate error is below* 10^{-14} *everywhere to the right of the intersection of the dashed line with the bottom of the graph. The accuracy of Newton's iteration is always limited by roundoff error, and the user must allow for this and terminate the iteration when the Newton corrections cease to diminish.*

continuation method is used as the predictor, and Newton's merely the corrector, as described in Chapter 8. Third, Newton's iteration can be the only rootfinder when provided with an analytic formula that is within the domain of convergence of a root in some finite part of the parameter space.

Definition 6.6 (Never-Failing Newton's Initialization (NFNI)). *An approximation to a root $x^{(0)}(\lambda_1, \lambda_2, \dots, \lambda_M)$ in an M-dimensional parameter space is a* never-failing Newton's

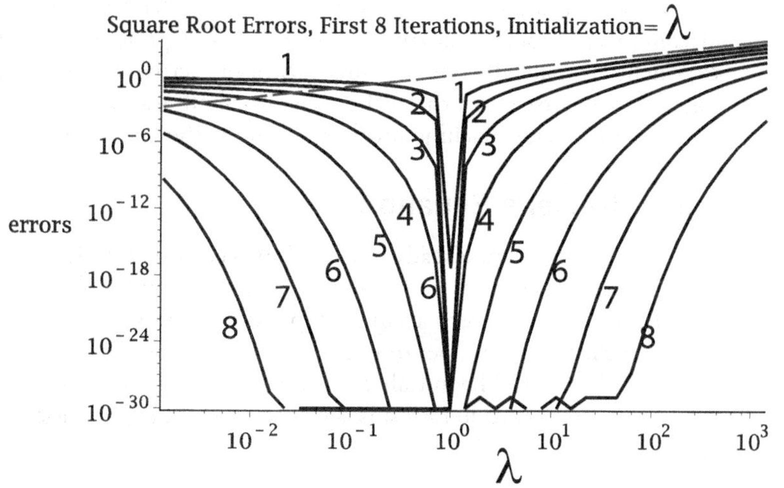

Figure 6.15. *Errors in the first 8 iterates for the square root of λ from the initialization $x^{(0)}(\lambda) = \lambda$, which is graphed as the dashed line, using 30-decimal-digit floating point arithmetic. $x^{(n+1)} = x^{(n)}/2 + \lambda/x^{(n)}$. The numbers on the curves are the iteration numbers n.*

initialization (NFNI) *in a continuous region* ⅂ *if the iteration always converges to a certain continuous solution branch everywhere within* ⅂.

Definition 6.7 (Uniform Never-Failing Newton's Initialization (UNFNI)). *An NFNI is* uniform *if the number of iterations n_{iter} required to reduce the error to less than* ℶ *for some small* ℶ *is bounded by a finite constant C over all of* ⅂.

A simple example will illuminate these concepts. The kth root of λ is the solution to the following equation with the corresponding Newton's iteration:

$$f(x,\lambda) \equiv x^k - \lambda \qquad \Rightarrow x^{(n+1)} = \frac{k-1}{k}x^{(n)} + \frac{\lambda}{k(x^{(n)})^{k-1}}. \tag{6.53}$$

The first guess,

$$x^{(0)}(\lambda) = \lambda, \tag{6.54}$$

is an NFNI for any real λ and any finite k. Consequently, Newton's iteration is the end of history for computation of kth roots; nothing else is used and nothing else is needed.

However, the initialization $x^{(0)}(\lambda)$ is not uniform, as illustrated in Fig. 6.15. For a UNFNI, the error contours would be horizontal or nearly so. When $k = 2$ (square roots) and $\lambda = 2^m$ for $m \gg 1$, the iteration reduces x by a factor of two at each iteration until close to x^m, at which time the usual digit-doubling convergence finally sets in. When $m \ll -1$, the same is observed except that $x^{(1)}$, the first iterate, is approximately $\lambda/(2x^{(0)})$. This geometric-then-digit-doubling rate of convergence is exactly the same as illustrated in section 6.5.

Accuracy is not necessary for an NFNI: a first guess of 1,000,000 for the square root of 10^6 is a horrible, thousand-times-too-big approximation to the square root, which is only 1,000. Nevertheless, 15 iterations suffice to find the square root to MATLAB/IEEE 16-digit accuracy.

A more accurate first guess can generate a UNFNI. For example, in floating point arithmetic, part of the information stored is the exponent, a string of binary digits. If this exponent is halved, one obtains—without actually calculating an exponential or logarithm—a first guess that is between one-half and double the true root. Fewer than 10 iterations yields machine precision even for the largest and smallest λ.

6.13.2 ▪ NFNI for Kepler's Equation

The Kepler equation, already analyzed in section 6.9, tells a similar story. Charles and Tatum [126] *proved* that $x^{CT} = \text{sign}(\lambda)\,\pi$ is an NFNI for all $\epsilon < 1$ and all λ. However, the initialization is not a very accurate approximation over most of the two-dimensional parameter space. The number of iterations can become arbitrarily large in the simultaneous limit $\lambda \to 0, \epsilon \to 1$. The initialization that solves a cubic polynomial is more complicated, but the ensuing third iteration is accurate to 16-digit precision over the entire parameter space.

6.13.3 ▪ Inverse Complete Elliptic Integral of the First Kind

The complete elliptic integral $K(m)$ arises in a thousand applications, and various approximations and series for it can be found in a variety of sources, such as the NIST Digital Library of Functions [284]. However, no simple form for the *inverse* is listed in any of these references. The complete elliptic integral of the first kind is

$$K(m) = \int_0^{\pi/2} \frac{dt}{\sqrt{1-t^2}\sqrt{1-mt^2}}. \tag{6.55}$$

It was shown by the author in [99] that an NFNI, derived from perturbation theory, is

$$\begin{aligned} x^{(0)} &= \exp(-\bar{\lambda}[d_0 + d_1\bar{\lambda}]/[1+d_2\bar{\lambda}] \\ &= \exp\left(-\bar{\lambda}[2.54+2.96\bar{\lambda}]/[1+1.48\bar{\lambda}]\right), \end{aligned} \tag{6.56}$$

where $\bar{\lambda} \equiv \lambda - \pi/2$,

$$d_0 = \frac{8}{\pi} \approx 2.54\ldots, \qquad d_1 = 4\frac{4-\pi}{\pi\,(\pi - 4\ln(2))} \approx 2.96, \qquad d_2 = d_1/2 \approx 1.48.$$

The maximum peak relative error is 0.081 at $x = 0.0223$, which is unimpressive in and of itself. However, accuracy at the start of an iteration is irrelevant except for the question: does the starting point lie in the convergence basin for the iteration? Careful numerical experiments show that (6.56) is an NFNI [99].

6.13.4 ▪ Inverse of the Incomplete Elliptic Integral of the Second Kind with Respect to Angle

The incomplete elliptic integral of the second kind is

$$\mathcal{E}(\sin(T), m) \equiv \int_0^T \sqrt{1 - m\sin^2(T')}\,dT', \tag{6.57}$$

where $m \in [0,1]$ is the elliptic modulus. This bivariate function can be inverted with respect to angle T by solving the transcendental equation $\mathcal{E}(\sin(T); m) - z = 0$. Boyd [93] showed that the first guess,

$$T^0(z,m) = \pi/2 + \sqrt{r}(\theta - \pi/2), \tag{6.58}$$

where

$$\zeta \equiv 1 - z/\mathscr{E}(1;m), \tag{6.59}$$
$$r = \sqrt{(1-m)^2 + \zeta^2}, \tag{6.60}$$
$$\theta = \arctan((1-m)/\zeta), \tag{6.61}$$

is an NFNI. The Newton's iteration is

$$T^{n+1} = T^n - \{\mathscr{E}(\sin(T);m) - z\} / \left\{\sqrt{1 - m\,\sin^2(T)}\right\}. \tag{6.62}$$

With the first guess above, the iteration always converged to $T(z;m) = \mathscr{E}^{-1}(z;m)$ within a relative error of less than 10^{-10} in three iterations or less and thus is a UNFNI. This initialization is an empirical, problem-specific choice; the reasoning behind the choice is given in [93] but is not reproduced here because it is so narrow.

The paper discusses three alternative initialization strategies: "homotopy" initialization $[T^0(z,m) \equiv (1-m)(z;0) + m\,T(z,m)(m;1)]$, perturbation series (in powers of m), and inversion of the Chebyshev interpolant of the incomplete elliptic integral. Although all work well and are general strategies applicable to a very wide range of problems, none of these three alternatives is as efficient as the empirical initialization, which is completely problem specific. This illustrates T. T. Wu's maxim "Usefulness is often inversely proportional to generality." (Harvard lecture, 1973.)

6.13.5 ▪ The Balanced Oscillator Gain Equation

The amplitude z and oscillator gain k of a balanced oscillator are connected by the transcendental equation $f(z,k) = 0$ [108]. It is easy to find $k(z)$ through numerical quadrature, but the inverse function $z(k)$ is nontrivial. Replacing the unknown z by x and the parameter k by λ yields

$$f(x,\lambda) \equiv (2/x) \int_0^1 \tanh(x\,\sin(2\pi t))\sin(2\pi t)\,dt - (1/\lambda), \tag{6.63}$$

where the parameter $\lambda \in [1,\infty]$ and the unknown $x \in [0,\infty]$. Newton's iteration requires the derivative, which is

$$\frac{df}{dx} = -\frac{2}{z^2}\int_0^1 \tanh(x\,\sin(2\pi t))\sin(2\pi t)\,dt + \frac{2}{x}\int_0^1 \operatorname{sech}^2(x\,\sin(2\pi t))\sin^2(2\pi t)\,dt. \tag{6.64}$$

Noting that one of the integrals for the derivative is also required for $f(x,\lambda)$, one must perform two numerical quadratures per iteration; theory shows that a roughly equal number of quadrature points is required for each. Thus, Newton's iteration is rather expensive, so a good first guess is desirable.

Boyd and Visser [108] show that

$$x^0 = \sqrt{(\lambda-1)(\lambda+3)} \tag{6.65}$$

is an NFNI for the entire parameter range, $\lambda \in [1,\infty]$. A better initialization blends the lowest order of the power series with the first two terms of the asymptotic expansion using error function ramps, as described later in this section:

$$x^0 \equiv \sqrt{1-\frac{1}{\lambda}\left\{1+\operatorname{erf}\left(3\left(\frac{3}{2}-\lambda\right)\right)\right\}+\frac{2}{\pi}\lambda\left\{1-\frac{\pi^4}{384\lambda^2}\right\}\left\{1+\operatorname{erf}\left(3\left(\lambda-\frac{3}{2}\right)\right)\right\}}. \tag{6.66}$$

The choice of the center of the blending region, $\lambda = 3/2$, is largely arbitrary but corresponds roughly to the radius of convergence of the power series. The blend is also an NFNI.

The two initializations are compared in Fig. 2 of [108]. Starting from the blended initialization, the error is less than 1.9×10^{-12} after just three iterations everywhere on $\lambda \in [1,\infty]$.

6.13.6 ▪ Equatorial Wave Dispersion Relation

The frequencies of the three equatorially trapped planetary-scale waves with latitudinal mode number n are the three real roots of

$$\omega^3 - (2n+1+k^2)\omega - k = 0, \qquad k \in [0,\infty],\ n \geq 1. \tag{6.67}$$

The *approximate* solutions, derived by perturbation theory in Chapter 11, are

$$\omega_{\text{Gravity}} \approx \pm(2n+1+k^2)^{1/2}, \tag{6.68}$$

where the plus sign gives the eastward-traveling gravity wave, and

$$\omega_{\text{Rossby}} \approx -k/(2n+1+k^2). \tag{6.69}$$

Boyd proved rigorous error bounds for these approximations [57]. The relative error in the Rossby wave approximation is never worse than 1 part in 60 for all n and k. The gravity wave approximations may be in error by as much as 10%, but the relative error can be reduced to less than 1% for all n,k by modifying (6.68) to be consistent with the Vieta root relations, Theorem 11.9, which for a cubic equation with a zero quadratic coefficient requires that, without approximation, the sum of the roots must be *zero*. This implies

$$\omega_{Gravity} \approx \pm(2n+1+k^2)^{1/2} - \frac{1}{2}\omega_{Rossby}, \tag{6.70}$$

which was previously derived by other means in [57].

These approximate solutions are NFNIs. One reason that a single formula (for each wave species) is successful is that for this geophysical application, the coefficients of the linear and constant terms in the cubic dispersion equations are always negative. The general reduced cubic equation has multiple regimes, most with only one real root, but the equatorial dispersion relation inhabits only the region in the parameter space of the general cubic equation where there are always three real roots.

Another reason that the approximate frequencies are NFNIs is that the approximations are very accurate. It cannot be said too strongly that sometimes very crude, inaccurate approximations are successful NFNIs. Nevertheless, the safest way to generate good initializations is to derive accurate approximations. Perturbation theory is very powerful in this context.

6.13.7 ▪ Lambert W-Function

The Lambert W-function is the solution $W(\lambda) \equiv x$ to

$$x \exp(x) = \lambda. \tag{6.71}$$

Computing $W(\lambda)$ is still an active research topic [181, 186, 31, 32]. The principle branch has the real-valued domain $\lambda \in [-\exp(-1), \infty]$. Boyd unsystematically devised an approximation that matches both the large and small λ asymptotics and showed his formula has a global absolute error of no worse than 4.7% [73]:[14]

$$\begin{aligned} W \approx & -1 + \{\log(y+10) - \log(\log(y+10))\} \tanh\left(\nu\sqrt{y}\right) \\ & \left\{1 + (1/10)(\log(y) - 1.4)\exp\left(-0.075(\log(y) - 1.4)^2\right)\right\}, \end{aligned} \tag{6.72}$$

where

$$y \equiv 1 + \exp(1)\lambda \qquad \text{and} \qquad \nu = \frac{\sqrt{2}}{\log(10) - \log(\log(10))}. \tag{6.73}$$

The rational Chebyshev series [73] and the piecewise analytical approximations of Barry et al. [31, 32] are much more accurate, but that is not important. Boyd shows that the approximation (6.72) is an NFNI, yielding a relative error smaller than 10^{-12} in four iterations for the entire branch. (The piecewise approximations of Barry et al. [31, 32] are also NFNIs.)

The local approximations that are combined to generate Boyd's global approximation and initialization are very different in form: for large y,

$$W \sim -1 + \log(y) - \log(\log(y)) + \frac{1 + \log(\log(y)}{\log(y)} + \frac{[\log(\log(y))]^2 - 1}{2\log^2(y)} + \cdots, \qquad y \to \infty. \tag{6.74}$$

For small y, near the limit point at $y = 0$,

$$W \approx -1 + \sqrt{2}\sqrt{y} - (2/3)y + \sqrt{2}(11/36)y^{3/2} + \cdots, \qquad y \ll 1. \tag{6.75}$$

Nevertheless, these can be merged into an approximation that is accurate everywhere.

Better approximations can be made by applying the systematic blending strategies described in the next section.

6.13.8 ▪ Blending and Partitions of Unity

Piecewise initializations are usually not good approximations in the neighborhood of the breaks; there, approximations for small λ and large λ are both being applied for intermediate λ where neither is accurate. An arbitrary number of pieces can be analytically blended by using smoothed step functions and top-hat functions (*box functions*) adjusted to have the partition of unity property.

Each local approximation is multiplied either by a smoothed box function (top hat) on the interior of the domain or by a "ramp" (smoothed step functions) at the ends.

[14] Errata: In Eq. (4) of this article, $11/36$ should be $11\sqrt{2}/36$.

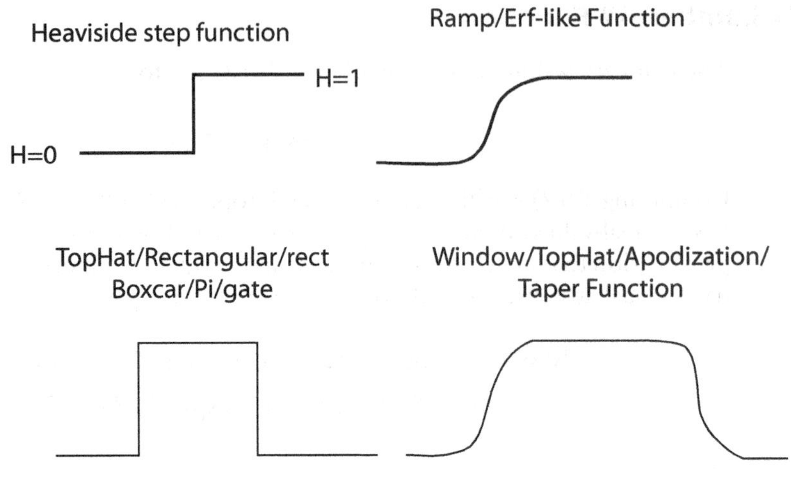

Figure 6.16. *A visual dictionary of ramps, bells, step function, and TopHat function. The discontinuous functions are on the left; their continuous counterparts are on the right. Note that the TopHat function on the bottom left and its smoothed counterpart on bottom right have multiple names, as indicated.*

Ramps and bells are schematically illustrated in Fig. 6.16. Suitable ramps and bells are the following:

$$\mathfrak{H}(x;S)=\frac{1}{2}\left\{1+\operatorname{erf}(Sx)\right\} \qquad \text{[Erf-Ramp]}, \tag{6.76}$$

$$\mathscr{H}(x;S,\Theta)\equiv(1/2)\{1+\mathscr{E}(x;S,\Theta)\} \qquad [C^{\infty}\text{-Ramp}], \tag{6.77}$$

where

$$\mathscr{E}(x;S,\Theta)=\begin{cases} -1, & x<-\Theta, \\ \operatorname{erf}\left(S\dfrac{x}{\sqrt{1-x^2/\Theta^2}}\right), & x\in[-\Theta,\Theta], \\ 1, & x>\Theta, \end{cases} \tag{6.78}$$

$$\mathfrak{T}(x;S,L)=\frac{1}{2}\left\{\operatorname{erf}(S[x+L])-\operatorname{erf}(S[x-L])\right\} \qquad \text{[Erf-TopHat]}, \tag{6.79}$$

$$\begin{aligned}\mathscr{T}(x;L,S,\Theta)&\equiv\frac{1}{2}\{\mathscr{H}([x+L+\Theta];S,\Theta)-\mathscr{H}([x-L-\Theta];S)\}\\ &=\begin{cases}\mathscr{H}([x+L+\Theta];S,\Theta), x\in[-2\Theta-L], \\ 1, \qquad x\in[-L,L], \\ \mathscr{H}(-[x-L-\Theta];S,\Theta), x\in[L,L+2\Theta].\end{cases}\end{aligned} \qquad [C^{\infty}\text{-TopHat}] \tag{6.80}$$

This TopHat tapers from one to zero on the intervals $x\in[-L-2\Theta,-L]$ and $x\in[L, L+2\Theta]$.

Definition 6.8 (Partition of Unity). *A set of N functions* $\mathfrak{W}_j$ *where* $N \geq 2$ *(and may be infinite) is a partition of unity on a domain* Ω *(in any number of dimensions) if and only if*

$$\sum_{j=1}^{N} \mathfrak{W}_j(x) = 1 \qquad \forall x \in \Omega. \tag{6.81}$$

The reason this property is useful is the following.

Theorem 6.9. *Suppose* $f(x)$ *is approximated on N different parts of the domain by local approximations* f_j *with errors* E_j. *Let the* $\mathfrak{W}_j(x)$ *denote a partition of unity. Then the blended approximation error is*

$$E_{blend} \equiv f(x) - \sum_{j=1}^{N} \mathfrak{W}_j(x) f_j \tag{6.82}$$

$$= \sum_{j=1}^{N} \mathfrak{W}_j(x) E_j(x). \tag{6.83}$$

In words, the error in the blended approximation comes entirely from the errors in the local approximations $E_j(x)$*; there is no error in partition-of-unity blending itself.*

Proof:

$$f(x) \approx \sum_{j=1}^{N} \mathfrak{W}_j(x) f_j \tag{6.84}$$

$$= \sum_{j=1}^{N} \mathfrak{W}_j(x)(f(x) - E_j(x)) \tag{6.85}$$

$$= f(x) \sum_{j=1}^{N} \mathfrak{W}_j(x) - \sum_{j=1}^{N} \mathfrak{W}_j(x) E_j(x) \tag{6.86}$$

$$= f(x) - \sum_{j=1}^{N} \mathfrak{W}_j(x) E_j(x). \tag{6.87}$$

■

Fig. 6.17 is an illustration of how two approximations, valid only for a small parameter and a large parameter, respectively, were blended together to create the NFNI for the oscillator gain equation of Boyd and Visser [108].

Fig. 6.18 shows a similar blend for the Lambert W-function. The large-y approximation is a bivariate series in log(y) and log(log(y)) and therefore is accurate only for rather large y. To extend the small-y approximation to the log/log(log) series, the blend employs a [3/3] Padé approximant formed from the first seven terms of the power series in $\sqrt{y}$:

$$W_{[3/3]} = \frac{(-2158\sqrt{2} - (3896/21)\sqrt{y} + (21865/12)\sqrt{2}y + (89131/135)y^{3/2})}{(2158\sqrt{2} + (94532/21)\sqrt{y} + (104225/84)\sqrt{2}y + (262861/1890)y^{3/2})}.$$

6.14 ▪ Maple and MATLAB Codes

Tables 6.3 and 6.4 are Maple and MATLAB code listings for Newton's iteration. Although not strictly speaking a part of the algorithm, a practical programmer needs to specify two

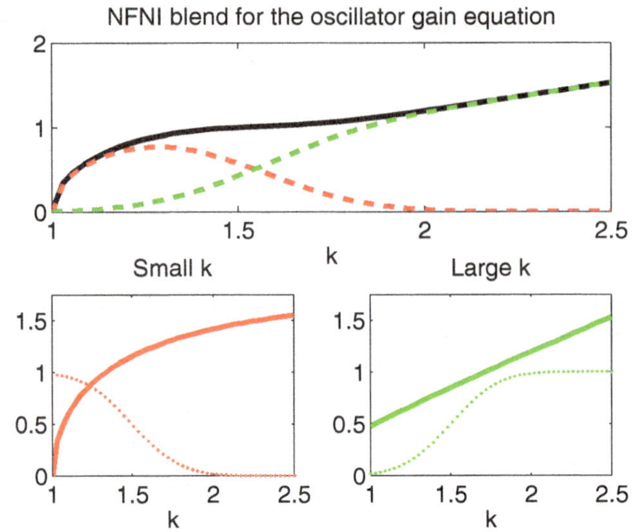

Figure 6.17. *Top: the thick solid curve is the blended NFNI from [108]. The red and green dashed curves are the two local approximations whose sum is the NFNI (solid curve). Bottom left: the solid curve is the approximation for small parameter k, which is actually the square root of a power series in $p \equiv 4(1-1/k)$; the dotted curve is the erf-ramp; the dashed red curve in the top panel is the product of the two red curves in the lower left frame. Bottom right: same but for the large-k asymptotic series; the dashed green curve in the top panel is the product of the two green curves in the lower right panel. The dotted green curve is the erf-ramp $(1/2)\{1+erf(3(k-\frac{3}{2}))\}$.*

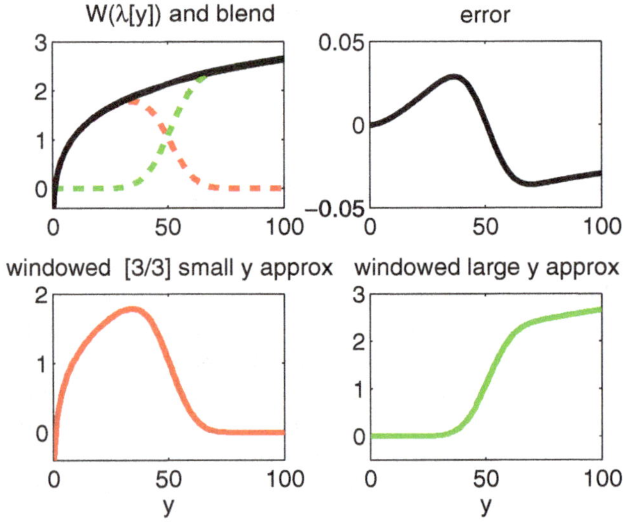

Figure 6.18. *Lambert W-function. Upper left: the solid black curve is the blended initialization. The red and green dashed curves are two local approximations, windowed by multiplication by an erf-ramp, whose sum is the solid curve. The coordinate y is connected to λ, the argument of the W-function, by $y \equiv 1+\exp(1)\lambda$. Top right: the error $W(\lambda[y]) - W_{blended}(y)$. Lower left: product of the erf-ramp $(1-erf((y-50)/12))/2$ with the $[3/3]$ Padé approximant formed from the power series in y about the limit point. Lower right panel: product of the erf-ramp $(1+erf((y-50)/12))/2$ with the large y approximation.*

Table 6.3. *Maple code for Newton's iteration.*

```
restart;
# Newton's iteration solves f(x) = 0;
# Newton's requires functions to evaluate f(x) and fx(x) <=> df/dx;
  f := x -> cos(Pi*x + x**2); fx := x -> -(Pi + 2*x)*sin(Pi*x + x**2);
# There are two user-choosable numerical parameters;
# "itermax" is the maximum number of iterations;
# "epsstop" is a stopping tolerance. This should be small compared;
# to one but large compared to machine epsilon;
  itermax := 10; epsstop := 1.0E-10;
  Digits := 16; # Number of decimal digits of precision;
# Newton's iteration requires an approximation to the root;
# the "initialization" or "first guess";
  x := 1;        # initialization
  delta := 1;        # initialize delta so that the while loop won't crash;
  for iter from 1 to itermax while abs(delta) > epsstop do
  delta := evalf( f(x)/fx(x) );        # Newton correction;
  x := evalf( x - delta);        # update to x;
  print(` iter = `,iter,` x = `,evalf[16](x),` delta = `,evalf[16](delta));
  od:
```

Table 6.4. *MATLAB code for Newton's iteration.*

```
% Newton's iteration solves f(x) = 0;
% Newton's requires functions to evaluate f(x) and fx(x) <=> df/dx;
  f = inline(' cos(pi*x + x^2)','x'); fx = inline(' -(pi + 2*x)*sin(pi*x + x^2)','x');
% There are two user-choosable numerical parameters;
% "itermax" is the maximum number of iterations;
% "epsstop" is a stopping tolerance. This should be small compared;
% to one but large compared to machine epsilon;
  itermax = 10; epsstop = 1.0E-10;
% Newton's iteration requires an approximation to the root;
% the "initialization" or "first guess";
  x = 1;
% initialization
  delta = 1; iter = 0;
% initialize delta, iter so that the while loop won't crash;
  while ( (abs(delta) > epsstop) & (iter <= itermax)), iter = iter + 1;
  delta = ( f(x)/fx(x) );
% Newton correction;
  x = x - delta;
% update to x;
  disp([' iter = ',int2str(iter),' x = ',num2str(x),' delta = ',num2str(delta)]);
end
```

numerical parameters. To prevent infinite loops, it is always desirable to specify a finite maximum number of iterations, "itermax." Furthermore, it is silly to keep iterating when the error has already been reduced below the user's needs, or the limits imposed by floating point precision; "epsstop" halts the iteration when the magnitude of the corrections

Table 6.5. *Newton corrections $\delta^{(n)}$, where n is the iteration number for various choices of the number of decimal digits of precision, "Digits" in Maple. The stopping tolerance was deliberately set to zero (not recommended) to illustrate what happens when it is omitted.*

n	Digits=10	Digits=16	Digits=64
1	−0.12488	−0.12488	−0.12488
2	0.016158	0.016158	0.016158
3	0.0000077855	0.0000077855	0.0000077855
4	4.8802×10^{-10}	1.13051×10^{-11}	1.13051×10^{-11}
5	4.8802×10^{-10}	2.65544×10^{-17}	2.3849×10^{-23}
6	4.8802×10^{-10}	2.65544×10^{-17}	1.0613×10^{-46}
7	4.8802×10^{-10}	2.65544×10^{-17}	1.00443×10^{-64}
8	4.8802×10^{-10}	2.65544×10^{-17}	1.00443×10^{-64}
9	4.8802×10^{-10}	2.65544×10^{-17}	1.00443×10^{-64}
10	4.8802×10^{-10}	2.65544×10^{-17}	1.00443×10^{-64}

is below this number. To achieve high accuracy, "epsstop" should be small. However, in floating point arithmetic, it is impossible to reduce the corrections to zero, as illustrated in Table 6.5; if the stopping parameter is too small, Newton's iteration will generate the same correction over and over until the iteration is stopped by "itermax."

If $f(x)$ is expensive to evaluate, it may be desirable to Newton-polish the roots of the truncated Chebyshev series f_N before a final iteration or two with $f(x)$. The Maple and MATLAB codes to sum a Chebyshev series and calculate the Chebyshev coefficients of df_N/dx from those of $f(x)$ are given in Table B.1 (Maple) and Table B.2 (MATLAB).

If the derivative of the transcendental function $f(x)$ is unavailable, or very expensive, an option is to use a quasi-Newton iteration in the final step by approximating df/dx by df_N/dx. The quasi-Newton corrections are then

$$\delta^{mixed} \equiv \frac{f(x)}{df_N/dx(x)}. \tag{6.88}$$

The final accuracy will be limited only by the accuracy in computing $f(x)$.

Chapter 7
Bifurcation Theory

"Proofs are to mathematics what spelling (or even calligraphy) is to poetry."
—Vladimir Igorevich Arnold (1937–2010) in [15]

7.1 ▪ Connecting the Points

Most science and engineering problems depend upon one or more parameters λ. Rootfinding's goal expands from mere solutions to the harder task of tracing branches of solutions and the interrelationships and connectivity of different branches. Older methods like Newton's iteration and newer strategies such as the Chebyshev-proxy rootfinder (CPR) yield zeros for discrete values of λ, but these point-solvers merely yield a collection of points in the (x, λ)-plane. Such a set of points is what computer scientists call a *point cloud.*

The continuation method, discussed in the next chapter, is a general strategy for tracing solution branches. Before we discuss continuation, it is advisable to first analyze how branches intersect and develop singularities.

The key theoretical guarantee that a solution branch is not singular is given by the following.

Theorem 7.1 (Implicit Function Theorem). *Suppose that $f(x, \lambda)$ is differentiable with finite derivatives in a neighborhood of a point (x_0, λ_0) such that $f(x_0, \lambda_0) = 0$ and $df/dx(x_0, \lambda_0) \neq 0$. Then there exists a branch $x(\lambda)$ in a finite neighborhood of (x_0, λ_0) that is holomorphic in that neighborhood* [224, 298].

When the first argument of a bivariate function is dependent upon the second argument, the total derivative of the function with respect to the second argument is given by the chain rule:

$$\frac{df}{d\lambda} = \frac{\partial f}{\partial \lambda} + \frac{\partial f}{\partial x}\frac{dx}{d\lambda} \qquad \text{[Total Derivative]}. \tag{7.1}$$

We shall refer to $df/d\lambda$ as the *total derivative* to distinguish it from the partial derivative of f with respect to λ. When $x(\lambda)$ is a solution branch, f is zero everywhere along the branch and therefore the total derivative is zero everywhere along the branch. This implies the differential equation

$$\frac{dx}{d\lambda} = -\frac{\partial f/\partial \lambda}{\partial f/\partial x} \qquad \text{[Davidenko Equation]}, \tag{7.2}$$

named after the mathematician who pointed out that this could be integrated as an initial value problem to trace solution branches [141, 142]. This equation also furnishes a proof of the implicit function theorem since it explicitly generates a solution branch precisely when the conditions of the theorem are satisfied.

7.2 ▪ Bifurcations and Folds

Definition 7.2 (Bifurcation Point). *A point in the space spanned by x and the parameters is a* bifurcation point *if two or more solution branches intersect at the point.*

The most common bifurcation point is the following.

Definition 7.3 (Limit Point/Fold/Turning Point). *If $f(x;\lambda) = 0$ is a single nonlinear equation with a solution branch $x(\lambda)$, then a* limit point, fold, *or* turning point *occurs wherever*

$$\frac{\partial f}{\partial x}(x[\lambda];\lambda) = 0; \qquad \frac{\partial f}{\partial \lambda}(x[\lambda];\lambda) \neq 0. \tag{7.3}$$

An equivalent definition is

$$\frac{dx}{d\lambda} \to \infty \qquad \textit{as} \qquad \lambda \to \lambda_{limit}. \tag{7.4}$$

Two distinct solution branches $x_1(\lambda)$ and $x_2(\lambda)$ meet at a limit point and exist only on one side of the limit point. That is, if $x_1(\lambda)$ and $x_2(\lambda)$ exist for $\lambda \leq \lambda_{limit}$, then they do not exist (as real solutions) for $\lambda > \lambda_{limit}$.

Newton's iteration has difficulties at and near a limit point because the iteration is attracted to both branches; these difficulties are described in detail in Chapter 6.

When $\vec{f}(\vec{x};\lambda)$ is a *system* of equations, (7.4) still applies. The interpretation of (7.3) is that (i) the determinant of the Jacobian matrix $\vec{f}_{\vec{x}}$ is zero and (ii) the column vector $\vec{f}(\vec{x};\lambda)$ does not lie in the range of $\vec{f}_{\vec{x}}$ so that the matrix equation $\vec{f}_{\vec{x}}\, d\vec{x}/d\lambda = -\vec{f}_\lambda$ has no bounded solution.

As an example, the simple algebraic equation

$$(x-1/2)^2 + \lambda^2 = 1 \tag{7.5}$$

has limit points at $\lambda = \pm 1$ where $x = 1/2$. This is merely the equation of a circle in the x-λ plane, as shown in Fig. 7.1, but nonetheless it is impossible to trace the entire curve in a single pass by marching in λ. By using a variable stepsize, we can come as close to the limit point as we want, but we can never turn the corner and march from the lower branch onto the upper branch using small steps in either λ or x.

However, it is trivial to write the equation of the circle in (7.5) in parametric form as

$$x = \cos(t) + 1/2; \qquad \lambda = \sin(t). \tag{7.6}$$

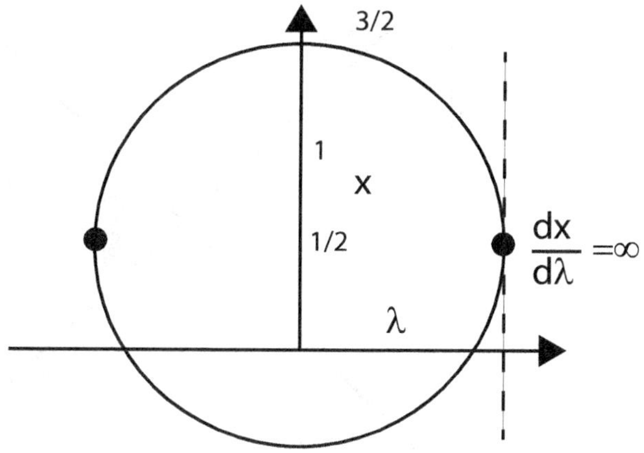

Figure 7.1. *Solution curve $x(\lambda)$ for a quadratic equation. The two limit points are marked by solid disks. There, as indicated on the right, the solution curve is tangent to a vertical line (dashed) and $dx/d\lambda = \infty$. For $-1 < \lambda < 1$, there are two solutions for a given λ. At the limit points (black disks), these upper and lower branches merge.*

Both x and λ are continuous, single-valued functions of t so that we can trace both branches of the solution in one continuous path if we march in t. The new continuation parameter t is the *arclength* along the solution curve.

It is usually impossible to *analytically* define the arclength, but H. B. Keller and various collaborators [222, 150, 125, 196, 223] have developed a "pseudoarclength continuation method" that can follow a solution smoothly around a limit point. The "globally convergent homotopy" [243] and "tangent vector continuation" methods [8] are very similar. We postpone discussion to the next chapter.

Limit points are generic in single-parameter problems in the sense that if a branch does not continue smoothly from $\lambda = -\infty$ to ∞, but instead exists only on one side or another of a finite, limited interval, then the branch cannot be holomorphic at the limit except by having a fold. One may construct exceptional $x(\lambda)$ that are analytic and vary monotonically along the whole real parameter axis in the parameter without folding, but limit points are the rule rather than the exception.

7.3 ▪ Bifurcation Points of Higher Codimension

If there are M parameters, then the zeros of $f(x;\lambda_1,\ldots,\lambda_M)$ live in a space of $N = M+1$ dimensions. With more than one parameter, the singularities of the solution branches need not be points but can be curves, surfaces, or hypersurfaces. The dimension of a singular surface is zero for a point, one for a curve, two for a surface or manifold parameterized by two variables, and so on. The difference between the dimension of the bifurcation structure and N, the dimension of the space spanning all the unknowns and parameters, is said to be the *codimension* of the bifurcation structure.

A fold is generically a bifurcation of codimension one. If there is only one parameter, the fold singularity is a point. When there are two parameters, the fold is a curve in the $\lambda_1 - \lambda_2$ plane, as illustrated for the two-parameter reduced cubic equation in Chapter 11. A *third* parameter generically allows a *surface* of folds, and so on.

More complicated bifurcations generically have higher codimensions.

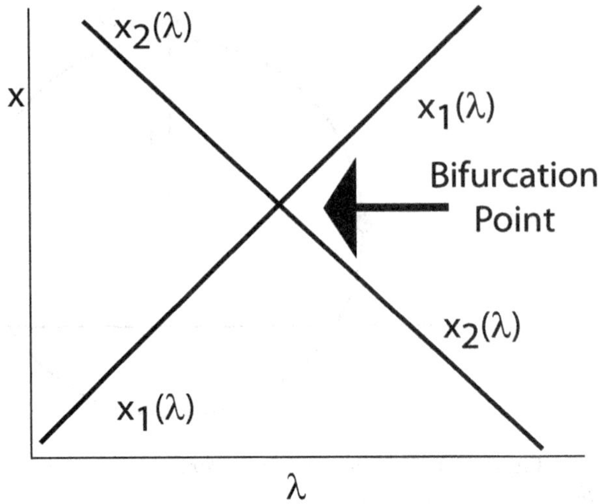

Figure 7.2. *Same as Fig. 7.1 except that the quadratic has been altered to $f(x,\lambda)=(x-\lambda)(1-x-\lambda)$ so that its solution branches have a* transcritical bifurcation *point instead of a pair of limit points. At a transcritical bifurcation point, two branches cross. As at a limit point, $df/dx \to 0$, which implies that Newton's correction is unbounded. Newton's iteration cannot converge to either one but instead diverges. Unlike a limit point, however, all branches have a finite slope $dx/d\lambda$ at a bifurcation point. A limit point is one where branches* merge; *a transcritical bifurcation point is one where branches* cross.

An important bifurcation of codimension two is the *transcritical bifurcation*:

Definition 7.4 (Transcritical Bifurcation/Branch-Crossing Bifurcation). *This is a point in parameter space where two branches intersect. The formal definition for a single equation $f(x;\lambda)=0$ is that a transcritical bifurcation point occurs whenever*

$$\frac{\partial f}{\partial x}(x[\lambda];\,\lambda)=0 \qquad \textit{and} \qquad \frac{\partial f}{\partial \lambda}(x[\lambda];\,\lambda)=0, \tag{7.7}$$

which differs from the definition of a limit point in that both *derivatives are zero instead of just one (Fig. 7.2).*

The domain of convergence of Newton's iteration shrinks to a point at the bifurcation point; unless it is very close to one branch and not so close to the other, the iteration literally cannot decide which branch of the two branches to converge to.

A fold is the solution of the simultaneous equations for (x,λ) ($f(x,\lambda)=0, f_x(x,\lambda)=0$), and solutions to two equations in two unknowns often have real roots. A transcritical bifurcation requires the simultaneous solution of *three* equations ($f=0, f_x=0$, $f_\lambda=0$). Generically, a trio of equations are solvable only in a three-dimensional space where the coordinates are the unknown x and two parameters. It is nevertheless possible to contrive transcritical bifurcations even when there is only a single parameter.

An example is the quadratic equation

$$x^2-2x-\lambda(\lambda-2)=0, \tag{7.8}$$

whose two roots are

$$x_1(\lambda)=\lambda \qquad \text{and} \qquad x_2(\lambda)=2-\lambda. \tag{7.9}$$

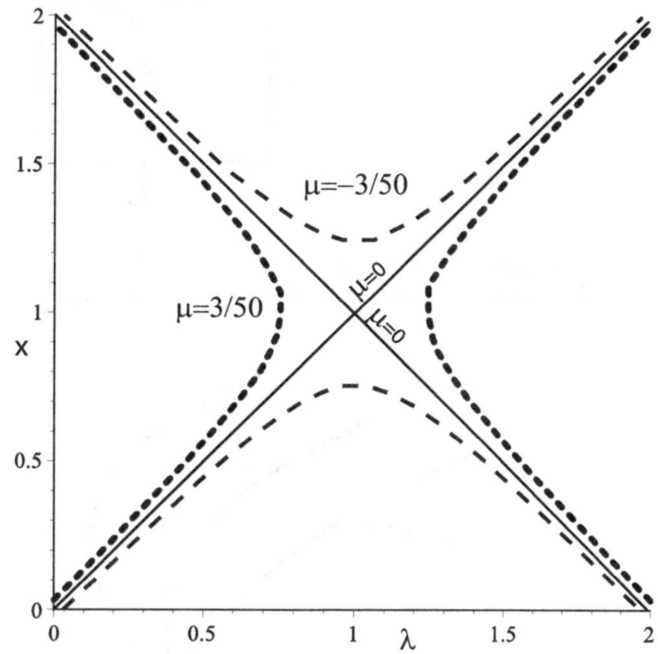

Figure 7.3. *Same as previous figure except that it also shows cross-sections at $\mu = 3/50$ (dots) and $\mu = -3/50$ (dashes) in the λ-x-μ volume; the pair of solid intersecting lines is the cross-section at $\mu = 0$, the only value of the second parameter μ where bifurcation occurs.*

The bifurcation point is located at $\lambda = 1$, where $x_1 = x_2$ and the two branches cross. The Davidenko ODE in λ for (7.8) is

$$\frac{dx}{d\lambda} = \frac{\lambda - 1}{x - 1}. \tag{7.10}$$

At the limit point, the denominator of (7.10) tends to 0—but so does the numerator. The slope $dx/d\lambda$ is always finite. Consequently, although we cannot compute $x(\lambda)$ *at* the bifurcation point, we can calculate either branch using a small, finite, fixed λ stepsize even at points very *near* $\lambda = \lambda_{\text{bifurcation}}$.

In the next section, we review techniques for continuation near and beyond a transcritical bifurcation.

One can see why the transcritical bifurcation is classified as codimension two by adding a second parameter. Let us generalize our example to

$$x^2 - 2x - \lambda(\lambda - 2) + \mu = 0, \tag{7.11}$$

where μ is the second parameter. The transcritical bifurcation occurs only at the point $(\lambda = 1, \mu = 0)$. Everywhere else in the three-dimensional λ-x-μ volume, the solution surfaces do not intersect, as illustrated in Fig. 7.3. The probability that a randomly selected two-dimensional cross-section will contain a transcritical bifurcation is a set of measure zero, i.e., vanishingly small.

In contrast, it is *generic* for folds to form curves in a three-dimensional x-λ-μ space.

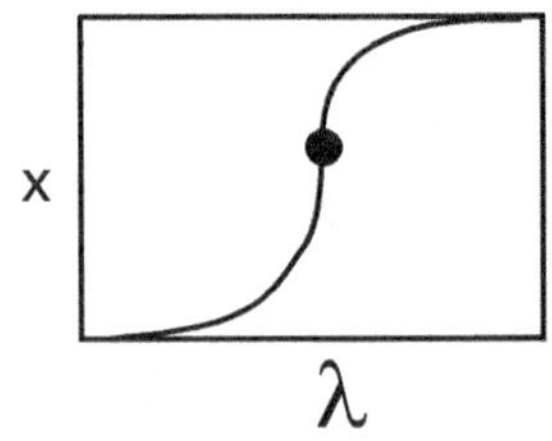

Figure 7.4. *The triple-point/hysteresis bifurcation point. The black disk marks the triple zero. This is structurally unstable, as illustrated in the next figure.*

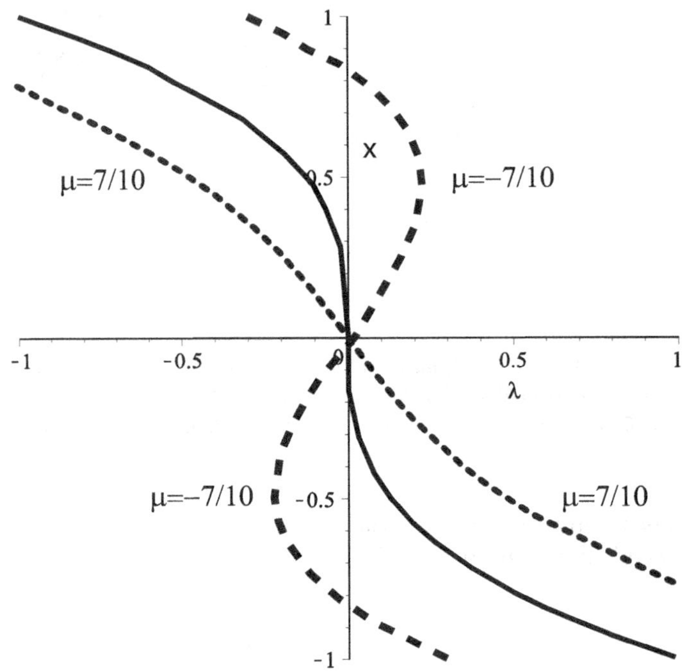

Figure 7.5. *The zero curve of* $x^3 + \mu x + \lambda$ *for three different cross-sections in the* λ*-*x*-*μ *volume at* $\mu = 7/10$ *(dots),* $\mu = -7/10$ *(dashes), and* $\mu = 0$ *(solid curve). Bifurcation occurs only at* $\mu = 0$.

Another type of codimension-two bifurcation is the following.

Definition 7.5 (Triple Point/Cusp/Inflection Point). *This bifurcation occurs when*

$$\frac{dx}{d\lambda} = 0 \qquad and \qquad \frac{\partial f}{\partial x} = \frac{\partial^2 f}{\partial x^2} = 0. \tag{7.12}$$

The triple-point bifurcation is so named because $f(x, \lambda)$ has a root of multiplicity three, as shown in Fig. 7.4. Generically, this bifurcation is an isolated point in a three-dimensional space spanned by the unknown and two parameters. A perturbation will replace the triple zero by three or one real-valued simple roots, as illustrated in Fig. 7.5.

7.4 ▪ Computing Bifurcation Points

The direct way to compute the bifurcation point is to solve the system

$$f(x,\lambda)=0, \qquad f_x(x,\lambda)=0 \tag{7.13}$$

as a bivariate system of two equations in two unknowns. General techniques for solving bivariate systems are given in Chapter 20. The Newton–Raphson iteration is

$$\mathscr{D}^{(n)} \equiv f_{xx}(x^n,\lambda^{(n)})f_\lambda(x^n,\lambda^{(n)}) - f_x(x^n,\lambda^{(n)})f_{x,\lambda}(x^n,\lambda^{(n)}), \tag{7.14}$$

$$x^{(n+1)} = x^{(n)} - \frac{f(x^n,\lambda^{(n)})f_{x,\lambda}(x^n,\lambda^{(n)}) - f_x(x^n,\lambda^{(n)})f_\lambda(x^n,\lambda^{(n)})}{\mathscr{D}}, \tag{7.15}$$

$$\lambda^{(n+1)} = \lambda^{(n)} - \frac{f(x^n,\lambda^{(n)})f_{xx}(x^n,\lambda^{(n)}) - \left(f_x(x^n,\lambda^{(n)})\right)^2}{\mathscr{D}}. \tag{7.16}$$

There is an interesting subtlety in applying Newton's iteration. When $\lambda - \lambda_{bif} \sim O(\epsilon)$, one might expect that the error after a single Newton's iteration would be $O(\epsilon^2)$, but if $(x^{(0)},\lambda_0)$ is a point on the solution curve, then the distance between x and its value at the bifurcation point is $O(\sqrt{\epsilon})$. After one iteration, both x and λ are still $O(\epsilon)$. Further iterations exhibit the usual digit-doubling convergence.

The bifurcation point can also be calculated indirectly. The pseudoarclength continuation described in the next chapter can smoothly round a fold. Passing through the singularity can be detected through the inevitable sign change in f_x at a fold. One can then employ inverse interpolation to recover the precise location of the zero of f_x. Indirect computations are common when attacking large systems of nonlinear equations because for large systems each Newton's iteration is expensive. In one dimension, however, Newton's iteration is usually the best option. Bivariate Cauchy or Muller iterations can also be employed using the technology of Chapter 20, though these are much more complex.

Of course, the goal is not merely to calculate the fold or transcritical point, but to track the multiple branches. Continuation strategies to do just that are the theme of the next chapter.

7.4 • Computing Bifurcation Points

Chapter 8
Continuation in a Parameter

"Ad augusta per angusta (towards big results by narrow paths)."

8.1 ▪ Introduction

Continuation, as noted earlier, is both a process and a family of algorithms. As a process, continuation is the tracing of solution branches. Equivalently, continuation is the systematic exploration of how zeros vary with the parameter λ. Branch-tracing is singular at bifurcation points, which by definition are points in the joint space of unknowns and parameters where two or more branches meet. We will describe how to continue beyond bifurcation points later in this chapter.

Continuation is also a family of algorithms in part because iterative methods have the vice that they must be initialized with an approximate solution. Continuation is a systematic strategy for furnishing a first guess; branch-tracing as a goal turns out to also provide essential concepts for the computations that achieve that goal.

The Chebyshev-proxy rootfinder (CPR) does not require a first guess but shares a different defect with Newton's iteration. Both are point-solvers in the sense that they find a single point on a solution branch. Repeated use of a point-solver will generate a point cloud, as computer scientists call it. But how does one connect the dots into continuous branch curves?

The answer is that continuation as the process of branch-tracing must be combined with continuation as algorithm, both to initialize iterations and also to connect the point clouds generated by point-solvers.

Two important issues in continuation are (i) the choice of a marching parameter λ and (ii) the initialization of the iteration at the first value of the parameter.

For many problems, the nonlinear equation becomes linear when the physical parameter $\lambda = 0$. The linear solution then initializes a march in the physical parameter. Sometimes, a low-order perturbation theory or other approximate solution for $\lambda = 0$ will be within the domain of convergence of an underrelaxed Newton's iteration, and the solution for $\lambda = 0$ is again easily obtained.

When a good initialization is not available, a common strategy is *artificial parameter homotopy*. Suppose that $f(x)$ is the problem, and suppose further that we can construct another function $g(x)$ whose solution $\check{x}$ is easily found. We can then define a so-called homotopy, which is an inflated problem that smoothly interpolates between $f(x)$ and $g(x)$ as an artificial parameter λ is varied between zero and one,

$$\mathscr{H}(x,\lambda) = \lambda f(x) + (1-\lambda)g(x). \tag{8.1}$$

When $\lambda = 0$, $\mathscr{H} \to g(x)$, with the known solution $x = \overline{x}$. We can then march in small steps in λ using the stratagems described in the rest of the chapter until we solve the target $f(x) = 0$. The continuation method is serenely indifferent as to whether the continuation parameter is real or artificial. Once $f(x) = 0$ is solved at the end of the march, one can then trace a solution branch by a second march, this time in the physical parameter λ.

However, this simple description of artificial parameter homotopy sloughs over many difficulties. How do we choose $g(x)$? How do we know that there is a smooth path connecting the artificial problem $g(x) = 0$ to the physical problem $f(x) = 0$? We shall return to homotopy in a later section to resolve these issues.

A second difficulty with continuation is that it will self-destruct at a limit point, also known as a fold. How do we pass from a lower branch of solutions around the fold singularity to an upper branch?

Three additional flavors of continuation have been developed to supplement straightforward, fail-at-a-fold continuation of $x(\lambda)$:

1. role-switching continuation,
2. pseudoarclength continuation, and
3. tangent vector continuation.

Role-switching means that λ becomes the unknown and x the parameter in the neighborhood of a fold. Continuation is agnostic as to which of the two variables in $f(x,\lambda)$ is the unknown and which is the parameter. It is perfectly legitimate to compute $\lambda(x)$ instead of $x(\lambda)$ everywhere, and it is advantageous near a fold where $dx/d\lambda = \infty$.

Role-switching continuation has the virtue of great simplicity; one can switch whenever $dx/d\lambda$ exceeds one in magnitude, for example. The drawback is that the end result is not an array of $x(\lambda_j)$ or a table of $\lambda(x_j)$ but a mixture of the two. Interpolation followed by refinement of the interpolated values by Newton's iteration is required to obtain a final answer for the unknown as a multibranched function of the parameter.

A popular strategy to avoid such role-switching is to make both x and λ functions of the arclength s along the solution curve. Because s monotonically increases along the solution curve even around a fold, $(x(s), \lambda(s))$ is singularity free even at a limit point. Arclength and pseudoarclength continuation are described at length in a later section.

There are difficulties with arclength continuation, however, so tangent vector continuation has become popular, as will also be explained below.

First, though, we must describe the basic predictor-corrector paradigm.

Full-length treatments of continuation are given in the books by Dankowicz and Schilder [139], Morgan [279], Allgower and Georg [8], Sommese and Wampler [337], and Bates, Hauenstein, Sommese, and Wampler [36].

8.1.1 ▪ Persistence Continuation

In meteorology, a *persistence forecast* is the prophecy that tomorrow's weather will be the same as today's. Analogously, we may label the simplest form of branch-tracing as "persistence continuation." The parameter λ is discretized as the set $\{\lambda_j\}$. The spacing $h_j \equiv \lambda_{j+1} - \lambda_j$ is entirely up to the user and need not be uniform. The persistence forecast is that the initialization of the computation of $x(\lambda_{j+1})$ is taken to be the last successfully computed solution, $x(\lambda_j)$. Assuming that Newton's iteration is used to refine the crude first guess to near machine precision, persistence continuation is the repetition, $j = 1, 2, \ldots$, of

$$x(\lambda_{j+1})^{(0)} \equiv x(\lambda_j), \tag{8.2}$$

$$x(\lambda_{j+1})^{(n+1)} = x(\lambda_{j+1})^{(n)} - \frac{f(x(\lambda_{j+1})^{(n)}, \lambda_{j+1})}{df/dx(x(\lambda_{j+1})^{(n)}, \lambda_{j+1})}. \tag{8.3}$$

If $|\lambda_{j+1} - \lambda_j|$ is sufficiently small, the persistence forecast will lie within the domain of convergence of Newton's iteration, which will very rapidly home in on the true point $x(\lambda_{j+1})$ on the solution branch.

8.2 ▪ The Predictor-Corrector Framework

Obviously, an iterative point-solver will converge more rapidly and reliably if the prediction of the next point on the curve is more accurate. Many predictors for initializing the iteration are feasible, just as Newton's iteration is only one of a large number of iteration schemes. All of these various methods can, however, be put in a common framework that we shall dub, by analogy with certain popular families of ODE methods, a "predictor-corrector" continuation.

Fig. 8.1 illustrates the predictor-corrector cycles. To generate a new point on the solution at $\lambda = \lambda_{n+1}$, the first task is to *predict* the root at λ from knowledge of the zeros at smaller λ. Then, one *corrects* this initialization by an iteration—Newton's, Halley's, Cauchy's, etc.

One dramatic contrast with predictor-corrector schemes for ODEs is that for ODEs, there is normally no way to avoid a slow, secular, steadily increasing accumulated error as time advances. For rootfinding, however, the corrector iteration purges away all the error accumulated on the march. The corrector of an ODE algorithm is really just another predictor. In contrast, the corrector in continuation is an iteration that enforces the transcendental equation, $f(x(\lambda), \lambda) = 0$, which refines the trajectory as precisely as one pleases at a point in λ, regardless of what has happened elsewhere on the branch of solutions.

Again, it is irrelevant whether the parameter λ is *physical*, such as the amplitude of a traveling wave or the strength of an external forcing, or is an *artificial* parameter chosen to allow us to make a good forecast where no good initial approximation is known.

There are two popular classes of predictors: polynomial/rational extrapolation and differential equation in the parameter (Davidenko) methods. We shall discuss each in turn in the next two sections.

Bates, Hauenstein, Sommese, and Wampler give a good discussion of path-following in [35, 36]. It is noteworthy that their Bertini system automatically shifts to multiprecision arithmetic as needed—a red flag that almost-intersecting paths are not unusual in applications. A good predictor is therefore important, as discussed next.

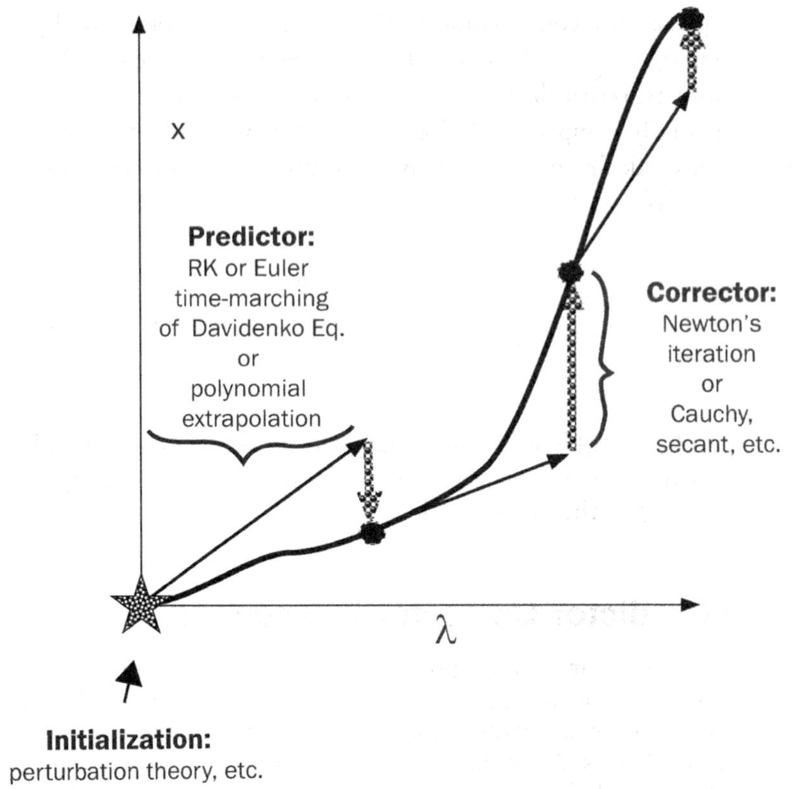

Figure 8.1. *Schematic of initialization/continuation/prediction cycle. The first step is to somehow—by perturbation theory, bold guess, exact solution for a special parameter value, artificial parameter homotopy, and so on—obtain at least an approximation to the solution for some parameter value $\lambda = \lambda_0$. (This special starting value is marked by the speckled star.) One may then predict the solution at the next discrete value of λ using* (i) *Euler forward or a Runge–Kutta integration of the Davidenko equation or* (ii) *persistence or higher-order polynomial extrapolation. (The grid spacing in λ is up to the user and can be adaptively varied.) Newton's iteration or a variant (vertical arrow with speckles) is the corrector, pulling the first guess back to the true solution branch.*

8.3 ▪ Polynomial and Rational Extrapolation

Polynomial predictors include the following:

$$x_{j+1} \approx x_j \qquad \text{[Constant Extrapolation (Persistence Forecast)]} \tag{8.4}$$

$$\approx x_j + \frac{\left(x_j - x_{j-1}\right)\left(\lambda_{j+1} - \lambda_j\right)}{\lambda_j - \lambda_{j-1}} \qquad \text{[Linear (Secant) Extrapolation]} \tag{8.5}$$

$$\approx \frac{x_j\left(\lambda_{j+1} - \lambda_{j-1}\right)\left(\lambda_{j+1} - \lambda_{j-2}\right)}{\left(\lambda_j - \lambda_{j-1}\right)\left(\lambda_j - \lambda_{j-2}\right)} + \frac{x_{j-1}\left(\lambda_{j+1} - \lambda_j\right)\left(\lambda_{j+1} - \lambda_{j-2}\right)}{\left(\lambda_{j-1} - \lambda_j\right)\left(\lambda_{j-1} - \lambda_{j-2}\right)}$$
$$+ \frac{x_{j-2}\left(\lambda_{j+1} - \lambda_j\right)\left(\lambda_{j+1} - \lambda_{j-1}\right)}{\left(\lambda_{j-2} - \lambda_j\right)\left(\lambda_{j-2} - \lambda_{j-1}\right)} \qquad \text{[Quadratic Extrapolation]} \tag{8.6}$$

$$\approx \sum_{m=1}^{N+1} x(\lambda_{j+1-m}) \frac{\prod_{k=1,k\neq m}^{N+1} (\lambda_{j+1} - \lambda_{j+1-k})}{\prod_{k=1,k\neq m}^{N+1} (\lambda_{j+1-m} - \lambda_{j+1-k})} \text{ [General Degree } N\text{].} \tag{8.7}$$

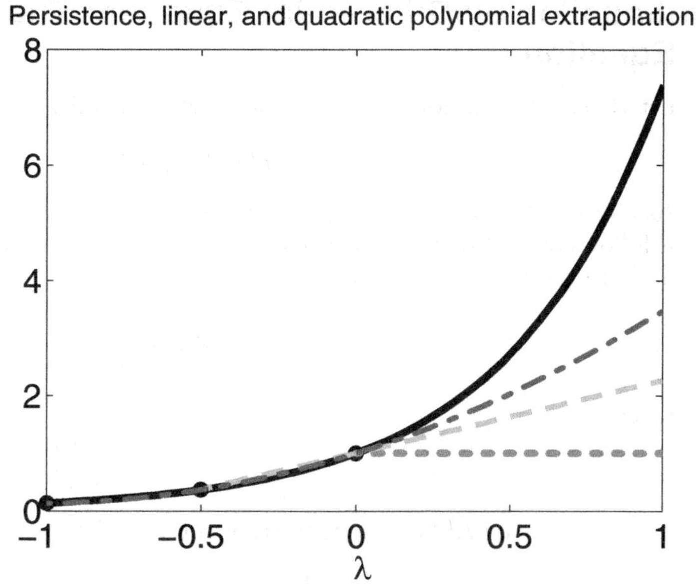

Figure 8.2. *Schematic of prediction by polynomial extrapolation. The uppermost curve (solid) shows the true variation of the function $x(\lambda)$ that the other curves are approximating. The disks show the interpolation points used to generate the polynomial extrapolations. Persistence (piecewise constant interpolation): horizontal dotted line. Linear interpolation: dashes. Quadratic interpolation: dot-dash curve. Persistence uses only the rightmost point, while linear extrapolation uses the middle and right points, and the quadratic extrapolation uses all three previously computed values.*

Polynomial interpolation predictors are illustrated in Fig. 8.2. Increasing the degree of the polynomial interpolant increases accuracy, but only up to a point. High-order interpolation on an evenly spaced grid is well known to be numerically unstable.

On the other hand, polynomial interpolation through a few previously computed values is very cheap. The strategy that underlies adaptive Runge–Kutta algorithms can be combined with polynomial interpolation, too. By computing approximations of, say, fourth and fifth order and comparing them, one can estimate the interpolation error. The software can then increase or decrease the step in λ so as to keep the interpolation error within a preset user tolerance without excessive cost.

Unfortunately, library software to implement adaptive polynomial extrapolation is scarce. However, it would be a fairly modest alteration of existing adaptive Runge–Kutta algorithms. High-order, adaptive polynomial extrapolation would seem to offer the best combination of low cost and reliability.

Many of the author's own studies have used a little subroutine that is called twice at each step in λ: once at the bottom of the loop to update the subroutine's arrays and once at the top of the loop to forecast $x(\lambda_n)$. The routine automatically switches from persistence to linear extrapolation to higher-order extrapolation as a sufficient number of previously computed solutions becomes available. Variable grid spacing in λ is easy; the Lagrangian formula for polynomial interpolation is well defined for an arbitrary distribution of interpolation points.

It is of course always feasible to fit a rational function instead of a polynomial to previously computed values. Recent variants of rational interpolation are constructed to be pole free on the approximation interval [41, 173].

8.4 ▪ Continuation by Differential Equation: The Davidenko Equation

Let $x(\lambda)$ be the solution to an algebraic equation that depends on a parameter λ:

$$f(x[\lambda]; \lambda) = 0. \tag{8.8}$$

As λ varies, the value of f along the curve $f(x(\lambda); \lambda)$ does not vary since it is always zero. It follows that $df/d\lambda = 0$ when x is chosen to vary with λ so as to be a root. This total derivative of f with respect to λ can be expressed in terms of partial derivatives as

$$\frac{df}{d\lambda} = f_\lambda + f_x \frac{dx}{d\lambda}, \tag{8.9}$$

where

$$f_\lambda = \left.\frac{\partial f}{\partial \lambda}\right|_{x=constant}, \qquad f_x = \left.\frac{\partial f}{\partial x}\right|_{\lambda=constant}. \tag{8.10}$$

The condition $df/d\lambda = 0$ along the root curve implies

$$\frac{dx}{d\lambda} = -\frac{1}{f_x} f_\lambda \qquad \text{[Davidenko Equation].} \tag{8.11}$$

This ODE in λ is called the Davidenko equation [141, 142].[15] This equation applies when f and x are either scalars or vectors. When x is a vector, the multiplication by the reciprocal of f_x should be interpreted as multiplication by the inverse of the Jacobian matrix, and the Davidenko equation is a system of ODEs.

The solution to the Davidenko equation is the *Davidenko flow*. In some ways, this is misleading because the algebraic equation $f(x) = 0$ usually has nothing to do with hydrodynamics. In other respects, the word "flow" is merited. In contrast to the static imagery of nineteenth-century rootfinding, where one or two numbers were extracted, through laborious pen-and-ink calculations, for fixed values of all parameters, the modern emphasis is on the continuous variation of roots as the parameter varies.

Example 1: Quadratic Equation
Define

$$f(x, \lambda) \equiv x^2 - 2\mu x + \lambda, \tag{8.12}$$

where μ is a second parameter. The exact roots are

$$x = \mu \pm \sqrt{\mu^2 - \lambda} \tag{8.13}$$

by the usual formula. These roots can also be calculated by solving the Davidenko differential equation.

By explicit partial differentiation,

$$f_x = 2x - 2\mu; \qquad f_\lambda = 1. \tag{8.14}$$

The Davidenko equation for this rootsolving problem is

$$\frac{dx}{d\lambda} = -\frac{f_\lambda}{f_x} = -\frac{1}{2(x-\mu)}. \tag{8.15}$$

[15] Zangwill and Garcia call this the homotopy differential equation, or HDE for short, but this label is no longer common [383].

At $\lambda = 0$, the roots are

$$x(0) = 2\mu, 0. \tag{8.16}$$

Integrating the Davidenko equation with the initial condition $x(0) = 2\mu$ generates the plus sign in (8.13); integrating from $x(0) = 0$ generates the root $x = \mu - \sqrt{\mu^2 - \lambda}$.

8.4.1 ▪ Virtues and Vices

One disadvantage of the Davidenko equation is that errors in solving it inevitably accumulate as one marches in λ. For this reason, the Davidenko equation is used mostly as a *predictor* in a predictor-corrector scheme. The Davidenko equation is integrated over a moderate interval in λ. The result is corrected by a Newton or quasi-Newton iteration. The accuracy of the correction is limited only by roundoff error; in the Newton iteration, the root is purged of all the error accumulated during the integration of the Davidenko differential equation.

Newton's method has the flaw that the iteration will fail if the first iterate is too far from the root. The Davidenko equation is a good partner for Newton's method because it allows one to generate an initialization of arbitrarily high accuracy for the corrector, neglecting roundoff.

Unfortunately, the Davidenko equation shares a vice of Newton's method: the Davidenko equation is singular at limit points or bifurcation points or wherever df/dx is zero.

An adaptive Runge–Kutta method, in which the fourth- and fifth-order methods are computed simultaneously, is relatively expensive because four additional evaluations of $f(x)$ per step are required. However, by comparing the fourth- and fifth-order approximations to estimate the error, the fourth/fifth-order Runge–Kutta scheme is able to adaptively vary the step in λ to maintain a given tolerance. Adaptive Runge–Kutta–Davidenko marching is extremely robust because of its inherent adaptivity.

In spite of the extra cost, the Davidenko equation is very powerful because all the theory, algorithms, and software for ODEs can be applied to the seemingly unrelated problem of solving a nonlinear algebraic system.

8.4.2 ▪ Continuation without Iteration: Examples of Pure Davidenko Rootfinding

Although we recommend against this in general, λ-marching the Davidenko equation is a fairly effective way to find roots even without a corrector. Fig. 8.3 illustrates this for the Kepler equation of celestial mechanics and also shows the rewards of applying Newton's method as a corrector to improve on the predictions of the initial value method. Wasserstrom successfully used pure initial value integration of the Davidenko equation to find all roots of a polynomial [363, 364].

The Davidenko method in particular and continuation in general can be applied with *artificial* parameters as well as physical ones. To illustrate this, observe that we can solve the quintic $x^5 + x - 1 = 0$ for the root nearest the origin by using any of three different artificial homotopies:

$$\begin{aligned} \lambda x^5 + x - 1 &= 0 \quad \text{[Weak Coupling]}, \\ x^5 + \lambda x - 1 &= 0 \quad \text{[Strong Coupling]}, \\ x^{1+4\lambda} + x - 1 &= 0 \quad \text{[Bender et al.]}, \end{aligned} \tag{8.17}$$

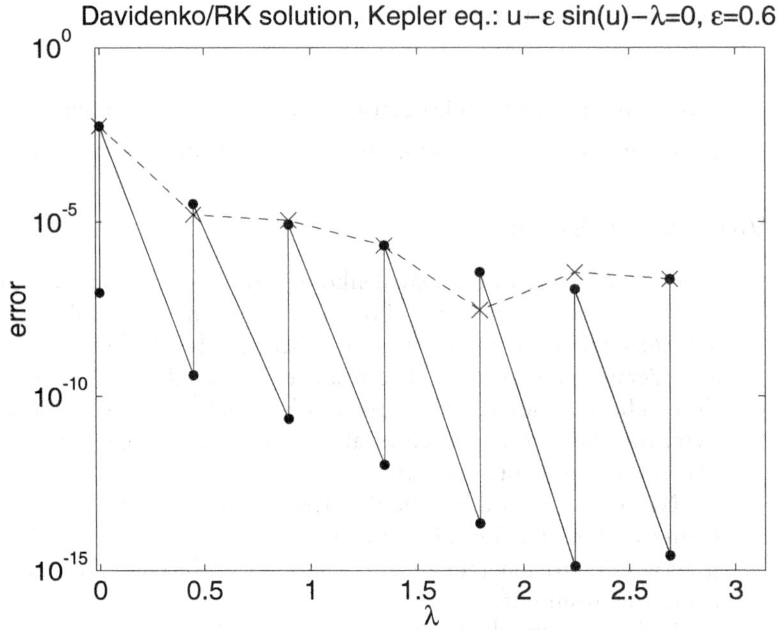

Figure 8.3. *The Kepler equation, $x - \epsilon \sin(x) - \lambda = 0$, is equivalent to the Davidenko equation (for fixed ϵ) $dx/d\lambda = 1/(1-\epsilon \cos(x))$. It can be verified that $x = \pi$ for $\lambda = \pi$ for all ϵ. The Kepler–Davidenko equation was solved twice in the* negative *λ direction from $\lambda = \pi$ to $\lambda = 0$. The dashed line with x's shows the errors in a fourth-order Runge–Kutta method in a pure initial value approach for $\epsilon = 3/5$. The errors (uncorrected) accumulate and grow as we march leftward. The zig-zag solid curve shows the errors when the Runge–Kutta method is used as the predictor and a single Newton's iteration is used as the corrector. (The bottom disk in each pair of vertically aligned disks is the error after one Newton's iteration.) The Euler predictor is cheaper per step, but experiments (not illustrated) showed that, with steps of $\pi/7$ in λ, it required several Newton's iterations per λ step, thus making it roughly as expensive as the Runge–Kutta method for the stepsize and equation illustrated.*

where the labels are borrowed from Bender, Milton, Pinsky, and Simmons [38]. These authors actually suggested these three homotopies for constructing perturbation series in powers of λ, not numerical solutions. This illustrates the principle: Perturbation parameters are good continuation parameters!

The Davidenko equation for the weak coupling homotopy is

$$\frac{dx}{d\lambda} = -\frac{x^5}{5x^4+1}, \qquad x(0)=1, \qquad \lambda \in [0,1]. \tag{8.18}$$

Note that *no approximation* is made in the Davidenko equation; except for time-marching errors in numerically solving the differential equation, the desired root is $x(\lambda = 1)$. The same is true, with different differential equations, for other homotopies as well. Fig. 8.3 shows a similar exercise for the Kepler equation.

The most dramatic way to demonstrate the power of continuation is to use the Davidenko equation *without correction* to solve nonlinear equations. Table 8.1 compares the errors in numerically integrating the Davidenko equation for each of the three homotopies described earlier for the quintic equation. All three, as well as a great diversity of other homotopies, generate very accurate roots even without Newton's iteration. From a practical standpoint, however, all homotopies are not equal: the *obvious* homotopy,

Table 8.1. *Errors in the numerical integration of the Davidenko equation. A comparison of pure Davidenko flow calculations (without Newton's iteration correction!) for three different homotopies for $x^5 + x - 1 = 0$ and two different time-marching schemes, fourth-order Runge–Kutta and first-order Euler. The stepsize in the parameter λ is h. The approximate roots are in columns two and four; the absolute errors are in columns three and five.*

Weak Coupling Homotopy				
h	RK4:Root	RK4:Error	Euler:Root	Euler: Error
1	0.718	−0.037	0	−0.75
1/2	0.7482	−6.7E−3	0.49	−0.27
1/4	0.75392	−9.6E−4	0.66	−0.094
1/8	0.75478	−9.8E−5	0.72	−0.039
1/16	0.75487	−7.4E−6	0.738	−0.017
1/32	0.754877	−4.8E−7	0.746	−0.0082
1/64	0.754878	−3.3E−8	0.7509	−0.0040
1/128	0.754878	−5.5E−9	0.7529	−0.0020
Strong Coupling Homotopy				
h	RK4:Root	RK4:Error	Euler:Root	Euler: Error
1	0.75520	3.2E−4	0.8	4.5E−2
1/2	0.754897	1.9E−5	0.781	2.6E−2
1/4	0.754879	1.1E−6	0.769	1.4E−2
1/8	0.754878	5.9E−8	0.762	7.4E−3
Bender et al. Homotopy				
h	RK4:Root	RK4:Error	Euler:Root	Euler: Error
1	0.746	−8.3E−3	1.19	0.44
1/2	0.75505	1.7E−4	0.91	0.16
1/4	0.754937	5.9E−5	0.83	0.070
1/8	0.754883	5.7E−6	0.788	0.033
1/16	0.754878	4.1E−7	0.771	0.016
1/32	0.754878	2.4E−8	0.763	0.0078

which is the weak coupling scheme, is less efficient than the other two in (8.17), which require more subtle justification.

However, such success is tricky and expensive. Note that a pure Davidenko integration must be initiated with either an exact solution or a very accurate approximate solution.

We recommend against using the Davidenko equation as a solo act, but instead advise a duet with a Newton corrector. The power of Newton's iteration is that it is self-correcting. It does not matter if the first guess for the Newton corrector has a few percent error; the iteration will purge it so long as the first guess lies within the domain of convergence of the iteration. It does not matter if df/dx is approximated by computing the derivative by finite differences. Newton's iteration will remove all traces of these approximations from the final answer.

Still, the fact that pure Davidenko computations are possible at all emphasizes that solving the Davidenko equation, especially by an adaptive high-order algorithm, is a very powerful and accurate predictor for continuation.

Table 8.2 summarizes the virtues and vices of a variety of predictors.

Table 8.2. *Predictors for continuation.*

Method	Comment
Persistence: $x(\lambda_{m+1}) \approx x(\lambda_m)$	simplest but least accurate
Higher-Degree Polynomial Extrapolation	easy to fit polynomial in λ through $\{x(\lambda_m), x(\lambda_{m-1}), x(\lambda_{m-2}), \ldots\}$; can accommodate changing stepsize in λ
Euler–Davidenko	Euler forward time integration of Davidenko eq.; cheap: df/dx, already available from computation of $x(\lambda_m)$, is used to predict $x(\lambda_{m+1})$
Adams–Bashforth–Davidenko	higher order than Euler and no more expensive but requires constant stepsize in λ
Runge–Kutta–Davidenko	extra evaluations of df/dx; therefore expensive; more accurate than Euler and allows variable λ-steps

8.5 ▪ Artificial Parameter Homotopy Initialization

The predictors have a flaw: they are useless until a good guess is obtained by some other means for the first point on the branch. Often, the problem can be solved analytically for a limiting value of the parameter. Equally often, it seems, no such solution is available. A useful strategy is to apply continuation in an artificial parameter, solving an inflated problem that interpolates between the target function $f(x)$ and a function $g(x)$ whose roots are known.

Definition 8.1 (Homotopy Method). *When an artificial parameter λ is added to a system of nonlinear equations to define an inflated problem such that* (i) $\lambda = 0$ *is analytically solvable and* (ii) $\lambda = 1$ *is the same as the original problem, the continuation method is usually called a* **homotopy method**. *The inflated system of equations is dubbed a* **homotopy** *of the original problem.*

Zangwill and Garcia define three popular choices of artificial homotopy, where the homotopy is denoted by $\mathscr{H}$ [383] and the unknown may be a vector:

$$\mathscr{H}(x;\lambda) \equiv f(x) - (1-\lambda) f(x^{(0)}) \text{ [Newton Homotopy]}, \tag{8.19}$$

$$\mathscr{H}(x;\lambda) \equiv (1-\lambda)(x - x^{(0)}) + \lambda f(x) \text{ [Fixed Point Homotopy]}, \tag{8.20}$$

$$\mathscr{H}(x;\lambda) \equiv \lambda f(x) + (1-\lambda)\, g(x) \text{ [Linear Homotopy]}, \tag{8.21}$$

where the linear homotopy contains an arbitrary function $g(x)$. However, there is an almost limitless range of possibilities.

To illustrate, suppose the goal is to solve $f(x) = x^3 - 2x + \sqrt{2} = 0$. Finding the zeros of a polynomial equation is messy and challenging in general, but an exception is the equation whose solutions are the N roots of unity:

$$x^N - 1 = 0 \quad \Leftrightarrow \quad x = \exp\left(2\pi i \frac{j}{N}\right), \qquad j = 0, 1, \ldots, (N-1). \tag{8.22}$$

Let us apply the linear homotopy with $g(x)$ equal to the $N = 3$ instance of this polynomial. The linear homotopy is then

$$h(x,\lambda) \equiv \lambda\left(x^3 - 2x + \sqrt{2}\right) + (1-\lambda)\left(x^3 - 1\right) = 0. \tag{8.23}$$

The trajectories of the roots for $\lambda \in [0,1]$ are illustrated in Fig. 8.4.

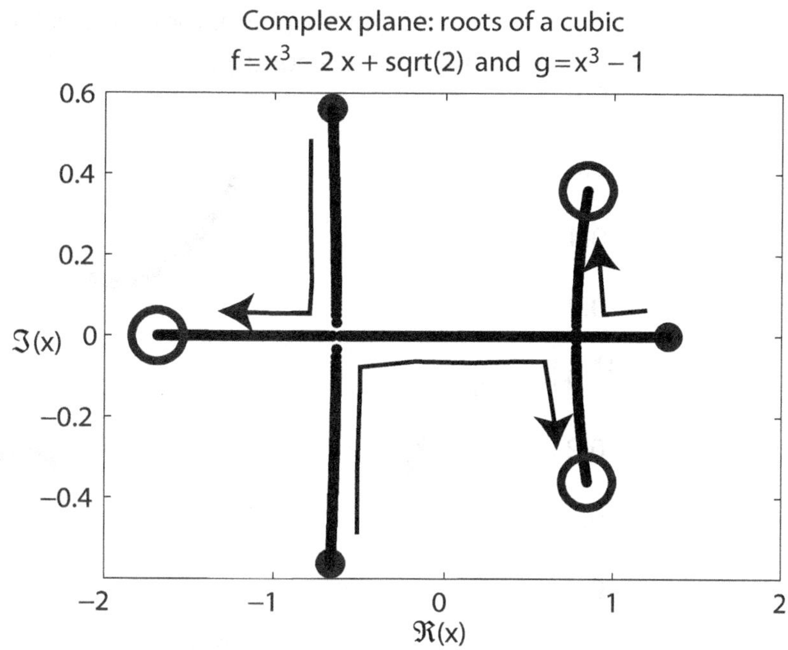

Figure 8.4. *One artificial parameter homotopy to solve* $f(x) = x^3 - 2x + \sqrt{2} = 0$ *is to use the auxiliary problem* $g(x) = x^3 - 1$ *whose roots are three roots of unity,* $(1, \exp(i(2/3)\pi), \exp(-i(2/3)\pi))$. *The three roots of* $f(x)$ *are the large hollow circles; the zeros of* $g(x)$ *are the disks. Unfortunately, the trajectories beginning at the complex roots of* $g(x)$ collide *on the real axis; one trajectory then moves left to become the real root of* $f(x)$. *The other moves right to a second collision, this time with the trajectory beginning at the real root of* $g(x)$*; these two trajectories then split to become the two complex zeros of the target,* $f(x)$.

Unfortunately, this homotopy is unsatisfactory because roots collide as λ is varied. Each of the collision points is a bifurcation point where $f_x = 0$, Newton's iteration fails, and the Davidenko equation is singular. This is a very common situation because f_x generally has many roots along the real λ-axis.[16] There is a good chance that the homotopy interval will include one of these roots, and then the homotopy fails. Pseudoarclength continuation around the fold is no help because it will merely turn us back toward smaller λ.

8.5.1 ▪ Complex Plane Continuation

The remedy is to create $g(x)$ so that its coefficients are *complex valued* and *random*. For example, we can replace $g(x)$ in our cubic polynomial example by

$$g \equiv \{x^3 - b\} = 0, \tag{8.24}$$

where, instead of one, b is a randomly chosen *complex* number. At $\lambda = 0$, the roots are simply the cube roots of b, i.e., if $b = r\exp(i\phi)$ for real r and ϕ, then the roots are $x = r^{1/3}\exp(i\phi/3 + i2\pi k/3)$, where $k = 0, 1, 2$. The roots are complex for all $\lambda \in [0, 1]$ except at $\lambda = 1$. This reduces the root-collision probability to infinitesimal (a set of

[16]For systems of many equations in many unknowns, the determinant of the Jacobian matrix is a scalar-valued function and so is equally prone to a root on $\lambda \in [0, 1]$.

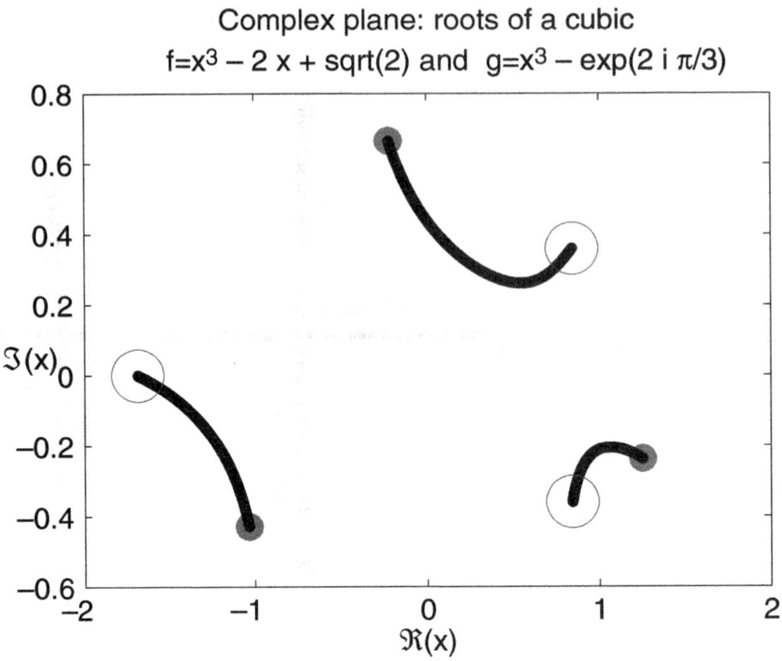

Figure 8.5. *Same as previous graph except that $g(x)$ was altered to $x^3 - \exp(i(2/3)\pi$ so that all three of its roots are complex. This change in $g(x)$ yields roots that smoothly move from the zeros of $g(x)$ to those of $f(x)$ without collisions.*

measure zero). At a collision, the derivative of $f(x)$ must be singular so that the implicit function theorem and its prediction of a single solution at a single point is broken. However, $df/dx = 0$ only at *isolated* points in the complex x-plane, whereas the trajectories are one-dimensional curves in a two-dimensional $(\Re(x), \Im(x))$-plane. Fig. 8.5 shows how this trick eliminates the trajectory collisions.

Metaphorically, a homotopy with polynomials whose coefficients are real is like a train system, confined to tracks and sometimes colliding head-on. Complex-valued coefficients liberate the trajectories to go off road like tanks or all terrain vehicles.

The technical jargon is that the complex-value randomized homotopy $f(x) + (1-\lambda)g(x)$ is a "probability-one homotopy" [243, 278]. What "probability-one" means is that if one picks the coefficients of $g(x)$ as random *complex-valued* numbers, the probability of root collision is infinitesimal, in theory. The random complex zeros for $g(x)$ ensure homotopy trajectories that are tank-like paths over the complex plane instead of train-like trajectories imprisoned on the real axis. In practice, *near*-collisions can cause difficulties in finite precision arithmetic, but this is unusual. Careful path-following helps, but the unfailing remedy is to repeat with a different $g(x)$.

Such complex plane continuation is a powerful tool for obtaining complete solutions to polynomials and systems of low degree. (Unfortunately, the relevant articles, which are mostly written by mathematicians, specify that the unknowns lie in the space of complex variables without emphasizing the importance of keeping the homotopy paths off the real axis in the space of the unknowns x.) Li [243] is a very readable and elementary introduction; Gelaro [190] offers a carefully worked example. Alexander and Yorke [7] seem to have originated this idea. Wasserstrom's review, Morgan's book and papers, and Henderson and Keller are good explications of complex plane continuation for a single polynomial equation [364], polynomial systems [279, 277, 276], and general systems of nonlinear equations [207], respectively.

8.6 ▪ Pseudoarclength and Tangent Vector Continuation

As noted earlier, a major problem for continuation in a physical parameter λ is the existence of limit points where $dx/d\lambda \to \infty$. In pseudoarclength continuation, both x and λ are taken as functions of a new parameter s that approximates arclength along the solution curve in the x-λ plane. The gradient $(dx/ds, d\lambda/ds)$ is always finite so there are no infinities.

The arclength is defined by approximating the solution curve $s(\lambda)$ by many short line segments. In the limit that the segments become infinitely short, the sum of the length of the segments between two points on the curve is the arclength between those points.

Let Δx and $\Delta\lambda$ denote changes in x and λ between the ends of the segment. As the segment becomes shorter and shorter, it becomes better and better approximated by a straight line segment of length Δs. By definition, Δs is the arclength of the segment. The changes in x and λ and the length of the line are connected by the Pythagorean theorem,

$$(\Delta s)^2 = (\Delta x)^2 + (\Delta\lambda)^2 \qquad \text{[Arclength Constraint]}. \tag{8.25}$$

In the 1970s, Herbert Keller devised a strategy he dubbed "pseudoarclength continuation" to smoothly follow solution curves around limit points. Calculating arclength accurately is unnecessary; fairly crude approximations—pseudoarclength—are quite sufficient [223]. Good descriptions can be found in [319] and Appendix D of [76]. However, pseudoarclength continuation has been largely replaced by a simpler strategy called tangent vector continuation [8]. When there is only a single unknown, the two strategies are effectively the same.

Like pseudoarclength continuation, the tangent vector scheme treats the unknown and the parameter on an equal basis. If t is any parameter for parameterizing the solution curve $(x(t), \lambda(t))$, the total derivative of $f(x(t); t)$ is

$$\frac{df}{dt} = f_x \frac{dx}{dt} + f_\lambda \frac{d\lambda}{dt} = 0, \tag{8.26}$$

which shows that the unit tangent vector to the solution curve is

$$\hat{t} = \sigma \frac{f_\lambda}{\sqrt{(f_x)^2 + (f_\lambda)^2}}, -\sigma \frac{f_x}{\sqrt{(f_x)^2 + (f_\lambda)^2}}, \tag{8.27}$$

where the sign $\sigma = \pm 1$ depends on whether the tangent vector is chosen to be oriented in the direction of increasing λ or in the opposite direction. (The denominators normalize the tangent vector to unit length.)

A small step of length τ in the tangent direction gives the predictor

$$\left| \begin{array}{c} x_{n+1} - x_n \\ \lambda_{n+1} - \lambda_n \end{array} \right| = \left| \begin{array}{c} \sigma \dfrac{f_\lambda}{\sqrt{(f_x)^2 + (f_\lambda)^2}} \\ -\sigma \dfrac{f_x}{\sqrt{(f_x)^2 + (f_\lambda)^2}} \end{array} \right| \tau, \tag{8.28}$$

where all functions on the right are evaluated at (x_n, λ_n). Comparison with the arclength condition then shows that $\tau = \delta s$, that is, this predictor is a step of arclength τ along the curve $x(\lambda)$.

If we continue to treat x and λ as independent parameters, then a first-order Taylor series expansion about the prediction (x_c, y_c) gives the Newton correction

$$f_x \delta^{(x)} + f_\lambda \delta^{(\lambda)} = -f(x_c, \lambda_c). \tag{8.29}$$

This defines a line, not a point, in the two-dimensional space spanned by $(\delta^{(x)}, \delta^{(\lambda)})$. To remove the ambiguity, we demand that the correction should be perpendicular to the tangent vector, thus taking the shortest path from the predicted point to the solution curve. This gives the system

$$\begin{vmatrix} f_x & f_\lambda \\ -f_\lambda & f_x \end{vmatrix} \begin{vmatrix} \delta^{(x)} \\ \delta^{(\lambda)} \end{vmatrix} = \begin{vmatrix} -f \\ 0 \end{vmatrix}. \tag{8.30}$$

The solution is

$$\begin{vmatrix} \delta^{(x)} \\ \delta^{(\lambda)} \end{vmatrix} = \frac{-f}{(f_x)^2 + (f_\lambda)^2} \begin{vmatrix} f_x \\ f_\lambda \end{vmatrix}. \tag{8.31}$$

When there are N unknowns, the so-called augmented Jacobian matrix is of dimension $N \times (N+1)$. The underdetermined Newton's iteration is, with $\vec{x} = (x_1, \ldots, x_N, \lambda)^T$,

$$\vec{x}^{(m+1)} = \vec{x}^{(m)} - \vec{\vec{J}}^{+} \vec{f}, \tag{8.32}$$

where $\vec{\vec{J}}^{+}$ is the Moore–Penrose pseudoinverse, defined by $\vec{\vec{J}}^{+} \equiv \vec{\vec{J}}^{T} (\vec{\vec{J}} \vec{\vec{J}}^{T})^{-1}$. When there is only one unknown, $\vec{\vec{J}}$ is the two-element vector (f_x, f_λ), $\vec{\vec{J}}\vec{\vec{J}}^{T} = (f_x)^2 + (f_\lambda)^2$, and $\vec{\vec{J}}^{+} = (f_x/\{(f_x)^2 + (f_\lambda)^2\}, f_\lambda/\{(f_x)^2 + (f_\lambda)^2\})^T$. Thus, the tangent vector Newton iteration is identical to (8.31). Allgower and Georg [8, p. 19] show that the Newton correction $-\vec{\vec{J}}^{+} \vec{f}$ (for any number of unknowns N) is, among the one-parameter family of vectors solving (8.30), the vector of smallest norm. It is by construction orthogonal to the tangent vector and is a movement from the predicted point to the *nearest* point on the solution curve.

Seydel is a good treatment of continuation, limit points, and bifurcation points that assumes only an undergraduate mathematical background: ODEs, partial derivatives, and matrix algebra [318]. Keller is an outstanding mix of review and research papers at a higher mathematical level [223]. Doedel, Keller, and Kernévez describe the theory behind the publicly available AUTO continuation method [157]. MatCont is a more recent software library described in [155]. Rheinboldt is a short historical review through the year 2000 [299]. His book [298] combines FORTRAN code and practical examples with advanced mathematical analysis.

8.7 ▪ Transcritical Bifurcation Point

There are two options for coping with a transcritical bifurcation point. One is to apply a Cauchy or Shafer iteration. Both iterations require solving quadratic equations; as the singularity is approached, both zeros of these quadratics will be good approximations to the intersecting branches. This is illustrated in Fig. 8.6. Because the slope of a branch, $dx/d\lambda$, remains finite, there is no obstacle to straightforward continuation in λ, and there is no difficulty in staying on the same branch from one side of the bifurcation point to the other. As shown in the figure, the second Cauchy root will converge to the second branch as the bifurcation point is approached; this second root can therefore be used as a first guess to switch to the other branch.

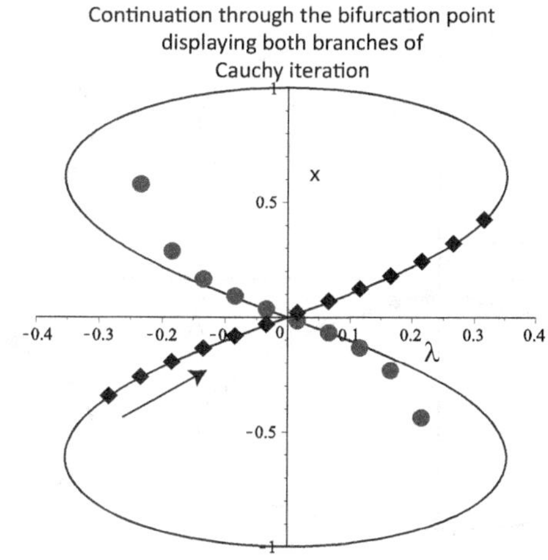

Figure 8.6. *The figure-eight curve is the zero isoline of* $f(x,\lambda) = (x^2+\lambda^2)^2 - (x^2-\lambda^2)$, *which is an instance of the lemniscate of Bernoulli. Marching in steps of* $1/20$ *in* λ *from* $(-\sqrt{2}/3, -1/3)$ *in the direction indicated by the arrow, the linear polynomial predictor had no trouble staying on a single branch even through the bifurcation point* $(0,0)$*; the converged values at each step are marked by diamonds. The second root of the Cauchy iteration at each point in* λ *is marked by the circles.*

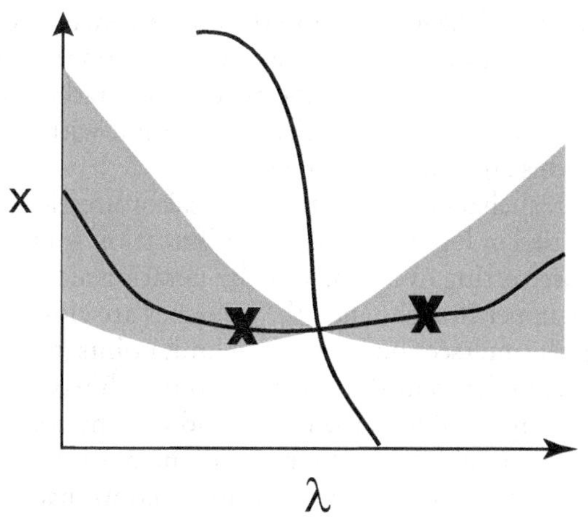

Figure 8.7. *Schematic of shooting a bifurcation point. Newton's iteration converges when the first guess is within the shaded region. Unfortunately, this convergence region shrinks to zero width at the bifurcation point. However, linear extrapolation from the left x will give a first guess as indicated by the right x. This lies within the convergence zone on the far side of the bifurcation point. Newton's iteration will converge from a guess of the point marked by the right x, and one may then apply continuation to trace out the whole solution branch to the right of the bifurcation point. The only complication is the issue of how far to proceed along the straight line when extrapolating beyond the bifurcation point so as to land within the Newton convergence domain (shaded); some trial and error may be needed.*

The second approach is to "shoot the bifurcation," H. Keller's phrase evocative of a river raft shooting the rapids, using only Newton's iteration (Fig. 8.7). Although the interval of convergence for Newton's iteration shrinks to zero at the bifurcation point, the smooth variation of $dx/d\lambda$ near the bifurcation point implies that a decent predictor can

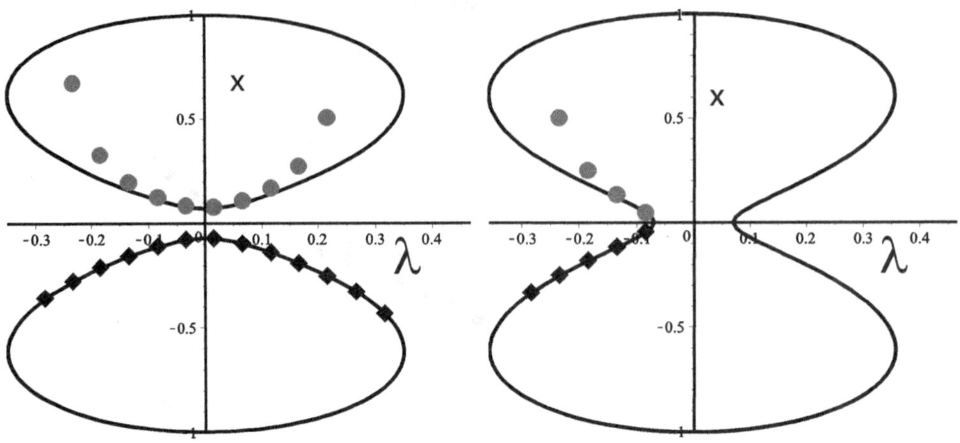

Figure 8.8. *Same as Fig. 8.6 except that $f(x,\lambda)$ for the lemniscate is perturbed. Left: two unconnected ovals result when the function f of the lemniscate of Bernoulli is perturbed by adding $1/200$ so that $f(x,\lambda)=(x^2+\lambda^2)^2-(x^2-\lambda^2)+1/200$. Right: subtracting the same small constant deforms the lemniscate into a vase-shaped curve defined implicitly by $f(x,\lambda)=(x^2+\lambda^2)^2-(x^2-\lambda^2)-1/200=0$; the transcritical bifurcation point has been replaced by a pair of folds.*

provide an initialization within the convergence region on the far side of the bifurcation from a point rather far from the bifurcation on the near side.

Cauchy iteration seems the better way, rather than Newton's. But the possibility of continuing past a transcritical bifurcation with Newton's iteration is a sign that such bifurcations are relatively benign.

The real challenge with transcritical bifurcations is their lack of structural stability, as illustrated in Fig. 8.8. When the bifurcation splits the intersecting branches into a pair of nonintersecting ovals, the Cauchy iteration along the lower branch still accurately predicts the upper branch when the branches are close. A perturbation of the opposite sign replaces the intersection by a pair of limit points (folds). The roots of the Cauchy iteration become complex valued if one tries to march past the fold through simple continuation in λ, but it is possible to use the second Cauchy root to continue onto the upper branch.

To jump to a different branch, all one needs is a single point on $x_2(\lambda)$; one may then trace the rest of the branch by ordinary continuation. Seydel and Keller are good references on coping with bifurcations of systems of N equations in N unknowns with $N>1$ [319, 223].

Part IV

Polynomials

Chapter 9

Polynomial Equations and the Irony of Galois Theory

"There is no problem that cannot be solved."
—François Vieta in *Introduction to the Analytical Art* (1591)

"None of the methods [Jenkins–Traub, Laguerre's modified algorithm, and companion (Frobenius) matrix methods] gives acceptable results for polynomials of degree higher than 50 .... If roots of high multiplicity exist, any ... method must be used with caution."
—S. Goedecker [197]

9.1 ▪ "The Great Art": The Quest for Solutions by Radicals

A univariate polynomial is

$$P(x) \equiv \sum_{j=0}^{N} a_j \, x^j, \tag{9.1}$$

where N is the degree of the polynomial. The solution of the quadratic, familiar to every high school student, was found by the Babylonians more than 2500 years ago. The cuneiform tablets, and the writings of Arab mathematicians including the poet Omar Khayyam (ca. 1100 A.D.), show methods for special classes of cubics ($N = 3$). The general solution to the cubic was found by the Italian mathematician Niccolo Fontana, alias Tartaglia (1500–1557), and published by Girolamo Cardano (1501–1576) in his *Ars Magna* (1545). The general quartic ($N = 4$) fell to Lodovico Ferrari (1522–1565) and was also published in *Ars Magna*, *"The Great Art."*

This success inspired a quarter-of-a-millennium quest for an algebraic solution to general polynomial equations of degree five or higher. An *algebraic solution* is defined to be an explicit, analytic formula that involves only a *finite* number of square roots, cube roots, fourth roots, etc., plus the usual four operations of arithmetic. Paolo Ruffini (1765–1822) and Niels Henrik Abel (1802–1829) proved that an algebraic solution to the general quintic was *impossible*; Evaristé Galois (1811–1832) proved that a finite number of algebraic operations would always fail, except for special cases, for any polynomial equation of degree $N > 4$. Analytic solutions for the quintic and higher degree can only be expressed in terms of *transcendental* functions, that is, functions that can only be evaluated exactly by an *infinite* series or an infinite sequence of iterations.

Charles Hermite (1822–1905) solved the general quintic in terms of elliptic modular functions. Camille Jordan (1838–1922) proved that polynomial equations of any order can be solved by modular functions (Jordan [218]; Umemura [356]).

From a modern engineering perspective, the "Great Art" of the Babylonians, Renaissance Italians, Abel, Galois, Hermite, and Jordan seems misguided, an evolutionary dead end like the Neanderthals. To be sure, Galois created group theory, a very valuable branch of mathematics, in proving his theorem, and the modular solutions of higher-degree equations advanced the theory of elliptic functions. For practical engineering, however, explicit polynomial solutions are always used for the quadratic equation, sometimes for the cubic, rarely for the quartic, and never at all for the quintic. Nevertheless, the "Great Art" had acquired such momentum over three centuries that it continued as a major branch of mathematics.

Numerical methods for polynomial rootfinding are afflicted with a similar enthusiasm. J. Michael McNamee's bibliography [265], which lists references by 29 categories with some papers appearing in multiple headings, ran 590 printed pages. Three supplements [266, 267, 268] culminated in his 2-volume treatise on polynomial equations, the second volume co-authored by Victor Y. Pan [269, 271].

Most of these articles are as impractical as the vast theoretical literature on polynomials, but three robust, black-box methods have emerged that are widely used in software libraries:

1. modified Laguerre method (Press et al. [297], which includes source listings),
2. Jenkins–Traub algorithm (Jenkins and Traub [216]), and
3. eigensolving the companion (Frobenius) matrix method.

These global, black-box library solvers require no input from the user except the coefficients of the polynomial. Further developments are reviewed in [270].

Ironically, the method of computing the roots by finding the eigenvalues of a matrix, implemented in the MATLAB "**roots**" command, is independent of the polynomial rootfinding literature! All parts of the companion matrix software were developed in the linear algebra community.

One reason for Goedecker's pessimistic remark quoted at the beginning of the chapter—that none of the 3 methods compared was reliable for polynomial degree N larger than 50—is that multiple roots are also *intrinsically* ill conditioned. Second, the representation of a polynomial as a sum of powers of the unknown is *numerically* ill conditioned and otherwise very unsatisfactory. Both difficulties are explained in turn in the next 2 sections.

The literature on polynomial rootfinding is vast. The magisterial but readable and unpretentious books by McNamee and McNamee and Pan [269, 271], totaling 1080 pages, are the best guide. Our chapter is a chrestomathy, a selection of the most important or most illuminating concepts.

9.2 ▪ Ill-Conditioning of Polynomial Roots, I: Multiple Roots

9.2.1 ▪ Definitions: Simple and Multiple Roots and Doublets

Definition 9.1 (Simple Root). *If x_* is a solution to $f(x)=0$, then the root is simple if and only if*

$$\frac{df}{dx}(x^*) \neq 0. \tag{9.2}$$

Definition 9.2 (Multiple Root). *If*

$$f(u) = (x - x_*)^m \, h(x), \tag{9.3}$$

where $h(x_*) \neq 0$, then x_* is said to be a multiple root *of* multiplicity m *of f. A simple root is a root of multiplicity $m = 1$.*

Nineteenth-century spectroscopists discovered that many spectral lines appear as closely spaced pairs (*doublets*) or a trio of closely spaced lines (*triplets*). Twentieth-century quantum mechanics showed that each spectral line is an eigenvalue of the matrix that discretizes the stationary Schrödinger equation in Heisenberg matrix mechanics. Equivalently, each spectral line has an energy that is the root of a polynomial, the characteristic polynomial of the matrix. The energy levels of the simplest species, the hydrogen atom, are highly degenerate in the sense that many roots are multiple. When the hydrogenic structure is perturbed through additional electrons, intermolecular bonds, or external magnetic or electric fields, the perturbation breaks the degeneracy, and a multiple root will split into simple roots that differ only slightly.

Since closely spaced simple roots are as numerically nasty as multiple roots, it is helpful to borrow the spectroscopic terminology:

Definition 9.3 (Doublet). *A* doublet *is a pair of closely spaced simple roots.*

Definition 9.4 (Triplet). *A* triplet *is a trio of closely spaced simple roots.*

These definitions are deliberately vague as to the meaning of "closely spaced." Neighboring simple zeros can wreck iterative algorithms and degrade the accuracy and rate of convergence even for successful methods; a root doublet is a numerical problem, as discussed below. However, often the parameters of an engineering or physics problem are known only within rather crude tolerances, and then a given pair of roots may be real and simple, a complex conjugate pair, or a double zero within the allowed parameter range; such doublets are closely spaced in an intrinsic or physics sense independent of numerical difficulties. Nevertheless, the intent is clear: roots are closely spaced whenever the separation between them is so small as to cause either numerical or engineering difficulties, or both.

9.2.2 ▪ The Multiple-Root Starburst

The polynomial of degree d,

$$P(x;\epsilon) \equiv (x - \chi)^d - \epsilon^d, \tag{9.4}$$

has, as easily shown by the change of variable $z \equiv x - \chi$, the exact roots

$$x_j = \chi + \epsilon \exp(2\pi i \, j/d), \qquad j = 0, 1, 2, \ldots, (d-1). \tag{9.5}$$

For simplicity, assume ϵ is real and nonnegative. When $\epsilon = 0$, there is a single dth-order root, $x = \chi$. As ϵ increases, the single high-order root explodes into a ring of simple roots, all on a circle of radius ϵ in the complex x-plane centered on $x = \chi$. The trajectories of the roots are rays that form a starburst pattern, as illustrated in the bottom panel of Fig. 9.1.

The reason that this "multiple-root starburst" is upsetting is that when d is large, a perturbation as small as ϵ^d where $\epsilon \ll 1$ will alter the roots by a complex ray of length ϵ.

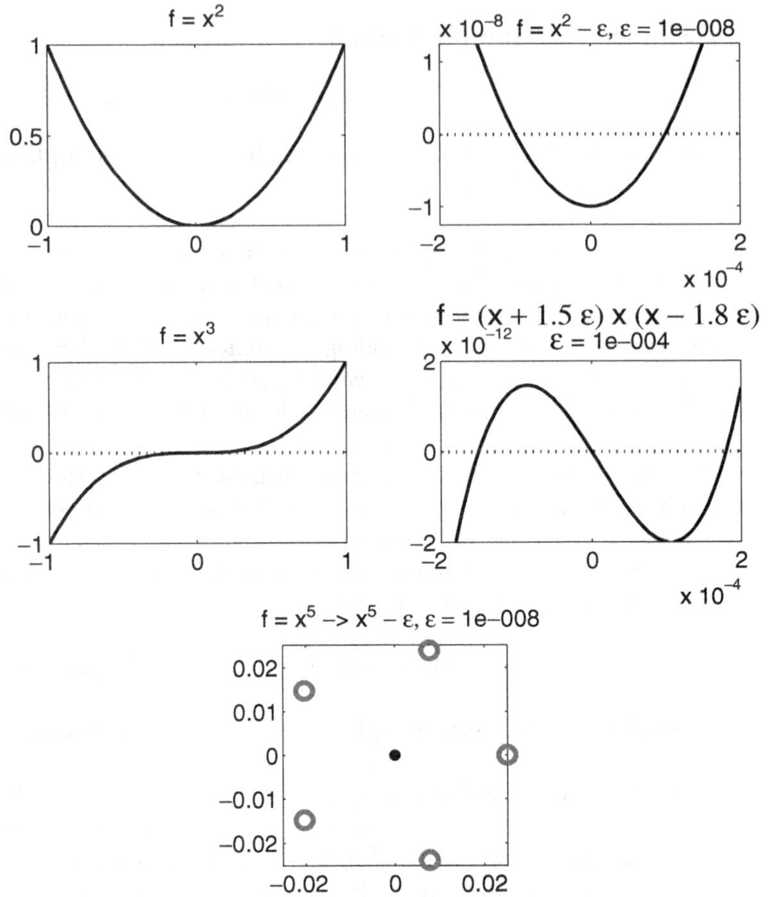

Figure 9.1. *Schematic illustrating the sensitivity of multiple roots to perturbations. Top: perturbing the parabola $f = x^2 \rightarrow x^2 - \epsilon$ converts a double root at the origin to a root doublet, that is, simple zeros at $x = \pm\sqrt{\epsilon}$. Middle: perturbation of a triple zero into three* real *zeros. $x^3 \rightarrow (x + 1.5\epsilon)x(x - 1.8\epsilon) = x^3 - \epsilon x^2 - 2.7\epsilon^2 x$. The separation between roots is $O(\epsilon)$, which does not seem alarming, but $f(x)$, on the interval spanned by these three simple zeros, oscillates within the very tiny range $O(\epsilon^3)$—10^{-12} when the perturbative change in the roots is only $O(10^{-4})$. Numerically, tracking these tiny oscillations, and thus computing the three simple roots accurately, is very challenging. Bottom: the multiple-root starburst for a fifth-order root. $x^5 \rightarrow x^5 - \epsilon$ converts the multiple root (solid black disk) into five simple zeros (circles) that form the vertices of a pentagon in the complex x-plane. The magnitude of the roots, $\epsilon^{1/5}$ is a million times the size of the perturbation, $\epsilon = 10^{-8}$.*

To put it another way, a perturbation in the dth digit will change the roots in the first decimal place.

The good news is that if a polynomial equation has M parameters, then the double roots will generically be found on submanifolds of dimension $(M - 1)$, triple zeros on surfaces of dimension $(M - 2)$, and so on within this M-dimensional parameter space. Thus, for the reduced cubic equation, discussed in Chapter 11, there are two parameters. A triple zero occurs only on a single zero-dimensional manifold, otherwise known as a point, within the two-dimensional plane spanned by the parameters. Double roots lie along one-dimensional curves in the same parameter space.

The reason that this behavior is generic, and not peculiar to the reduced cubic equation, is that a dth-order root must not only solve the polynomial but also simultaneously solve the polynomials that are its first $(d-1)$ derivatives with respect to x. If we regard the parameters as unknowns, then we have d equations in $(M+1)$ unknowns. When $M=d-1$, the number of unknowns matches the number of equations, and the solutions occur at discrete points in the M-dimensional parameter space. When $M>(d-1)$, the roots occur on submanifolds of dimension $(M+1-d)$ within the M-dimensional parameter space. It is of course possible to construct exceptions, but the statement that the dimensionality of the surface of d-fold roots is of dimension $(M+1-d)$ is generic; the exceptions require imposing constraints upon the equation.

9.2.3 ▪ Pseudoresult

Because of roundoff error, amplified by the ill-conditioning problems just described, a computed root is always really a *pseudoresult*, as formally defined by the following:

Definition 9.5 (Pseudoresult). *$\tilde{x}$ is a* pseudoresult *for the equation system $\vec{f}(x;\lambda)=0$ whose coefficients (or parameters) λ are known within a tolerance ϵ if $\tilde{x}$ is the exact solution of $f(x,\tilde{\lambda})=0$ for some set of approximate coefficients $\tilde{\lambda}$ whose difference from the original lies within the specified tolerances. The set of $\tilde{x}$ for all possible coefficients $\tilde{\lambda}$ that lie within the specified tolerances of λ is the* pseudoresult set.

In ensemble forecasting in meteorology, the same forecast is repeated with slightly different initial conditions and the ensemble of predictions is then compared to evaluate the reliability of the forecast and the spread of errors. In the same spirit, one can approximate the pseudoresult set by repeatedly solving a polynomial equation with randomly perturbed coefficients. Roots that change little under the perturbations are reliable; zeros with a big pseudoresult set are untrustworthy.

A root is given by some function $x_*(\lambda)$ that is not known explicitly. However, the Jacobian of $x_*(\lambda)$ as a function of λ can be calculated from the Davidenko equation as

$$\frac{dx}{d\lambda}=-\frac{\partial f/\partial\lambda}{\partial f/\partial x}. \tag{9.6}$$

One can then apply the following.

Theorem 9.6 (Bound on Pseudoresult Set). *Let $\upsilon(\lambda,\epsilon)$ denote a pseudoresult set. Then*

$$diam(\upsilon(\lambda,\epsilon))\leq 2\left|\frac{\partial f/\partial\lambda}{\partial f/\partial x}\right||\epsilon| \tag{9.7}$$

(Stetter [343]).

The diameter of a set, *diam*, is defined below.

Definition 9.7 (Diameter of a Set).

$$diam(M)=\max_{m_1,m_2\in M}||m_1-m_2||, \tag{9.8}$$

where m_1,m_2 are any two vectors in the set.

9.2.4 ▪ Bounds on Roots

Theorem 9.8 (Perturbation of a Simple Root). *Suppose x_* is a root of a function $f(x)$, where f need not be a polynomial. Suppose that f is perturbed to*

$$\tilde{f} \equiv f(x) + \epsilon g(x). \tag{9.9}$$

If the root is a simple root, then it is perturbed to $\tilde{x}$, where in the limit $\epsilon \to 0$,

$$\tilde{x} = x_* - \epsilon \frac{g(x_*)}{df/dx(x_*)} + O(\epsilon^2). \tag{9.10}$$

If $g(x_*)$ is finite and $df/dx(x_*) \neq 0$, then an $O(\epsilon)$ perturbation in f produces only an $O(\epsilon)$ change in the root.

Proof: Use the Newton–Raphson iteration. ■

There are a couple of subtleties in applying this theorem. First, if $g(x)$ is a high power of x such as x^{10}, then the perturbation $\epsilon g(x)$ is *never small* for *sufficiently large* $|x|$. This is why Wilkinson was able to find his spectacular example of the sensitivity of roots to tiny changes in the coefficients. However, only the large roots are ill conditioned. The roots where $|x| \sim 1$ are well conditioned to slight changes in the coefficients of powers of x.

The second caveat is that at a multiple root, the theorem fails. (When two roots are very close so that we have a near-double root, the theorem still technically applies but perhaps only for such tiny ϵ as to be useless for all practical purposes.) We need a different relation, which is given as follows.

Theorem 9.9 (Perturbation of a Multiple Root). *Suppose x_* is an m-fold root of a function $f(x)$ where f need not be a polynomial. This implies that the first $m-1$ derivatives of f are also zero at $x = x_*$. Suppose that f is perturbed to*

$$\tilde{f} \equiv f(x) + \epsilon g(x). \tag{9.11}$$

If the root is perturbed to $\tilde{x}$, then in the limit $\epsilon \to 0$,

$$\tilde{x} = x_* + \epsilon^{1/m} \left\{ \frac{-g(x_*)m!}{\frac{d^m f}{dx^m}(x_*)} \right\}^{1/m} + O(\epsilon^{(m+1)/m}). \tag{9.12}$$

Proof:

$$\tilde{f}(\tilde{x}) = \epsilon g(x_*) + \tilde{f}(x_*) + \frac{1}{m!}\frac{d^m f}{dx^m}(x_*)(\tilde{x} - x_*)^m + O\left\{(\tilde{x} - x_*)^{m+1}\right\} \tag{9.13}$$

where $\tilde{f}(x_*) = 0$ because x_* is a zero of $\tilde{f}$ by definition. Solving for $\tilde{x}$ proves the theorem. ■

This theorem is consistent with the multiple-root starburst phenomenon: m-fold roots are altered as $O(\epsilon^{1/m})$ by a perturbation of $O(\epsilon)$.

More elaborate theorems with rigorous upper bounds (true even for finite ϵ) can be found in Winkler [375] and Gautschi [184].

9.2.5 ▪ Square-Free Polynomials and All That

Multiple roots are very annoying because they are poorly conditioned. In some circumstances, they can be suppressed so as to create a *square-free* polynomial $f_{squarefree}(x)$ that has the same roots as the original polynomial $f(x)$ except that all the roots of $g(x)$ are simple. The formula is

$$f_{squarefree}(x) = \frac{f(x)}{gcd(f, df/dx)}, \tag{9.14}$$

where gcd denotes the greatest common denominator.

The good news is that the gcd can be calculated by the so-called Euclidean algorithm in a finite number of steps without approximation. Furthermore, a routine to do so is built into computer algebra packages like Maple and Mathematica.

The bad news is that computing the square-free part of a polynomial makes sense only in *exact arithmetic*. The reason is that in floating point arithmetic, roundoff error will prevent the gcd from exactly dividing f: $f_{squarefree}(x)$ will be a rational function, not a polynomial, with poles of tiny residue at each of the multiple roots of f. In addition, roundoff effects in computing $f(x)$ may perturb the multiple root to a doublet or triplet.

When $f(x)$ is a polynomial with integer or rational coefficients, it may well have multiple roots, and the square-free part of f will then be an exact and useful simplification. In floating point arithmetic, though, the concept of a multiple root must be replaced by the vaguer notion of "double root or doublet," and similarly for higher multiplicity. It is impossible to precisely distinguish between a double root and a doublet in finite precision arithmetic unless one has side information—i.e., constraints or theorems about the physics—that narrows the possibilities.

9.2.6 ▪ Kahan's Pejorative Manifolds for Multiple Roots

In work that was never formally published, William Kahan observed that multiple roots were not necessarily as ill conditioned as the above theorems would indicate. The reason is that if one is given as additional side information that the polynomial has multiple roots of a certain multiplicity, it is possible to devise special rootfinding algorithms that preserve the multiplicity.

Wilkinson observed that for a very wide class of problems and numerical algorithms, the numerical solution is the exact solution to a slightly perturbed problem. For polynomial rootfinding, for example, the *backward error* is the difference between the original and perturbed polynomials f, the *forward error* is the difference between x_* and $x_{numerical}$, and the condition number is the backward error divided by the forward error. For multiple roots, the condition number is usually rather large because even a very tiny, $O(\epsilon)$ error in either the computation of f or the iteration scheme will split a double root into two single roots separated by $O(\sqrt{\epsilon})$.

Kahan showed, however, that if both the original and perturbed polynomials are constrained to have roots of the same multiplicity, then the numerical error is drastically reduced. He defined polynomials that have the same multiplicity as the original polynomial as members of the *pejorative manifold* in the space defined by the coefficients of the polynomial. For example, the pejorative manifold of the quadratic equation is defined by

$$f(x; \lambda) \equiv (x-\lambda)^2 = x^2 - 2\lambda x + \lambda^2. \tag{9.15}$$

As long as the linear coefficient is -2λ and the constant in the polynomial is λ^2 for some parameter λ, the quadratic polynomial will have a double root. If we define a two-parameter space with coefficients a_1 and a_0, then the pejorative manifold is a parabola in this space. The full parameter space has two dimensions, but the manifold of polynomials with a double root is a one-dimensional submanifold. The double root can be computed as half the negative of the linear coefficient in the quadratic or as the square root of the constant in the quadratic; the difference between the two is the sensitivity of the zero to the errors in the coefficients of the quadratic polynomial.

Dayton, Li, and Zeng [147] showed that it is possible to develop algorithms which first compute the multiplicity by calculating the gcd of f and df/dx and second apply special algorithms to approximate roots only on the pejorative manifold. He showed that multiple roots can be computed to nearly full machine precision by his publicly available MATLAB software, Multroot. This is in contrast to the theorems given earlier, which show that for a *general* perturbation, the roots can only be calculated to roughly the d-fold root of machine epsilon.

The rub is that in most engineering and scientific situations, the multiplicity is not known a priori. A perturbation that splits a double root into a doublet of simple roots is just as plausible as a perturbation that preserves the double root. Consequently, we shall not discuss the details of Zeng's complicated but effective algorithms.

9.3 ▪ Ill-Conditioning of Polynomial Roots, II: Vices of the Power Form (Monomial Basis)

"When $[N] \geq 7$ or 8, however [where N is the degree of the fitted polynomial], one begins to hear strange rumblings of discontent in the computing laboratory; the gist of the unhappiness is that each method selected to solve the system (11) [normal equations of least-squares fitting] fail somehow for the larger values of $[N]$."

George Forsythe on p. 77 of [174], replacing his k by our N.

9.3.1 ▪ Power Form: Linear Independence and Nonuniformity

Algebraic geometry almost universally employs $x^j y^k z^m \ldots$—monomials—as basis functions. In one dimension, the monomial basis implies the following representation.

Definition 9.10 (Power Form of a Polynomial). *A polynomial is in* power form *or* monomial basis form *when it is written as a sum of powers of the unknown, i.e.,*

$$P(x) \equiv \sum_{j=0}^{d} a_j x^j. \tag{9.16}$$

The power form is the way middle school students are first exposed to polynomials; it seems almost absurd to bother to even define it. Numerically, though, this familiar form is poison. Numerical routines for least-squares polynomial fitting have avoided the power form for at least half a century; in the 32-bit precision of the first generation of computers, disasters began for N as small as 6. Similarly, algorithms for solving differential and integral equations by high-degree polynomials (*spectral methods*) use the power form only in exact arithmetic in an algebraic manipulation.

The reason for these numerical disasters is two-fold. First, the powers of x, although always linearly independent in exact mathematics, easily become numerically dependent in finite precision arithmetic. For example, all powers of x greater than 16 are smaller than machine epsilon in MATLAB/IEEE double precision (2×10^{-16}) on $x \in [0, 1/10]$.

Second, the powers of x are very nonuniform in the sense that powers are very tiny near the origin but grow hugely for $|x| > 1$.

9.3.2 ▪ The Perils of Conversion to Power Form

In engineering and physics, high-degree polynomials in power form are usually the result of conversion from some other form. For example, if $\vec{A}$ is a matrix and $\vec{I}$ is the identity matrix, then the eigenvalues λ of $\vec{A}$ are the roots of the determinant $|\vec{A} - \lambda\vec{I}|$. This can always be converted into a polynomial, but this is very ill conditioned and can vastly amplify the effects of roundoff errors. This conversion is strongly deprecated in all articles on eigenvalue solving.

Indeed, the sound procedure is to go in the opposite direction. A useful strategy to solve polynomial equations is to compute the matrix whose *characteristic polynomial* is the target polynomial and then find the roots by computing the eigenvalues.

The best strategy for avoiding the perils of ill-conditioning in polynomial equations is thus to solve the polynomial in its original nonpower form. If $f(x)$ is a polynomial created by a Chebyshev interpolation, then solve it as a sum of Chebyshev polynomials rather than a series of powers. If $f(x)$ is the characteristic polynomial of a matrix, solve the matrix.

Walter Gautschi showed that the condition number of a polynomial of degree N in the monomial form is proportional to $(1+\sqrt{2})^N = (2.414)^N$ [185].

There are two exceptions to this deprecation of the monomial basis:

1. When the degree N of the polynomial is *small.*

2. When the coefficients of a high-degree polynomial can be computed *exactly*, as is often possible in an algebraic manipulation language such as Mathematica or Maple.

9.4 ▪ The Ultimate Oracle: The Fundamental Theorem of Algebra

The factorization theorem and the fundamental theorem of algebra are probably already familiar to the reader but are so important that they are worth a reminder.

Theorem 9.11 (Factorization). *A polynomial of degree N can always be written as a product of N factors that are linear in the unknown x:*

$$P(x) \equiv \sum_{j=0}^{N} a_j\, x^j = a_N \prod_{j=1}^{N} (x - x_j), \tag{9.17}$$

where x_j is a zero of the polynomial. This factorization is unique, modulo permutations of the order of the factors.

Definition 9.12 (Multiple Root: Second, Equivalent Definition). *A zero of multiplicity k is a root x_j that appears k times in the factorization equation* (9.17). *A root of multiplicity one is a* simple *root whereas zeros of higher multiplicity are* multiple zeros.

Theorem 9.13 (Fundamental Theorem of Algebra). *If the zeros are counted according to their multiplicity so that a root of multiplicity three is counted as equivalent to three distinct roots, then a polynomial of degree N has exactly N zeros, some or all of which may be complex valued.*

The fundamental theorem of algebra may be justly described as the "ultimate oracle" because it makes a precise prediction about the number of roots. Once all N roots have been identified, then a rootfinding algorithm may safely terminate.

9.5 ▪ Discriminant: Identifying Multiple Roots

Theorem 9.14 (Multiple Roots of a Polynomial). 1. *All multiple roots of $P(x)$ are also roots of its derivative, dP/dx.*

2. *All multiple roots of $P(x)$ are zeros of the* discriminant polynomial,

$$Disc(P(x);x) \equiv \frac{1}{a_d} Resultant(P, dP/dx; x), \tag{9.18}$$

where $Resultant(P, dP/dx; x)$ denotes the resultant of $P(x)$ with its first derivative and a_d is the coefficient of x^d in $P(x)$.[17]

3. *The discriminant is a homogeneous polynomial of degree $2d-2$ in the coefficients $\{a_0, a_1, \ldots, a_d\}$ of $P(x)$.*

9.6 ▪ Oracles for Polynomial Equations: Prophecies for the Number of Roots in an Interval or Domain

9.6.1 ▪ Oracles for Polynomial and Nonpolynomial Equations

Theorem 9.15 (Bolzano $f(a)f(b) < 0$ Oracle). *If, whether $f(x)$ is a polynomial or a general function that is analytic on the interval $x \in [a,b]$ and*

$$f(a)f(b) < 0, \tag{9.19}$$

then f has at least *one root on the interval $x \in [a,b]$. More precisely, there is an* odd *number of roots on the interval.*

The weakness of the Bolzano oracle is that it is blind to the presence of an *even* number of roots on an interval. Still, it is extremely useful because of its simplicity. A much longer discussion is given in section 19.2.

The Kronecker–Picard oracle, described in section 19.5, counts the number of roots by evaluating an integral whose value is approximately equal to the number of roots (excluding their multiplicity).

9.6.2 ▪ Descartes's Rule of Signs

Theorem 9.16 (Descartes's Rule of Signs). *Let $P(x)$ denote a polynomial:*

$$P(x) \equiv \sum_{j=0}^{N} a_j x^j. \tag{9.20}$$

There is a sign alternation *if*

$$a_j a_{j+1} < 0. \tag{9.21}$$

[17] The resultant is defined and discussed in Section 20.5.

The number n_{pos} *of* positive real roots *of* $P(x)$ *is* bounded from above *by the number of sign alternations. That is, if the number of sign changes is* n_{change} *and the number of positive real roots is* $n_{positive/real}$, *then*

$$n_{positive/real} \leq n_{change}. \tag{9.22}$$

Descartes's rule can test for the number of negative roots by applying it to $P(-z)$ [347].

Corollary 9.17 (Real Roots for a Polynomial with m Terms). *A polynomial with* m *nonzero terms has at most* $(2m-1)$ *real roots, and this bound is the best possible (Sturmfels* [345]*).*

Descartes's rule and its corollary are especially useful for sparse polynomials. For example, the rule implies that the polynomial $P(x) = x^{200} - x^{157} + 8x^{101} - 23x^{61} + 1$, which has 200 roots, has at most four positive real roots; the actual number of positive roots is two.

9.6.3 ▪ Budan–Fourier Oracle: Real Roots on an Interval

Theorem 9.18 (Budan–Fourier). *Let* $n_{interval}(a,b)$ *denote the number of real zeros of* $P(x)$ *on the interval* $x \in [a,b]$ *where* (i) *neither* $P(a) = 0$ *nor* $P(b) = 0$ *and* (ii) *zeros are counted according to their multiplicity so that a single double root would give* $n_{interval}(a,b)$ $= 2$, *for example. Let* V_a *denote the number of sign variations in the sequence* $P(a), dP/dx(a), \ldots, d^N P/dx^N(a)$, *and similarly* V_b *denote the number of sign variations in the sequence* $P(b), dP/dx(b), \ldots, d^N P/dx^N(b)$. *Then*

$$n_{interval}(a,b) = V_a - V_b - 2k, \tag{9.23}$$

where k *is a positive integer or zero. The theorem is true regardless of the polynomial basis (Chebyshev or monomial).*

This oracle is imperfect in the sense that the number of real roots could be less than $V_a - V_b$. However, if $V_a = V_b$, then we can be sure that the interval is zero free.

Floating point error is worse and worse as the order of the derivatives increases, but this is not a problem in exact rational arithmetic in a computer algebra system.

9.6.4 ▪ Sturm's Oracle

Definition 9.19 (Sturm Sequence). *The Sturm sequence of a polynomial* $P(x)$ *is defined by*

$$p_0 = P(x), p_1 = \frac{dP}{dx}, \qquad p_j = -rem\left(\frac{p_{j-2}(x)}{p_{j-1}(x)}\right), j \geq 2, \tag{9.24}$$

where rem denotes the remainder of the division of the numerator polynomial by the denominator polynomial and where the last nonzero polynomial in the sequence is denoted $p_m(x)$ *for some finite* m.

Theorem 9.20 (Sturm Sequence Oracle). *If* $a < b$ *in* $\mathbb{R}$ *and* $P(a) \neq 0$ *and also* $P(b) \neq 0$, *the number of* real zeros *in the interval* $x \in [a,b]$ *is the number of sign changes in the Sturm sequence evaluated at* $x = a$, *i.e.,* $\{p_0(a), p_1(a), \ldots, p_m(a)\}$, *minus the number of sign changes in the Sturm sequence evaluated at* $x = b$, $\{p_0(b), p_1(b), \ldots, p_m(b)\}$.

The Sturm oracle is more complicated than Descartes's rule of signs since it is necessary to first calculate the Sturm sequence. Once these auxiliary polynomials,

$$\{p_0(x), p_1(x), \ldots, p_m(x)\},$$

are known, however, the oracle can be applied repeatedly on different intervals. Note that unlike the rule of signs, the Sturm oracle gives the *exact number* of *real roots* on the interval, $\{p_0(a), p_1(a), \ldots, p_m(a)\}$. Note that it is possible to compute polynomial division with remainder, and thence Sturm sequences, entirely in the Chebyshev basis [25]; conversion to the monomial basis is unnecessary for the Sturm sequence oracle.

9.6.5 ▪ Bernstein Form Oracle

Theorem 9.21 (Positive Bernstein Coefficients Oracle). *A polynomial can be converted without approximation to a series of Bernstein basis functions,*

$$f(x) \equiv \sum_{j=0}^{N} g_j \, \mathfrak{B}_j(x;N), \tag{9.25}$$

where

$$\mathfrak{B}_j(x;N) \equiv \binom{N}{j} x^j (1-x)^{N-j}. \tag{9.26}$$

Note that this is not *a Bernstein quasi interpolant such that $g_j = f(x/j)$. Rather, the g_j must be computed by standard interpolation so that when $f(x)$ is a polynomial of degree N, the g_j reproduce the exact polynomial from the Bernstein basis series. Then if the g_j are all of the* same sign, *the polynomial $f(x)$ is free of zeros on $x \in [0,1]$.*

The theorem can be generalized to transcendental functions by expanding $f(x)$ as

$$f(x) = \sum_{j=0}^{N} a_j T_j(2x-1) + E(x;N) = f_N(x) + E(x;N) \tag{9.27}$$

and then re-expanding the truncated Chebyshev series $f_N(x)$ as a Bernstein polynomial series. Suppose that $|E(x;N)|$ can be bounded by ϵ on the interval $x \in [0,1]$. Let the g_j denote the coefficients of the Bernstein basis form of the Chebyshev polynomial approximation, $f_N(x)$. If

$$g_j > \epsilon \qquad \forall j = 0, 1, \ldots, N \qquad \text{or} \qquad g_j < -\epsilon, \tag{9.28}$$

then the interval $x \in [0,1]$ contains no zeros of $f(x)$.

9.7 ▪ Whittaker's Root Series

Whittaker showed that the root of a polynomial of *smallest magnitude*, that is, the root closest to the origin, can be computed by a geometrically converging series [370, 369]. This expansion is now rarely used because Newton's iteration, begun from a first guess of $x = 0$, is usually more efficient. Whittaker's expansion is still worth a paragraph just to show that series expansions for roots are possible.

Denote the polynomial by

$$P(x) \equiv \sum_{j=0}^{N} a_j x^j. \tag{9.29}$$

Then the Whittaker series is

$$x_{smallest} = \sum_{j=1}^{\infty} r_j, \tag{9.30}$$

$$r_1 = -\frac{a_0}{a_1}, \qquad r_2 = -a_0^2 \frac{a_2}{a_1 \left\{ a_1^2 - a_0 a_2 \right\}}, \tag{9.31}$$

$$r_j = -a_0^j \det(\vec{\vec{D}}_{top}) / \left\{ \det(\vec{\vec{D}}_A) \det(\vec{\vec{D}}_B) \right\}, \tag{9.32}$$

where det denotes the usual matrix determinant. The elements of these matrices for the jth-order correction are specified by the following. Note that a_j with a negative index j is replaced by 0:

$$D_{top,ik} = a_{2-i+k}, \qquad i,k = 1,2,\ldots,(j-1), \tag{9.33}$$

$$D_{A,ik} = a_{1-i+k}, \qquad i,k = 1,2,\ldots,(j-1), \tag{9.34}$$

$$D_{B,ik} = a_{1-i+k}, \qquad i,k = 1,2,\ldots,j. \tag{9.35}$$

The Whittaker series converges geometrically with

$$|r_{j+1}| = \frac{|x_{smallest}|}{|x_{second\ smallest}|} |r_j|, \tag{9.36}$$

where $x_{smallest}$ is the root of smallest magnitude (i.e., the sum of the series) and $x_{second\ smallest}$ is the root whose absolute value is the smallest of all the roots except $x_{smallest}$. Thus, the series will *fail* if the two smallest roots have equal magnitude, i.e., $x_{second\ smallest} = -x_{smallest}$. However, if there is any separation in magnitude between the target root and the second-smallest root, then the Whittaker series will converge geometrically.

The method is more powerful than it first appears. If one wants to find a root that is thought to be near x_0, even if this is not the root of smallest magnitude, one can define a new variable

$$z \equiv x - x_0 \tag{9.37}$$

and rewrite the polynomial in terms of this new coordinate. The Whittaker series will then converge to the root closest to x_0.

Similarly, if the Whittaker series does not converge rapidly (or at all!), one can apply a translation, and it will succeed.

Lastly, if the function is not a polynomial, one can apply Whittaker's series to a truncated Taylor or Chebyshev series of $f(x)$.

The only drawback of Whittaker's method is that it converges more slowly than Newton's. (However, on modern workstations, the issue of numerical efficiency for a single equation in a single unknown has become irrelevant, unless one needs to solve the equation millions of times with different parameters as part of some larger calculation.)

Fig. 9.2 shows the Whittaker corrections for a problem considered by Whittaker himself. The polynomial is $P = 20 - 321x - 4x^2 + x^3$ with the exact roots $(0.0622, -16.06, 20)$. Six corrections suffice to give the smaller root with an error approaching machine epsilon.

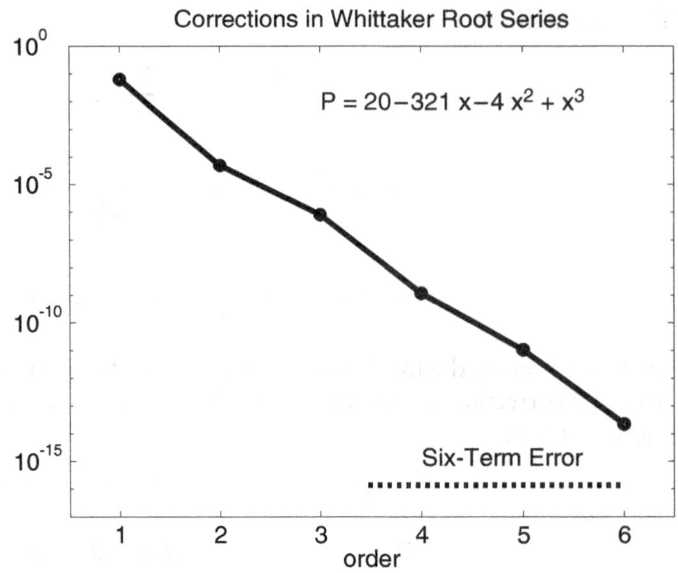

Figure 9.2. *The disks show the successive terms r_j in the Whittaker root series for the smallest root of the polynomial $P = 20 - 321x - 4x^2 + x^3$. The horizontal dashed line shows the error in the six-term approximation.*

9.8 ▪ When Radicals Fail: Transcendental Formulas for Roots of Any Degree

In later chapters, the cubic and quartic equations are discussed in detail. In this section, we briefly explain the good and the bad about polynomial equations of degree five and higher. The good news is that it is possible to derive *analytic* solutions for the roots of equations of any order. The bad news is that when the degree is higher than four, the analytic solution does not involve only square roots, cube roots, and higher radicals, but rather transcendental functions so complicated that the solution is effectively useless.

Camille Jordan showed in 1870 [218] that any polynomial equation can be solved using modular functions. Umemura discusses in detail a theorem that allows one to express the solution of any equation of any degree as the product of so-called hyperelliptic theta constants [356]. Better yet, it is unnecessary to use preliminary transformations to bring the algebraic equation into a special form. The bad news is that evaluating the necessary theta constants is hideous: a modestly difficult numerical problem has been recast into a much more difficult numerical problem. The periods of the theta constants, analogous to the elliptic modulus, must be determined by evaluating a set of hyperelliptic integrals, more integrals as the degree of f increases. No known examples of explicit calculations of roots for sextic and higher-degree equations via theta constants are known to us.

On a happier note, Puiseux series can be applied at any degree, as described in the next section.

9.9 ▪ Puiseux Series for Roots of Polynomial Equations of Any Degree

Sturmfels [346, 347] gives a good review of the theory of fractional power series (Puiseux series) for the roots of polynomial equations of any order. The most general theory, employing so-called $\mathscr{A}$-hypergeometric series, allows for all of the coefficients of the polynomial to be indeterminates—as general as possible!

Here, however, only the simpler case where the exponentials are rational functions of a parameter ϵ will be discussed.

Definition 9.22 (Puiseux Series). *A Puiseux series in a variable ϵ is an expansion in powers of ϵ with* rational exponents. *The set of rational exponents must satisfy the twin constraints of* (i) *being bounded from below and* (ii) *having a common denominator.*

Theorem 9.23 (Puiseux's Theorem). *If $P(u,\epsilon)$ is a polynomial of degree d whose coefficients are rational functions of ϵ, then the polynomial has d roots in the field of Puiseux series. Another way of saying the same thing is that the whole neighborhood of any point x of an algebraic curve may be uniformly represented by a finite number of convergent Puiseux series (Sturmfels [347], Weisstein [367]).*

The theorem is proved algorithmically, that is to say, by an explicit construction for all the Puiseux series. The **"algcurves"** package in Maple has a **"puiseux"** command for computing these series to any desired order.

The formalism is straightforward. Suppose that the coefficient of x^j in $P(x,\epsilon)$ has a Laurent series in ϵ whose leading term is

$$a_j \sim c_j\,\epsilon^{E_j} + \cdots . \tag{9.38}$$

Each Puiseux series has the leading term

$$x(\epsilon) \sim \mu\,\epsilon^{\tau} + \cdots , \tag{9.39}$$

where τ is as yet unknown. Substituting the leading order of the Puiseux series into the polynomial gives

$$c_d\,\epsilon^{E_j + d\,\tau} + c_{d-1}\,\epsilon^{E_{j-1}+(d-1)\tau} + \cdots + c_1\,\epsilon^{E_1+\tau} + c_0\,\epsilon^{E_0} + \text{higher-order terms} = 0. \tag{9.40}$$

In general, the terms displayed in (9.40) are of different orders, but there must be at least two terms of the same smallest order so that these dominant terms can cancel. If the lowest term is proportional to, say, $O(\epsilon^{3/2})$, and all the other terms are proportional to higher powers of ϵ, then the polynomial cannot have a root. Thus, τ is determined by the condition that

$$\min\left\{E_j + d\,\tau, E_{j-1} + (d-1)\,\tau, \ldots, E_1 + \tau, E_0\right\} \tag{9.41}$$

is attained at least twice.

Once the exponent τ of the leading term has been determined, it is straightforward to determine the rest of the series by substitution into $P(x,\epsilon) = 0$. It is important to note that the exponent τ is usually not the same for all d series. A more formal method to determine exponents constructs the *Newton polytopes* of the polynomials. Sturmfels provides a semipopular account in [345] and high-level treatments in [346, 347].

Puiseux series can arise in solving nonpolynomial equations by perturbation theory, too, as discussed in section 14.3.

9.10 ▪ Roots of the Quintic

Two different approaches to the roots of the quintic have been developed. The Hermite scheme uses a sequence of Tschirnhausen transformations that involve only square roots and cube roots and the solving of lower-degree equations to reduce the general cubic to the so-called Bring form

$$x^5 - x - \lambda = 0. \tag{9.42}$$

Hermite showed that the five roots of this equation are

$$B\Phi(\tau),\ B\Phi(\tau+16),\ B\Phi(\tau+32),\ B\Phi(\tau+48),\ B\Phi(\tau+64), \tag{9.43}$$

where

$$B = (1/2)5^{-3/4}k^{-1/4}(1-k^2)^{-1/2} \tag{9.44}$$

and where Φ is a function defined in terms of the elliptic modular function ϕ by

$$\Phi(\tau) \equiv \left\{\phi(5\tau)+\phi\left(\frac{\tau}{5}\right)\right\}\left\{\phi\left(\frac{\tau+16}{5}\right)-\phi\left(\frac{\tau+64}{5}\right)\right\}\left\{\phi\left(\frac{\tau+32}{5}\right)-\phi\left(\frac{\tau+48}{5}\right)\right\}. \tag{9.45}$$

Its argument τ is expressed in terms of the elliptic modulus k by

$$\tau \equiv \frac{iK(\sqrt{1-k^2})}{K(k)}, \tag{9.46}$$

where k is the elliptic modulus[18] and $K(k)$ is the usual elliptic integral. The modulus k is determined from λ by first computing the auxiliary quantity

$$\alpha(\lambda) = \text{arcsine}\left(\frac{16}{25\sqrt{5}\,\lambda^2}\right). \tag{9.47}$$

Then one can choose k to be any of the four options

$$k = \tan(\alpha/4),\ k = \tan(\alpha/4+\pi/2),\ k = \tan(\pi/4-\alpha/4),\ k = \tan((3/4)\pi-\alpha/4). \tag{9.48}$$

Davis [143] gives a good description of the traditional numerical way to apply these complicated formulas, which is to expand $\Phi(\tau)$ in a series that converges rapidly when $|\lambda| \gg 1$. However, it is far easier in this regime to bypass all of Hermite's elliptic function machinery to apply the methods of perturbation theory (Chapter 14) to derive

$$x \approx \frac{1}{\epsilon} + \frac{1}{5}\epsilon^3 - \frac{1}{25}\epsilon^7 + \frac{1}{125}\epsilon^{11} - \frac{21}{15625}\epsilon^{19} + \cdots, \tag{9.49}$$

where

$$\epsilon = 1/\lambda^{1/5}. \tag{9.50}$$

This perturbative series applies to *all five* roots with appropriate choice of the fifth root of λ in the definition of the small parameter ϵ. For Davis's example of $\lambda = 2$, the exact real root is 1.267168, whereas the perturbation series through the terms above gives 1.267136, an absolute error of only 0.000031.

King and Canfield [226, 227] have implemented an alternative algorithm due to Kiepert. This has the virtue that only elliptic functions, rather than the more complicated elliptic modular functions, are required. However, the preliminary steps are at least as complicated as those of the Hermite–Bring scheme. Indeed, as King reports amusingly [225], Kiepert's scheme, originally proposed in 1878, was apparently never used to actually calculate numerical roots until King and Canfield more than a century later!

[18]The "m modulus" is also in widespread use, where $m = k^2$—be careful!

The original paper of Kiepert was so riddled with errors—because it had never been used to solve anything, not a single example—that King and Canfield had to independently recreate many of the steps in a kind of mathematical archaeology [225].

Because of their complexity, the various quintic formulas have never apparently been put to any practical use. Yet they stand as the end of a line of mathematical research that stretched over several centuries.

Chapter 10

The Quadratic Equation

"Facts are only the steam which obscures the mirror of truth."
—Msgr. Ronald Knox [228, p. 113]

10.1 ▪ Quadratic Equation

Theorem 10.1 (Analytic Solution of the Quadratic). *The general solution of the quadratic equation*

$$Ax^2 + Bx + C = 0 \tag{10.1}$$

is given by

$$x = -\frac{B}{2A} \pm \frac{1}{2A}\sqrt{B^2 - 4AC} \tag{10.2}$$

or equivalently by

$$x = \frac{2C}{-B \pm \sqrt{B^2 - 4AC}}. \tag{10.3}$$

When $B^2 \gg AC$, one can avoid cancellations by expressing the roots as a combination of these two forms: If $sign(B) > 0$, use

$$x_1 = -\frac{B}{2A} - \frac{1}{2A}\sqrt{B^2 - 4AC}, \qquad x_2 = -\frac{2C}{B + \sqrt{B^2 - 4AC}}. \tag{10.4}$$

For negative B,

$$x_1 = -\frac{B}{2A} + \frac{1}{2A}\sqrt{B^2 - 4AC}, \qquad x_2 = \frac{2C}{-B + \sqrt{B^2 - 4AC}}. \tag{10.5}$$

The discriminant is

$$\Delta \equiv B^2 - 4AC. \tag{10.6}$$

Its significance is

$$\Delta > 0, \qquad \textit{two real roots;} \tag{10.7}$$

$$\Delta = 0, \qquad \textit{two identical real roots;} \tag{10.8}$$

$$\Delta < 0, \qquad \textit{two complex-valued roots.} \tag{10.9}$$

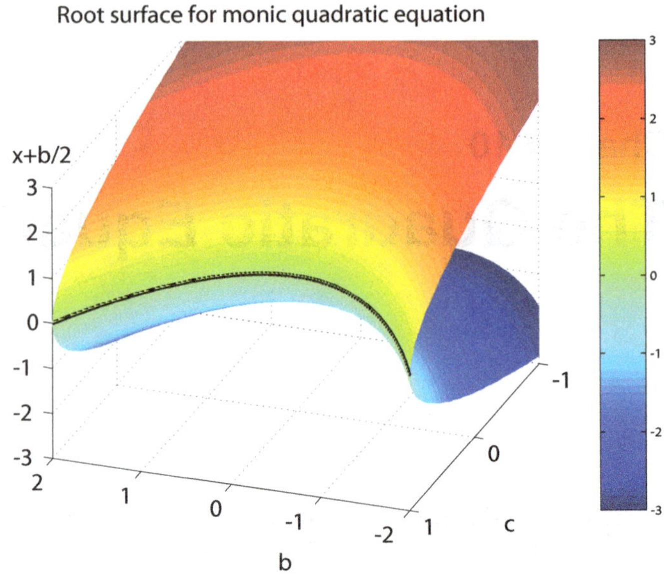

Figure 10.1. *Roots of the monic quadratic equation* $x^2 + bx + c = 0$. *The vertical axis is the* shifted *root,* $x + b/2$. *The black curve is the discriminant, the curve where the upper and lower surfaces merge. This is the parabola* $b^2 - 4c = 0$, *plotted in the plane* $x + b/2 = 0$.

10.2 ▪ Geometry of the Root Surface of the Monic Quadratic

By dividing out the coefficient of the quadratic term, every quadratic can be transformed into the *monic quadratic*, which depends on two parameters, (b, c):

$$x^2 + bx + c = 0. \tag{10.10}$$

Theorem 10.2 (Geometry of Monic Quadratic). 1. *The discriminant (where roots merge) is the* parabola

$$b^2 - 4c = 0 \tag{10.11}$$

in the b-c plane.

2. *The root surface is a* parabola *in the x-c plane for a fixed value of b:*

$$(x + b/2)^2 + (c - b^2/4) = 0. \tag{10.12}$$

3. *The parameter c is a single-valued function of x and the parameter b:*

$$c = \frac{1}{4}b^2 - (x + b/2)^2. \tag{10.13}$$

Fig. 10.1 shows that the two branches of the solution form a continuous surface. The branches merge along the line $x = -b/2$; the graph employs $x + b/2$ as the vertical axis. The surface for x itself is similar to the plot except that the surface is displaced upward (by $-b/2$) to the left of $b = 0$ and twisted downward by the same amount for positive b.

The discriminant in the b-c plane is illustrated in Fig. 10.2.

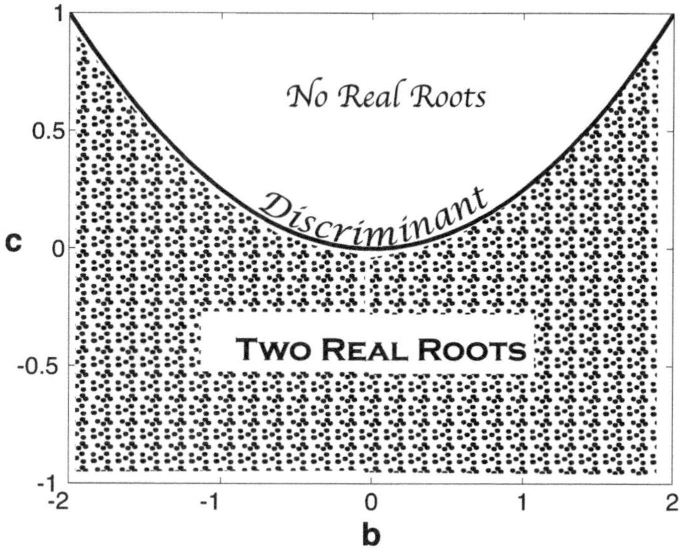

Figure 10.2. *The black thick parabola is the discriminant, $b^2-4c=0$, of the monic equation $x^2+bx+c=0$. The discriminant is the boundary between the region where there are two real roots (stippled) and the region where there are two complex-valued roots (unshaded).*

10.3 ▪ Reduced Quadratic

Theorem 10.3 (Reduction by Translation Quadratic). *The linear term in the quadratic may be eliminated by defining the new unknown*

$$y=x+b/2. \tag{10.14}$$

The reduced quadratic *is*

$$y^2+(c-b^2/4)=0 \qquad \leftrightarrow \qquad y=\pm\sqrt{\frac{b^2}{4}-c}. \tag{10.15}$$

10.4 ▪ Vieta's Theorem for the Quadratic

Theorem 10.4 (Vieta's Root Theorem: Quadratic). *The sum and products of the roots of $Ax^2+Bx+C=0$ satisfy the exact relationships*

$$x_1+x_2=-B/A, \qquad x_1x_2=C/A. \tag{10.16}$$

The general solution of the quadratic equation was known to the Babylonians, although they did not understand complex-valued roots [230, p. 88].

Chapter 11

Roots of a Cubic Polynomial

"[Of a difficult cosmic ray experiment] in order to make it work you had to spit on the wire on some Friday evening in Lent."
—Patrick Blackett, Nobel Laureate [165, p. 214]

11.1 ▪ Introduction

11.1.1 ▪ Reduction to a Monic Polynomial

The general cubic equation,

$$\tilde{P} \equiv a_3 x^3 + a_2 x^2 + a_1 x + a_0, \tag{11.1}$$

contains four parameters. However, it is trivial to reduce the cubic to a monic polynomial, that is, one whose leading coefficient is one, by dividing the polynomial by a_3, a transformation that does not alter the roots:

$$P \equiv x^3 + a x^2 + b x + c, \tag{11.2}$$

where $a = a_2/a_3$, $b = a_1/a_3$, and $c = a_0/a_3$.

11.1.2 ▪ The Reduced Cubic: Reduction by Translation

The number of parameters in the cubic can be reduced from three to two by making a translation or shift of the unknown. Define

$$x = y - a/3 \qquad \leftrightarrow \qquad y = x + a/3. \tag{11.3}$$

The cubic is transformed to the so-called reduced cubic

$$y^3 - \tau y + \xi = 0, \tag{11.4}$$

where

$$\tau = \frac{1}{3} a^2 - b, \tag{11.5}$$

$$\xi = \frac{2}{27} a^3 - \frac{1}{3} a b + c. \tag{11.6}$$

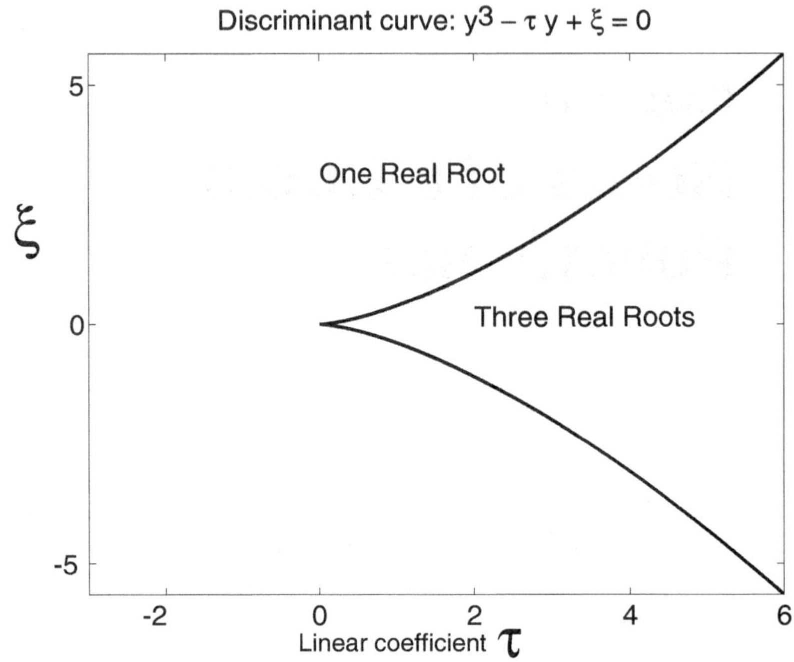

Figure 11.1. *Discriminant of the reduced cubic equation in the plane spanned by the coefficient of the linear term τ and the constant in the reduced cubic, ξ.*

11.1.3 ▪ Discriminant of the Reduced Cubic

The *discriminant* of a polynomial defines an implicit surface in the space of coefficients on which the polynomial has multiple roots. For the cubic, the discriminant is

$$\Delta = 4\tau^3 - 27\xi^2. \tag{11.7}$$

The meaning of this is that if the coefficients of the cubic are all real valued, then when

$$\Delta > 0, \quad \text{three real roots;} \tag{11.8}$$
$$\Delta = 0, \quad \text{three real roots; at least two are equal;} \tag{11.9}$$
$$\Delta < 0, \quad \text{one real root.} \tag{11.10}$$

The triple root occurs at the point $\tau = \xi = 0$.

The curve where $\Delta = 0$ (and double or triple roots occur) is a semicubical parabola. It is illustrated in Fig. 11.1.

11.2 ▪ Explicit Solutions

11.2.1 ▪ Trigonometric and Hyperbolic Solutions to the Reduced Cubic

Theorem 11.1 (Analytic Solution to the Reduced Cubic). *Define the cubic polynomial*

$$P \equiv y^3 - \tau y + \xi = 0. \tag{11.11}$$

The discriminant of the cubic is

$$\Delta = 4\tau^3 - 27\xi^2. \tag{11.12}$$

If the coefficients of the cubic (τ, ξ) *are real valued, then* Δ *controls the number of roots as follows:*

$$\Delta > 0, \quad \textit{three real roots}; \tag{11.13}$$

$$\Delta = 0, \quad \textit{double or triple root}; \tag{11.14}$$

$$\Delta < 0, \quad \textit{one real root}. \tag{11.15}$$

Note that if $\tau < 0$*, then there is always only a single real root because then from its definition,* Δ *is negative, too.*

The triple root is the single point $\xi = \tau = 0$.

Three Real Roots Case: $\boldsymbol{\Delta \geq 0}$: *Define the parameter*

$$\rho = \sqrt{\frac{4}{3}\tau}. \tag{11.16}$$

Note that $\tau \geq 0$ *whenever there are three real roots and therefore* ρ *is real and positive. The zeros are*

$$y = \rho \cos\left(\frac{1}{3}\arccos\left(-4\frac{\xi}{\rho^3}\right) + \frac{2\pi}{3}k\right), \tag{11.17}$$

where $k = 0, 1, 2$ *and each value of* k *generates a different root. To avoid difficulties for small* ξ*, Vignes* [360] *recommends the alternative*

$$y_1 = \rho \cos\left(\frac{1}{3}\omega + \frac{\pi}{6}\right), \tag{11.18}$$

$$y_2 = -\rho \cos\left(\frac{1}{3}\omega - \frac{\pi}{6}\right), \tag{11.19}$$

$$y_3 = \rho \sin\left(\frac{1}{3}\omega\right), \tag{11.20}$$

where

$$\omega = \arctan\left(9\frac{\xi}{\sqrt{12\tau^3 - 81\xi^2}}\right), \qquad -\frac{\pi}{2} < \omega < \frac{\pi}{2}. \tag{11.21}$$

One Real Root Case: $\tau < 0 \Rightarrow \boldsymbol{\Delta < 0}$: *Negative* τ *implies the discriminant is always negative, too.*

Define the parameter

$$\rho = \sqrt{-\frac{4}{3}\tau}. \tag{11.22}$$

Then the roots are, defining

$$\phi = \operatorname{arcsinh}(-4\xi/\rho^3), \tag{11.23}$$

$$y = \rho \sinh(\phi/3), \qquad -(\rho/2)\sinh(\phi/3) \pm i\,\rho(\sqrt{3}/2)\cosh(\phi/3). \tag{11.24}$$

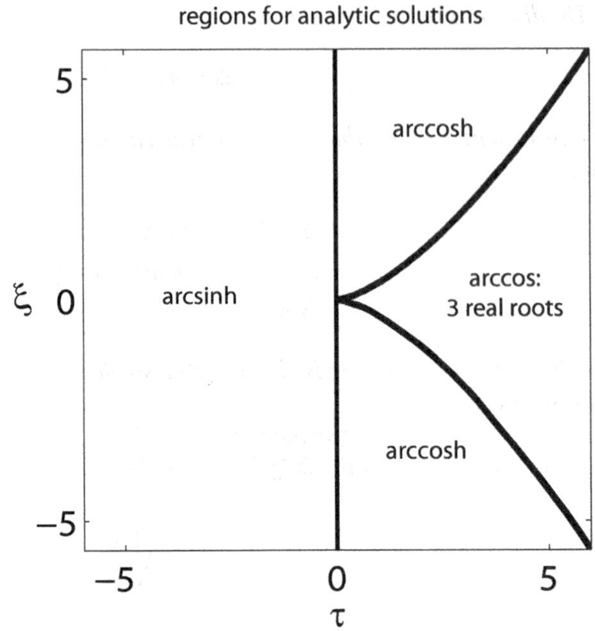

Figure 11.2. *Regions of validity of the three analytic solutions of the reduced cubic equation using inverse trigonometric or hyperbolic functions. The explicit solution in terms of radicals for the real root* (11.28) *is valid everywhere there is only a single real root, that is, everywhere outside the wedge labeled "arccos:* 3 *real roots."*

One Real Root Case: $\tau > 0$ ***and*** $\Delta < 0$*: Define*

$$\rho = \sqrt{\frac{4}{3}\tau}. \tag{11.25}$$

Then the roots are, defining

$$\phi = \operatorname{arccosh}(-4\xi/\rho^3), \tag{11.26}$$

$$y = \rho\cosh(\phi/3), \qquad -(\rho/2)\cosh(\phi/3) \pm i\,\rho(\sqrt{3}/2)\sinh(\phi/3). \tag{11.27}$$

If the coefficients are real and $\Delta < 0$*, then it may be more convenient to compute the single real zero by Press et al.* [297, p. 146]*:*

$$y = -\operatorname{sign}(\xi)\left[\left\{|\xi/2| + \sqrt{\frac{\xi^2}{4} - \frac{\tau^3}{27}}\right\}^{1/3} + \frac{\tau}{3}\Big/\left\{|\xi/2| + \sqrt{\frac{\xi^2}{4} - \frac{\tau^3}{27}}\right\}^{1/3}\right]. \tag{11.28}$$

Fig. 11.2 *shows the regions where each of these analytic solutions applies.*

11.2.2 ▪ Cardano–Tartaglia Formula (Deprecated)

The *NBS Handbook of Mathematical Functions*, edited by Abramowitz and Stegun [1], gave the cubic solution only in the Cardano–Tartaglia form. Unfortunately, the book omitted all discussion of roundoff error and the proper choice of branches for these formulas. The floating point accuracy of this solution and its competitors is tested later in this section.

Theorem 11.2 (Cardano–Tartaglia Solution to the Reduced Cubic). *The reduced cubic* $y^3 - \tau y + \xi = 0$ *is solved by*

$$D = \frac{\xi^2}{4} - \frac{\tau^3}{27}, \tag{11.29}$$

$$s_1 = -\left\{\frac{\xi^2}{2} + \sqrt{D}\right\}^{1/3}, \tag{11.30}$$

$$s_2 = -\left\{\frac{\xi^2}{2} - \sqrt{D}\right\}^{1/3}, \tag{11.31}$$

$$y_1 = s_1 + s_2, \tag{11.32}$$

$$y_2 = -\frac{1}{2}(s_1 + s_2) + i\frac{\sqrt{3}}{2}(s_1 - s_2), \tag{11.33}$$

$$y_3 = -\frac{1}{2}(s_1 + s_2) - i\frac{\sqrt{3}}{2}(s_1 - s_2), \tag{11.34}$$

throughout the τ*-*ξ *plane.*

Note that the cube root must be defined as $Q^{1/3} = \mathit{sign}(Q)|Q|^{1/3}$.

11.2.3 ■ Symmetrized Solution by Radicals (Deprecated)

The NIST Digital Library has replaced Cardano's formula by the following.

Theorem 11.3 (Symmetrized Solution). *The reduced cubic* $y^3 - \tau y + \xi = 0$ *is solved by*

$$\varphi = \exp([2/3]i\pi), \tag{11.35}$$

$$\Phi = \frac{1}{2} - \xi + \sqrt{\xi^2 - \frac{4}{27}\tau^3}, \tag{11.36}$$

$$u_1 = \Phi^{1/3}, \qquad v_1 = \frac{\tau}{3\,\Phi^{1/3}}, \tag{11.37}$$

$$u_2 = \varphi\Phi^{1/3}, \qquad v_2 = \varphi^2\frac{\tau}{3\Phi^{1/3}}, \tag{11.38}$$

$$u_3 = \varphi^2\Phi^{1/3}, \qquad v_3 = \varphi\frac{\tau}{3\Phi^{1/3}}, \tag{11.39}$$

$$y_1 = u_1 + v_1 = \Phi^{1/3} + \frac{\tau}{3\Phi^{1/3}}, \tag{11.40}$$

$$y_2 = u_2 + v_2 = \varphi\Phi^{1/3} + \frac{\tau}{3\,\varphi\Phi^{1/3}}, \tag{11.41}$$

$$y_3 = u_3 + v_3 = \varphi^2\Phi^{1/3} + \frac{\tau}{3\,\varphi^2\Phi^{1/3}}, \tag{11.42}$$

throughout the τ*-*ξ *plane.*

The NIST Digital Library abandoned the Cardano formulas in favor of this alternative. This may be dubbed the "symmetrized solution" because a different branch for the cube root merely *permutes* the roots without changing the numbers.

It is said that Gauss sometimes stayed up all night wondering where to put the minus signs. Choosing the proper branch of the multibranched function is equally maddening. The symmetrized solution has the great advantage that any consistent choice of the square root and the cube root will always yield the same trio of correct roots.

11.2.4 ▪ Numerical Tests of Cubic Solutions: Experimental Design

To test the limits of the formulas in floating point arithmetic is a little tricky. Points in the parameter space where one or both parameters are very large or very small can severely tax a formula. The cubic equation has folds (bifurcations) where two branches merge at

$$\xi_{fold}(\tau) = \sqrt{\frac{4}{27}\tau^3}. \tag{11.43}$$

It is also necessary to test the formulas very, very close to this singular curve. Lastly, incorrect or inaccurate formulas sometimes work for special values, such as integers. We therefore randomly perturbed the parameter values; an entry in the tables below for $\xi = 10^{-4}$ is only the order of magnitude of a number that may actually be 0.788×10^{-4} or 2.984×10^{-4}.

One simplification is that the symmetry theorem proved below shows that it is sufficient to test the formulas for positive ξ only, since the errors are symmetric about $\xi = 0$.

For $\tau < 0$, we tested accuracy on a grid with a uniform *logarithmic* spacing:

$$10^{-8}, 10^{-4}, 1, 10^{4}, 10^{8}.$$

For $\tau > 0$ and $\xi > \xi_{fold}(\tau)$ (one real root), the grid was logarithmic with a shift, that is, $\xi - \xi_{fold}$ was a uniform logarithmic grid.

In the region of three real roots, the range was $\xi \in [0, \xi_{fold}(\tau)]$, with points clustered logarithmically near both endpoints.

Each root from a formula was refined by five Newton iterations, whose result was taken as the "exact" root. When the root from a formula is sufficiently bad, Newton's iteration will not converge. The Newton/formula difference may be larger than the true error, but the failure is not thereby disguised.

11.2.5 ▪ Experimental Results

Using MATLAB/IEEE double precision (machine epsilon $\approx 2.2 \times 10^{-16}$), all three methods gave maximum relative errors in the *casus irreducibilis* no larger than 3×10^{-15}.

Tables 11.1 and 11.2 show a more complicated story.

The Cardano–Tartaglia formulas are inaccurate, sometimes unacceptably so, for both very small $|\tau|$ and very large $|\tau|$ in both regions where there is but one real root. Reruns with higher precision showed that the maximum errors at any of the test points did not exceed 10^{-36} in 64-digit arithmetic.

The symmetrized formulas fail utterly when $|\tau| \ll 1$, for either sign of τ, unless ξ is also very small, because the difference $-\xi + \sqrt{\xi^2 - \frac{4}{27}\tau^3}$ may become smaller than machine epsilon, triggering overflow since $1/(-\xi + \sqrt{\xi^2 - \frac{4}{27}\tau^3})$ appears in the formulas for v_j and therefore in the roots themselves. The symmetrized formulas are very inaccurate

Table 11.1. *Base-10 logarithm of the errors for various explicit formulas for the roots of the reduced cubic equation for negative τ (i.e., positive linear coefficient in the reduced cubic equation). Overflows and errors larger than 10^{-6} are both shown in boldface.*

		Cardano			
τ ξ	10^{-8}	-10^{-4}	1	-10^{4}	10^{8}
-10^{-8}	−10.7	**−5.81**	−8.48	−11.1	−13.8
-10^{-4}	−14.3	−13.6	−7.95	−7.14	−9.81
−1	−9.33	−12.3	−15.0	−13.5	**−5.69**
-10^{4}	−2.38	−7.20	−10.1	−14.2	−14.3
-10^{8}	0.	**0.330**	−4.62	−8.80	−13.0
		Symmetrized			
τ ξ	10^{-8}	-10^{-4}	1	-10^{4}	10^{8}
-10^{-8}	**−6.48**	**NaN**	**NaN**	**NaN**	**NaN**
-10^{-4}	−13.8	−11.0	**−3.48**	**NaN**	**NaN**
−1	−7.79	−11.7	−15.0	−10.2	**NaN**
-10^{4}	**−2.38**	**−5.51**	−9.81	−14.2	−11.2
-10^{8}	**4.20**	**0.569**	**−3.34**	−8.21	−11.6
		Sinh			
τ ξ	10^{-8}	-10^{-4}	1	-10^{4}	10^{8}
-10^{-8}	−15.0	−15.0	−14.9	−14.6	−14.5
-10^{-4}	−15.1	−15.7	−15.0	−14.8	−14.3
−1	−15.0	−15.3	−16.2	−14.7	−15.0
-10^{4}	−15.0	−15.5	−15.2	−15.1	−15.2
-10^{8}	−15.0	−15.0	−15.2	−15.6	−15.4

when $|\tau| > \xi$, but only for negative τ. Experiments with higher-precision arithmetic showed that with 64-digit precision, the maximum errors at any of the test points did not exceed 10^{-23}.

The trigonometric and hyperbolic solutions are therefore recommended. The cosh formula for $\tau > 0$ and one real root is relatively inaccurate when τ is huge and ξ is tiny, but this difficulty is mild compared to the errors of the solutions by radicals:

1. The trigonometric and hyperbolic solutions are the clear winners. Relative errors are $O(10^{-15})$ everywhere except for the cosh solutions very close to the fold (curve of zero discriminant) for $\tau \sim O(10^8)$ and $\xi \sim O(10^{11})$, where the relative error rises to $O(10^{-8})$.

2. Cardano's formula is equally accurate when there are three real roots. For $\tau \sim O(1)$ and $\xi \sim O(10^8)$, the error is $O(10^{-8})$

Lebedev [241], Vignes [360], and Chapter 12 of [271] provide further detailed discussions of the numerical properties of various explicit cubic root formulas.

11.3 ▪ Symmetry Theorem for the Reduced Cubic

Theorem 11.4. *If $y_j, j = 1,2,3$, are the roots of the reduced cubic $P(y;\tau,\xi) \equiv y^3 - \tau y + \xi$, then $-y_j, j = 1,2,3$, are the roots of $P(y;\tau,-\xi)$. In words, flipping the sign of the constant in the reduced cubic changes the sign of all roots of the cubic.*

Table 11.2. *Same as previous table but for positive τ, one real root: base-10 logarithm of errors in explicit formulas for roots of the reduced cubic equation. Overflows and errors larger than 10^{-6} are both shown in boldface.*

Cardano					
τ ξ	10^{-8}	-10^{-4}	1	-10^4	10^8
-10^{-8}	−9.9	−5.81	−8.48	−11.1	−13.8
-10^{-4}	−15.8	−12.7	−7.95	−7.14	−9.81
−1	0	−15.4	−15.1	−10.4	−5.7
-10^4	0	0	0	−15.1	−14.0
-10^8	**NaN**	0	0	−12.1	−14.3
Symmetrized					
τ ξ	10^{-8}	-10^{-4}	1	-10^4	10^8
-10^{-8}	**−6.98**	**NaN**	**NaN**	**NaN**	**NaN**
-10^{-4}	−14.7	−10.9	**−3.48**	**NaN**	**NaN**
−1	−16.8	−15.4	−15.2	−6.52	**NaN**
-10^4	−15.4	−16.0	−15.2	−15.1	−11.2
-10^8	−15.3	−15.4	−15.5	−15.7	−14.3
Cosh					
τ $\xi-\xi_{cusp}(\tau)$	10^{-8}	-10^{-4}	1	-10^4	10^8
-10^{-8}	−14.9	−14.8	−15.0	−14.6	−14.5
-10^{-4}	−14.4	−15.3	−15.0	−14.7	−14.5
−1	−11.3	−13.3	−15.3	−14.6	−14.9
-10^4	−8.31	−10.3	−12.2	−14.5	−15.2
-10^8	−7.59	−7.63	−15.5	−11.8	−13.7

Proof:

$$P(-y;\tau,-\xi)\equiv(-y)^3-\tau(-y)+(-\xi) \tag{11.44}$$

$$\equiv(-1)(y)^3-(-1)\tau(y)+(-1)\xi \tag{11.45}$$

$$=y^3-\tau y+\xi \tag{11.46}$$

$$=P(y;\tau,\xi). \tag{11.47}$$

In words, the reduced cubic is invariant to the simultaneous substitution $y\to -y$, $\xi\to-\xi$. ■

The theorem reduces the problem of the cubic to understanding the roots for $\xi\geq 0$ because the roots for negative ξ follow trivially from the roots in the upper half of the τ-ξ plane.

In most of what follows, we shall show the entire parameter plane for clarity, but the symmetry should always be kept in mind.

11.4 ▪ Dilation Theorems: Reducing the Parameters in the Reduced Cubic

Theorem 11.5 (Dilation of a Reduced Cubic). *Let $y(\tau,\xi)$ be a solution to the reduced cubic $y^3-\tau y+\xi=0$, and let s be an arbitrary constant. Then for any s,*

$$y(\tau,\xi) = s y\left(\frac{\tau}{s^2}, \frac{\xi}{s^3}\right) \tag{11.48}$$

is a solution to the cubic equation. In other words, given a solution $y(\tau,\xi)$ to the reduced cubic, we can generate a new solution, for any positive constant s, by simultaneously reducing the linear coefficient τ by the square of s, diminishing the constant in the reduced cubic polynomial by s^3, and multiplying the root by s.

Proof: Define a function $z(\tau,\xi,s)$ by

$$z \equiv s\, y\left(\frac{\tau}{s^2}, \frac{\xi}{s^3}\right), \tag{11.49}$$

which is the right-hand side of (11.48). Then z/s must solve

$$\left(\frac{z}{s}\right)^3 - \frac{\tau}{s^2}\frac{z}{s} + \frac{\xi}{s^3} = 0. \tag{11.50}$$

Multiplying this by s^3 gives

$$z^3 - \tau z + \xi = 0. \tag{11.51}$$

This demands that $z = y(\tau,\xi)$, which is the left side of (11.48), thus proving the theorem. ■

Theorem 11.6 (Doubly Reduced/Single-Parameter Cubic). *Define the parameter*

$$\epsilon \equiv \xi/\tau^{3/2} \tag{11.52}$$

and the new unknown

$$w(\epsilon) \equiv y(\tau,\xi)/\sqrt{\tau}. \tag{11.53}$$

The w satisfies the doubly reduced, one-parameter cubic

$$w^3 - w + \epsilon = 0. \tag{11.54}$$

It is possible to reproduce the entire three-parameter solution space of the monic cubic from the one-parameter family of solutions $w(\epsilon)$. That is, given $w(\epsilon)$, $y(\xi,\tau) = \sqrt{\tau}w(\xi/\tau^{3/2})$. Similarly, the solution to the general, unreduced cubic is

$$x = -\frac{a}{3} + \sqrt{\frac{a^2}{3} - b}\; w(\xi/\tau^{3/2}). \tag{11.55}$$

Proof: The dilation theorem for the special case $s = \sqrt{\tau}$ is

$$y(\tau,\xi) = \sqrt{\tau} y\left(1, \frac{\xi}{s^3}\right). \tag{11.56}$$

The theorem follows by applying the definition of the parameter ϵ and then invoking the definition of w. ■

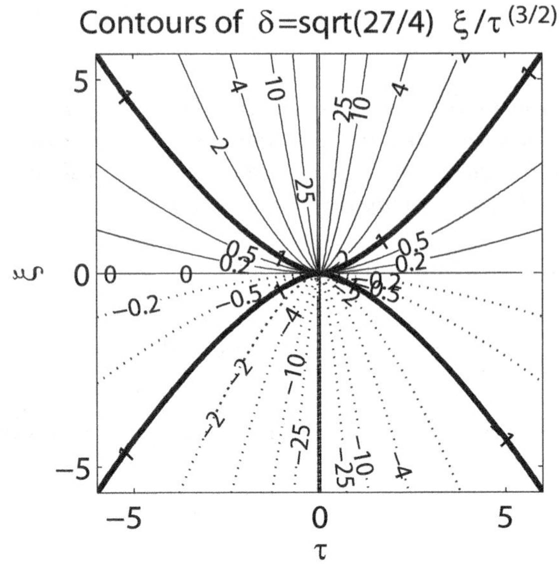

Figure 11.3. *The contours of the key parameter,* $\delta = \sqrt{27/4}\,\epsilon$, *where* $\epsilon = \xi/\tau^{3/2}$. *Note that real values of* δ *are plotted in the right half of the figure where* $\tau \geq 0$; *negative-valued contours are dashed. The left of the figure* ($\tau < 0$) *shows the contours of the imaginary part of* δ, *which is pure imaginary for real* ξ *and negative real* τ. *A curve of the form* $\xi^2 = \epsilon^2\tau^3$ *is known variously as a semicubical parabola, isochrone, or Neile's semicubical parabola; all the isolines of* δ *(and* ϵ*) are semicubical parabolas. The thick contours are the isolines where* $|\delta| = 1$.

11.4.1 ▪ Rescaling the Single Parameter

Roots merge along the discriminant curve $\Delta = 0$ in the τ-ξ plane where

$$\Delta = 4\tau^3 - 27\xi^2 = 27\tau^3\left(\frac{4}{27} - \epsilon^2\right). \tag{11.57}$$

Thus, there are big changes where $|\epsilon| = \sqrt{4/27}$. It is therefore convenient to rescale ϵ to a new parameter δ so that the discriminant curve of the single-parameter doubly reduced cubic is at $\delta = \pm 1$:

$$\delta \equiv \sqrt{\frac{27}{4}}\epsilon = \sqrt{\frac{27}{4}}\frac{\xi}{\tau^{3/2}}. \tag{11.58}$$

The isolines of δ as illustrated in Fig. 11.3 are the equiconvergence contours for the various perturbation schemes derived in the next section in the sense that each approximation gives an equally large or small error everywhere along a curve $\delta = \text{constant}$.

11.5 ▪ The Trigonometric Solution: Derivation and Connection with Chebyshev Polynomials

The Chebyshev polynomials are defined by the interesting identity

$$T_n(\cos(t)) \equiv \cos(nt), \qquad n = 0, 1, 2, \ldots, \tag{11.59}$$

where n is the degree of the nth Chebyshev polynomial. In words, the trigonometric function $\cos(nt)$ turns into an ordinary polynomial of degree n in the variable $x \equiv \cos(t)$. The special case of $n = 3$ is a cubic polynomial:

$$T_3(x) \equiv 4x^3 - 3x. \tag{11.60}$$

The Chebyshev polynomials are relevant to the cubic equation because the special cubic

$$T_3(x) = \lambda \tag{11.61}$$

can thus be solved by making the change of variable

$$x = \cos(t) \tag{11.62}$$

to transform it to

$$4\cos^3(t) - 3\cos(t) = \lambda. \tag{11.63}$$

The fundamental Chebyshev/trigonometric identity implies that these two powers of $\cos(t)$ can be combined into a single cosine to transform the equation to

$$\cos(3t) = \lambda. \tag{11.64}$$

Taking the inverse cosine of both sides gives the exact solution

$$t = \frac{1}{3}\{\arccos(\lambda) + 2\pi k\} \qquad \longleftrightarrow \qquad x = \cos\left(\frac{1}{3}\arccos(\lambda) + \frac{2\pi}{3}k\right), k = 0, 1, 2, \tag{11.65}$$

where k is an arbitrary integer that arises because the inverse cosine is a multivalued function.

This trick only works for a cubic equation with particular coefficients. However, without loss of generality, we can transform the general reduced cubic into this special form by using dilation.

11.5.1 ▪ Dilation into the Chebyshev Cubic

Define a new unknown via

$$y \equiv \rho z. \tag{11.66}$$

Multiplying the reduced cubic by $(4/\rho^3)$, it becomes

$$4z^3 - \left(4\frac{\tau}{\rho^2}\right) z + \left(4\frac{\xi}{\rho^3}\right) = 0. \tag{11.67}$$

This is of the form $T_3(z) = s$ if

$$\rho = \sqrt{\frac{4}{3}\tau}, \tag{11.68}$$

$$s \equiv -4\frac{\xi}{\rho^3}. \tag{11.69}$$

It follows from the previous subsection that the *general* solution y to the reduced cubic is

$$y = \rho\cos\left(\frac{1}{3}\arccos\left(-4\frac{\xi}{\rho^3}\right) + \frac{2\pi}{3}k\right), \tag{11.70}$$

thus proving (11.17).

11.6 ▪ Perturbation Theory

The reduced cubic polynomial

$$y^3 - \tau y + \xi = 0 \tag{11.71}$$

can be solved by perturbation theory with the assumption that one of the terms is small compared to the other two. Since there are three possible pairwise balances in the three-term reduced cubic, it follows that there will be three distinct cases, each requiring a different treatment.

11.6.1 ▪ First Pairwise Balance: Linear and Constant Terms Dominate

If we neglect the cubic term, then the cubic polynomial becomes

$$-\tau y + \xi \approx 0 \qquad \Leftrightarrow \qquad y \approx \xi/\tau. \tag{11.72}$$

Is this a consistent balance? Yes, if ξ/τ is sufficiently small, because then the cubic term we have neglected is $O([\xi/\tau]^3)$, which is smaller by a factor of ξ^2/τ^3 than the constant in the cubic, ξ. Thus, the effective perturbation parameter is

$$\epsilon^2 \equiv \xi^2/\tau^3. \tag{11.73}$$

To make this more explicit, factor out the lowest-order approximation by defining a new unknown:

$$y = \frac{\xi}{\tau} v. \tag{11.74}$$

The cubic equation becomes

$$\epsilon^2 v^3 - v + 1 = 0. \tag{11.75}$$

A two-parameter problem has been reduced without approximation to a single-parameter problem.

Substituting a perturbation series in the form

$$v = 1 + \sum_{j=1}^{\infty} a_j \epsilon^{2j} \tag{11.76}$$

and matching powers of ξ gives

$$y = \frac{\xi}{\tau}\Big\{1 + \epsilon^2 + 3\epsilon^4 + 12\epsilon^6 + 55\epsilon^8 + 273\epsilon^{10} + 1428\epsilon^{12} + 7752\epsilon^{14} + 43263\epsilon^{16}$$
$$+ 246675\epsilon^{18} + 1430715\epsilon^{20} + 8414640\epsilon^{22} + 50067108\epsilon^{24} + \cdots\Big\} \text{[Series I]}.$$

What is the radius of convergence of this series? As discussed earlier, two roots merge at the discriminant curve, $\Delta = 0$. The discriminant is

$$\Delta = 4\tau^3 - 27\xi^2 = 27\tau^3\left(\frac{4}{27} - \epsilon^2\right). \tag{11.77}$$

This suggests that the radius of convergence is $|\epsilon| < \sqrt{4/27}$ or equivalently $|\delta| < 1$. This conjecture can be confirmed by noting that the radius of a power series is

$$\epsilon_c = 1/\limsup_{n\to\infty}(a_n/a_{n-1}). \tag{11.78}$$

Fig. 11.4 strongly suggests that the conjecture is true.

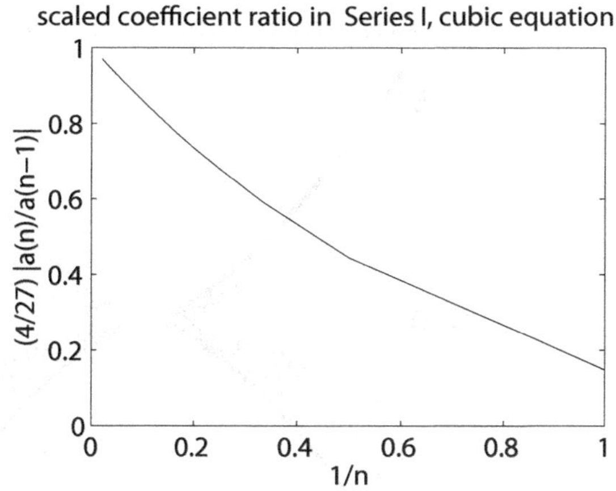

Figure 11.4. *The scaled ratio of magnitudes of terms in the perturbation series, $(27/4)|a_n|/|a_{n-1}|$, plotted versus $1/n$. The usual power series convergence theory asserts that $|a_n/a_{n-1}|$ should converge to the radius of convergence, and thus the scaled ratio should converge to one if the radius of convergence is $4/27$, as conjectured. Since this series is an expansion in ϵ^2, the convergence of the scaled ratio to one as $n \to \infty \leftrightarrow 1/n \to 0$ demonstrates that the radius of convergence is indeed $|\epsilon|^2 < 4/27$ or, equivalently, $\epsilon_c = \sqrt{4/27}$.*

When $\tau < 0$ (so that there is only one real root for real ξ), this series converges to the real root, but only for $\xi^2/|\tau|^3 < 4/27$. When $\tau > 0$, there are three real roots for real ξ and $\tau > 0$ whenever the product $\xi\tau > 0$. This same series converges to the root of *intermediate* magnitude, but only out to the discriminant curve, $\xi^2/\tau^3 < 4/27$, where this root merges with one of the other two (Fig. 11.6). This merger (fold) again limits the radius of convergence to $|\epsilon| < \sqrt{4/27}$.

We can normalize the radius of convergence to unity in a new parameter by defining

$$\delta \equiv \sqrt{\frac{27}{4}}\epsilon = \sqrt{\frac{27}{4}}\frac{\xi}{\tau^{3/2}}. \tag{11.79}$$

The perturbation series is then

$$\begin{aligned} y = \frac{\xi}{\tau}\Big\{ & 1 + 0.1481481481\,\delta^2 + 0.06584362140\,\delta^4 + 0.0390184423\,\delta^6 + 0.0264940040\,\delta^8 \\ & + 0.0194824595\,\delta^{10} + 0.0150975185\,\delta^{12} + 0.0121419196\,\delta^{14} + 0.0100389074\,\delta^{16} \\ & + 0.00847990879\,\delta^{18} + 0.00728644015\,\delta^{20} + 0.00634883498\,\delta^{22} + 0.00559638046\,\delta^{24} \\ & + 0.00498164668\,\delta^{26} + 0.00447175728\,\delta^{28} + 0.00404328066\,\delta^{30} \\ & + 0.00367911312\,\delta^{32} + \cdots \Big\} \text{ [Series I]}. \end{aligned}$$

Isolines of errors are shown in Fig. 11.5.

11.6.2 ▪ Second Pairwise Balance: Cubic and Linear Terms Balance

The previous series gives only a single root, but the cubic has three. To find the other two, we must assume a different pairwise balance. If $\xi \ll 1$, it is plausible to look for a

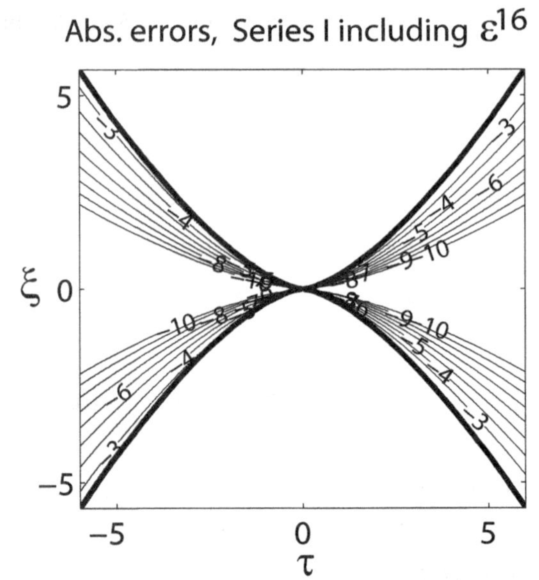

Figure 11.5. *Contours of the logarithm, base* 10, *of the absolute error in approximating one root of the cubic equation by perturbation Series* I, *in powers of* ϵ *or* δ, *up to and including* δ^{16}. *The thick curve is the discriminant (and the reflection of the discriminant about* $\tau = 0$*); these curves are the limit of convergence for this series.*

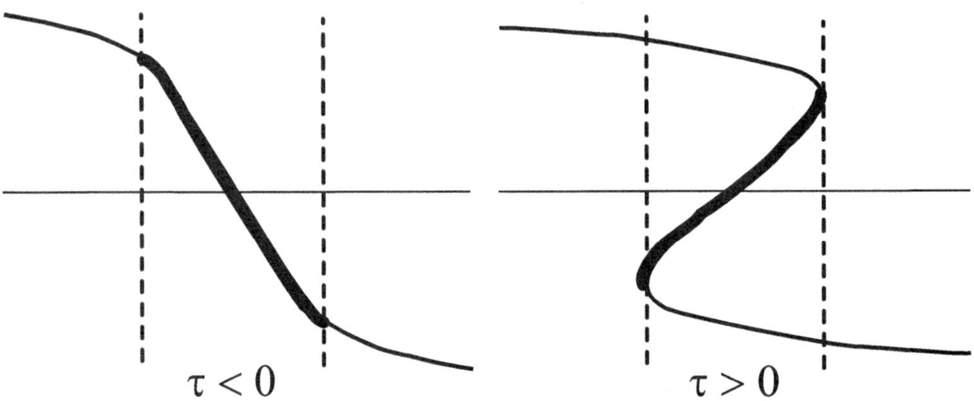

Figure 11.6. *The thick solid lines, bracketed by vertical dashed lines at the limits of convergence, show what the first perturbation series converges to: the sole root if* $\tau < 0$ *and the intermediate zero if* $\tau > 0$.

second pairwise balance by neglecting the constant in favor of the cubic and linear terms, yielding approximately

$$y^3 - \tau y = 0 \qquad \Leftrightarrow \qquad y \approx \pm\sqrt{\tau}. \tag{11.80}$$

(The trivial root, $y \approx 0$, was approximated by a different series in the previous section.) If we factor out the lowest-order dependence by introducing

$$y = \sqrt{\tau} w, \tag{11.81}$$

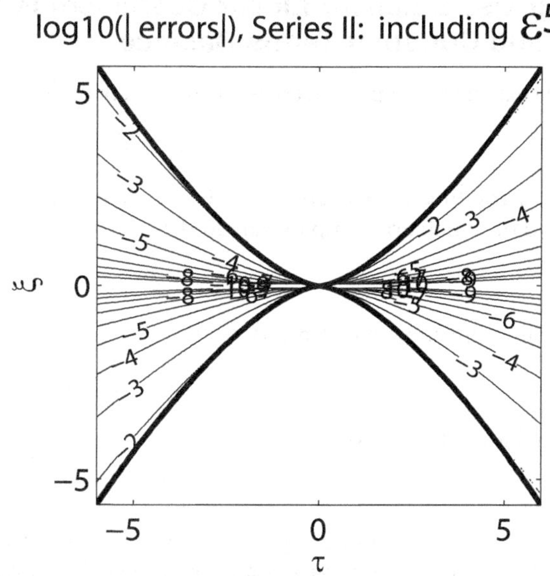

Figure 11.7. *Contours of the logarithm, base 10, of the absolute error in approximating one root of the cubic equation by the second perturbation series, in powers of ϵ or δ, up to and including ϵ^5. The thick curve is the discriminant (and the reflection of the discriminant about $\tau = 0$); these curves are the limit of convergence for this series.*

the full cubic becomes

$$w^3 - w + \epsilon = 0, \tag{11.82}$$

where

$$\epsilon = \frac{\xi}{\tau^{3/2}}. \tag{11.83}$$

Expanding

$$w = 1 + \sum_{j=1}^{\infty} b_j \epsilon^j \tag{11.84}$$

and matching powers of ϵ gives

$$y_2 = \sqrt{\tau}\left\{1 - \frac{1}{2}\epsilon - \frac{3}{8}\epsilon^2 - \frac{1}{2}\epsilon^3 - \frac{105}{128}\epsilon^4 - \frac{3}{2}\epsilon^5 - \frac{3003}{1024}\epsilon^6 - 6\epsilon^7 - \frac{415701}{32768}\epsilon^8\right\},$$

$$y_3 = \sqrt{\tau}\left\{-1 - \frac{1}{2}\epsilon + \frac{3}{8}\epsilon^2 - \frac{1}{2}\epsilon^3 + \frac{105}{128}\epsilon^4 - \frac{3}{2}\epsilon^5 + \frac{3003}{1024}\epsilon^6 - 6\epsilon^7 + \frac{415701}{32768}\epsilon^8\right\}.$$

These series are identical except for sign differences and can be combined as

$$y_\pm = \sqrt{\tau}\left\{\pm 1 - \frac{1}{2}\epsilon \mp \frac{3}{8}\epsilon^2 - \frac{1}{2}\epsilon^3 \mp \frac{105}{128}\epsilon^4 - \frac{3}{2}\epsilon^5 \mp \frac{3003}{1024}\epsilon^6 - 6\epsilon^7 \mp \frac{415701}{32768}\epsilon^8\right\} \text{ [Series II]}.$$

Fig. 11.7 shows the errors in the series for y_2. The graph is not quite symmetrical about the ξ-axis because the roots swap identities across the axis in the sense $y_2(-\epsilon) = -y_3(\epsilon)$; for small $|\epsilon|$, $y_2 \approx 1$ on both sides of the axis. It is unnecessary to plot the errors for y_3 because this graph is just the mirror image of Fig. 11.7, reflected about the ξ-axis.

11.6.3 ▪ Third Pairwise Balance: Linear Coefficient Is Small so That the Cubic and Constant Terms Balance

Suppose that the small parameter is the *linear* coefficient τ in the cubic

$$y^3 - \tau y + \xi = 0. \tag{11.85}$$

Neglecting the linear term gives $y \approx (-\xi)^{1/3}$, which gives approximations to all three roots, two of which are complex valued if ξ is real. It is then convenient to write

$$y = \xi^{1/3} z, \tag{11.86}$$

transforming the cubic equation to

$$z^3 - \mu z + 1 = 0, \tag{11.87}$$

where we have introduced

$$\mu \equiv \frac{\tau}{\xi^{2/3}}. \tag{11.88}$$

(Although $\mu = \epsilon^{-2/3}$, computing it in terms of τ, ξ ensures the correct choice of branch.) The lowest approximation to z is any of three cube roots of -1, which we shall denote by the symbol s:

$$s = -1, \exp(i\pi/3), \exp(-i\pi/3). \tag{11.89}$$

Matching powers of μ in the residual, we have

$$z \approx s + \frac{1}{3}\frac{\mu}{s} - \frac{1}{81}\frac{\mu^3}{s^5} + \frac{1}{243}\frac{\mu^4}{s^7} - \frac{4}{6561}\frac{\mu^6}{s^{11}} + \frac{5}{19683}\frac{\mu^7}{s^{13}} + \cdots \text{ [Series III]},$$

which applies to all three roots, with s chosen to be one of the cube roots of -1.

Fig. 11.8 shows the error of this series for the real root, which is $s = -1$. The graphs for the two complex roots are not shown because they are almost indistinguishable from Fig. 11.8.

Collectively, the three perturbation series cover the entire τ-ξ plane except in the neighborhood of the discriminant curve and the reflection of the discriminant curve about $\tau = 0$ as shown in Fig. 11.9.

11.7 ▪ Perturbative Expansions around the Discriminant

11.7.1 ▪ Upper Branch: $\delta \approx 1 \longleftrightarrow \xi^2 \approx (4/27)\tau^3$

The previous series all fail in the vicinity of the discriminant curve where $\delta = \pm 1$. However, it is possible to derive series that are accurate in the vicinity of these branch points where pairs of roots merge.

The dilation theorem shows that by defining

$$y = \sqrt{\tau}\, w, \tag{11.90}$$

the full cubic becomes

$$w^3 - w + \sqrt{4/27}\,\delta = 0. \tag{11.91}$$

At $\delta = 1$, for example, the double root is $w = \sqrt{3}/3$. If we write

$$\delta = 1 - \nu^2, \tag{11.92}$$

$$w = u + \sqrt{3}/3, \tag{11.93}$$

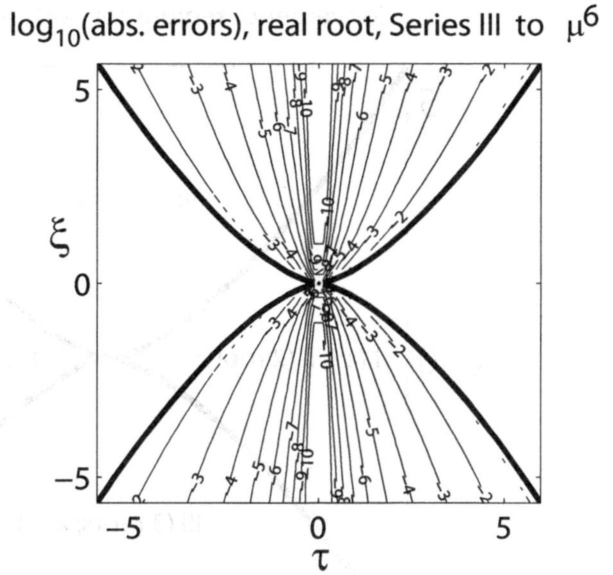

Figure 11.8. *Errors in the third perturbation series, in this case for the single real root in the sectors where this series is accurate.*

then the cubic in u lacks a linear term:

$$u^3 + u^2\sqrt{3} - 2/9\sqrt{3}v^2 = 0. \tag{11.94}$$

This shows that u has a double root, $u = 0$, at $v = 0$. For small, finite v, $u^2 \approx (2/9)v^2$. Expanding in powers of v gives

$$\begin{aligned} y_+ = \sqrt{\tau}\Bigg\{ & \frac{\sqrt{3}}{3} + \frac{\sqrt{2}}{3}v - \frac{\sqrt{3}}{27}v^2 + \frac{5}{324}\sqrt{2}v^3 - \frac{4}{729}\sqrt{3}v^4 + \frac{77}{23328}\sqrt{2}v^5 \\ & - \frac{28}{19683}\sqrt{3}v^6 + \frac{2431}{2519424}\sqrt{2}v^7 - \frac{80\sqrt{3}}{177147}v^8 + \frac{1062347}{3265173504}\sqrt{2}v^9 \\ & - \frac{2288}{14348907}v^{10}\sqrt{3} + \frac{14003665}{117546246144}\sqrt{2}v^{11} - \frac{23296\sqrt{3}}{387420489}v^{12} \\ & + \frac{1168767425\sqrt{2}}{25389989167104}v^{13} - \frac{82688\sqrt{3}}{3486784401}v^{14} + \frac{50657719535\sqrt{2}}{2742118830047232}v^{15} \\ & - \frac{909568\sqrt{3}}{94143178827}v^{16} + \frac{6022308892955\sqrt{2}}{789730223053602816}v^{17} - \frac{30764800\sqrt{3}}{7625597484987}v^{18} \\ & + \frac{823154536579165\sqrt{2}}{255872592269367312384}v^{19} - \frac{117877760\sqrt{3}}{68630377364883}v^{20} + \cdots \Bigg\}. \end{aligned} \tag{11.95}$$

The series for the second root is identical except for a sign reversal of every odd power of v:

$$y_- = \sqrt{\tau}\left\{ \frac{\sqrt{3}}{3} - \frac{\sqrt{2}}{3}v - \frac{\sqrt{3}}{27}v^2 - \frac{5}{324}\sqrt{2}v^3 - \frac{4}{729}\sqrt{3}v^4 - \frac{77}{23328}\sqrt{2}v^5 + \cdots \right\}.$$

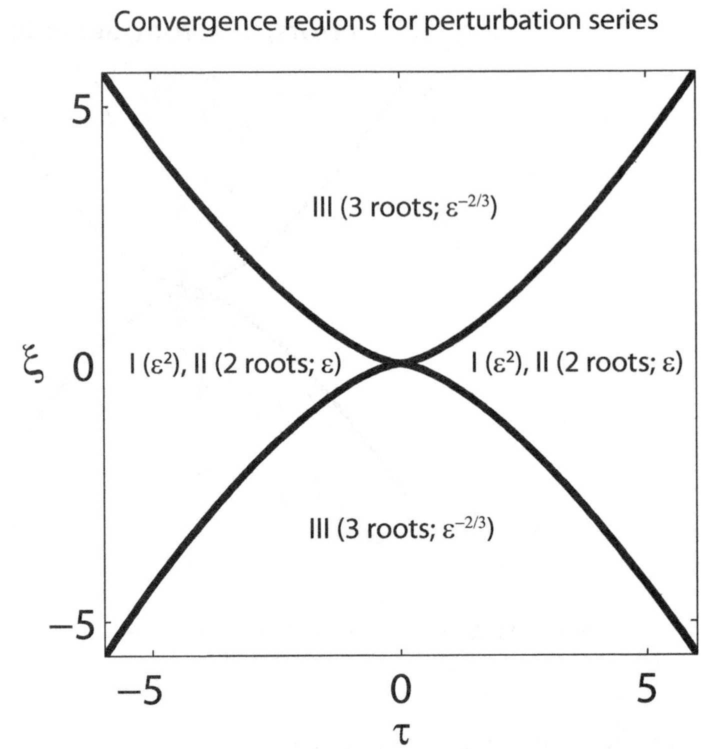

Figure 11.9. *The discriminant curve and its reflection about $\tau = 0$, shown as the heavy curves, partition the τ-ξ parameter plane into four quasi-triangular sectors. Each perturbation series converges within two of the sectors. The first perturbation series (I) in even powers of ϵ gives a single root in the left and right sectors. The second perturbation series gives two roots in the same sectors. The third perturbation series in powers of $1/\epsilon^{2/3}$ converges in the top and bottom sectors.*

The two series can be written together as

$$y_{\pm}^{upper} = \sqrt{\tau}\left\{ \frac{\sqrt{3}}{3} \pm \frac{\sqrt{2}}{3}\nu - \frac{\sqrt{3}}{27}\nu^2 \pm \frac{5}{324}\sqrt{2}\nu^3 - \frac{4}{729}\sqrt{3}\nu^4 \cdots \right\} \text{ [Series IV]}.$$

What is the radius of convergence of these expansions? When $\nu = \sqrt{2}$, $\delta = -1$, which is the other branch of the discriminant. Because roots merge (in different combinations) at $\delta = -1$, the series must diverge at that point. Numerically, the ratio test shows that $d_n/d_{n-1} \approx 1/\sqrt{2}$, which implies that the radius of convergence in ν is indeed $\sqrt{2}$. When we multiply d_n by $(-1)^n(\sqrt{2})^n n^{3/2}$, we find that the scaled coefficients rapidly asymptote to about 0.39, implying that

$$d_n \sim 0.39\frac{(-1)^n}{n^{3/2}\,2^{n/2}}, \qquad n \gg 1. \tag{11.96}$$

Translating the radius of convergence in ν into the parameter δ, one finds the series converges for $\delta \in [-1,3]$ and also $\Im(\delta) \in [-\sqrt{3}, \sqrt{3}]$.

The isolines of δ are also isolines of ν because of the definition of ν, $\delta = 1 - \nu^2$. The equiconvergence contours, that is, curves in the complex plane where the exponential rate of convergence is the same, are circles in the complex ν-plane as for any power series in ν. When mapped to the τ-ξ plane, these equiconvergence contours are semicubical parabolas radiating from the $(\tau = 0, \xi = 0)$ plane. Thus, we obtain the error contours and sectors of convergence illustrated in Fig. 11.10.

The third root can be approximated by writing

$$\delta = 1 + \beta, \tag{11.97}$$

$$\begin{aligned} y_3^{upper} = \sqrt{\tau}\Big\{ & -2/3\sqrt{3} - \frac{2}{27}\sqrt{3}\beta + \frac{8}{729}\sqrt{3}\beta^2 - \frac{56}{19683}\sqrt{3}\beta^3 + \frac{160}{177147}\sqrt{3}\beta^4 \\ & - \frac{4576}{14348907}\sqrt{3}\beta^5 + \frac{46592}{387420489}\sqrt{3}\beta^6 - \frac{165376}{3486784401}\sqrt{3}\beta^7 \\ & + \frac{1819136}{94143178827}\sqrt{3}\beta^8 - \frac{61529600}{7625597484987}\sqrt{3}\beta^9 + \frac{235755520}{68630377364883}\sqrt{3}\beta^{10} \\ & - \frac{2752522240}{1853020188851841}\sqrt{3}\beta^{11} + \frac{32551567360}{50031545098999707}\sqrt{3}\beta^{12} \\ & - \frac{129705476096}{450283905890997363}\sqrt{3}\beta^{13} + \frac{14082308833280}{109418989131512359209}\sqrt{3}\beta^{14} \\ & - \frac{171221451538432}{2954312706550833698643}\sqrt{3}\beta^{15} + \frac{698693987729408}{26588814358957503287787}\sqrt{3}\beta^{16} \\ & - \frac{8604771410378752\sqrt{3}}{717897987691852588770249}\beta^{17} + \frac{319605795242639360\sqrt{3}}{58149737003040059690390169}\beta^{18} \\ & - \frac{1325250203602124800\sqrt{3}}{523347633027360537213511521}\beta^{19} \\ & + \frac{16555433312691159040\sqrt{3}}{14130386091738734504764811067}\beta^{20} + \cdots \Big\} \text{ [Series V]}. \end{aligned} \tag{11.98}$$

Numerically, the coefficients rapidly asymptote to

$$p_n \sim 0.19 \frac{(-1)^n}{n^{3/2}\, 2^n}, \qquad n \gg 1. \tag{11.99}$$

The series converges in the sector $\delta \in [-3, 1]$ (Fig. 11.11).

11.7.2 ▪ Lower Branch of the Discriminant: $\delta \approx -1$

To expand about the lower branch of the discriminant where $\delta = -1$, it is convenient to write

$$\delta = -(1 - \tilde{\nu}^2), \tag{11.100}$$

which differs only by a multiplicative minus sign from the analogous expression for the series about the upper branch. One finds for Series VI (not graphed)

$$\begin{aligned} y_{\pm}^{lower} = \sqrt{\tau}\Big\{ & -\frac{\sqrt{3}}{3} \pm \frac{\sqrt{2}}{3}\tilde{\nu} + \frac{\sqrt{3}}{27}\tilde{\nu}^2 \pm \frac{5}{324}\sqrt{2}\tilde{\nu}^3 + \frac{4}{729}\sqrt{3}\tilde{\nu}^4 \pm \frac{77}{23328}\sqrt{2}\tilde{\nu}^5 \\ & + \frac{28}{19683}\sqrt{3}\tilde{\nu}^6 + \cdots \Big\} \text{ [Series VI]}, \end{aligned} \tag{11.101}$$

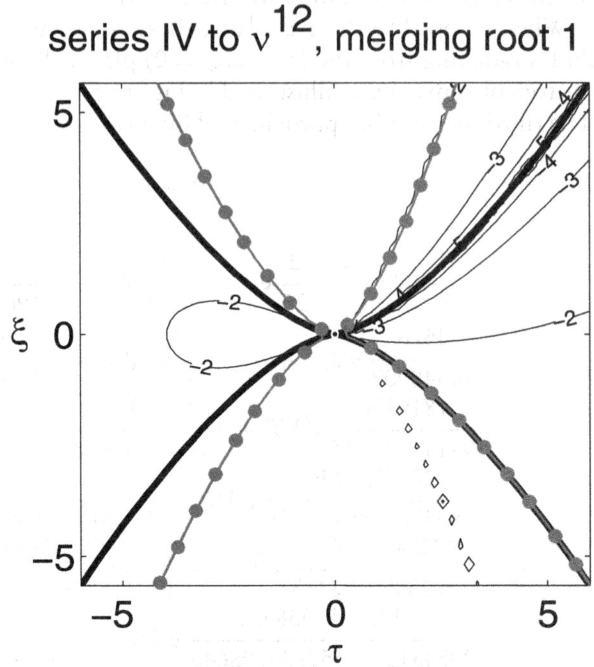

Figure 11.10. *Base-10 logarithm of the errors in the fourth perturbation series, the expansion of one of the two roots that merge at the discriminant. The series is in* ν *where* $\delta = 1-\nu^2$. *The thick curves are the discriminant, as in earlier figures. The curves with disks bound the convergence domain for the series.*

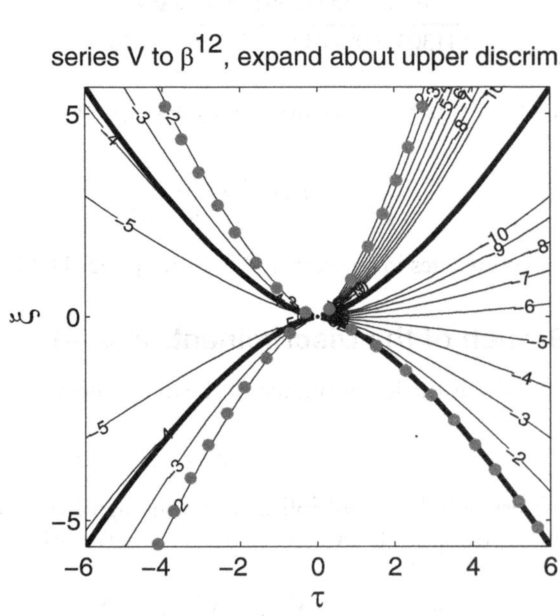

Figure 11.11. *Errors in the fifth perturbation series, the expansion of one of the two roots that merge at the discriminant. The series is in* β, *where* $\delta = 1+\beta$.

which is of exactly the same form as the expansion about the upper branch of the discriminant except that (i) $\nu \to \tilde{\nu}$ and (ii) all the signs are reversed: $y_{\pm}^{lower}(\tilde{\nu}) = -y_{\mp}^{upper}(\tilde{\nu})$.

The expansion about the lower discriminant converges for $\delta \in [-3,1]$ and also $\Im(\delta) \in [-\sqrt{3}, \sqrt{3}]$.

The third root can be approximated by writing

$$\delta = 1 + \beta, \tag{11.102}$$

$$y_3^{lower} = \sqrt{\tau}\left\{ 2/3\sqrt{3} - \frac{2}{27}\sqrt{3}\beta - \frac{8}{729}\sqrt{3}\beta^2 - \frac{56}{19683}\sqrt{3}\beta^3 - \frac{160}{177147}\sqrt{3}\beta^4 \right.$$
$$- \frac{4576}{14348907}\sqrt{3}\beta^5 - \frac{46592}{387420489}\sqrt{3}\beta^6 - \frac{165376}{3486784401}\sqrt{3}\beta^7$$
$$\left. - \frac{1819136}{94143178827}\sqrt{3}\beta^8 - \frac{61529600}{7625597484987}\sqrt{3}\beta^9 + \cdots \right\} \text{[Series VII]}.$$

Note that $y_3^{lower}(\beta) = y_3^{upper}(-\beta)$. Numerically, the coefficients rapidly asymptote to

$$p_n \sim \frac{0.19}{n^{3/2}\,2^n}, \qquad n \gg 1. \tag{11.103}$$

The series converges in the sector $\delta \in [-1, 3]$.

11.7.3 ▪ Perturbation Series about the Reflection of the Discriminant: $\delta = \pm i$

The doubly reduced cubic is

$$w^3 - w + \sqrt{4/27}\,\delta = 0. \tag{11.104}$$

Define

$$\delta = i(1 + \alpha). \tag{11.105}$$

We expand

$$w = w_0 + \sum_{n=1}^{\infty} w_j\, \alpha^n, \tag{11.106}$$

where w_0 is in turn each of three roots for $\delta = i$, generating three series whose coefficients are given in Table 11.3.

The asymptotic ratio test shows that all three series converge for $|\alpha| \leq \sqrt{2}$, or, in other words, $\Im(\delta) \in [1 - \sqrt{2}, 1 + \sqrt{2}]$.

11.8 ▪ Surface of the Roots

In elementary algebra and also in the trigonometric formula above, the roots of the cubic are presented as discrete, almost unrelated numbers. In reality, the roots at different points in the coefficient (τ, ξ)-plane are not unrelated, but rather form a single, connected surface both in the region where there is one real root and also, without break or discontinuity, in the region where there are three real roots, as illustrated in Fig. 11.12. The surface becomes multivalued—i.e., there are three real roots at a given point in the τ-ξ plane—when the surface folds over on itself. Although the transition from one root to three roots is seemingly a discontinuous jump—there are never two-and-a-half real roots, for example—the surface formed by connecting all the real roots is always smooth and continuous.

Table 11.3. *Coefficients of the perturbation series about the reflection of the discriminant in powers of* α^n*, Series* VIII, IX, *and* X.

$\Re(w_n)$ first root	$\Im(w_n)$ first root	$\Re(w_n)$ second root	$\Im(w_n)$ second root	$\Im(w_n)$ third root
1.043468e+00	−1.720711e−01	−1.043468e+00	−1.720711e−01	3.441421e−01
7.024772e−02	−1.419980e−01	−7.024772e−02	−1.419980e−01	2.839960e−01
1.142329e−02	3.071975e−02	−1.142329e−02	3.071975e−02	−6.143950e−02
−1.091518e−02	−4.841504e−03	1.091518e−02	−4.841504e−03	9.683008e−03
4.970991e−03	−1.951797e−03	−4.970991e−03	−1.951797e−03	3.903594e−03
−1.054098e−03	2.442148e−03	1.054098e−03	2.442148e−03	−4.884295e−03
−5.258184e−04	−1.314746e−03	5.258184e−04	−1.314746e−03	2.629492e−03
7.254685e−04	3.094154e−04	−7.254685e−04	3.094154e−04	−6.188308e−04
−4.209517e−04	1.698926e−04	4.209517e−04	1.698926e−04	−3.397852e−04
1.044921e−04	−2.466256e−04	−1.044921e−04	−2.466256e−04	4.932512e−04
6.059361e−05	1.493304e−04	−6.059361e−05	1.493304e−04	−2.986608e−04
−9.075015e−05	−3.828864e−05	9.075015e−05	−3.828864e−05	7.657728e−05
5.647976e−05	−2.299913e−05	−5.647976e−05	−2.299913e−05	4.599826e−05
−1.480041e−05	3.518087e−05	1.480041e−05	3.518087e−05	−7.036174e−05
−9.111747e−06	−2.231995e−05	9.111747e−06	−2.231995e−05	4.463991e−05
1.415206e−05	5.941154e−06	−1.415206e−05	5.941154e−06	−1.188231e−05
−9.106785e−06	3.724644e−06	9.106785e−06	3.724644e−06	−7.449287e−06
2.452801e−06	−5.852064e−06	−2.452801e−06	−5.852064e−06	1.170413e−05
1.559319e−06	3.807047e−06	−1.559319e−06	3.807047e−06	−7.614094e−06
−2.472161e−06	−1.034859e−06	2.472161e−06	−1.034859e−06	2.069718e−06
1.622215e−06	−6.652032e−07	−1.622215e−06	−6.652032e−07	1.330406e−06
−4.442284e−07	1.062295e−06	4.442284e−07	1.062295e−06	−2.124590e−06
−2.881222e−07	−7.019788e−07	2.881222e−07	−7.019788e−07	1.403958e−06
4.628646e−07	1.933972e−07	−4.628646e−07	1.933972e−07	−3.867944e−07
−3.076501e−07	1.263710e−07	3.076501e−07	1.263710e−07	−2.527421e−07
8.518782e−08	−2.040268e−07	−8.518782e−08	−2.040268e−07	4.080536e−07
5.601323e−08	1.362747e−07	−5.601323e−08	1.362747e−07	−2.725495e−07
−9.081658e−08	−3.789617e−08	9.081658e−08	−3.789617e−08	7.579234e−08
6.091256e−08	−2.505103e−08	−6.091256e−08	−2.505103e−08	5.010207e−08
−1.700133e−08	4.076397e−08	1.700133e−08	4.076397e−08	−8.152794e−08
−1.129051e−08	−2.744001e−08	1.129051e−08	−2.744001e−08	5.488002e−08
1.843034e−08	7.683251e−09	−1.843034e−08	7.683251e−09	−1.536650e−08
−1.244535e−08	5.122950e−09	1.244535e−08	5.122950e−09	−1.024590e−08
3.494465e−09	−8.385709e−09	−3.494465e−09	−8.385709e−09	1.677142e−08
2.338246e−09	5.678260e−09	−2.338246e−09	5.678260e−09	−1.135652e−08
−3.836810e−09	−1.598307e−09	3.836810e−09	−1.598307e−09	3.196613e−09
2.604422e−09	−1.072828e−09	−2.604422e−09	−1.072828e−09	2.145656e−09
−7.346986e−10	1.764225e−09	7.346986e−10	1.764225e−09	−3.528451e−09
−4.945310e−10	−1.200180e−09	4.945310e−10	−1.200180e−09	2.400361e−09
8.148225e−10	3.392325e−10	−8.148225e−10	3.392325e−10	−6.784651e−10
−5.554052e−10	2.289140e−10	5.554052e−10	2.289140e−10	−4.578280e−10

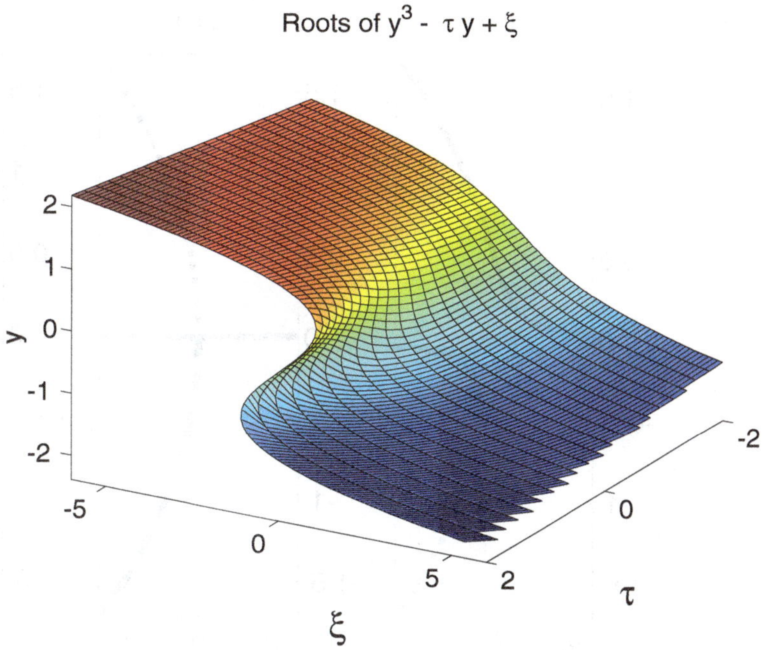

Figure 11.12. *Surface of real-valued roots of the reduced cubic equation in the plane of the linear and constant coefficients of the polynomial. Note that the surface folds for* $\tau > 0$*; there is an ever-widening region about the axis, bounded by the semicubical parabola that is the discriminant in the* τ-ξ *plane, where the surface is triple valued. Three cross-sections are given in Fig.* 11.13.

11.8.1 ▪ The Root Surface of the Cubic and Catastrophe Theory

Catastrophe theory is a branch of mathematical physics that tries to understand qualitative changes in behavior by looking at the big picture. If a one-dimensional potential energy function is a quartic polynomial, for example, the equilibrium positions of a particle in this potential can be found by solving a cubic equation. In some regions of parameter space, there will be just one equilibrium; in others, there will be three. Catastrophe theory asks and answers the question: What can we learn about the transition from one equilibrium to three by studying the topology of the root surface?

For the special case of a quartic potential/cubic for the equilibria, the answer is that the transition occurs when the root surface develops a fold or a pleat as τ increases. The region of three equilibria is the area of the τ-ξ plane that is enclosed by the discriminant of the cubic equation. Since the discriminant has a cusp at $\tau = 0$ where both the upper and lower branches of the semicubical parabola intersect and are simultaneously tangent to one another, this cubic for the equilibria governs the *cusp catastrophe*. It is one of seven elementary catastrophes, as described by Saunders [307] and Arnold [14].

11.9 ▪ Cubic Equation and the One-Dimensional Advection Equation

Another connection of the mathematics of the cubic equation with the physics of something seemingly far removed is that the root surface of the cubic is also a solution to the so-called one-dimensional advection equation, also known as the inviscid Burgers equation.

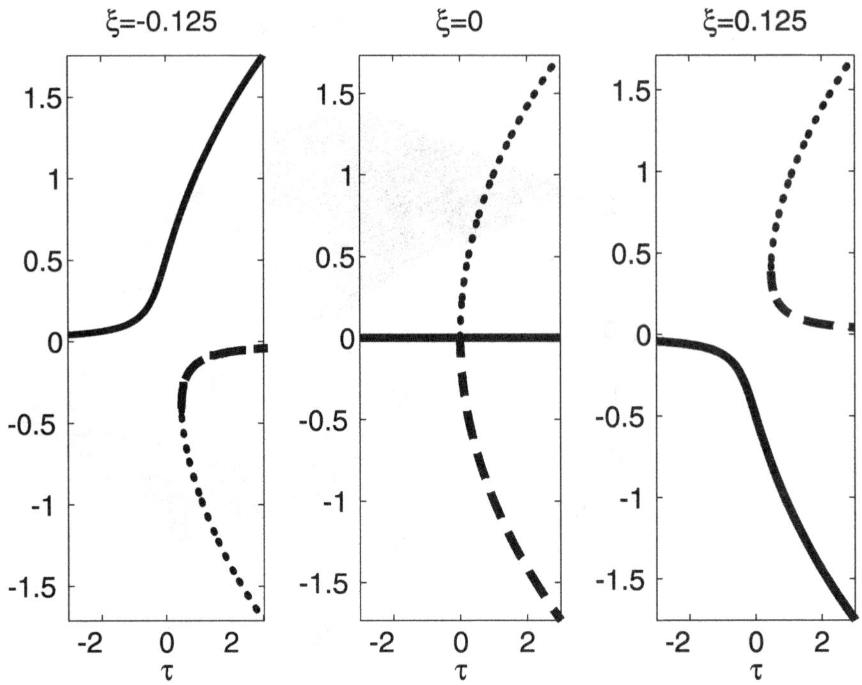

Figure 11.13. *Slices through the root surface of the reduced cubic equation, $y^3 - \tau y + \xi = 0$, for three different fixed values of ξ and various τ. The continuation of the root from negative τ to positive τ is shown as the heavy solid line; straightforward continuation in τ misses the two additional roots (dashed and dotted), which are real valued only for positive τ.*

Note that in this section only, the unknown is u instead of x so that x can have its usual hydrodynamic meaning as the spatial coordinate, and the parameter is t (time) instead of λ.

The reduced cubic can be written in the form

$$f(u;t,x) = u^3 - tu + x = 0. \tag{11.107}$$

The condition that the total derivative with respect to t is zero, $Df/Dt = 0$, gives

$$\frac{\partial f}{\partial t} + \frac{\partial f}{\partial x}\frac{\partial u}{\partial t} = 0 \quad \Rightarrow \quad \frac{\partial u}{\partial t} = \frac{u}{3u^2 - t}. \tag{11.108}$$

Similarly, $Df/Dx = 0$ requires

$$\frac{\partial u}{\partial x} = -\frac{1}{3u^2 - t}. \tag{11.109}$$

Elimination gives

$$\frac{\partial u}{\partial t}\bigg/\frac{\partial u}{\partial x} = -u \tag{11.110}$$

or, equivalently,

$$\frac{\partial u}{\partial t} + u\frac{\partial u}{\partial x} = 0. \tag{11.111}$$

In other words, the function $u(x,t)$ that solves the reduced cubic equation also solves the one-dimensional advection equation.

This is consistent with the fact that the isolines of u in the x-t plane are straight lines of slope u.

11.10 ▪ Phase-Plane Analysis and the Root Surface

It is easy to use the trigonometric solution or the built-in polynomial solvers in MATLAB, Maple, and other software packages to compute the roots at an arbitrary point in the τ-ξ plane. How does one connect these discrete points into a smooth but folded and triple-valued surface? Perhaps the simplest way to graph sections of the surface at given values of τ is to use a *phase-plane* analysis.

The idea is to replace the Davidenko equation (Chapter 8), in which the marching variable is one of the parameters, by a pair of ODEs in which the marching variable is a dummy parameter introduced to parameterize what otherwise would be multivalued curves.

Define

$$f(x,\tau,\xi) \equiv x^3 - \tau x + \xi. \tag{11.112}$$

The partial derivatives are

$$f_x = 3x^2 - \tau; \qquad f_\xi = 1. \tag{11.113}$$

The Davidenko equation in ξ for fixed τ would be

$$dx/d\xi = -f_\xi/f_x = -1/(3x^2 - \tau). \tag{11.114}$$

This fails at a limit point where $dx/d\xi \to \infty$ because two roots have merged; this happens at the discriminant curve in the τ-ξ plane. However, if we introduce a new variable T, we can write [264, pp. 314–315]

$$d\xi/dT = f_x, \qquad dx/dT = -f_\xi. \tag{11.115}$$

These are identical to the equations of tangent vector continuation except for the omission of denominators of $(f_x)^2 + (f_\lambda)^2$, which normalize the continuation parameter so that T approximates s, the arclength. However, this normalization is strictly optional; the choice here is to leave the phase-plane equations unnormalized.

For the cubic equation, these phase-plane equations can be solved *analytically* to give

$$x = x_0 - T, \qquad \xi = \xi_0 + \left(3x_0^2 - \tau\right)T - 3x_0 T^2 + T^3. \tag{11.116}$$

There is no error in these parametric equations for the root surface; however, one must solve the cubic (usually at large $|\xi|$ where there is only a single real root) so that the parameter x_0 that appears in the phase-plane solution is a solution to the cubic at $\xi = \xi_0$ for the chosen τ.

11.11 ▪ Contours

The isolines or contours of the real-valued roots of the reduced cubic, $y(\tau,\xi)$, have a fascinating property: even though they are defined by a *nonlinear* equation, these curves of constant value of the root are *linear*, as expressed formally by the following theorem.

Theorem 11.7 (Contours of Cubic Roots). *The isoline where $y = U$, for any constant U, is given by the straight line*

$$\xi = U\tau + \xi_0 \tag{11.117}$$

in the plane spanned by the parameters (τ, ξ) of the reduced cubic.

Proof: Along the contour $y = U$, the reduced cubic is

$$U^3 - \tau U + \xi(\tau) = 0. \tag{11.118}$$

Rewrite the linear expression for ξ given in the theorem, $\xi = U\tau + \xi_0$, in the equivalent form obtained by adding $U\tau_0$ to the constant term and subtracting it from the linear term:

$$\xi = U(\tau - \tau_0) + \xi_1, \qquad \xi_1 = \xi_0 + U\tau_0, \tag{11.119}$$

where τ_0 is arbitrary.

If we replace ξ by the linear expression $\xi(\tau)$, then the cubic becomes

$$U^3 - \tau U + \{U(\tau - \tau_0) + \xi_1\} = 0. \tag{11.120}$$

We find that the terms in τ cancel. It follows that if U satisfies the cubic equation at the point (τ_0, ξ_1), then it will also satisfy the cubic everywhere on the line of slope U through that point, as asserted by the theorem. It is not necessary that the solution to the cubic be single valued nor that the contour line exist for $\tau = 0$. ■

Fig. 11.14 is a graphical proof of the theorem. Note that three sets of contour lines intersect in the region where there are three real roots.

Lalanne's 1843 nomograph used a contour plot like this to graphically solve the cubic equation [203]. (Plots that are graphical calculating devices are nomographs). His calculator by plot and ruler was so successful that it inspired a century of vigorous nomography that is not quite extinct, even now [205, 163, 164].

11.11.1 ▪ Contour Intersections with the Discriminant

Theorem 11.8 (Intersections with the Discriminant). *The contours of constant root, $y = U$, that exist for negative τ are tangent to the discriminant at the point*

$$\tau_t = 3U^2, \qquad \xi_t = 2U^3. \tag{11.121}$$

When $U > 0$, the tangency is with the upper branch of the discriminant. Tangent *means that the contour touches the discriminant curve at a point where the slopes of both the contour and the discriminant are equal. The point of tangency is the only point of intersection between the contour line and the discriminant.*

Proof: For simplicity, assume $U > 0$. This means that we can take the upper branch of the discriminant curve,

$$\xi_d(\tau) = \pm\sqrt{\frac{4}{27}}\,\tau^{3/2}. \tag{11.122}$$

The curve of the contour is

$$\xi_c(\tau) = U\tau - U^3. \tag{11.123}$$

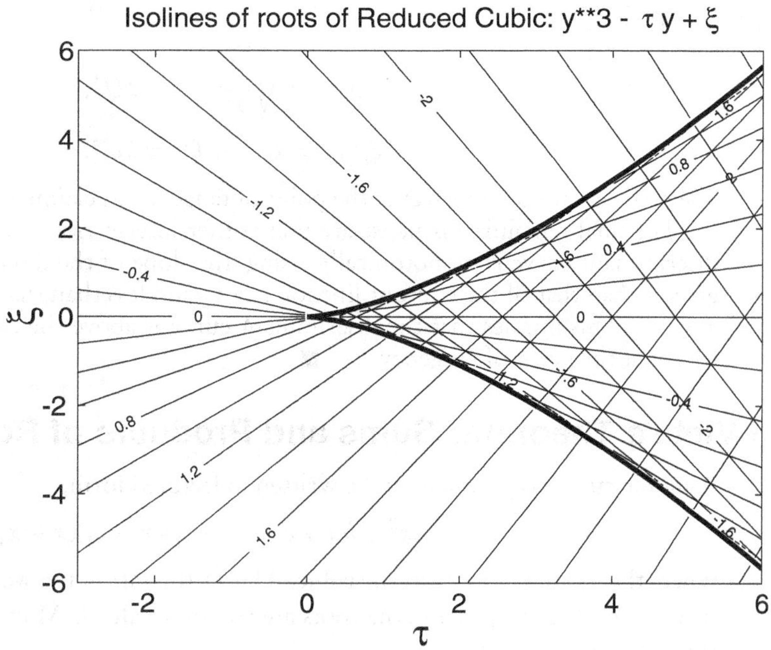

Figure 11.14. *Contours of real-valued roots of the reduced cubic equation in the plane of the linear and constant coefficients of the polynomial. In* 1843, *Louis Lalanne published a similar plot as a nomogram: to solve the reduced cubic graphically for a given* (τ, ξ), *find the given point on the contour plot, identify the isolines of y that bracket the point, and interpolate between those two contours to estimate x. The graphical approximation is crude (one or two decimal places), but this is always sufficient for rapid convergence of Newton's iteration. In the triple-root region, the same strategy must be applied to each of the three families of contour lines separately, but is still successful (Grattan-Guinness,* [203]*).*

The choice of the constant $\xi_0 = -U^3$ ensures that the reduced cubic equation is satisfied by $y = U$ at $\tau = 0$.

The slopes of the curves are, by explicit differentiation,

$$\frac{d\xi_d}{d\tau} = \sqrt{\frac{4}{27}\frac{3}{2}}\,\tau^{1/2} = \frac{1}{3^{1/2}}\tau^{1/2}, \tag{11.124}$$

$$\frac{d\xi_c}{d\tau} = U, \quad \text{a constant.} \tag{11.125}$$

The slope of the discriminant is zero at $\tau = 0$ and then grows without bound, whereas the slope of the contour line is always a positive constant. It follows that there must be a point τ_t where the slopes match. Equating the slopes gives

$$\frac{1}{3^{1/2}}\tau^{1/2} = U \qquad \leftrightarrow \qquad \tau_t = 3U^2. \tag{11.126}$$

One remaining issue is: do the values of the curves themselves match at the points where their slopes are equal?

We find

$$\xi_d(\tau_t) = \sqrt{\frac{4}{27}\tau_t^{3/2}} = 2U^3, \tag{11.127}$$

$$\xi_c(\tau_t) = U\tau_t - U^3 = 2U^3. \tag{11.128}$$

Thus, the curves do intersect at the point of tangency as claimed.

The final remaining issue is: are there other intersections? Because the slope of the discriminant curve is monotonically rising, the slope of the discriminant curve must be greater than that of the contour line for $\tau > \tau_t$ and less than that of the contour line for $\tau < \tau_t$. This implies that the discriminant curve is above the contour line everywhere except at the point of tangency. ■

11.12 ▪ Vieta's Theorem: Sums and Products of Roots

A monic cubic polynomial can be written in factored form,

$$x^3 + ax^2 + bx + c = (x - x_1)(x - x_2)(x - x_3), \tag{11.129}$$

where the x_j are the roots of the cubic. This factorization is always valid even when two or more roots are equal and the roots are complex valued. Multiplying the factors gives the following theorem.

Theorem 11.9 (Vieta's Root Identities: Cubic).

$$x_1 + x_2 + x_3 = -a, \qquad x_1x_2 + x_2x_3 + x_3x_1 = b, \qquad x_1x_2x_3 = -c. \tag{11.130}$$

Historically, these relationships were very useful in looking for roots that are integers. For example, if $c = 8$ and all three roots are integers, the only possible roots are the integral factors of 8: $\pm 1, \pm 2, \pm 4, \pm 8$. Algebraic manipulation systems like Maple and Mathematica are very good at factoring polynomials if the roots are integers, however, so trial-and-error searches for integer zeros are obsolete.

Nevertheless, these relationships are sometimes still useful for physical interpretation. For example, the dispersion relation for equatorial waves in oceanography is a cubic polynomial in the frequency ω,

$$\omega^3 - \left(2n + 1 + k^2\right)\omega - k, \tag{11.131}$$

where n is the *latitudinal mode number*, an integer, and k is the east-west wavenumber. In the oceanographic literature, the roots are usually written approximately as

$$\omega_{Rossby} \approx -\frac{k}{2n + 1 + k^2}, \qquad \omega_{gravity} \approx \pm\sqrt{2n + 1 + k^2}. \tag{11.132}$$

This implies that the two "gravity" waves travel at equal but opposite velocities. The root sum relationship tells us, however, that this cannot be so; since the quadratic coefficient a is zero, the sum of the roots must be zero, too. It follows that the correct relationship between the frequencies of the gravity waves must be

$$\omega_{gravity,1} = -\omega_{gravity,2} - \omega_{Rossby}. \tag{11.133}$$

The next perturbative order gives a *Vieta-compatible* approximation

$$\omega_{gravity} \approx \pm\sqrt{2n + 1 + k^2} + \frac{1}{2}\frac{k}{2n + 1 + k^2}. \tag{11.134}$$

Chapter 12

Roots of a Quartic Polynomial

"A series or explicit expression sprawling over many pages can often be replaced by an iteration filling half a line."
—John P. Boyd (unpublished)

12.1 ▪ Introduction: Reduction by Translation

The general quartic equation in monic form is

$$P \equiv x^4 + ax^3 + bx^2 + cx + d = 0. \tag{12.1}$$

Lodovico Ferrari (1522–1565) found a general analytic (literal) solution that was published in Cardano's *Ars Magna* in 1545, but it is extremely messy. However, one parameter can be eliminated by a simple translation without changing the qualitative character of the roots. Define

$$x \equiv y - a/4. \tag{12.2}$$

Then the *reduced quartic* is

$$P = y^4 + P_2 y^2 + P_1 y + P_0, \tag{12.3}$$

where

$$P_2 = b - \frac{3}{8}a^2, \tag{12.4}$$

$$P_1 = -\frac{1}{2}ab + \frac{1}{8}a^3 + c, \tag{12.5}$$

$$P_0 = -\frac{1}{4}ca + d + \frac{1}{16}a^2 b - \frac{3}{256}a^4. \tag{12.6}$$

The next section describes the analytic (literal) solution of the reduced quartic.

12.2 ▪ Analytic Solution of the Reduced Quartic

Theorem 12.1 (Analytic Solution of the Reduced Quartic). *The four roots of the reduced quartic,*

$$y^4 + P_2 y^2 + P_1 y + P_0 = 0, \tag{12.7}$$

are given by

$$y = \pm\frac{1}{6}\frac{\sqrt{\mathscr{C}}}{\mathscr{B}^{1/6}} \pm \frac{1}{6\mathscr{B}^{1/6}\mathscr{C}^{1/4}}\Big\{\sqrt{\mathscr{C}}\left[-12P_2\mathscr{B}^{1/3} - 9\mathscr{B}^{2/3} - 12P_0 - P_2^2\right] \\ -3P_1\sqrt{6}\sqrt{-72P_0P_2 + 27P_1^2 + 2P_2^3 + 3\mathscr{A}}\Big\}^{1/2}, \quad (12.8)$$

$$\mathscr{A} \equiv \sqrt{3}\sqrt{-256P_0^3 + 128P_0^2P_2^2 - 16P_0P_2^4 - 144P_0P_2P_1^2 + 27P_1^4 + 4P_1^2P_2^3}, \quad (12.9)$$

$$\mathscr{B} \equiv -\frac{4}{3}P_0P_2 + \frac{1}{2}P_1^2 + \frac{1}{27}P_2^3 + \frac{1}{18}\mathscr{A}, \quad (12.10)$$

$$\mathscr{C} \equiv -6P_2(\mathscr{B})^{1/3} + 9(\mathscr{B})^{2/3} + 12P_0 + P_2^2. \quad (12.11)$$

Multiple roots occur whenever the discriminant $\Delta = 0$, *where the discriminant is defined by*

$$\Delta = 256P_0^3 - 128P_0^2P_2^2 + 16P_0P_2^4 + 144P_0P_2P_1^2 - 27P_1^4 - 4P_1^2P_2^3. \quad (12.12)$$

12.3 ▪ Dilation: Doubly Reduced Quartic

Because the explicit solution to the reduced quartic is very complicated, it is highly advantageous to employ dilation (i.e., rescaling the magnitude of the roots) to eliminate another parameter. Define

$$y \equiv \rho z. \quad (12.13)$$

After dividing by a common factor, the reduced quartic becomes

$$P(y) = z^4 + \frac{P_2}{\rho^2}z^2 + \frac{P_1}{\rho^3}z + \frac{P_0}{\rho^4}. \quad (12.14)$$

To avoid complex values, which are inconvenient when all the coefficients of the quartic are real valued, choose (when $P_2 \neq 0$)

$$\rho = \sqrt{|P_2|}. \quad (12.15)$$

The reduced quartic simplifies to

$$P(z) = z^4 + \sigma z^2 + p_1 z + p_0, \quad (12.16)$$

where the three values of σ, each corresponding to a distinct case, are

$$\sigma = \begin{cases} 1, & P_2 > 0, \\ 0, & P_2 = 0, \\ -1, & P_2 < 0, \end{cases} \quad (12.17)$$

and where

$$p_1 = \frac{1}{\rho^3}P_1, \qquad p_0 = \frac{1}{\rho^4}P_0. \quad (12.18)$$

12.4 ▪ The Discriminant and the Swallowtail Catastrophe

In catastrophe theory, the discriminant of a quartic polynomial is the surface of the *swallowtail catastrophe*, one of the seven elementary catastrophes. By custom, the discriminant

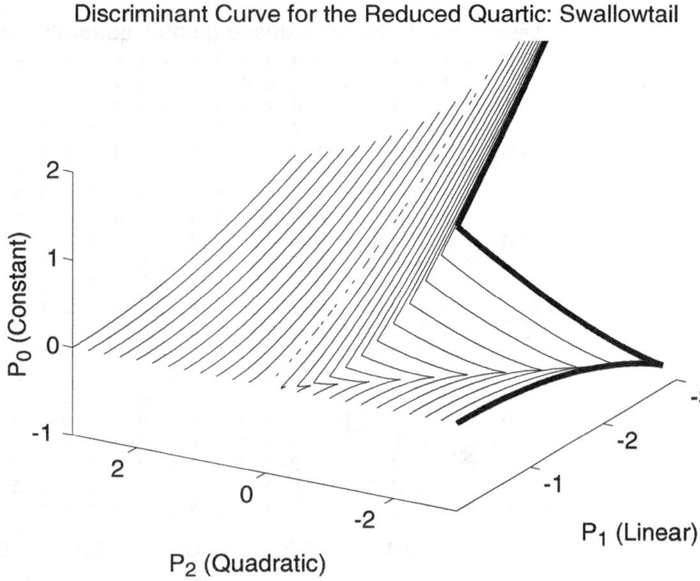

Figure 12.1. *The swallowtail surface (discriminant of the reduced quartic equation), plotted as a function of the quadratic, linear, and constant coefficients, P_2, P_1, P_0. Because the surface is symmetric with respect to the linear (P_1) axis, only negative P_1 is shown. The surface is single valued when the quadratic coefficient is positive, i.e., $P_2 > 0$, but triple valued in the constant (vertical, P_0) coordinate for fixed values of the other two coefficients. The curve that is the $P_2 = -2$ slice of the discriminant surface is shown as a thick solid line to make it easier to visualize the swallowtail shape.*

surface is usually illustrated for the once-reduced, three-parameter quartic, as in Fig. 12.1 here. A plot of the discriminant in the three-dimensional space spanned by the quadratic, linear, and constant coefficients of the reduced quartic allows us to see all three distinct cases of the doubly reduced quartic as a single, continuous surface.

The discriminant surface by definition is where multiple roots occur so that the number of complex-valued roots changes in crossing from one side of the surface to the other. In catastrophe theory, such changes in the number of roots are changes in the character of the equilibrium states of a dynamical system that has a one-dimensional potential energy that can be described (or approximated) by a fifth-degree polynomial in the spatial coordinate. The discriminant surface is the *bifurcation set* for the dynamical system because the number of equilibrium states changes as the surface is crossed.

To visualize the number of complex-valued roots at each point, it is convenient to use the doubly reduced quartic. Each case can then be represented by a two-dimensional graph that is equivalently a cross-section at $P_2 = \pm 1$ through the surface in three-dimensional coefficient space illustrated in Fig. 12.1. Figs. 12.2, 12.3, and 12.4 illustrate the two nontrivial cases. For the third case, $\sigma = 0$ (and therefore the quadratic coefficient $P_2 = 0$), the discriminant surface is analytically given by $p_0 = (3/8)2^{1/3} p_1^{4/3}$.

12.4.1 ▪ Analysis of the Discriminant of the Rescaled, Reduced Quartic

The discriminant of

$$P(z) = z^4 - z^2 + p_1 z + p_0 \tag{12.19}$$

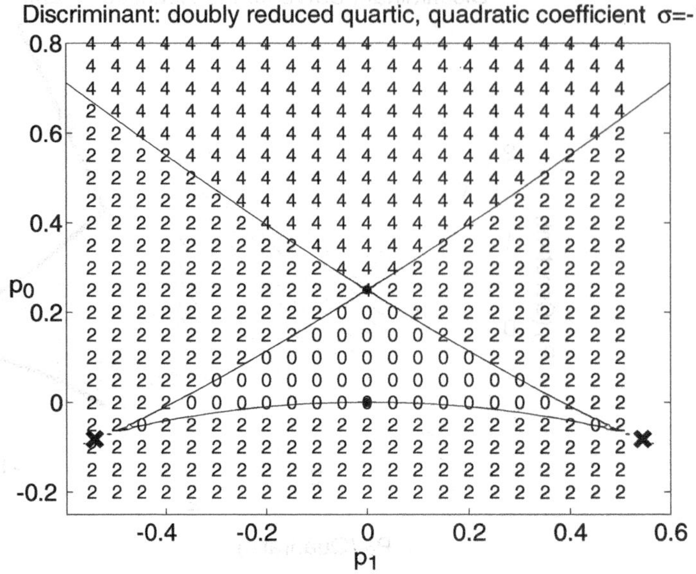

Figure 12.2. *The discriminant surface for the doubly reduced quartic $z^4 + \sigma z^2 + p_1 z + p_0$ for $\sigma = -1$. (This is identical to a slice through the three-dimensional parameter space of the reduced quartic for $P_2 = -1$.) The horizontal axis is the coefficient of the linear term in the quartic, p_1. The vertical coordinate is the constant p_0 in the fourth-degree polynomial. The numbers indicate the number of complex-valued roots at each point in the p_1-p_0 plane. The black disks are the analytically determined points where the swallowtail crosses the p_1-axis. The x's are the analytically determined tips of the tail.*

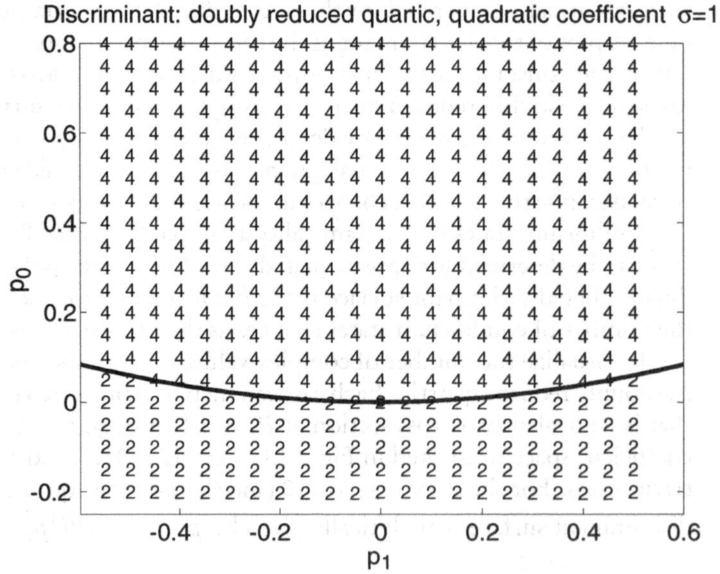

Figure 12.3. *Same as the previous figure but for $\sigma = 1$, that is, when the coefficient of the quadratic term is positive. The numbers indicate the number of complex-valued roots at each point in the p_0-p_1 plane: four complex-valued roots for p_0 above the discriminant curve, two real and two complex zeros when p_0 is below the discriminant.*

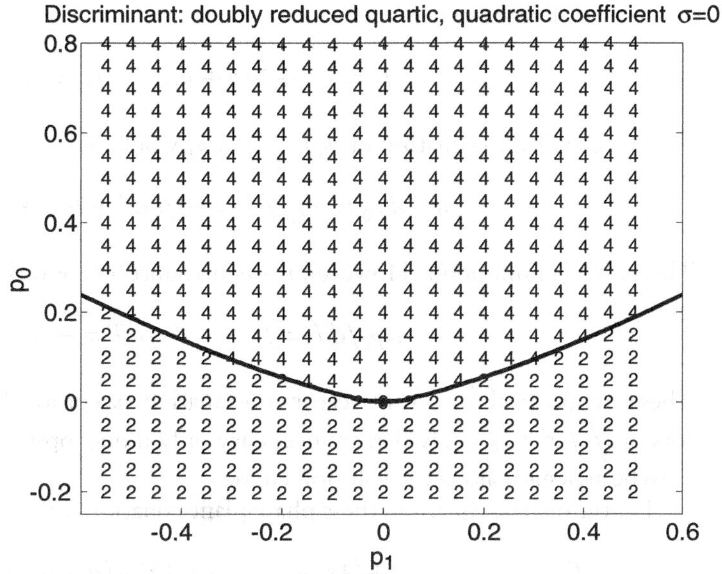

Figure 12.4. *Same as the previous figure but for $\sigma = 0$, that is, when the coefficient of the quadratic term is positive. The numbers indicate the number of complex-valued roots at each point in the p_0-p_1 plane: four complex-valued roots wherever p_0 is above the discriminant curve; two real and two complex zeros when p_0 is below the discriminant.*

is

$$\Delta = 4p_1^2 - 27p_1^4 + 16p_0 - 128p_0^2 - 144p_1^2 p_0 + 256p_0^3. \tag{12.20}$$

This can be solved as a quadratic equation in p_1^2 to give

$$p_1^2 = \frac{2}{27} - \frac{8}{3}p_0 \pm \frac{2}{27}(1 + 12p_0)^{3/2}. \tag{12.21}$$

One finds that when $p_1 = 0$, then $p_0 = 0, 1/4$, where the second value is actually a double root. Thus, we can analytically determine where the swallowtail crosses the p_0-axis (black disks in Fig. 12.2).

The argument of the square root is zero when

$$p_0 = -1/12, p_1 = \pm\sqrt{8/27}, \tag{12.22}$$

which analytically gives the cusps of the curve, marked by a pair of x's in Fig. 12.2.

As p_0 increases toward large positive values,

$$p_1 \sim \pm \frac{4}{3} 3^{1/4} p_0^{3/4}, \tag{12.23}$$

so that the upper branch of the swallowtail is unbounded.

12.5 ▪ Phase-Plane Analysis and the Root Surface

Define

$$f(z, p_1, p_0) \equiv z^4 + \sigma z^2 + p_1 z + p_0. \tag{12.24}$$

The partial derivatives are

$$f_z = 4z^3 + 2\sigma z + p_1, \qquad f_{p_0} = 1. \tag{12.25}$$

The Davidenko equation in p_0 for fixed p_1 would be

$$dz/dp_0 = -f_{p_0}/f_z = -1/(4z^3 + 2\sigma z + p_1). \tag{12.26}$$

This fails at a limit point. However, if we introduce a new variable T, we can write

$$dp_0/dT = f_z, \qquad dz/dT = -f_{p_0}. \tag{12.27}$$

These are identical to tangent vector continuation except for the omission of denominators of $\sqrt{(f_z)^2 + (f_{p_0})^2}$, which are necessary only if one (optionally) wishes to normalize T to be arclength along the solution curve.

For the quartic equation, these phase-plane equations can be solved *analytically* to give

$$z = z_0 - T, \qquad p_0 = p_0(0) + \left(4z_0^3 + 2\sigma z_0 + p_1\right)T - \left(\sigma + 6z_0^2\right)T^2 + 4z_0 T^3 - T^4. \tag{12.28}$$

There is no error in these parametric equations for the root surface; however, one must solve the quartic at $p_0 = p_0(0)$ for some constant $p_0(0)$ to determine z_0.

12.6 ▪ Contours of Constant Root for the Doubly Reduced Quartic

The contours of the real-valued roots of the quartic, like those of the reduced cubic, are *linear*, as expressed formally by the following theorem, illustrated in Fig. 12.5.

Theorem 12.2 (Linearity of Contours of Quartic Roots). *The isoline where $z = Z$, for any constant Z, is given by the straight line*

$$p_0 = -Z\,p_1 + p_0(0) \tag{12.29}$$

in the plane spanned by the parameters (p_1, p_0) of the doubly reduced quartic.

Proof: Define a line in the p_0-p_1 plane by $p_0 = S p_1 + p_0(0)$ for some constant S. Along this line, $\partial f/\partial p_1$, which is the numerator of the Davidenko equation, is

$$\partial f/\partial p_1 = z + S. \tag{12.30}$$

If we choose $S = -z$, then $\partial f/\partial p_1 = 0$ along the line. It follows that z is unchanging on this line, which is therefore a contour curve of constant z. Repeating this argument for arbitrary S shows that all such lines are contours with $z = -S$. ■

12.7 ▪ Sums and Products of Roots: Vieta's Theorem

A monic quartic polynomial can be written in factored form,

$$x^4 + ax^3 + bx^2 + cx + d = (x - x_1)(x - x_2)(x - x_3)(x - x_4), \tag{12.31}$$

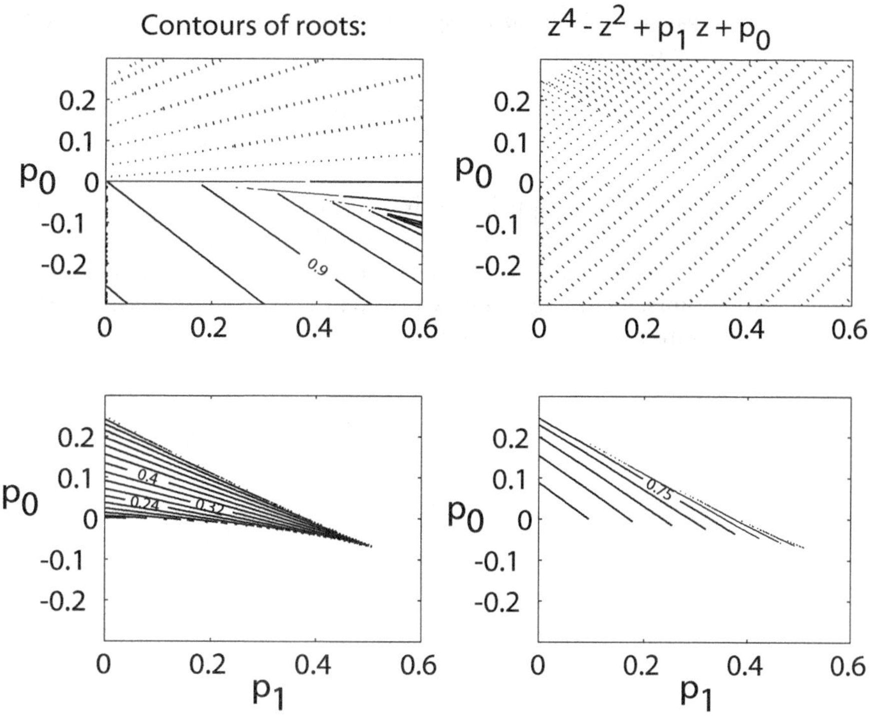

Figure 12.5. *Isolines of the real-valued roots of the doubly reduced quartic. Each plot is a different zero. Solid contours indicate positive roots; negative isolines are dashed. There are two real roots through the entire p_1-p_0 plane when the coefficient of the quadratic coefficient is negative (top panels). The contours span only a curvilinear triangle in the lower two graphs because there are four real roots only within the triangle.*

where the x_j are the roots of the quartic. This factorization is always valid even when two or more roots are equal and the roots are complex valued. Multiplying the factors gives the following theorem.

Theorem 12.3 (Vieta's Root Identities: Quartic).

$$x_1 + x_2 + x_3 + x_4 = -a, \tag{12.32}$$

$$x_1x_2 + x_1x_3 + x_1x_4 + x_2x_3 + x_2x_4 + x_3x_4 = b, \tag{12.33}$$

$$x_1x_2x_3 + x_1x_2x_4 + x_1x_3x_4 + x_2x_3x_4 = -c, \tag{12.34}$$

$$x_1x_2x_3x_4 = d. \tag{12.35}$$

These relationships are sometimes still useful for physical interpretation, as illustrated for cubic equations in the previous chapter. They are also useful in checking the choice of signs for the square roots and cube roots in the (messy!) explicit solution to the quartic.

12.8 ▪ The Reduced Quartic Is Equivalent to Intersecting Conic Sections

$$P = y^4 + P_2y^2 + P_1y + P_0 \tag{12.36}$$

is equivalent to the bivariate system of two equations in the new unknowns (u, v), where $u = y^2$ and $v = y$ [168]:

$$u^2 + p_2 u + p_1 v + p_0 = 0, \tag{12.37}$$

$$u - v^2 = 0. \tag{12.38}$$

The top equation implicitly defines a conic section (ellipse, hyperbola, or parabola); the second curve is always a parabola through the origin. Solving the system is equivalent to finding the intersections of the two curves. Two conic sections can intersect in at most four points, including complex-valued intersections, reflecting the fact that a quartic polynomial may have at most four roots.

Part V

Analytical Methods

Chapter 13
Methods for Explicit Solutions

"Often, usefulness is inversely proportional to generality; the wider the range of a method, the smaller its value for your particular problem. Special tricks that work only for a single problem may be the only way to unlock that problem."
—Tai Tsun Wu (1933–), Harvard lecture (1973)

"The opposite of a correct statement is a false statement. But the opposite of a profound truth may well be another profound truth."
—Niels Bohr (1885–1962)

13.1 ▪ Introduction

It is rather unusual to find closed-form, explicit solutions to a nonlinear equation or a system of equations. Nevertheless, there is a fairly wide range of problems that are amenable to analytic, nonperturbative techniques.

Wu's assertion that special, narrow methods—we shall dub them "stenarithms" from the Greek word for "narrow"—are the most useful may seem to denigrate the quest in most of this book to describe methods of broad generality. However, Bohr's Zen-like quotation is also true. Very broad and very narrow methods may both be powerful. The search for explicit solutions is a pursuit where Wu-like narrow tricks are sometimes the only option.

Even so, there are several broad strategies for pursuing literal (analytic) formulas for zeros, including the following:

1. Apply a computer algebra system.
2. If the problem involves special functions, find a mathematician who is an expert in special functions.
3. Look for special cases: often these have an explicit solution when the general case does not.
4. If the function is a polynomial, trigonometric, or hyperbolic polynomial, or involves exponentials, there are special techniques described in the rest of the chapter.
5. Graph some solutions; the visual has ever been an inspiration to the analytical.

Computer algebra systems like Maple and Mathematica have "solve" commands that return explicit solutions to surprisingly complicated problems. The good news is that these systems are steadily getting "smarter" as new expertise from the symbolic computing research algebra is baked into the software. The bad news is that there are vast numbers of explicitly solvable equations that the software cannot yet solve. But Maple is nonetheless a very painless first step.

Transcendentals like Bessel functions, confluent hypergeometric functions, and so on, satisfy a vast number of relationships; an electrical engineer (as a hobby) wrote an entire book about the identities and formulas of the dilogarithm function [242]. SIAM has a Special Interest Group on Orthogonal Polynomials and Special Functions: an organic interactive archive of arcane but useful lore to be tapped when local expertise is inadequate.

Solutions for special cases often illuminate properties of the general case. More important, special-case explicit solutions are very useful to initialize a numerical continuation to calculate the rest of the solution branch.

13.2 ▪ Polynomials

13.2.1 ▪ Univariate Polynomials

Excluding special cases, a single polynomial equation in a single unknown can be solved explicitly if and only if the degree is four or less. The quadratic, cubic, and quartic formulas are given in Chapters 10, 11, and 12, respectively. Explicit solutions in terms of elliptic modular functions or multidimensional theta functions formally exist for quintic and higher equations. However, in practice, these solutions are too complicated to be useful. Note, though, that it is often possible to find explicit solutions for *particular* polynomial equations of fifth or higher degree, even though this is not possible for the *general* equation of this order.

The best way to find simple, explicit solutions of higher-degree polynomial equations is to apply a good computer algebra system like Maple or Mathematica. "Under the hood," powerful algorithms are applied to look for a simple factorization. If the search is successful, then one need not know the underlying Galois theory/substitution/algebraic geometry methods to understand the answer.

A pair of bivariate quadratic polynomials, whose roots are the intersection of conic sections, can always be reduced to a single quartic univariate polynomial.

13.3 ▪ Surds and Radicals

An equation with radicals ("surds") can often be converted into ordinary polynomial equations by introducing a new unknown. For example,

$$x^2 + 5\sqrt{x} + 2 = 0 \tag{13.1}$$

is transformed by $y \equiv \sqrt{x}$ into

$$y^4 + 5y + 2 = 0. \tag{13.2}$$

The slightly more complicated case of

$$x^2 + 5\sqrt{x+3} + 2 = 0 \tag{13.3}$$

requires the new variable

$$y \equiv \sqrt{x+3} \qquad \longleftrightarrow \qquad y^2 = x + 3 \qquad \longleftrightarrow \qquad x = y^2 - 3 \tag{13.4}$$

to become the polynomial equation

$$y^4 - 6y^2 + 5y + 11 = 0. \tag{13.5}$$

Some problems require the introduction of a new variable that does not replace the old, but instead becomes an *extra unknown*. For example,

$$x^2 + 5\sqrt{x+3} + \sqrt{x+5} = 0 \tag{13.6}$$

cannot be converted into a single polynomial equation by a change of variables because there are two radicals. We therefore define *two* new unknowns:

$$y \equiv \sqrt{x+3}, \qquad w \equiv \sqrt{x+5}. \tag{13.7}$$

The original equation is transformed to the polynomial *system* of *two* equations in *two* unknowns:

$$y^4 - 6y^2 + 5y + 9 + w = 0, \qquad w^2 - y^2 - 2 = 0. \tag{13.8}$$

Substituting $w = -(y^4 - 6y^2 + 5y + 9)^2$ yields the univariate polynomial

$$y^8 - 12y^6 + 10y^5 + 54y^4 - 60y^3 - 84y^2 + 90y + 79 = 0.$$

This yields four real roots: −2.6519887,−2.465761,−1.059048,−0.795764. Substitution into the original yields the corresponding w-values

$$-3.005502,\ 2.8425305,\ 1.766800,\ -1.6227263.$$

13.4 ▪ Equations Involving Trigonometric Functions

There are three methods for transforming a *trigonometric* polynomial rootfinding problem into an ordinary polynomial zero calculation:

1. complex-valued substitution $z = \exp(ix)$,
2. tan half-angle substitution, and
3. cosine and sine pair substitution: $c = \cos(x)$, $\sin(x) = \sqrt{1-c^2}$.

All have both virtues and flaws, and therefore all three are in widespread use.

Jan Gullberg's *Mathematics from the Birth of Numbers*, a thousand-page love poem from an amateur mathematician to his passion, contains the best elementary discussion of explicit solutions to trigonometric equations in one or two unknowns [204, pp. 516–526].

13.4.1 ▪ Roots of a Trigonometric Polynomial: $z = \exp(ix)$ Conversion

Our goal is to find the real zeros on the canonical interval $x \in [-\pi, \pi]$ of the trigonometric polynomial

$$f(x;N) = \frac{1}{2}a_0 + \sum_{n=1}^{N} a_n \cos(nx) + \sum_{n=1}^{N} b_n \sin(nx) + \frac{1}{2}a_{N+1}\cos([N+1]x). \tag{13.9}$$

In this subsection, the strategy is to convert the trigonometric polynomial into an ordinary polynomial and then apply standard polynomial software.

Note that if

$$z \equiv \exp(ix) \qquad \leftrightarrow \qquad x = \text{angle}(z), \tag{13.10}$$

then

$$\cos(nx) = \frac{1}{2}\left(z^n + z^{-n}\right), \qquad \sin(nx) = -i\frac{1}{2}\left(z^n - z^{-n}\right). \tag{13.11}$$

The polynomial becomes

$$f(x;N) = \frac{1}{2}a_0 + \frac{1}{2}\sum_{n=1}^{N} a_n\left(z^n + z^{-n}\right) \\ - i\frac{1}{2}\sum_{n=1}^{N} b_n\left(z^n - z^{-n}\right) + \frac{1}{4}a_{N+1}\left(z^{N+1} + z^{-(N+1)}\right). \tag{13.12}$$

This can be rewritten as

$$f(x;N) = \frac{1}{2}z^{-(N+1)}\left\{a_0 z^{N+1} + \sum_{n=1}^{N} a_n\left(z^{n+N+1} + z^{N+1-n}\right)\right. \\ \left. - i\sum_{n=1}^{N} b_n\left(z^{N+1+n} - z^{N+1-n}\right) + \frac{1}{2}a_{N+1}\left(z^{2N+2} + 1\right)\right\}. \tag{13.13}$$

Define the polynomial g as

$$f = (1/2)z^{-(N+1)}\, g(z) \qquad \Leftrightarrow \qquad g \equiv 2z^{N+1} f(z). \tag{13.14}$$

Then

$$g(z) = \frac{1}{2}a_{N+1} + \frac{1}{2}a_{N+1}z^{2N+2} + a_0 z^{N+1} + \sum_{n=1}^{N} a_n\left(z^{n+N+1} + z^{N+1-n}\right) \\ - i\sum_{n=1}^{N} b_n\left(z^{N+1+n} - z^{N+1-n}\right). \tag{13.15}$$

This can be rewritten as

$$g(z) = \sum_{k=0}^{2N+2} g_k z^k, \tag{13.16}$$

where

$$g_0 = g_{2N+2} = \frac{1}{2}a_{N+1}, \qquad g_{N+1} = a_0, \tag{13.17}$$

$$g_k = a_{-k+N+1} + i\, b_{-k+N+1},\ k = 1, 2, \ldots, N, \tag{13.18}$$

$$g_k = a_{k-(N+1)} - i\, b_{k-(N+1)}, k = N+2, 2, \ldots, 2N+1. \tag{13.19}$$

The real roots in x on $x \in [-\pi, \pi]$ are the *angles* of the roots of $g(z)$ on the *unit circle*, that is,

$$x_j^{root} = \text{angle}(z_j^{root})\ \forall\, z_j^{root} \text{ such that } \left|z_j^{root}\right| = 1. \tag{13.20}$$

For example, the trigonometric polynomial

$$f = a_0 + a_1 \cos(x) + b_1 \sin(x) + (1/2) a_2 \cos(2x) \tag{13.21}$$

is converted to

$$g = \frac{1}{2} a_2 z^4 + (a_1 - i\, b_1) z^3 + 2 a_0 z^2 + (i b_1 + a_1) z + \frac{1}{2} a_2. \tag{13.22}$$

For the particular case

$$f = 1 + \frac{3}{2}\cos(x) + \frac{1}{2}\sin(x) + \frac{1}{16}\cos(2x), \tag{13.23}$$

the corresponding polynomial in g is

$$g = \frac{1}{16} z^4 + \left(\frac{3}{2} - i\frac{1}{2}\right) z^3 + 2z^2 + \left(\frac{3}{2} + i\frac{1}{2}\right) z + \frac{1}{16}. \tag{13.24}$$

The roots of g are $z = -22.786 + 8.419i, -0.858 + 0.514i, -0.317 - 0.948i, -0.0386 + 0.0143i$. The first and fourth roots are far from the unit disk and therefore irrelevant to the real roots of $f(x)$. Taking the argument or angle of the second and third roots (which *are* on the unit disk) gives

$$\begin{aligned} x_1^{root} &= \text{angle}(-0.858 + 0.514i) = 2.60155, \\ x_2^{root} &= \text{angle}(-0.317 - 0.948i) = -1.89365, \end{aligned} \tag{13.25}$$

which are indeed the two roots of $f(x)$ on $x \in [-\pi, \pi]$.

13.4.2 ▪ New Variable: $c = \cos(x)$, $\sin(x) = \sqrt{1 - c^2}$

Trigonometric functions satisfy many identities, such as

$$\cos^2(x) + \sin^2(x) = 1. \tag{13.26}$$

Define a new variable c via

$$c \equiv \cos(x), \qquad \sin(x) = \sqrt{1 - c^2}. \tag{13.27}$$

Gander and Gruntz [182] comment that "It is a bad idea to substitute $\cos(x)$ by a variable c and then $\sin(x)$ by $\sqrt{1-c^2}$, because new roots are introduced which make the equation more difficult to solve. Furthermore, if it is solvable, new solutions are introduced, which are not solutions of the problem, because squaring is necessary to get rid of the roots."

This isn't the whole story; in complicated problems, the square root factors that are the spoor of the sine may be eliminated by further algebraic manipulations. Solving nonlinear algebraic equations is hard; one must be prepared to try many strategies.

This convert-trig-to-algebra strategy is ubiquitous in robotics, where determining the correct angle in a movement of a robot arm is a problem in trigonometric rootfinding, usually in more than one unknown [278, 361, 337, 36].

13.4.3 ▪ Tan Half-Angle Substitution: Rationalizing Trigonometric Functions

Define

$$t \equiv \tan(x/2) \qquad \longleftrightarrow \qquad x = 2\arctan(t). \tag{13.28}$$

Then the sin and cosine function are *rationalized*:

$$\sin(x) = \frac{2t}{1+t^2}, \qquad \cos(x) = \frac{1-t^2}{1+t^2}, \qquad \tan(x) = \frac{2t}{1-t^2}. \tag{13.29}$$

By multiplying through by powers of $(1+t^2)$ and/or $(1-t^2)$, a trigonometric equation can thus be converted into an ordinary algebraic polynomial equation. The roots of the trigonometric polynomial are twice the arctangent of each root of the algebraic polynomial.

For example, consider

$$f(x) = 1 + (1/2)\cos(x) - \sin^2(x) \qquad \longleftrightarrow \qquad g(t) = 3 - 4t^2 + t^4. \tag{13.30}$$

The roots are

$$t = \pm 1, \pm\sqrt{3} \qquad \rightarrow \qquad x = \pm\frac{\pi}{2} + 2\pi m, \; \pm\frac{\pi}{3} + 2\pi m, \tag{13.31}$$

where m is an arbitrary integer.

Schweikard gives explicit expressions for the coefficients of the algebraic polynomials in terms of the coefficients of the trigonometric polynomials [311, 312].

13.4.4 ▪ Parity Symmetries

When the trigonometric polynomial is composed of only a subset of a general Fourier polynomial because of parity symmetries with respect to the origin and perhaps $x = \pi/2$, both numerical and explicit methods benefit. Polynomials have parity if they belong to any of the following six classes:

1. only cosines (plus the constant),
2. only sines,
3. only even cosines $\{\cos(2nx), n = 0, 1, 2, \ldots\}$,
4. only even sines $\{\sin(2nx), n = 0, 1, 2, \ldots\}$,
5. only odd cosines $\{\cos([2n-1]x), n = 1, 2, \ldots\}$,
6. only odd sines $\{\sin([2n-1]x), n = 1, 2, \ldots\}$.

Parity is discussed at length in section 4.6, so we shall only illustrate two cases in sections 13.4.7 and 13.4.8.

13.4.5 ▪ Linear Trigonometric Polynomial

The general linear trigonometric polynomial is

$$f_1(t) \equiv a_0 + a_1\cos(t) + b_1\sin(t). \tag{13.32}$$

Define

$$S \equiv \sqrt{b_1^{\,2}\left(a_1^{\,2}-a_0^{\,2}+b_1^{\,2}\right)}, \qquad R \equiv a_1^{\,2}+b_1^{\,2}. \tag{13.33}$$

The roots of f_1 are $t^{\pm}$:

$$t^{+} = -\arctan\left(\frac{a_0 b_1^{\,2}+a_1 S}{R b_1}, \frac{-a_0 a_1 + S}{R}\right), \tag{13.34}$$

$$t^{-} = -\arctan\left(\frac{a_0 b_1^{\,2}-a_1 S}{R b_1}, \frac{-a_0 a_1 - S}{R}\right), \tag{13.35}$$

using the two-argument form of the arctangent function. These solutions were derived using the half-angle tangent substitution described above.

13.4.6 ▪ Quadratic ($N = 2$) Trigonometric Solutions: Sine, Cosine, and General

The quadratic trigonometric polynomial is

$$f_2(t) = a_0 + a_1\cos(t) + a_2\cos(2t) + b_1\sin(t) + b_2\sin(2t). \tag{13.36}$$

Theorem 13.1 (Number of Roots of a Real Quadratic Trigonometric Polynomial). *A quadratic trigonometric equation with real coefficients must have zero, two, or four real-valued roots on $t \in (-\pi, \pi]$ when roots are counted according to their multiplicity.*

Proof: Apply the transformation $z = \exp(ix)$ described above to convert the trigonometric polynomial to a quartic algebraic polynomial and apply the fundamental theorem of algebra. ∎

Theorem 13.2 (Other Properties of Quadratic Trigonometric Polynomials). *The general quadratic trigonometric polynomial is*

$$f(t) \equiv \sum_{j=0}^{2} a_j \cos(jt) + \sum_{j=1}^{2} b_j \sin(jt). \tag{13.37}$$

Let t_j denote its four roots, counted according to their multiplicity, on the fundamental strip, $|\Re(t)| \le \pi$. Then

1.

$$f(t) = \exp(-i2t)\frac{1}{2}(a_2 - ib_2)\prod_{k=1}^{4}(\exp(it) - \exp(it_k)). \tag{13.38}$$

2.

$$f(t) = (a_2 - ib_2)\left\{\prod_{k=1}^{4}\exp\left(i\frac{t_k}{2}\right)\right\}\prod_{k=1}^{4}\sin\left(\frac{t-t_k}{2}\right). \tag{13.39}$$

3. *If the polynomial has real coefficients, then*

$$f(t) = \sqrt{(a_2)^2 + (b_2)^2}\cos(\Lambda)\prod_{k=1}^{4}\sin\left(\frac{t-t_k}{2}\right), \tag{13.40}$$

where

$$\Lambda = \text{angle}(a_2 - i b_2) + \frac{1}{2}\sum_{k=1}^{4} t_k \tag{13.41}$$

is always an integral multiple of π (including zero), even when some of the roots are complex valued.

This is a special case of the general factorization theorem given in section 4.7.
The explicit solutions for f_2 are

$$t = -\arctan\left(\frac{a_0 - a_2 + a_1 Z + 2a_2 Z^2}{b_1 + 2b_2 Z}, Z\right), \tag{13.42}$$

where Z is one of the four roots of the quartic

$$\left(4b_2^2 + 4a_2^2\right)Z^4 + (4b_1 b_2 + 4a_1 a_2)Z^3 + \left(a_1^2 - 4b_2^2 - 4a_2^2 + 4a_0 a_2 + b_1^2\right)Z^2$$
$$+(-2a_1 a_2 - 4b_1 b_2 + 2a_0 a_1)Z + a_0^2 + a_2^2 - 2a_0 a_2 - b_1^2 = 0. \tag{13.43}$$

This quartic polynomial in Z was derived by (i) the $\tilde{s} = \tan(t/2)$ transformation to convert the trigonometric polynomial to an algebraic polynomial in $\tilde{s}$, (ii) solving the quartic polynomial, and (iii) taking the arctan function to convert the roots in $\tilde{s}$ to zeros in t.

The Maple "solve" function did all these steps automatically—a testimony to the ever-growing competence of computer algebra systems—but considerable manual simplification and numerical checking was required.

This fourth-degree polynomial equation is no real improvement over the companion matrix method because the default strategy for calculating the zeros of an algebraic polynomial in MATLAB, for example, is to find the eigenvalues of a Frobenius companion matrix, which for (13.43) is the *same size* as the Fourier companion matrix. The explicit solution requires further processing with an arctangent function; the companion matrix strategy requires postprocessing of eigenvalues with the arccosine function.

An alternative to solving (13.43) is to compute the eigenvalues of the $N = 2$ Fourier multiplication matrix

$$\begin{bmatrix} 0 & 2 & 0 & 0 \\ -\frac{a_0}{a_2} + 1 & -\frac{a_1}{a_2} & -\frac{b_1}{a_2} & -\frac{b_2}{a_2} \\ 0 & 0 & 0 & 1 \\ M_{4,1} & M_{4,2} & M_{4,3} & M_{4,4} \end{bmatrix}, \tag{13.44}$$

where

$$M_{4,1} = \frac{-a_0 a_2 b_1 - b_2 a_1 a_2 + a_0 b_2 a_1 - a_2^2 b_1}{\left(b_2^2 + a_2^2\right)a_2}, \tag{13.45}$$

$$M_{4,2} = -\frac{a_1 a_2 b_1 + 2b_2 a_2^2 + 2b_2 a_0 a_2 - a_1^2 b_2}{\left(b_2^2 + a_2^2\right)a_2}, \tag{13.46}$$

$$M_{4,3} = -\frac{2a_2^2 a_0 + b_1^2 a_2 - b_1 b_2 a_1 - 2a_2^3}{\left(b_2^2 + a_2^2\right)a_2}, \tag{13.47}$$

$$M_{4,4} = -\frac{-b_2^2 a_1 + 2b_2 b_1 a_2 + a_2^2 a_1}{\left(b_2^2 + a_2^2\right)a_2}. \tag{13.48}$$

13.4.7 ▪ $N = 2$ Cosine Polynomial

If $f(t)$ is a quadratic cosine polynomial,

$$f(t) = a_0 + a_1 \cos(t) + a_2 \cos(2t), \tag{13.49}$$

then the transformation $x = \cos(t)$ converts this to

$$P(x) = a_0 + a_1 x + a_2(2x^2 - 1). \tag{13.50}$$

The roots are

$$t_k = \arccos\left\{-\frac{a_1}{4a_2} \pm \frac{1}{4a_2}\sqrt{a_1^2 - 8a_2 a_0 + 8a_2^2}\right\} + 2\pi k, \qquad k = \text{any integer}. \tag{13.51}$$

All functions with the property $f(-t) = f(t)$ have even parity with respect to the origin and can be approximated by a cosine polynomial. When $N = 2$, parity makes $f(t)$ equivalent to a *quadratic* algebraic polynomial rather than to a *quartic* algebraic polynomial.

13.4.8 ▪ $N = 2$ Sine Polynomial

If $f(t)$ is a quadratic sine polynomial,

$$f(t) = b_1 \sin(t) + b_2 \sin(2t), \tag{13.52}$$

the identity $\sin(2t) = 2\sin(t)\cos(t)$ and division by $\sin(t)$ gives

$$t_k = \left\{0, \pi, \arccos\left(-\frac{b_1}{2b_2}\right)\right\} + 2\pi k, \qquad k = \text{any integer}. \tag{13.53}$$

13.5 ▪ Equations with Hyperbolic Functions

If the function only depends on hyperbolic functions without powers of x, then a good procedure is as follows:

1. Replace the hyperbolic functions by exponentials.
2. Replace $\exp(x)$ by a new variable y.
3. Solve the resulting *polynomial* equation.

For example, the inverse of the hyperbolic secant function is obtained by solving

$$f(x) = \operatorname{sech}(x) - \lambda = \frac{2}{\exp(x) + \exp(-x)} - \lambda. \tag{13.54}$$

The substitution $y \equiv \exp(x)$ yields the algebraic equation

$$f(y[x]) = \frac{2}{y + 1/y} - \lambda. \tag{13.55}$$

Multiplying through by $y + 1/y$ yields a quadratic polynomial divided by y. Since $y = 0$ is equivalent to $u = -\infty$, we can safely multiply through by y to obtain a quadratic polynomial. Taking the logarithm of a root in y gives

$$x = \log\left\{\frac{1}{\lambda} + \sqrt{\frac{1}{\lambda^2} - 1}\right\} = \operatorname{sech}^{-1}(\lambda). \tag{13.56}$$

Table 13.1. *A sampler of equations solvable by the Lambert W-function.*

Equation	Solution
$x\exp(x)=\lambda$	$x=W(\lambda)$
$x\exp(-x)=\lambda$	$x=-W(-\lambda)$
$x\exp(bx)=\lambda$	$x=\frac{1}{c}W(\lambda c)$
$x\exp(cx^2)=\lambda$	$x=\pm\frac{1}{\sqrt{2c}}\sqrt{W(2c\lambda^2)}$
$ax+b\log(x)=\lambda$	$\frac{b}{a}W\left(\frac{a}{b}\exp(\lambda/b)\right)$
$x\exp(-q/x)=\lambda$	$x=q/W(q/\lambda)$
$x^2\exp(x)=\lambda$	$x=2W\left(\pm[1/2]\sqrt{\lambda}\right)$
$x^2\exp(-x)=\lambda$	$x=-2W\left(\pm[1/2]\sqrt{\lambda}\right)$
$x^2\exp(-q/x)=\lambda$	$x=(q/2)/W\left\{\pm(1/2)\sqrt{q^2/\lambda}\right\}$
$\sqrt{x}\exp(x)=\lambda$	$x=\lambda^2/\left\{\exp\left([1/2]W(2\lambda^2)\right)\right\}^2$
$\exp(-x)-\lambda-\mu x=0$	$x=W(\exp(\lambda/\mu)/\mu)-\lambda/\mu$

Although the quadratic equation has two roots, one is always negative and therefore its logarithm is complex valued; only the real-valued logarithm is shown above.

13.6 ▪ Lambert W-Function

To match the notation that is standard for the W-function, in this section, the unknown x will be denoted by W and the parameter λ will be replaced by x. The Lambert function $W(x)$ is then the solution of

$$W\exp(W)=x. \tag{13.57}$$

Many transcendental equations are solved by the Lambert W-function. An exhaustive list is impossible, but Table 13.1 suggest the possibilities.

The Lambert W-function is also the zeroth solution for many interesting perturbative approximations. For example, Gil, Segura, and Temme [195, sect. 10.2] thoroughly work the example of asymptotic inversion of the complementary error function. Dubinov and Dubinova [160] review applications in plasma physics, including the electric charge of a solitary body in plasma, nonlinear ion-acoustic waves, the equilibrium configuration of a neutralized electric beam, the dynamics of shaping virtual cathodes in a medium with viscous friction, and the dispersion equation for electron plasma waves in the framework of generalized Boltzmann kinetics. Veberic reviews other physics applications [359]. Pickett and Millev use the Riemann–Hilbert formalism to derive a "ladder" of log-of-log-of-log asymptotic approximations for the Lambert W-function (although they do not use this appellation) [295].

The Wright ω function is closely related to the W-function [380]. The formal definition is

$$\omega(z)=W_K(\exp(z)), \tag{13.58}$$

where the unwinding number $K(z)=\lceil\frac{\Im(z)-\pi}{2\pi}\rceil$ is defined so $z=\log(\exp(z))+2\pi K(z)$ for all z, and where $\lceil x\rceil$ is the usual ceiling function, the smallest integer whose absolute

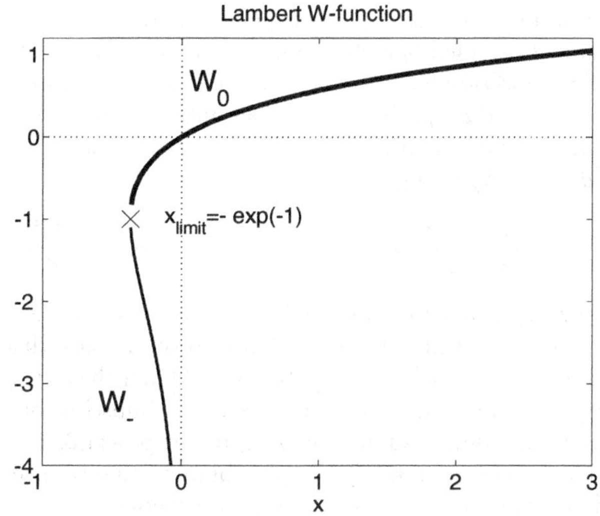

Figure 13.1. *The two real-valued branches of the Lambert W-function. The upper branch $W_0(x)$ is the thick solid line; $W\_(x)$ is the thinner, lower curve. The limit point at $x = x_{limit} = -\exp(-1)$, $W = -1$ is marked by the X.*

value is greater than x. The ω function solves [327]

$$y+\log(y)=z \qquad \rightarrow \qquad y=\begin{cases} \omega(z), & z\neq t\pm\pi i, t\leq -1, \\ \omega(z), \omega(z-2\pi i), & z=t+\pi, t\leq -1, \\ \text{no solutions in } \omega, & z=t-\pi, t\leq -1. \end{cases} \tag{13.59}$$

It satisfies the differential equation

$$\frac{d\omega}{dz}=\frac{\omega}{1+\omega}. \tag{13.60}$$

The Lambert and Wright functions each have their uses. The Wright function has the virtue that it is single valued and analytic everywhere except on the two half-lines $z=t\pm\pi i$ where it is discontinuous.

Numerical algorithms for the Lambert function (other than those discussed here) are explained in Fukushima [181], Fritsch, Shafer, and Crowley [178], Siewert and Burniston [327] and Barry et al. [31, 30, 29] and for the Wright function in Lawrence, Corless, and Jeffrey [238].

13.6.1 ▪ Properties of the Lambert W-Function

The Lambert W-function has been intensively studied, beginning with Lambert in 1758 and Euler in 1779. Corless et al. [135] provide a comprehensive review of both applications and theory with 77 references. For present purposes, it is sufficient to concentrate on the 2 real-valued branches, illustrated in Fig. 13.1.

The upper branch, W_0, is real valued for $x\in[-\exp(-1),\infty]$ and has the property that $W_0(x)\geq -1$ for all x in this range. The lower branch is real valued only for $x\in[-\exp(-1),0]$, and $W_{-1}(x)\leq -1$ everywhere it is real valued. The two branches meet at the limit point $x = x_{limit} = -\exp(-1)$; there are no real-valued branches for $x < x_{limit}$ but two real-valued branches immediately to the right of the limit point.

The condition for a double root for an arbitrary function $f(W)$ is that the conditions $f(W) = 0$ and $\partial f/\partial W = 0$ hold simultaneously. When the solution depends on a parameter x, the fact that $f(W(x);x) = 0$ everywhere on the solution curve implies that $df/dx = 0$. Applying the chain rule shows

$$\frac{df}{dx} = \frac{\partial f}{\partial x} + \frac{\partial f}{\partial W}\frac{dW}{dx} = 0 \qquad \longleftrightarrow \qquad \frac{dW}{dx} = -\frac{\partial f/\partial x}{\partial f/\partial W}. \tag{13.61}$$

For the particular function $f(W;x) \equiv W\exp(W) - x$, $\partial f/\partial x = -1$. It follows that $\partial f/\partial W = 0$ not only implies a double root, but also that $dW/dx \to \infty$. Explicit differentiation gives $\partial f/\partial W = \exp(W)[W+1]$, which confirms that the only possible double roots/limit points are where $W = -1, -\infty$. Substitution into $W\exp(W) = x$ shows that the corresponding x-values are $x_{limit} = -\exp(-1), 0$, as is already evident from Fig. 13.1.

The first expansion for the upper branch is a power series about the origin, which can be derived by using regular perturbation theory:

$$W_0(x) = \sum_{n=1}^{\infty} \frac{(-n)^{n-1}}{n!} x^n. \tag{13.62}$$

This has a radius of convergence exactly equal to

$$x_c = \exp(-1) \approx 0.36 \tag{13.63}$$

because of the limit point at $x = -x_c$ where the upper and lower branches merge.

An asymptotic approximation is derived in section 15.4 as a double series in the variables

$$\sigma = \frac{1}{\log(x)}, \qquad \tau \equiv \frac{\log(\log(x))}{\log(x)}. \tag{13.64}$$

Then

$$W \approx \frac{1}{\sigma} - \frac{\tau}{\sigma} + \tau + \frac{1}{2}\tau^2 - \tau\sigma + \frac{1}{3}\tau^3 - \frac{3}{2}\tau^2\sigma + \tau\sigma^2 + \frac{1}{4}\tau^4 - \frac{11}{6}\tau^3\sigma + 3\tau^2\sigma^2 - \tau\sigma^3$$
$$+ (1/5)\tau^5 - (25/12)\tau^4\sigma + (35/6)\tau^3\sigma^2 - 5\tau^2\sigma^3 + \tau\sigma^4 + \cdots. \tag{13.65}$$

The accuracy of a given truncation can be improved by Padé approximants as explained in section 15.4.3.

A half-power Puiseux series about the limit point is derived in section 14.14 using the new variable

$$\rho \equiv \sqrt{1 + \exp(1)x} \qquad \longleftrightarrow \qquad x = \exp(-1)(\rho^2 - 1). \tag{13.66}$$

The equation to solve becomes

$$f(W;\rho) \equiv W\exp(W) - \exp(-1)(\rho^2 - 1), \tag{13.67}$$

whence perturbation theory yields the alternating series

$$W = -1 + \sqrt{2}\rho - (2/3)\rho^2 + (11/36)\sqrt{2}\rho^3 - (43/135)\rho^4 + (769/4320)\sqrt{2}\rho^5 + \cdots. \tag{13.68}$$

For $\rho > 0$, this series approximates the upper branch $W_0(x)$. For negative ρ, however, all the terms in this series are negative and all the partial sums to all orders are smaller than -1, implying that the series approximates the *lower* branch $W_{-}(x)$.

Table 13.2. *Roots x_k for $f(x,\lambda) = \lambda\tan(x) + \tanh(x)$ from the Burniston–Siewert method. Formulas for the roots are given in the upper half of the table; the root expressions depend on a wide variety of auxiliary expressions given in the lower half of the table.*

Roots	
x_0	$\lambda \in [-1,0]$, $\text{arctanh}(\|\lambda\|/\|z_0\|)$, real root with $\|x_0\| \le \pi/2$
x_0	$\lambda \in [-\infty,-1]$, $i\,\text{Tan}^{-1}(\|\lambda\|/\|z_0\|)$, imaginary root with $\|x_0\| \le \pi/2$
x_k	$\lambda \in [-\infty,\infty]$, $\left(k - \frac{\lambda}{2\|\lambda\|}\right)\pi + \frac{\|\lambda\|}{\lambda}\text{Tan}^{-1}(\|y_k\|)$ $k = 1,2,\ldots,$ real roots with $\|x_k\| > \pi/2$
x_k^{im}	$\lambda \in [-\infty,\infty]$, $i x_k(1/\lambda)$
Auxiliary Functions	
y_k	$\lambda + \frac{1+\lambda}{k\pi} + \frac{1}{2\pi}\int_0^1 \log(M_k(x))dx - \frac{\|\lambda\|}{\pi}\int_{-1}^1 \phi_k(\|\lambda\|x)dx,$
$R(x)$	$\text{arctanh}(x) + \frac{\|\lambda\|}{\lambda}\left[\text{sgn}(x)\pi - 2\text{Tan}^{-1}(x/\|\lambda\|)\right]$
$M_k(x)$	$\dfrac{R^2(x) + (2k-1)^2\pi^2}{R^2(x) + (2k+1)^2\pi^2}$
$\theta_k(x)$	$\arctan\left(\dfrac{-2\pi R(x)}{R^2(x) + \pi^2(4k^2-1)}\right)$
$\phi_k(x)$	$\arctan\left(\dfrac{(\|\lambda\|/\lambda)\pi}{I_k(x)}\right)$
$I_0(x)$	$-2\frac{\|\lambda\|}{\lambda}\text{arctanh}\left(\frac{x}{\|\lambda\|}\right) + 2\text{Tan}^{-1}(x) - \text{sgn}(x)\pi$
$I_k(x)$	$I_0(x) + 2k\pi, \quad k = 1,2,\ldots$
$\phi_0(x)$	$\arctan\left(\dfrac{-\pi}{I_0(x)}\right)$
$\theta_0(x)$	$\arctan\left(\dfrac{-\pi}{R(x)}\right)$
z_0^2	$\frac{2}{\pi}\int_0^{\|\lambda\|}\phi_0(y)dy - \frac{2}{\pi}\int_0^1 \theta_0(y)dy - \frac{(1-\lambda^3)}{3(1+\lambda)}$

13.7 ▪ Reduction to Quadratures through Riemann–Hilbert Problems

Employing ideas that first appeared in the work of Muskhelishvili [282], Burniston and Siewart showed that for many transcendentals, roots can be given in terms of exact quadratures by solving so-called Riemann-Hilbert problems [118]. This spawned a cottage industry [119, 326, 327, 328, 330, 323, 331, 332, 324, 295]. Leathers and McCormick provide a brief but useful review [240].

For example, Siewert [325] showed that

$$f(x) \equiv (5-x)\exp(x) - 5 \tag{13.69}$$

has the root

$$x = 4\exp\left\{-\frac{1}{\pi}\int_0^\infty \left[\arctan\left(\frac{\pi}{\log(5) - 5 - t - \log(t)}\right) - \pi\right]\frac{dt}{t+5}\right\} \approx 4.96511.$$

Similarly, an infinite number of zeros of

$$\lambda \tan(x) + \tanh(x) = 0 \tag{13.70}$$

are summarized in Table 13.2 [332].

There is of course a philosophical and semantic issue as to whether an answer in the form of an integral that cannot be evaluated in closed form is "explicit," especially for formulas as complicated as those in the table.

Chapter 14

Regular Perturbation Methods for Roots

"The battle cry of the engineer in the 1990s and beyond should be...
Do asymptotic now, save CPU time later!"
—David C Wilcox [371, p. 189]

14.1 ▪ Introduction

When $x(\epsilon)$ is known for the special case that the parameter $\epsilon = 0$, it is possible to generate an expansion for this root for $\epsilon \ll 1$. If the dependence on ϵ is sufficiently well behaved as $\epsilon \to 0$, then an ordinary power series in powers of ϵ will suffice. When the function is not well behaved as $\epsilon \to 0$, it is often still possible to obtain an expansion. However, one must apply *singular perturbation theory*, and the approximation may involve both logarithms and powers of ϵ, or something still more complicated.

In this chapter, we describe *regular* perturbation theory for roots.

14.2 ▪ Power Series in ϵ for a Root of a Scalar-Valued Function

The problem is to find a zero of an ordinary function $f(x,\epsilon)$. Suppose that for $\epsilon = 0$, we know a root x_0. (If a root is known for some nonzero value λ_0 of a parameter λ, then we can apply the formalism below by defining the perturbation parameter $\epsilon \equiv \lambda - \lambda_0$.)

Expand the root as a power series in ϵ:

$$x = x_0 + \epsilon x_1 + \epsilon^2 x_2 + \epsilon^3 x_3 + \cdots. \tag{14.1}$$

By expanding $f(x(\epsilon),\epsilon)$ and matching powers of ϵ, one finds

$$x_1 = -\frac{f_\epsilon}{f_x}, \tag{14.2}$$

$$x_2 = -\frac{1}{f_x}\left\{\frac{1}{2} f_{\epsilon\epsilon} + f_{x\epsilon}\, x_1 + \frac{1}{2} f_{xx}\, x_1^2\right\}, \tag{14.3}$$

$$x_3 = -\frac{1}{f_x}\left\{\frac{1}{6} f_{\epsilon\epsilon\epsilon} + \frac{1}{2} f_{x\epsilon\epsilon}\, x_1 + \frac{1}{2} f_{xx\epsilon}\, x_1^2 \frac{1}{6} f_{xxx}\, x_1^3 + f_{x\epsilon} x_2 + f_{xx}\, x_1\, x_2\right\}. \tag{14.4}$$

Table 14.1. *Maple program for regular perturbation theory for a single unknown.*

```
# User must replace the next line with his/her own choice
# of f(x,epsilon), x0, and N;
  f:= cos(x) + epsilon*sin(x); x0 := Pi/2; N := 5;
  x := x0; for j from 1 by 1 to N do x := x + x||j*epsilon**j: od:
  r := series(f,epsilon,N + 1); # Expand f as epsilon-series;
  for j from 1 by 1 to N do eq||j := coeff(r,epsilon,j); x||j := solve(eq||j,x||j); od:
```

The subscripted variables denote partial differentiation with respect to that variable, and all derivatives of f are evaluated at $x = x_0$, $\epsilon = 0$; thus, $f_{x\epsilon} \equiv \partial^2 f / (\partial x \, \partial \epsilon)(x_0, 0)$.

It is easy to write a program in a symbolic manipulation language like Maple or Mathematica to calculate the perturbation series for an arbitrary $f(x,\epsilon)$ to arbitrary order N, as shown in Table 14.1.

Example 1:

$$f(x,\epsilon) \equiv \epsilon x^2 + x - 1; \qquad x_0 = 1. \tag{14.5}$$

The exact root such that $x \approx 1$ is[19]

$$x = \frac{2}{1 + \sqrt{1 + 4\epsilon}}, \tag{14.6}$$

$$f_x = 2\epsilon x + 1 \Big|_{x=1,\epsilon=0} = 1, \; f_\epsilon = x^2 \Big|_{x=1,\epsilon=0} = 1, \tag{14.7}$$

which suffice to deduce that the $O(\epsilon)$ correction (14.2) is $-(f_\epsilon / f_x)\epsilon = -\epsilon$. Continuing to higher order yields

$$x \approx 1 - \epsilon + 2\epsilon^2 - 5\epsilon^3 + 14\epsilon^4 - 42\epsilon^5 + 132\epsilon^6 - 429\epsilon^7 + \cdots. \tag{14.8}$$

Example 2:

$$f(x,\epsilon) = \cos(x) + \epsilon \sin(x); \qquad x_0 = \pi/2. \tag{14.9}$$

Expanding $f(x(\epsilon),\epsilon)$ in ϵ gives

$$\begin{aligned} f(x,\epsilon) &= (-1 + x_1)\epsilon - x_2\epsilon^2 + \left\{ -x_3 + \frac{1}{6}x_1^3 - \frac{1}{2}x_1^2 \right\} \epsilon^3 \\ &+ \left\{ -x_4 - x_1 x_2 + \frac{1}{2}x_1^2 x_2 \right\} \epsilon^4 + \left\{ -x_5 - x_1 x_3 + \frac{1}{2}x_1 x_2^2 + \frac{1}{2}x_1^2 x_3 + \frac{1}{24}x_1^4 - \frac{1}{120}x_1^5 - \frac{1}{2}x_2^2 \right\} \epsilon^5. \end{aligned} \tag{14.10}$$

The exact root is $x = \pi/2 + \arccos\left(1/\sqrt{1+\epsilon^2}\right)$. Order by order in ϵ, zeroing the expansion of $f(x,\epsilon)$ gives

$$x = \pi/2 + \epsilon - (1/3)\epsilon^3 + (1/5)\epsilon^5 + \cdots. \tag{14.11}$$

[19] The usual quadratic solution $x = \left(-1 + \sqrt{1+4\epsilon}\right)/(2\epsilon)$ has an apparent pole; the alternative quadratic solution in (14.6) is explicitly nonsingular. The two forms of the solution to the general quadratic are discussed in Chapter 10.

14.3 ▪ Perturbation Theory in Fractional Powers of ϵ (Puiseux Series)

For many problems, the preceding algorithm must be generalized by computing an expansion in *fractional* powers of ϵ. Such expansions are called *Puiseux series*. Sturmfels's prize-winning essay explains how one can compute the "Newton polytope" of the equation, and thence the fractional exponents [346]. A later article gives all 2^{N-1} distinct Puiseux series expansions of the solution of a general, single polynomial equation of degree N [347]. Unfortunately, his work requires too much background to be repeated here. But much can be learned by rather elementary means.

The reduced cubic polynomial, $P(x) = x^3 - px + q$, is a good illustration. This furnishes several different cases, depending upon which coefficients are assumed to be small in the perturbation parameter. However, the overall procedure is the same for all cases.

First, write

$$x = \epsilon^{\beta} v, \tag{14.12}$$

where the exponent β is as yet unknown and must be determined later, and the new unknown v is assumed to be $O(1)$. Second, substitute (14.12) into the polynomial and examine pairwise balances. The third step is to choose β to be equal to those values that give sensible balances in the sense that the two largest terms are of equal magnitude as $\epsilon \to 0$ so that this choice of β generates one or more roots. The cubic has three roots; it is usually necessary to choose *different* exponents β for different roots.

Case 1: Suppose that the small parameter is the *constant* in the cubic so that we wish to solve

$$x^3 - px + \epsilon = 0 \tag{14.13}$$

for small $|p| \sim O(1)$ and $|\epsilon| \ll 1$. The cubic becomes, after dividing by ϵ,

$$\epsilon^{3\beta-1} v^3 - p\epsilon^{\beta-1} v + 1 = 0. \tag{14.14}$$

We see that if $\beta = 1$, then the second and third terms are both $O(1)$, while the cubic term is smaller by $O(\epsilon^2)$:

$$\epsilon^2 v^3 - pv + 1 = 0. \tag{14.15}$$

This is a possible balance, which yields the lowest-order root $v \sim 1/p$ or, in others words, $x \approx (1/p)\epsilon$.

Because the perturbation is a function of ϵ^2 rather than ϵ after we have rescaled the cubic, it is not surprising that the expansion involves only *even* powers of ϵ. If we write a formal expansion with symbolic coefficients,

$$v = 1/p + \epsilon^2 v_2 + \epsilon^4 v_4 + +\epsilon^6 v_6 + \cdots, \tag{14.16}$$

then substitution into the cubic gives the residual

$$R = \left(\frac{1}{p^3} - pv_2\right)\epsilon^2 + \left(3v_2\frac{1}{p^2} - pv_4\right)\epsilon^4 + \left(+3v_2^2\frac{1}{p} + 3v_4\frac{1}{p^2} - pv_6\right)\epsilon^6. \tag{14.17}$$

Solving for v_2 so as to make the coefficient of ϵ^2 equal to zero, then computing v_4 so that the fourth-order term in R is zero, and so on, gives

$$v = \frac{1}{p} + \frac{1}{p^4}\epsilon^2 + \frac{3}{p^7}\epsilon^4 + \frac{12}{p^{10}}\epsilon^6 + \cdots \text{ [Case 1: 1st root]}. \tag{14.18}$$

However, this is only one root, and the cubic has three. If we return to the scaled cubic, the first and third terms are both $O(1)$ if $\beta = 1/3$, but then the second term is $O(\epsilon^{-2/3})$ and therefore huge with nothing to cancel it. It follows that this is not a possible balance. If we instead choose β so that the first and second terms are equal, then the requirement that $3\beta - 1 = \beta - 1$ demands that $\beta = 0$. The first and second terms are then $O(1/\epsilon)$ larger than the last term, so this is a consistent balance. The lowest-order approximation to the second and third roots is the solution to $v^2 - p = 0$, which implies that the roots are $x = \pm\sqrt{p}$. The expansion for these two roots involves all powers of ϵ, not just the even ones:

$$x_{\pm} = \pm\sqrt{p} + x_1\epsilon + x_2\epsilon^2 + x_3\epsilon^3 + \cdots. \tag{14.19}$$

The residual is

$$R_{\pm} \equiv x^3 - px + \epsilon = (2px_1 + 1)\epsilon + \left(2px_2 \pm \sqrt{p}3x_1^2\right)\epsilon^2 + \cdots. \tag{14.20}$$

Again, solving for x_1 from the coefficient of ϵ in R, computing x_2 from the $O(\epsilon^2)$ term in R, etc., gives

$$x_{\pm} = \pm\sqrt{p} - \frac{1}{2p}\epsilon + \left(\mp\sqrt{p}\frac{3}{8}\frac{1}{p^3}\right)\epsilon^2 + \left(-\frac{1}{2}\frac{1}{p^4}\right)\epsilon^3 + \cdots \text{ [Case 1: 2nd and 3rd roots].} \tag{14.21}$$

Case 2: Suppose that the small parameter is the *linear* coefficient in the cubic so that we wish to solve

$$x^3 - \epsilon x + q = 0. \tag{14.22}$$

Neglecting the linear term gives $x \approx (-q)^{1/3}$, which gives approximations to all three roots, two of which are complex valued if q is real. Introducing the parameter $s \equiv (-q)^{1/3}$, one obtains by matching powers in the residual

$$x = s + \left(\frac{1}{3s}\right)\epsilon + 0\epsilon^2 + \left(\frac{-1}{81s^5}\right)\epsilon^3 + \cdots \text{ [Case 2: all roots],} \tag{14.23}$$

which applies to all three roots, with s chosen to be each of the cube roots of $(-q)$ in turn.

Case 3: The parameter is the coefficient of the cubic term

$$\epsilon x^3 - x + q = 0. \tag{14.24}$$

One root is approximated by neglecting the cubic perturbation to give $x \approx q$. Matching powers in the residual, $R \equiv \epsilon x^3 - x + q$, gives

$$x = q + q^3\epsilon + 3q^5\epsilon^2 + 12q^7\epsilon^3 + \cdots \text{ [Case 3: 1st root].} \tag{14.25}$$

The other two roots exist only because of the cubic perturbation, which raises a linear polynomial to a cubic and therefore adds two additional zeros. This implies that the cubic term must be as important as one or both of the other terms. This is possible if and only if $x = v/\epsilon^{1/2}$, where $v \sim O(1)$ as $\epsilon \to 0$. The lowest-order approximation to these roots is $x = \pm 1/\sqrt{\epsilon}$. The rescaled cubic in terms of v is

$$v^3 - v + q\epsilon^{1/2} = 0. \tag{14.26}$$

However, this is just (14.13) with $p=1$ and ϵ replaced by $q\epsilon^{1/2}$. Quoting the earlier series with the appropriate substitutions gives

$$v_{\pm}=\pm 1-\frac{1}{2}q^{1/2}\epsilon^{1/2}+\left(\mp\frac{3}{8}q\right)\epsilon+\left(-\frac{1}{2}q^{3/2}\right)\epsilon^{3/2}+\cdots \quad \text{[Case 3: 2nd and 3rd roots]}. \tag{14.27}$$

Thus, even for the very simple case of a cubic equation, we find that it is necessary to assume different scalings and different exponents for different roots in different parameter regimes. Our expansions were in $\sqrt{\epsilon}$, ϵ, or ϵ^2. When the perturbation parameter was the linear coefficient of the cubic, a single scaling and expansion simultaneously approximated all three roots. For the other two cases, however, one root was wildly different in magnitude from the other pair, and so the pair and the other root required different expansions in different powers of ϵ.

The methodology for expansions in fractional powers of ϵ is identical to that of the previous section, however. If we need an expansion in ϵ^{β}, we can simply rewrite the problem in terms of $\delta \equiv \epsilon^{\beta}$ and expand the root as an ordinary power series in the new perturbation parameter δ. It is very common in applications, however, to find that $\beta \neq 1$, so one must always analyze the problem before blindly applying perturbation theory.

Additional perturbation series for the reduced cubic equation, together with many graphs and convergence analysis, are described in Chapter 11.

14.3.1 ▪ Breaking Trail with Newton's Iteration

If the form of a perturbation expansion is not obvious, but a lowest-order approximation can be found, then Newton's iteration can be used to determine the form of the perturbation series. It is easy to apply the Newton–Raphson iteration in an algebraic manipulation language such as Maple. One can use the Maple "series" command, or its equivalent in other languages, to expand the (usually very messy) Newton's iterate as a series in ϵ. This expansion will show the form of the series and also provide the lowest term or two in the expansion—more terms when we expand a higher iterate. (We shall return to discuss the exact relationship between Newton's iterates and perturbation series later in this chapter.)

14.4 ▪ Perturbation Theory in Inverse Powers

14.4.1 ▪ Modon Dispersion Relation

When a parameter is large, expansions can often be found in inverse powers of the parameter. For example, vortex pairs in the atmosphere and ocean known as *modons* travel at a phase speed that solves the *modon dispersion relation* [53] for x as a function of the parameter ϵ in the equation, where the parameter is the nondimensional width of the vortex:

$$f \equiv J_2(x)\,\epsilon K_1(\epsilon)+x J_1(x) K_2(\epsilon). \tag{14.28}$$

The Bessel functions have the asymptotic expansions

$$K_0(\epsilon)\sim\sqrt{\pi/(2\epsilon)}\exp(-\epsilon)\left\{1-\frac{1^2}{1!\,8\epsilon}+\frac{(1\bullet 3)^2}{2!\,(8\epsilon)^2}+\cdots\right\}, \tag{14.29}$$

and similarly for K_1. The $\exp(-\epsilon)\sqrt{\pi/(2\epsilon)}$ is a common factor that can be canceled out. As a result, the root has an expansion in inverse powers of ϵ:

$$x\sim x_0+x_1\frac{1}{\epsilon}+x_2\frac{1}{\epsilon^2}+\cdots. \tag{14.30}$$

The lowest-order residual is proportional to $J_2(x_0)$. It follows that x_0 must be one of the countable infinity of roots of this Bessel function. The physically interesting mode is associated with the smallest root, $j_{2,1} = 5.135622$. However, it is easy to keep x_0 as a symbol and compute higher-order terms. We thus obtain a perturbation series that simultaneously approximates an infinite number of roots.[20] Matching powers of $1/\epsilon$ and solving order by order gives

$$x \sim j_{2,m}\left\{1 - \frac{1}{\epsilon} + \frac{1}{\epsilon^2} + \left\{\frac{1}{3}(j_{2,m})^2 - \frac{3}{8}\right\}\frac{1}{\epsilon^3} - \left\{\frac{4}{3}(j_{2,m})^2 + \frac{3}{2}\right\}\frac{1}{\epsilon^4} + \cdots\right\}, \tag{14.31}$$

where for the mth branch of the solution, $j_{2,m}$ denotes the mth root of $J_2(x)$.

For the lowest mode in particular,

$$x \sim 5.13562 - 5.13562\frac{1}{\epsilon} + 5.13562\frac{1}{\epsilon^2} + 43.224\frac{1}{\epsilon^3} - 188.30\frac{1}{\epsilon^4} - 289.35\frac{1}{\epsilon^5}. \tag{14.32}$$

Note that because the Bessel expansions are *asymptotic* but *divergent*, one would expect that the series for the roots is also *divergent*. Contrary to expectations, Fig. 14.1 suggests that the large-ϵ series is *convergent*, with a radius of convergence of roughly $\epsilon > 5.5$.

However, this is probably a fluke peculiar to this example. Since many special functions have divergent asymptotic expansions for large arguments, it follows that algebraic equations involving these transcendentals will often have divergent perturbation series in $1/\epsilon$ for the roots.

The small-ϵ expansion of the solutions of the modon dispersion relation is a log-and-power series, which will be discussed in the next chapter.

14.4.2 ▪ Roots of $x\tan(x) - 1$

The goal is to find the zeros of

$$f(x) \equiv x\tan(x) - 1. \tag{14.33}$$

Note that if x_* is a root, then so is $-x_*$, as follows from the property that

$$f(-x) = -x\tan(-x) - 1 = f(x), \tag{14.34}$$

because $\tan(-x) = -\tan(x)$.

For large x, $x\sin(x) \gg \cos(x)$ except very close to the roots of the sine function. It follows that for large x, the roots must be *near* those of the sine function, $x = n\pi$, where n is any integer. Let us write

$$x_n = n\pi + \delta(n), \qquad |\delta| \ll 1, \tag{14.35}$$

for sufficiently large n. Using the identities $\sin(x) = \sin(n\pi + \delta) = \cos(n\pi)\sin(\delta)$ and $\cos(x) = \cos(n\pi + \delta) = \cos(n\pi)\cos(\delta)$, $f(x) = 0$ becomes

$$\cos(\delta)(n\pi + \delta) - \sin(\delta) = 0. \tag{14.36}$$

[20] Although Maple has a built-in function to evaluate the Bessel roots, this explicit computational solution does not violate our earlier assertion that all but a few of the zeros are best approximated by asymptotic expansions. Hidden from the user, asymptotic expansions are precisely what most library software employs, albeit mostly to initialize Newton's iteration to compute large Bessel function zeros.

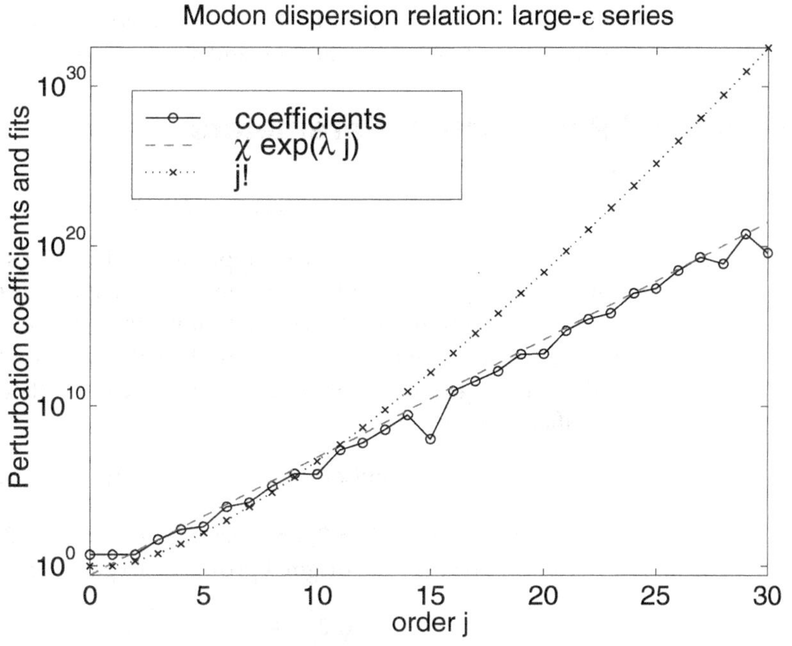

Figure 14.1. *The solid curve with circles shows the magnitude of the coefficients of the large-ϵ expansion of the first branch of the modon dispersion relation. The dashed curve is the exponential fit $\chi \exp(\lambda j)$ where $\chi = 0.27$ and $\exp(\lambda) = 5.45$; this tightly bounds the oscillating coefficients. The dashed line with x's shows $j!$. This dashed-with-x curve has been included because divergent x asymptotic series typically grow at a factorial rate. Note that if the coefficients satisfy a bound $a_n \leq constant \exp(-\mu n)$ for some positive μ (such as the dashed line), then the series is geometrically convergent. A graph is not as good as a proof because the latter can take limits whereas the graph necessarily has a finite range, but the graph certainly suggests that the series is geometrically convergent.*

Table 14.2. *Zeros of* $x\tan(x) - 1$.

n	Exact	Approximate	Fifth-order error
1	3.428446592068	3.42561845948172	0.0028
2	6.4373258101015	6.43729817917194	2.76E−5
3	9.52933610260433488	9.52933440536196	1.70E−6
4	12.64528745442429106	12.6452872238566	2.31E−7
5	15.77128492356986261 1	15.7712848748158	4.88E−8

Expanding δ as a series of negative powers of n and matching powers of n allows the coefficients to be calculated to whatever order is desired:

$$x_n = n\pi + \frac{1}{\pi}\frac{1}{n} - \frac{4}{3\pi^3}\frac{1}{n^3} + \frac{53}{15\pi^5}\frac{1}{n^5} + O\left(\frac{1}{n^7}\right). \tag{14.37}$$

Table 14.2 shows that the expansion through n^{-5} gives even the first root to an absolute error of only 0.28%. Accuracy can easily be improved by calculating more terms in the series or by a couple of Newton's iterations.

The zeroth positive root requires a different approach, which is described in the next section.

A generalization, $f(x) \equiv x \tan(x) - 1/x$, is analyzed with seven different approximations throughout Chapter 7 of Gil, Segura, and Temme [195].

14.5 ▪ Artificial Perturbation Parameters

The function

$$g(x) = x \sin(x) - \cos(x) \tag{14.38}$$

has a root near the origin that cannot be approximated by the series in inverse powers of n given in the previous section because this root corresponds to $n = 0$. For small x, we can expand the trigonometric functions in powers of x. To second order, this gives $g(x) \approx -1 + (3/2)x^2 + \cdots$, which correctly predicts $x \approx \sqrt{2/3} = 0.816$.

To calculate higher approximations analytically, we can inflate the problem by introducing the artificial parameter ϵ:

$$h(x;\epsilon) \equiv \epsilon(x \sin(x) - \cos(x)) + (1-\epsilon)\left(-1 + (3/2)x^2\right). \tag{14.39}$$

This is constructed so that for $\epsilon = 0$, an exact root is $x = \sqrt{2/3} = 0.816$; when $\epsilon = 1$, $h(x;\epsilon) = g(x)$ and we recover the original problem. Expanding

$$x = \sqrt{2/3} + p_1\epsilon + p_2\epsilon^2 + \cdots, \tag{14.40}$$

substituting this in $h(x;\epsilon)$, and matching powers of ϵ yields

$$x = \sqrt{2/3} + 0.0366425\epsilon + 0.0056531\epsilon^2 + 0.0011675\epsilon^3 + \cdots. \tag{14.41}$$

It is obvious that at $\epsilon = 1$, the only value of interest, the terms of the series are converging rapidly. Including the terms up to and including ϵ^3 gives 0.8603335 versus the true root of 0.8599599062 with an error of 0.00037.

14.6 ▪ Extending the Range: Padé Approximants

Power series are a ubiquitous topic of lower-level college mathematics courses. This popularity obscures the fact that for real applications, Taylor series per se are often useless. The real power of power series is in combination with other ideas. Newton's iteration (power series plus iteration of the linear approximation) and finite differences (multiple Taylor expansions, each on its own subinterval and linked) are familiar examples.

The $[M/N]$ Padé approximant to a function $f(x)$ is a polynomial of degree M divided by a polynomial of degree N that is chosen so that the leading terms of the power series of the approximant match the first $(M+N+1)$ terms of the power series of $f(x)$. One might suppose that the approximant would be restricted to the same domain of convergence as the Taylor expansion whence it came. In reality, the Padé approximant will usually converge on the entire real axis if $f(x)$ is free of singularities on the real axis [20, 21, 39].

By matching powers in the equation $Q(x)f_T(x) = P(x) + O(x^{M+N+1})$, where f_T is the power series of $f(x)$ and $P(x)$ and $Q(x)$ are the numerator and denominator polynomials of the Padé approximant, one can derive a set of *linear* equations to determine the coefficients of the approximant. Recurrence relations such as the *q-epsilon* algorithm do the same job even more efficiently than solving a matrix problem. Because Padé-finding algorithms are built-in utilities in most algebraic manipulation languages, such as Maple and Mathematica, and are also available in most FORTRAN and C software libraries, we shall assume that we have access to such utilities. The theory and practice of Padé approximants are described in many texts, such as [20, 21, 39].

One useful piece of folklore is that the most accurate approximant is usually the *diagonal* approximant, that is, from all combinations of M and N such that $M+N=N_{trunc}$ where N_{trunc} is the power series truncation, it is usually best to choose $M=N$.

Unfortunately, Padé approximants have quirks that greatly impair their reliability. In particular, the denominator may have real zeros, which will cause the approximant to have *poles* in the desired interval of approximation. More confusing, these poles may appear only in a subsequence of the family of approximants. Thus, for example, the [2/2] and [3/3] approximants may be accurate over a large interval, whereas the [4/4] has poles in the same interval and thus is almost useless. Pairs of poles and zeros at *almost* the same location, alias *Froissart doublets*, are another ubiquitous difficulty. Furthermore, even when the approximants are free of poles, convergence with increasing degrees M and N may not be monotonic. Many examples of both success and failure, often for the same function at different orders, are given in [20, 21, 39]. We shall illustrate this phenomenon for two-point Padé approximants in the next section.

Recently, Gonnet, Güttel, and Trefethen have removed much of this weirdness [199]. The coefficients of the numerator and denominator of the approximant are the solution to a matrix problem but are usually calculated by the much faster epsilon algorithm. Speedwise, Gonnet, Güttel, and Trefethen [199] do the opposite by computing the SVD of the matrix. When the SVD reveals rank-deficiency, their algorithm automatically reverts to a lower-degree but nonsingular approximation. A request for a [10/10] approximation may return an answer that is only [3/3], but the trustworthiness of the result is vastly improved.

14.7 ▪ Two-Point Padé Approximants

When one has a power series in a parameter ϵ for small ϵ and also a series in inverse powers of ϵ for $\epsilon \gg 1$, it is possible to fit a rational function of degree p over a polynomial of degree q so that it matches both expansions up to a given order, not necessarily the same, for each series (McCabe [261], McCabe and Murphy [262], and Frost and Harper [180]). The resulting approximation is usually called a *two-point Padé* approximant.

The method is straightforward. If the approximant is

$$x_{[M/N]} \equiv P(\epsilon)/Q(\epsilon), \tag{14.42}$$

then multiplication gives

$$x_{[M/N]}\, Q(\epsilon) \equiv P(\epsilon). \tag{14.43}$$

The only difference from the computation of ordinary (single-point) Padé approximants is that this equation must be expanded twice: once in powers of ϵ (and then matched up to and including $\epsilon^{\tilde{M}}$) and a second time in $1/\epsilon$ (and matched through $\epsilon^{\tilde{N}}$). We must have

$$\overline{M}+\overline{N}+2=M+N+1, \tag{14.44}$$

so that the number of matching conditions is equal to the number of degrees of freedom in the rational approximation, but otherwise the integers $M, N, \overline{M}, \overline{N}$ are user choosable and unconstrained. (As for single-point Padé approximants, however, diagonal approximants $[M=N]$ are usually the most accurate for a given sum of $p+q$, and diagonal approximants are the only ones that have a finite limit as $\epsilon \to \infty$.)

Again as for ordinary Padé methods, the matching conditions give a set of *linear* equations for the unknown coefficients of the numerator and denominator polynomials in $x_{[M/N]}$ even though the approximation itself is nonlinear.

The linear matrix problem for the coefficients can be somewhat ill conditioned, however, both for one-point and two-point approximants. Symbolic algebra systems such as Maple and Mathematica are very efficient both for computing the perturbative expansions and for computing Padé approximations using exact rational or user-chosen-precision arithmetic, thereby eliminating the roundoff error bugaboo.

The modon dispersion relation is a good illustration of both the positive and negative aspects of Padé approximants. The large-ϵ expansion was derived as (14.31) above; the small-ϵ series was obtained as (15.61) in the next chapter. Although the small-ϵ series has logarithms at fourth order, the first three orders involve powers only.

Boyd derived the [2/2] two-point approximant that matches the small-ϵ series up to and including ϵ^2 and the large-ϵ expansion up to and including $1/\epsilon$ [53]. For the first branch of solutions,

$$x_{[2/2]}^{2,1} \approx \frac{3.83170 + 1.51030\epsilon + 0.51395\epsilon^2}{1 + 0.39416\epsilon + 0.10008\epsilon^2}, \tag{14.45}$$

where the superscript denotes the highest power matched in the small ϵ (first superscript) and large ϵ (second superscript) series. The $[2/2]^{2,1}$ approximant [53] has a maximum *relative* error of 1 part in 700 and a maximum *absolute* error of 0.007 over the entire interval $\epsilon \in [0, \infty]$. This is quite extraordinary accuracy for an approximant that is merely the ratio of two quadratic polynomials.

Unfortunately, the $x_{[3/3]}^{2,3}$ approximant has a maximum error no smaller than that of $x_{[2/2]}^{2,1}$. The approximant $x_{[4/4]}^{3,4}$ has poles at 0.4212 and 11.0144 and thus is completely unacceptable as a "global" approximation. We are thus left in the odd position of cheering a low-order approximation while booing two approximations of (nominally) higher order.

This strange experience is, alas, rather common with Padé approximants; the SVD remedy of [199] is very helpful here. Like dynamite, Padé approximants are both useful and dangerous.

It should be noted that for this problem, there are complications in matching the small-ϵ series beyond third order because logarithms of ϵ appear at fourth order. As explained in the next chapter, one can usually compute such logarithms by pretending that the root is a function of two parameters, not one, and expanding as a double series in ϵ and $\nu \equiv \log(1/\epsilon)$. In principle, one can then apply bivariate Padé approximants.

This generalization fails here for two-point approximants because including a term in $\log(\epsilon)$ yields an approximation that is either infinite or zero in the limit $\epsilon \to \infty$. This is unacceptable because the modon root tends to a finite nonzero limit.

Thus, Padé approximants are a wonderful tool for increasing the power of perturbation series—*some* of the time. Thoughtful use of SVD to compute the approximants can greatly increase the success rate [199].

14.8 ▪ Hermite–Padé Approximants

Ordinary Padé approximants help to accelerate summation and improve the accuracy over optimal truncation in many cases, but they can also miss important information. For example, they are incapable of representing multivalued functions or giving any hints that the function that corresponds to the real-valued divergent series actually has a nonzero imaginary part.

A family of generalizations called *Hermite–Padé approximants* has been successfully used in quantum mechanics, the theory of water waves, series analysis of multivalued functions, and other applications [358, 194, 316, 348, 317]. (The quadratic Shafer form, the heart of the Shafer iteration discussed in Chapter 6, is a special case.) On the theoretical side, many illuminating and useful theorems have been proved about convergence properties of the Hermite–Padé approximants. However, the accuracy required in some applications can make a problem intractable for modern-day computers, or at least make the approximants less efficient than other methods.

The quadratic Hermite–Padé approximant $f[K/L/M]$ is defined to be the solution of the quadratic equation [320]

$$P(f[K/L/M])^2 + Qf[K/L/M] + R = 0, \tag{14.46}$$

where the polynomials P, Q, and R are of degrees K, L, and M, respectively. These polynomials are chosen so that the power series expansion of $f[K/L/M]$ agrees with that of f through the first $N = K + L + M + 1$ terms. The constant terms in P and Q can be set arbitrarily without loss of generality since these choices do not alter the root of the equation, so the total number of degrees of freedom is as indicated. Like ordinary Padé approximants, the coefficients of the polynomials can be computed by solving a matrix equation, and the most accurate approximations are obtained by choosing the polynomials to be of equal degree, again called *diagonal* approximants. An example is given in section 2.19.

14.9 ▪ Newton's Iteration and Perturbation Theory

Just like perturbation theory, iterative schemes can be applied when $f(x;\epsilon)$ depends on a small parameter and the root x_0 is known for $\epsilon = 0$. The digit-doubling convergence of Newton's method implies the following.

Theorem 14.1 (Newton's Iteration with a Small Parameter). *Suppose that Newton's iteration is applied to find the roots of $f(x;\epsilon)$, where f and x are either scalars or vectors. If the first guess, $x^{(0)}$, is the exact root of $f(x;0) = 0$ and if $x(\epsilon)$ has a power series about $\epsilon = 0$, then*

$$|x(\epsilon) - x^{(j)}(\epsilon)| \sim O\left(\epsilon^{2j}\right), \qquad \epsilon \to 0, \qquad j = 1, 2, \ldots. \tag{14.47}$$

In other words, the expansion of the first, second, and third Newton's iterates in powers of ϵ will reproduce the perturbation series for the root up to and including the terms in ϵ, ϵ^3, and ϵ^7, and so on.

Example:

$$f \equiv \left\{x^3 - 7x^2 + 14x - 8\right\} + \epsilon\left\{x^3 - 5x^2 + 2x + 8\right\}. \tag{14.48}$$

One root is

$$x = \frac{1-\epsilon}{1+\epsilon} = 1 + 2\sum_{j=1}(-1)^j \epsilon^j. \tag{14.49}$$

The Newton's iterates, with $x^{(0)} = x(\epsilon = 0) = 1$, are rational functions. The power series of these rational functions are given below with a box around those terms that agree with the corresponding terms for the series of $x(\epsilon)$:

$$x^{(1)} = \frac{3 - 11\epsilon}{3 - 5\epsilon} = \boxed{1 - 2\epsilon} - \frac{10}{3}\epsilon^2 + \cdots, \tag{14.50}$$

$$x^{(2)} = \frac{81-270\epsilon+828\epsilon^2-2514\epsilon^3+1363\epsilon^4}{(3-5\epsilon)(27+9\epsilon+165\epsilon^2-137\epsilon^3)} = \boxed{1-2\epsilon+2\epsilon^2-2\epsilon^3} - \frac{970}{27}\epsilon^4+\cdots, \tag{14.51}$$

$$\begin{aligned} x^{(3)} = \{&-1594323+3720087\epsilon-18068994\epsilon^2+61529058\epsilon^3-198463689\epsilon^4 \\ &+605166957\epsilon^5-1821482316\epsilon^6+5467065516\epsilon^7-10280709621\epsilon^8 \\ &+20651592897\epsilon^9-36811441170\epsilon^{10}+38184200658\epsilon^{11} \\ &-19577230862\epsilon^{12}+3869933531\epsilon^{13}\} \\ &/\{(3-5\epsilon)(27+9\epsilon+165\epsilon^2-137\epsilon^3) \\ &(-19683-19683\epsilon-87480\epsilon^2+58320\epsilon^3-894726\epsilon^4+1011042\epsilon^5 \\ &-3371328\epsilon^6+9429336\epsilon^7-9192735\epsilon^8+2824793\epsilon^9)\} \\ =& \boxed{1-2\epsilon+2\epsilon^2-2\epsilon^3+2\epsilon^4-2\epsilon^5+2\epsilon^6-2\epsilon^7} - \frac{4189930}{2187}\epsilon^8+\cdots. \end{aligned} \tag{14.52}$$

The example shows a defect of analytical applications of Newton's method: even when f is a simple cubic polynomial, the complexity of the Newton's iterates grows very rapidly with iteration number j. The power series derived from the Newton's iterates are much simpler. However, the domain of accuracy in ϵ may be greatly reduced by power series expansion of the rational iterates since the iterates may have singularities closer to $\epsilon = 0$ than does the exact root.

Still, the theorem shows that iteration followed by expansion may be a very efficient way to rapidly generate the terms of a power series. This is especially true when one does not know a priori whether the expansion will proceed in *fractional* or *integral* powers of ϵ. If one can devise an approximation to the root in the limit $\epsilon \to 0$, one can initialize Newton's method with this approximation and let the algorithm do the thinking. Table 14.3 is a Maple implementation.

Table 14.3. *A Maple program for perturbation theory from Newton's method.*

```
# User must replace the next line with his/her own choice;
# of f(x,epsilon), x0, and N;
  f:=x**5-x-1/epsilon**5;   x0:=1/epsilon;   maxNewt:=5;   Npert:=20;
  newt:=x-f/diff(f,x);
  for j from 1 by 1 to maxNewt do x||j:=subs(x=x||(j-1),newt);
  xpert||j:=series(x||j,epsilon,Npert); od:
# Compare powers in x||maxNewt, x||(maxNewt-1) to see how many coefficients
  have converged to perturbation series coefficients;
```

14.10 ▪ Nonperturbative Series: Chebyshev Expansions of Root Branches

In many problems, the goal isn't merely to generate numerical values of a root. Rather, the needs of a larger research project may make it convenient to have an analytical approximation that can replace expensive Newton's iterations with the summation of a series. Perturbation series have the virtue that they always generate such a cheap-to-evaluate analytical approximation, but they have the vice that this approximation is highly nonuniform in ϵ. The error of a power series (for any given truncation of the series) rises exponentially fast as ϵ increases.

Chebyshev series have the drawback that they require that the values of the roots must first be found by some other method at a discrete set of values of the parameter ϵ. These nonperturbative expansions have the virtue that once the required numerical table has been constructed, the Chebyshev series can replace the table (and the Newton's iteration or other algorithm used to construct it) by an expansion that is highly uniform in ϵ over the entire interval of interest.

The mechanics of adaptive Chebyshev interpolation on an interval $\epsilon \in [a,b]$ are thoroughly discussed in Chapter 3. Recall that the Chebyshev polynomials $T_j(\epsilon)$ (and their cousins, the rational Chebyshev functions) are the images of the cosine functions of an ordinary Fourier series under a change of coordinate:

$$T_j(\epsilon) \equiv \cos(j[\arccos(\epsilon)]), \qquad j = 0, 1, 2, \ldots \tag{14.53}$$

or, equivalently, $T_j(\cos(t)) \equiv \cos(jt)$. The jth Chebyshev polynomial is a polynomial of degree j. A root can be approximated by a Chebyshev series on an arbitrary, user-chosen interval in the form

$$x(\epsilon) \approx \sum_{j=0}^{N} a_j \, T_j\left(\frac{2}{b-a}\left\{\epsilon - \frac{b+a}{2}\right\}\right), \qquad \epsilon \in [a,b]. \tag{14.54}$$

The *truncation* N is user choosable; the larger the N, the better the approximation, but also the more expensive to compute the series coefficients and sum the series.

The Chebyshev strategy can also be extended to semi-infinite or infinite domains. The rational Chebyshev functions TL_j, which are the images of the Chebyshev polynomials under a change of coordinate, can be employed to expand a function on the interval $\epsilon \in [0,\infty]$; the formulas are also given in section 2.14.1.

An expansion over an unbounded domain is obviously a more impressive feat than an approximation over a finite interval, so we shall illustrate the potential of Chebyshev expansions for roots by showing the TL_j coefficients for the first branch of the modon dispersion relation in Fig. 14.2. The coefficients decrease very fast even though (i) the expansion interval is unbounded and (ii) $x(\epsilon)$ is logarithmically singular at $\epsilon = 0$, as evident from the form of the small-ϵ perturbation theory derived in section 15.5.

Severely truncated rational Chebyshev expansions can mimic Padé approximants. For example, for the modon dispersion relation,

$$x \approx 4.288 + 0.6457\frac{\epsilon-2}{\epsilon+2} + 0.20615\frac{\epsilon^2 - 12\epsilon + 4}{(\epsilon+2)^2} = \frac{15.3938 + 14.6782\epsilon + 5.1399\epsilon^2}{(\epsilon+2)^2}. \tag{14.55}$$

The maximum relative error is only 0.44%. This is a factor of three worse than the two-Padé approximant of the same degree. However, unlike Padé approximants, which require high-order perturbation series and often develop poles in a subsequence of approximants, Chebyshev expansions are very robust. A reliable theory [76] guarantees that the error will fall as the truncation N increases, so long as we can reliably compute the roots at the Chebyshev interpolation points.

Other parametric Chebyshev expansions of root branches (or eigenvalue branches, which are branches of the zeros of determinants) include [49, 77].

In applications, it is unfortunately not uncommon for $x(\lambda)$ to be singular on part of the range of interest, often at an endpoint of the range of the parameter λ. Extensions of the Chebyshev method to cope with endpoint singularities are discussed in [85, 108]. Simultaneous approximation of multiple branches that meet at a limit point is analyzed in [77, 88].

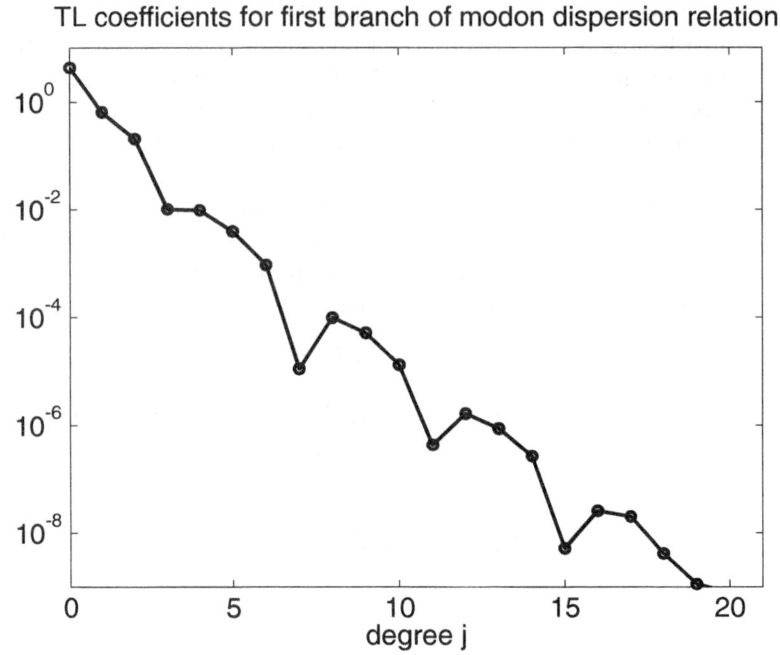

Figure 14.2. *Expansion of the first branch of the modon dispersion relation for the semi-infinite range $\epsilon \in [0,\infty]$ as a series of rational Chebyshev functions, $TL_j(\epsilon, L)$. Experimentation showed that the best choice of map parameter for the Chebyshev series was $L = 2$.*

14.11 ▪ Rescaling and Transformation

Before we apply the perturbative techniques discussed earlier, it is often highly useful to transform the problem to reduce the number of parameters. For example, the problem [290]

$$f(x) \equiv ax + bx^{7/8} - c = 0 \tag{14.56}$$

contains three parameters, (a, b, c), which are known from the physics to be positive constants. Because $c > 0$, we lose no generality by dividing every term in the equation to obtain the modified problem

$$\tilde{f}(x) \equiv Ax + Bx^{7/8} - 1 = 0, \tag{14.57}$$

which contains only two parameters, where $A \equiv a/c$ and $B = b/c$. If we define the new unknown

$$w \equiv Ax, \tag{14.58}$$

the equation becomes

$$w + BA^{-7/8} w^{7/8} - 1 = 0. \tag{14.59}$$

If we define a new parameter $\lambda = BA^{-7/8}$, we finally recast the problem so that it contains but a single parameter:

$$w + \lambda w^{7/8} - 1 = 0. \tag{14.60}$$

Because only one parameter remains, it is easy to expand w as a perturbation series in various parameter regimes. For small λ, for example,

$$w \approx 1 - \lambda + \frac{7}{8}\lambda^2 - \frac{91}{128}\lambda^3 + \frac{35}{64}\lambda^4 + \cdots, \qquad \lambda \ll 1, \tag{14.61}$$

while for large λ,

$$w \approx \lambda^{-8/7} - \frac{8}{7}\lambda^{-16/7} + \frac{68}{49}\lambda^{-24/7} - \frac{600}{343}\lambda^{-32/7}, \qquad \lambda \gg 1. \tag{14.62}$$

14.12 ▪ Choice of Perturbation Parameter

14.12.1 ▪ 3.99 Dimensions, the 1/*N* Expansion, and All That

Sometimes a very nonobvious choice of perturbation parameter is best. In the theory of critical phenomena, an important tool is perturbation in a small parameter ϵ that is the difference between the dimension d and 4, which led to a famous paper with the amusing title "Phase transitions in 3.99 dimensions" by Wilson and Fisher [374, 373]. In the words of the physicist and Fields Medalist Ed Witten, "After decades in which the study of critical phenomena was thwarted by the absence of an expansion parameter, Kenneth Wilson and Michael Fisher suggested that to introduce a parameter, one should regard the number of spatial dimensions not as a fixed number, three, but as a variable parameter. They showed that critical phenomena are simple in four dimensions and that in $4-\epsilon$ dimensions critical phenomena can be understood by perturbation theory in ϵ. Even at $\epsilon = 1$, the original three-dimensional problem, this perturbation theory is quite successful" [379].

A number of other problems simplify when the dimension N is infinite and the perturbation parameter is the reciprocal of N. As Witten explains, the $1/N$ series was first invented in atomic physics, but the Nobel Laureate Gerald t'Hooft extended it to quantum chromodynamics (QCD) in 1974 by pretending that there are N quarks instead of the three quarks of standard QCD.

Similarly, Bender, Milton, Pinsky, and Simmons treated the exponent of nonlinearity, which in most applications is restricted to an integer, as a continuous small parameter [38]. Thus, a polynomial equation may be rewritten as

$$x^5 + ax + b = 0 \qquad \rightarrow \qquad x^{1+\epsilon} + ax + b = 0 \tag{14.63}$$

and solved as a perturbation series in powers of ϵ. This article has been cited more than a hundred times and has spawned many applications, including differential equations and algebraic equations. One of their examples is discussed in more detail in section 8.4.2.

J. He [206] has noted that when an equation is inflated to a one-parameter family of equations by introducing an artificial homotopy parameter λ, the homotopy parameter can be taken as a perturbation parameter. For this strategy to be successful, one must choose a zeroth-order problem $g_0(x) = 0$ that is as close as possible to the desired problem $g(x) = 0$—but still explicitly solvable. One then expands x in powers of ϵ and applies perturbation theory in the usual way to

$$f(x;\epsilon) \equiv (1-\epsilon)\, g_0(x) + \epsilon\, f(x) = 0 \tag{14.64}$$

or any other homotopy such that $f(x;\epsilon = 0) = g_0$ and $f(x;\epsilon = 1) = g(x)$. The desired solution to $g(x) = 0$ is obtained by summing the perturbation series with $\epsilon = 1$. The idea is the same whether there is one unknown or many.

14.12.2 ▪ Case Study: Paterson's Problem

The problem

$$w+\lambda w^{7/8}-1=0 \tag{14.65}$$

is a good example. (This is the simplified by rescaling version of the chemical engineering equation of the previous section [290].) If we generalize this to

$$w+\lambda w^{1-\epsilon}-1=0, \qquad \lambda\in[0,\infty], \tag{14.66}$$

then the problem for $\epsilon=0$ is trivial to solve:

$$w(\epsilon=0)=\frac{1}{1+\lambda}. \tag{14.67}$$

To simplify the perturbative series, it is convenient to introduce

$$\mu\equiv\frac{1}{1+\lambda}, \qquad \lambda=\frac{1-\mu}{\mu}, \qquad \mu\in[0,1]. \tag{14.68}$$

We find

$$w=\mu-(\mu-1)\mu\log(\mu)\epsilon+\frac{1}{2}\mu\log(\mu)\left\{\log(\mu)(2\mu-1)(\mu-1)+2(\mu-1)^2\right\}\epsilon^2+\cdots. \tag{14.69}$$

Fig. 14.3 shows the errors in approximating $w(\mu[\lambda])$ by the first three orders for $\epsilon=1/8$, which corresponds to the chemical engineering problem we started with. The left panel shows that the two higher approximations are graphically indistinguishable from the exact root. On the right, one sees that the maximum *relative* error in the three-term approximation is only 0.55%.

Thus, this oddball choice of perturbation parameter—a small correction to the exponent of the nonlinear term—has generated a good literal approximation for *all* $\lambda\in[0,\infty]$, which is equivalent to $\mu\in[0,1]$.

14.12.3 ▪ Bender et al. Polynomial Equation

Bender, Milton, Pinsky, and Simmons [38] solved

$$x^5+x+1=0, x_*=0.75487767\ldots \tag{14.70}$$

using three different series. They calculated the radius of convergence ρ_c of the power series, the relative error in the perturbation series up to and including ϵ^6, and the relative error in the [3/3] Padé approximant obtained from the sixth-order series (Table 14.4). The best result comes from the Padé approximant of their series, but the strong coupling series—without Padé acceleration—is almost as good. Two of the three series fail without Padé acceleration, but the Padé treatment of the strong coupling series doubles the error (albeit to only 1 part in 600).

14.13 ▪ Perturbative Expansion about a Limit Point

When a solution exists only on one side of $x=x_0$, then x_0 is said to be a *fold* or *limit point*. Because an upper and lower branch meet at x_0, naive perturbation theory must fail.

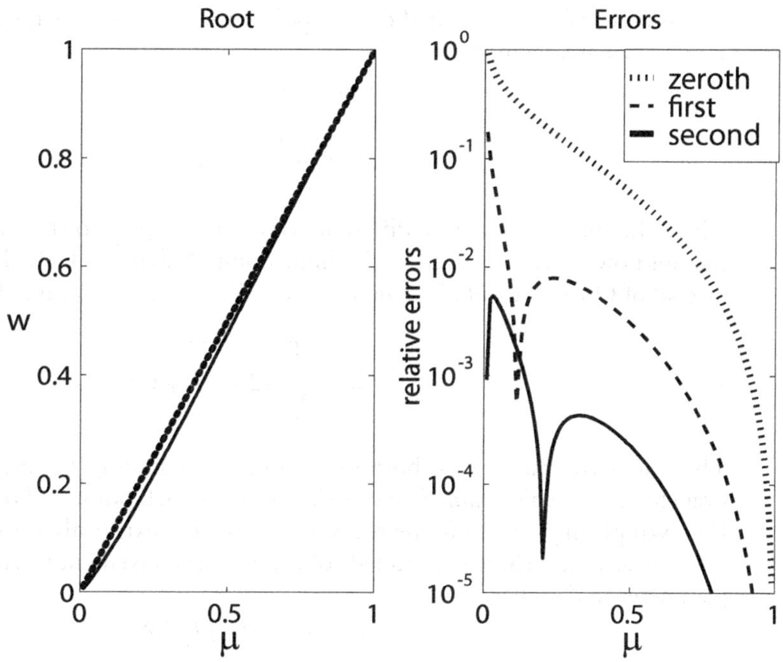

Figure 14.3. *The root w of the transformed Paterson problem, $w + \lambda w^{1-\epsilon} - 1 = 0$ for $\epsilon = 1/8$ (left), and the errors in the zeroth-, first-, and second-order approximations from a power series in powers of ϵ. The dependence on λ is plotted using the dependent parameter μ, where $\mu \equiv 1/(1+\lambda)$ and $\lambda \in [0, \infty]$ is equivalent to $\mu \in [0, 1]$. The errors are relative errors defined by $|w_{exact} - w_{approx}|/w_{exact}$. The left panel compares the approximate roots with the exact roots, but the graphs (except for the zeroth-order approximation, $w_0 = \mu$ [dotted line]) are visually indistinguishable; the right panel displays the errors of different perturbative orders versus the parameter μ.*

Table 14.4. *Comparison of different perturbation schemes to solve $x^5 + x + 1 = 0$.*

Form	Name	ρ_c	Sixth-Order Error at $\epsilon = 1$	[3/3] Padé error at $\epsilon = 1$
$\epsilon x^5 + x - 1 = 0$	weak coupling	0.08192	huge	1.20%
$x^5 + \epsilon x - 1 = 0$	strong coupling	1.64938	0.07%	0.16%
$x^{1+4\epsilon} + x - 1 = 0$	Bender et al.	1/4	huge	0.05%

The perturbation expansion is single valued, but there are two branches: the algorithms respond to this plurality of branches by converging to neither. Nevertheless, it is still possible to derive a perturbative expansion when $\epsilon = 0$ is a limit point.

Definition 14.2 (Limit Point at $\epsilon = 0$). *The point $(x = x_0, \epsilon = 0)$ is a limit point for $f(x; \epsilon)$ if and only if the following two conditions are met:*

$$\frac{\partial f}{\partial x}(x = x_0, \epsilon = 0) = 0, \qquad \frac{\partial f}{\partial \epsilon}(x = x_0, \epsilon = 0) \neq 0. \tag{14.71}$$

(Note that if both partial derivatives are zero, then $\epsilon = 0$ is a transcritical bifurcation *point.)*

This is just a restatement of Definition 7.3.

Because $\partial f/\partial x = 0$ at the limit point, the lowest-order terms in the power series about $\epsilon = 0$ are, defining

$$x = x_0 + v, \qquad |v| \ll 1, \tag{14.72}$$

$$f \approx \frac{1}{2} f_{xx} v^2 + f_\epsilon \epsilon, \tag{14.73}$$

where the subscripts denote differentiation with respect to the subscripted variable and the unknowns are evaluated at the limit point. Solving (14.73) shows that $v \sim O(\epsilon^{1/2})$ instead of $O(\epsilon)$, as would be true when $\epsilon = 0$ is not a limit point. Explicitly,

$$v = \pm\sqrt{-2\frac{f_\epsilon}{f_{xx}}\epsilon} + O(\epsilon). \tag{14.74}$$

There are two branches for both signs of ϵ, but, assuming f_{xx} and f_ϵ are real valued, the branches are *complex valued* on one side of the branch point and real valued on the other. For example, if $f_\epsilon/f_{xx} > 0$, the real-valued branches exist only for $\epsilon < 0$.

To track only these real-valued solutions, it is convenient to change the perturbation parameter to δ, where

$$\epsilon = -\text{sign}(f_\epsilon/f_{xx})\,\delta^2. \tag{14.75}$$

The unknown is expanded as

$$x = x_0 + x_1\delta + x_2\delta^2 + \cdots, \tag{14.76}$$

which is equivalent to an expansion in *half-integral* powers of ϵ. Define

$$x_1 \equiv \sqrt{|2f_\epsilon/f_{xx}|}, \tag{14.77}$$

where the derivatives are evaluated at the limit point, $(x = x_0, \epsilon = 0)$, and f_ϵ is computed before making the change $\epsilon \to \pm\delta^2$. The upper branch (of the two real-valued branches merging at the limit point) is obtained by evaluating the perturbative approximation for $\delta > 0$; the lower branch (such that $x(\delta) < x_0$) is the sum of the same series for negative δ.

Table 14.5 is a Maple script for computing such a perturbative expansion. Fig. 14.4 shows the accuracy of the expansion of various orders for the problem

$$f(x,\epsilon) \equiv (x-1)^2 - \epsilon(1+x^3) \;\rightarrow\; x = 1+\sqrt{2}\delta+\frac{3}{2}\delta^2+\frac{33}{16}\sqrt{2}\delta^3+\frac{11}{2}\delta^4+\frac{4095}{512}\sqrt{2}\delta^5+24\delta^6. \tag{14.78}$$

14.14 ▪ Another Example of a Half-Power Expansion about a Limit Point: Lambert W-Function

The Lambert function $W(x)$ is the solution of

$$W\exp(W) = x. \tag{14.79}$$

This arises in many applications and has been intensively studied, as explained in depth in section 13.6. Note that here to follow conventional notation, W is the root, and x is the parameter instead of λ.

Table 14.5. *Maple program: Perturbation theory at a limit point.*

```
# Regular perturbation theory when x0 is a limit (fold) point;
# User must replace next line with his/her own choice of f, x0,N;
  f := (x − 1)**2 − epsilon*(1 + x**3); x0 := 1; N := 12;
  fe := subs(epsilon = 0,x = x0,diff(f,epsilon));
  fxx := subs(epsilon = 0,x = x0,diff(f,x,x));
  if (fe/fxx) < 0 then epsilon := delta*delta; x1 := sqrt(−2*fe/fxx);
  else epsilon := −delta*delta; x1 := sqrt(2*fe/fxx); fi:
  x := x0 + delta*x1; for j from 2 by 1 to N do x := x + x||j * delta**j; od:
  r := series(f,delta,N + 2); # Expand f as series in delta;
  for j from 2 by 1 to N do eq||j := coeff(r,delta,j + 1); x||j := solve(eq||j,x||j); od:
```

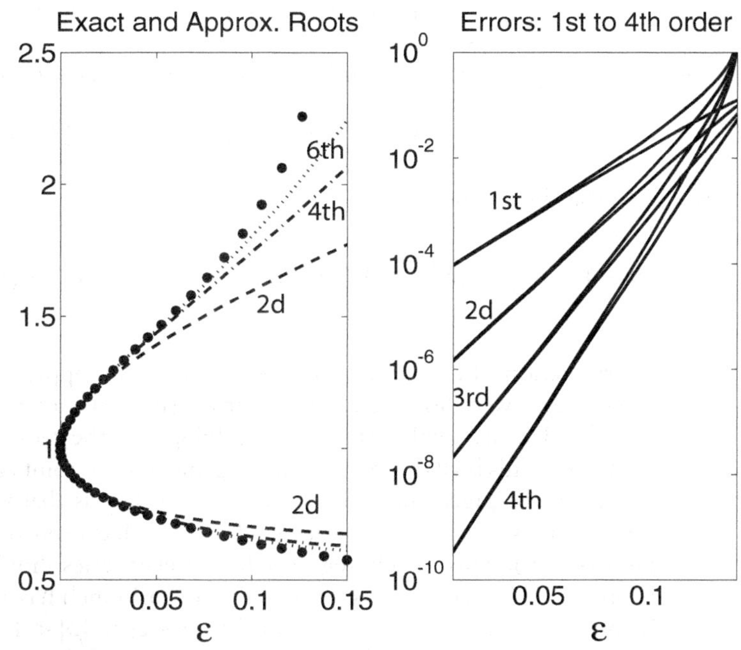

Figure 14.4. *Left panel: the black disks show the exact solution of* $f(x,\epsilon) \equiv (x-1)^2 - \epsilon(1+x^3)$*; the dashed, dash-dotted, and dotted curves show the sums of the perturbative series up to and including orders* δ^2, δ^4, *and* δ^6. *Right panel: the errors in the approximations up to and including* δ^j *for* $j = 1,2,3,4$ *in descending order. The perturbation series has a radius of convergence of approximately* $|\delta| = 0.4$, *or equivalently* $|\epsilon| < 0.16$. *For a given order, each error curve is double because the errors for the upper and lower branches are the same only as* $\epsilon \to 0$.

The two real-valued branches, $W_0(x)$ and $W_{-1}(x)$, meet at a limit point at $x = -\exp(-1)$ where $W = -1$.

Near a limit point x_{lim}, a function $f(W,x)$ can be approximated as

$$f(W,x) \approx f(W(x_{lim}),x_{lim}) + \frac{\partial f}{\partial W}(W - W[x_{lim}]) + \frac{1}{2}\frac{\partial^2 f}{\partial W^2}(W - W[x_{lim}])^2 + \frac{\partial f}{\partial x}(x - x_{lim}) + \cdots . \quad (14.80)$$

At a limit point, the first and second terms vanish. The remaining terms must cancel so that $f(W(x),x)=0$. Therefore,

$$+\frac{1}{2}\frac{\partial^2 f}{\partial W^2}(W-W[x_{lim}])^2+\frac{\partial f}{\partial x}(x-x_{lim})=0 \Rightarrow$$
$$W \approx W[x_{lim}] \pm \sqrt{\frac{\partial f/\partial x}{(1/2)\partial^2 f/\partial W^2}}\sqrt{x-x_{lim}}. \tag{14.81}$$

This shows that near the limit point, the two branches that merge are functions of the *square root* of $x-x_{lim}$ rather than $x-x_{lim}$ itself.

This suggests the change of variable

$$\rho \equiv \sqrt{1+\exp(1)x} \qquad \leftrightarrow \qquad x=\exp(-1)(\rho^2-1). \tag{14.82}$$

One can then apply *regular* perturbation theory in powers of ρ to

$$f(W;\rho) \equiv W\exp(W)-\exp(-1)(\rho^2-1) \tag{14.83}$$

to obtain the alternating series

$$W=-1+\sqrt{2}\rho-(2/3)\rho^2+(11/36)\sqrt{2}\rho^3-(43/135)\rho^4+(769/4320)\sqrt{2}\rho^5+\cdots. \tag{14.84}$$

The interpretation of this series is subtle. For $\rho>0$, it approximates the upper branch $W_0(x)$. For negative ρ, however, all the terms in this series are negative and all the partial sums to all orders are smaller than -1, implying that the series approximates the *lower* branch $W_{-}(x)$. This is all the same as for the general limit point and its pair of expansions, as discussed in the preceding subsection. The subtlety is that the upper branch has no finite singularities except at $x=-\exp(-1)$, which has been resolved by the change of variable $x \to \rho(x)$. This naively suggests the ρ power series should have an *infinite* radius of convergence. However, this is wrong. The lower branch has a limit point at $\rho=1 \leftrightarrow x=0$, and this limits the convergence of the ρ-series to $|\rho|<1$. Thus, the branch point on the second Riemann sheet limits the convergence of the series for the upper branch (first Riemann sheet).

14.14.1 ▪ Padé Approximants for the Lambert Series

Padé approximants work exceptionally well for the Lambert series. First, the approximants have coefficients that are rationals multiplied by the surd, $\sqrt{2}$, left as *symbolic*, and thus are calculated in Maple without error. For example,

$$W_{[3/3]}=\frac{-8157240\sqrt{2}-701280\sqrt{\rho}+6887475\sqrt{2}\rho+2495668\rho^{3/2}}{8157240\sqrt{2}+17015760\sqrt{\rho}+4690125\sqrt{2}\rho+525722\rho^{3/2}}. \tag{14.85}$$

Floating point errors arise only when a floating point number is substituted for y. Second, the approximants converge geometrically far beyond the radius of convergence of the Puiseux series whence it came: at $y=100$—one hundred times the radius of convergence—the relative error of the $[15/15]$ approximation is only 4.0×10^{-8} (Fig. 14.5).

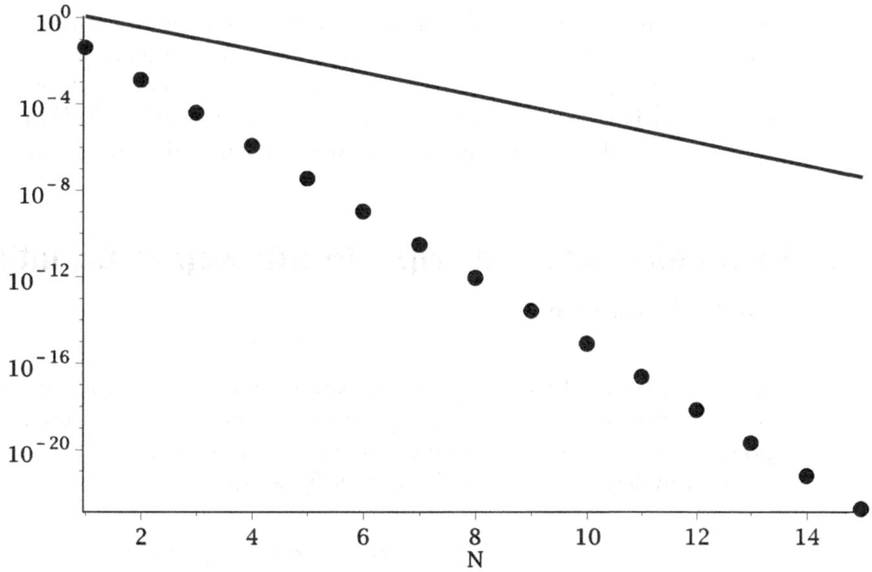

Figure 14.5. *Errors in the diagonal Padé approximants formed from the limit power series. The solid circles show the errors at $\rho = 1$, the radius of convergence of the power series. The solid line shows the errors at $\rho = 100$.*

14.14.2 ▪ Inferiority of Perturbation Series: Strengths of the Chebyshev Series in ρ

The success of perturbation theory for the Lambert W-function is encouraging. However, it is true for both regular and nonpower perturbation theory that nonperturbative methods can often yield better analytical approximations.

For example, the expansion of $W(x)$ on $x \in [-\exp(-1), 0]$ using shifted Chebyshev polynomials in the variable $\rho \equiv \sqrt{1+\exp(1)x}$ gives a rapidly convergent expansion on $\rho \in [0,1]$. In contrast, the perturbation series in powers of ρ (i.e., half-integer powers of $(x+\exp(-1))$) diverges at $\rho = 1$. Standard theory shows that with a singularity at $\rho = -1$, the coefficient of $T_n^*(\rho)$ for expansion on $\rho \in [0,1]$ is proportional to $(0.17)^n$. In other words, each additional term reduces the error *uniformly* over the whole interval by more than a factor of five. In contrast, the error in the ρ power series decreases by a factor ρ^n, which is much slower except for very small ρ.

The quintic Chebyshev approximation is

$$\begin{aligned} W_{Cheb,5} \equiv & -0.4593694\, T_0(2\rho-1) + 0.4954710\, T_1(2\rho-1) - 0.0400573\, T_2(2\rho-1) \\ & + 0.0044537\, T_3(2\rho-1) - 0.0005648\, T_4(2\rho-1) + 0.0000740\, T_5(2\rho-1) \\ = & -0.9999905 + 1.4133479\rho - 0.6542483\rho^2 + 0.3700897\rho^3 - 0.1671150\rho^4 \\ & + 0.0379233\rho^5. \end{aligned} \tag{14.86}$$

Its maximum absolute and relative pointwise errors on $\rho \in [0,1]$ are only 0.000024.

Furthermore, the Chebyshev series can be extended to a larger range in ρ—far enough to join up with the log/log(log) asymptotic approximation—at the price of a slower but still exponential rate of convergence.

The perturbation theory could be extended by converting the series into a Padé approximant in ρ, as done in the previous subsection. (For alternating series, the accuracy at $\rho = \rho_1$ is usually very similar to a Chebyshev approximation of the same order on the interval $\rho \in [0, \rho_1]$, where *order* for a Padé approximant is defined to be the sum of the degrees of the numerator and denominator polynomials.) However, pure perturbation theory is a dramatic tradeoff compared to perturbation/Padé or Chebyshev approximations: simplicity at the price of a narrower range of accuracy and much less uniform accuracy.

14.15 ▪ Expansion about a Triple Point: Kepler Equation

The Kepler equation is

$$f(x) = x - \epsilon \sin(x) - \lambda, \tag{14.87}$$

where $\lambda \in [-\pi, \pi]$ and $\epsilon \in [0, 1]$ are parameters. An expansion in integral powers of ϵ or λ is useless in the corner of the parameter plane where $1 - \epsilon$ and λ are both small: the point $\epsilon = 1, \lambda = 0$ is a *triple* point where three roots merge.

We can derive a series for this corner by writing

$$\epsilon = 1 - \mu\lambda \qquad \Leftrightarrow \qquad \mu = \frac{1-\epsilon}{\lambda}, \tag{14.88}$$

where μ is $O(1)$. When $\lambda \ll 1$, $x \ll 1$ also. Expanding the $\sin(x)$ as a power series in x and neglecting $O(x^5)$ gives

$$f = \mu\lambda x + \frac{1}{6}x^3 - \lambda. \tag{14.89}$$

If the last two terms balance, then

$$x \approx (6\lambda)^{1/3}. \tag{14.90}$$

The $\lambda\mu x$ term is then $O(\lambda^{4/3})$, whereas the second and third terms are both $O(\lambda)$. Thus, the lowest-order approximation is consistent.

If we substitute the expansion

$$x = (6\lambda)^{1/3} + x_1 \lambda^{2/3} + x_2 \lambda + \cdots, \tag{14.91}$$

we find by matching powers of $\lambda^{1/3}$ that

$$\begin{aligned} x = {} & (6\lambda)^{1/3} - \frac{1}{3}6^{2/3}\mu\lambda^{2/3} + \frac{1}{10}\lambda \\ & + 6^{1/3}(\mu/5 + (2/27)\mu^3)\lambda^{4/3} + \frac{1}{56700}(243 - 9450\mu^2 + 1400\mu^4)6^{2/3}\lambda^{5/3} + O(\lambda^2). \end{aligned} \tag{14.92}$$

14.16 ▪ Replacing $f(x)$ by Its Asymptotic Expansion: When the Perturbation Parameter Is Implicitly the Ordinal of the Zero

The error function, erf(x), has a root at the origin plus a countable infinity of complex-valued roots. Because of the known symmetries of the error function in the complex plane, it is sufficient to find the roots only in the *upper right quadrant*, that is, the roots whose real and imaginary parts are both positive.

A good preliminary strategy for finding complex roots is to make a contour plot of the *logarithm* of the *absolute values*, the MAG/complex plane algorithm discussed in section 5.11. Fig. 14.6 shows that, except for the trivial root at the origin, the roots in the

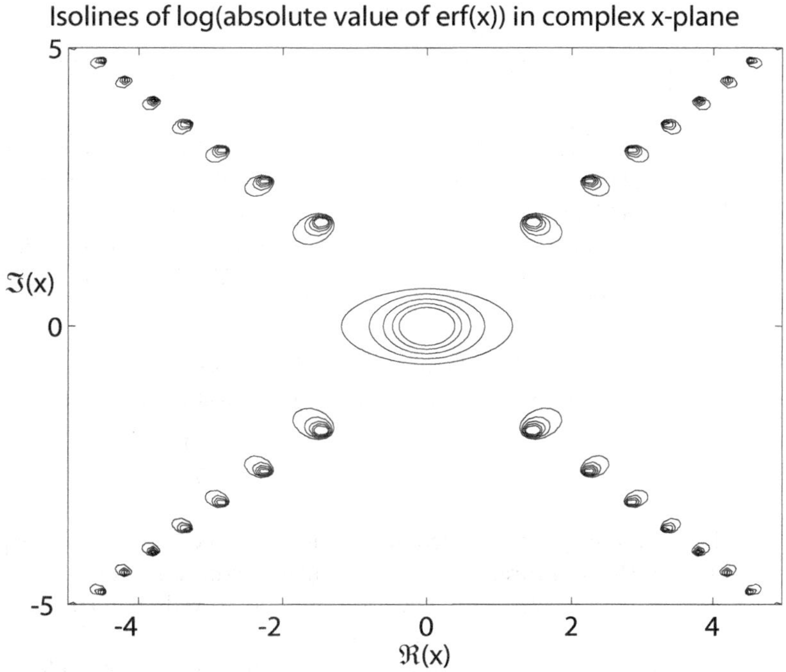

Figure 14.6. *The contour lines of* $\log(|erf(x)|)$ *in the complex x-plane. The isolines whose values are* 0.1, 0.3, ..., 0.9 *are plotted.*

upper right quadrant of the complex x-plane are clustered around the line $\Re(x) = \Im(x)$. The isoline plot of magnitudes also shows the four-fold symmetry already mentioned:

$$\operatorname{erf}(x_n) = 0 \rightarrow \operatorname{erf}(-x_n) = 0 \ \& \ \operatorname{erf}(\pm x_n^*) = 0, \tag{14.93}$$

where the asterisk denotes complex conjugation.

An important clue is that the four roots nearest the origin have a magnitude of about 2.4. If we number the roots so that the zero nearest the origin in the upper right quadrant is $m = 1$, the second-nearest is $m = 2$, and so on, then clearly $|x_m|$ is increasing with m. It follows that for large m, it should be a good approximation to replace erf(x) by its large-$|x|$ asymptotic approximation,

$$\operatorname{erf}(x) \sim 1 - \frac{\exp(-x^2)}{\sqrt{\pi}\, x} \left\{ 1 - \frac{1}{2}\frac{1}{x^2} + \frac{3}{4}\frac{1}{x^4} - \frac{15}{8}\frac{1}{x^6} + \cdots + \frac{(-1)^n (2n)!}{n! 2^{2n}} \frac{1}{x^{2n}} \right\}, |x| \gg 1, \tag{14.94}$$

where this expansion is valid everywhere in the upper right quadrant.

To apply this expansion to find roots requires two conceptual leaps. First, the effective perturbation parameter is not any quantity that appears *explicitly* in the definition of $f(x)$—indeed, the error function contains *no explicit parameters*. Instead, the small parameter is *implicitly*

$$\epsilon \equiv 1/m, \qquad \epsilon \ll 1, \qquad m \equiv \text{root index}. \tag{14.95}$$

Second, the expansion is not a simple matching of powers of ϵ but rather is the asymptotic expansion for $f(x)$ itself.

Table 14.6. *Roots of the error function.*

m	$x_m^{Lambert} = \sqrt{0.5W(2m,2/\pi)}$	Exact	Error in $x_m^{Lambert}$
1	$1.436154318 + 1.868751686i$	$1.450616163 + 1.880943000i$	0.01891489090
2	$2.239971792 + 2.612606940i$	$2.244659274 + 2.616575141i$	0.006141588286
3	$2.837288409 + 3.173502016i$	$2.839741047 + 3.175628100i$	0.003245869118
4	$3.333904121 + 3.644799380i$	$3.335460735 + 3.646174376i$	0.002076935518
5	$3.767908741 + 4.059714388i$	$3.769005567 + 4.060697234i$	0.001472757120

If we truncate the asymptotic expansion after the first nonconstant term, then this implies that the error function will have zeros at all large x such that

$$\exp(x^2) = \frac{1}{\sqrt{\pi}\, x}. \tag{14.96}$$

This is a messy transcendental equation in which both x and exponentials of x appear. Indeed, this is a foretaste of the singular perturbation expansions described in the next chapter. However, this equation can be solved in terms of the so-called Lambert W-function, whose properties are described at great length in section 13.6. What is important for present purposes is that, denoting the root of (14.96) by $x_m^{Lambert}$,

$$x_m \sim x_m^{Lambert} = \sqrt{W(2m,2/\pi)/2}, \qquad m = 1,2,\ldots, \tag{14.97}$$

where $W(k;x)$ denotes the kth branch of this infinitely many branched function. The two real-valued branches, $k = -1, 0$, give roots that are too small to be consistent with the accuracy of the large-x asymptotics; these roots are legitimate zeros of the approximate equation (14.96) but are spurious roots for the error function itself. The odd branches, as may be verified by direct substitution, do not give legitimate error functions roots either; the reason is a subtle consequence of the square root enclosing the W-function. The even-numbered roots are accurate, however; Table 14.6 shows that the approximation is accurate to better than 2% even for the lowest root.

The error can be reduced still further by a modified Newton's iteration. If we denote the lowest approximation by $x^{(0)}$, then

$$x^{(1)} = x^{(0)} - \frac{\operatorname{erf}\left(x^{(0)}\right)}{(2/\sqrt{\pi})\exp(-[x^{(0)}]^2)}. \tag{14.98}$$

If the Lambert solution is the first approximation, then $x^{(1)}$ can be simplified by (i) invoking the transcendental equation solved by $x^{(0)} \equiv x_m^{Lambert}$ and (ii) replacing the error function by the first three terms of its asymptotic approximation, yielding

$$x^{(1)} = x_m^{Lambert} - \frac{1}{4[x_m^{Lambert}]^3}. \tag{14.99}$$

The disadvantage of these approximations is that both involve the Lambert function. Although the W-function is native to Maple and other software libraries, it is still useful to note that it can be removed by invoking the asymptotic approximation for its $2m$th branch:

$$W(2m,2/\pi) \sim \log(2/\pi) + 4\pi m i - \log(\log(2/\pi) + 4\pi m i), m \gg 1. \tag{14.100}$$

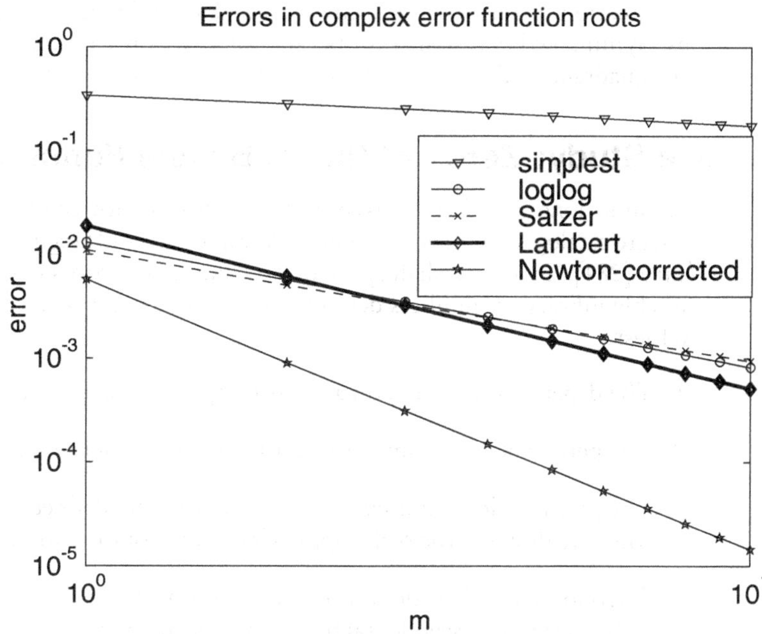

Figure 14.7. *Errors in approximation of the complex roots of the error function versus m, the index of the root. The "simplest" approximation is (14.102), $x_m^{simplest} \sim \sqrt{2\pi m}\exp(i\pi/4)$. $x_m^{Lambert}$ is labeled as the "Lambert" approximation; "Newton-corrected" is the (approximate) first Newton correction of $x_m^{Lambert}$, $x^{(1)} = x_m^{Lambert} - 1/(4[x_m^{Lambert}]^3)$.*

This defines the cruder but simpler approximation

$$x_m^{loglog} \sim \sqrt{(1/2)\log(2/\pi) + 2\pi m i - (1/2)\log(\log(2/\pi) + 4\pi m i)}. \tag{14.101}$$

Neglecting the logarithms gives

$$x_m^{simplest} \sim \sqrt{2\pi m}\exp(i\pi/4). \tag{14.102}$$

The simplest approximation is useful because it shows explicitly that the mth root is $O(m^{1/2})$. Inspecting $x^{(1)}$ shows that the error in $x_m^{Lambert}$ is $O(m^{-3/2})$. The *log-log* approximation can be expanded in inverse powers of $1/\sqrt{m}$ to get an expression similar to that of Salzer [305], which is included in the table of error function zeros in Abramowitz and Stegun [1]:

$$x_m^{Salzer} \sim (\sigma_m - \tau_m) + i(\sigma_m + \tau_m), \tag{14.103}$$

$$\sigma_m = (1/2)\sqrt{\pi(4m - 1/2)}, \ \tau_m = \frac{\log(\pi\sqrt{2m - 1/4})}{2\sqrt{\pi(4m - 1/2)}}. \tag{14.104}$$

Fig. 14.7 compares the errors in the various approximations. The simplest approximation has an error that falls off as $1/\sqrt{m}$. Salzer's approximation is better than the log-log and Lambert approximations for the lowest root but inferior to these approximations for $m > 3$.

The important point is that by treating the index m of the root as the reciprocal of an implicit perturbation parameter and approximating $f(x)$ by its asymptotic, large-$|x|$

approximation, we have successfully obtained a whole hierarchy of analytic approximations. Symmetry is valuable, too: because of it, we only need to explicitly find the roots in one quadrant, and the others are immediately known by reflection in one or both axes.

14.17 ▪ Case Study: Zeros of the J_0 Bessel Function

MacMahon long ago derived asymptotic expansions for the roots of Bessel functions, which are cataloged for general order in Abramowitz and Stegun [1] and the *NIST Digital Library* [284]. Here, we shall specialize to finding the roots of $J_0(x)$. These zeros are a countable infinity of numbers denoted by $j_{0,k}$, where $k = 1, 2 \ldots$. This example has four novelties:

1. The dependence of $f(x)$ on x must be approximated by an asymptotic series.
2. The perturbation parameter $\epsilon \ll 1$ is the *inverse* of the zero index k.
3. The perturbation parameter ϵ is mathematically defined for all ϵ, but because the roots are discrete, the series approximates a root only for a discrete set of ϵ.
4. Trigonometric identities must be used to transform the asymptotic approximation of J_0 so every transcendental function after the transformation has a *small argument* and can be expanded in ϵ.

For large argument, the J_0 function has the asymptotic approximation

$$J_0(x) \sim \sqrt{\frac{2}{\pi x}} \left[\cos(x - \pi/4) \left\{ 1 - \frac{9}{128x^2} + \frac{3675}{32768\,x^4} + \cdots \right\} \right. \\ \left. \sin(x - \pi/4) \left\{ \frac{1}{8x} - \frac{225}{3072\,x^3} + \cdots \right\} \right] + \cdots. \tag{14.105}$$

The leading term is proportional to $\cos(x - \pi/4)$, so for sufficiently large x, the Bessel function must have a root very close to $x_0 = \pi(k - 1/4)$, where k is an integer.

The obvious choice of small parameter is $1/k$, where k is the order of the zero, and this choice is legitimate and gives an asymptotic expansion. However, MacMahon recognized that an even better choice is the reciprocal of the lowest-order approximation, that is,

$$\epsilon \equiv 1/(\pi(k - 1/4)). \tag{14.106}$$

The reason is that the combination $k - 1/4$, rather than k alone, arises as we chug through the orders of perturbation theory. The inverse series in $1/k$ contains all integer powers, whereas the expansion in ϵ has nonzero coefficients for *odd powers only*. This example shows that subtle changes in the choice of perturbation parameter can greatly simplify a perturbation series.

As noted earlier, the series for $x(\epsilon)$ is defined for all real ϵ, but it only approximates the kth root of J_0 when $k(\epsilon)$ is an *integer*.

The next step is to substitute $x = 1/\epsilon + a_0 + a_1\epsilon + \cdots$ into the asymptotic expansion of $f(x) \equiv \sqrt{\pi x/2}J_0$, where the factor multiplying J_0 is chosen to eliminate an overall multiplicative factor in the asymptotic series of J_0. However, there is a complication: the argument of the sine and cosine factors in (14.105) is *large* rather than small. The trigonometric functions therefore cannot be approximated by expanding in powers of ϵ.

The way around this dilemma is to write

$$x = 1/\epsilon + v, \tag{14.107}$$

where v turns out to be $O(\epsilon)$ and is therefore small. However, the trigonometric identities $\cos(x_0+v) = \cos(x_0)\cos(v)-\sin(x_0)\sin(v)$ and $\sin(x_0+v) = \sin(x_0)\cos(v)+\cos(x_0)\sin(v)$ imply, replacing $1/\epsilon$ by its definition,

$$\cos\left(\left[\pi\left(k-\frac{1}{4}\right)+v\right]-\frac{\pi}{4}\right) = \cos(k\pi)\sin(v), \tag{14.108}$$

$$\sin\left(\left[\pi\left(k-\frac{1}{4}\right)+v\right]-\frac{\pi}{4}\right) = -\cos(k\pi)\cos(v), \tag{14.109}$$

which have been simplified because $\sin(k\pi) = 0$. $\cos(k\pi)$ is a common factor to every term of the asymptotic expansion and so does not appear in the series for the roots. The only transcendental functions remaining are $\cos(v)$, $\sin(v)$, whose arguments are $O(\epsilon)$. We can now expand these trigonometric functions and the inverse powers of x in ϵ. Next, collect powers of ϵ and solve order by order, thus forcing each term in the ϵ power series of the residual to vanish. This yields MacMahon's expansion:

$$\begin{aligned} x = 1/\epsilon &+ (1/8)\epsilon - (31/384)\epsilon^3 + (3779/15360)\epsilon^5 - (6277237/3440640)\epsilon^7 \\ &+ (2092163573/82575360)\epsilon^9 - (8249725736393/14533263360)\epsilon^{11} \\ &+ (423748443625564327/22671890841600)\epsilon^{13} \\ &- (3697238718425172707533/4353003041587200)\epsilon^{15} + \cdots \end{aligned} \tag{14.110}$$

or, in terms of floating point numbers,

$$\begin{aligned} x &= 1/\epsilon + 0.125\epsilon - 0.08072916667\epsilon^3 + 0.2460286458\epsilon^5 - 1.824438767\epsilon^7 \\ &+ 25.33641480\epsilon^9 - 567.6444121\epsilon^{11} + 18690.47653\epsilon^{13} - 849353.5803\epsilon^{15} + \cdots. \end{aligned} \tag{14.111}$$

The rapid increase in the coefficients a_j as $j \to \infty$ suggests that the series is asymptotic but divergent, just like the $1/x$ series for J_0 whence it came.

Fig. 14.8 shows the errors in the expansion, truncated at different orders, for the first three zeros. As typical of asymptotic series, for a given ϵ, there is a minimum error at some finite order, and then further corrections only worsen the approximation. As the order of the zero k increases, the minimum moves to higher order and smaller error. It is remarkable that although the rationale for the approximation is large-order k, the minimum relative error for the very first zero is only 1 part in 1500.

14.18 ▪ Other Perturbative Methods: Lagrange's Inversion and Adomian's Decomposition

Lagrange's method for reversion of a series is discussed in Chapter 18.

Regular perturbation series sometimes have other names. The *Adomian decomposition* method is an unmodified regular perturbation series in an artificial parameter [5, 6, 221]. This method has been applied to hundreds of problems, mostly redoing known solutions. Because of its simplicity, Adomian decomposition has sadly become a path to tenure at third-tier universities, where the number of publications is essential for survival but quality is not. Some journals now refuse papers using Adomian decomposition.

This unfortunate academic sociology does not change the fact that perturbation series are very useful, at least when applied to novel problems.

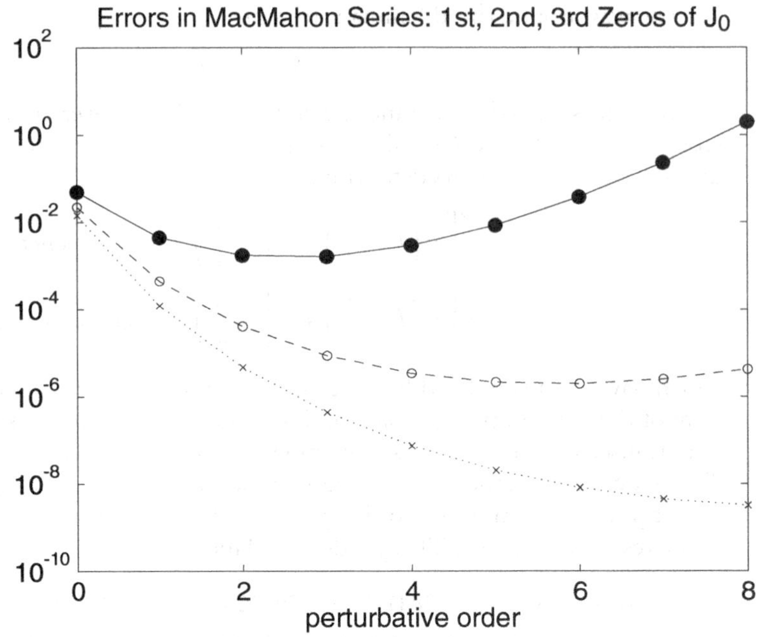

Figure 14.8. *Errors in the lowest three roots of J_0 as given by MacMahon's series, plotted versus order. (Zeroth order is the term in ϵ^{-1}, first order is $O(\epsilon)$, second order is $O(\epsilon^3)$, and so on.) The upper curve (solid disks) is the approximation to the first zero ($k = 1$), exact value 2.405...; $\epsilon = 0.4244$. The smallest error is third order, including terms up to and including ϵ^5, with a minimum error of 0.0016. The middle curve is for the $k = 2$ zero, which is minimum at sixth order (includes $O(\epsilon^{11})$) with an error of 2.0×10^{-6} ($\epsilon = 0.1818$). The bottom curve is for the third zero; the minimum error may be greater than eighth order, off the axis, and the minimum error is no larger than 3.2×10^{-9} with $\epsilon = 0.1157$.*

14.19 ▪ Case Study: The Zeros of the Stationary Point Equation for the Method of Steepest Descent for Integrals

The problem is to derive an approximation series for an integral of the form

$$I(z) \equiv \int_{-\infty}^{\infty} \exp(z\Psi(t))dt \tag{14.112}$$

when the parameter z is large. As $|z|$ increases, the integrand will become more and steeply peaked about points where $\Psi(t)$ has a local maximum. These *stationary points*, t_s, satisfy

$$\frac{d\Psi(t)}{dt}(t_s) = 0. \tag{14.113}$$

It is often necessary to deform the contour of integration in the complex t-plane so that it passes through stationary points off the real axis. A case-by-case analysis is necessary to determine which stationary points are on the deformed integration path and which stationary points are off-paths and can be ignored. It is possible for more than one stationary point to be on the integration path. Such refinements are thoroughly discussed in [75], [74], and [273]. Each stationary point is treated independently of all the others, so there is no loss of generality in treating a single stationary point.

The key step is to rewrite the integral without approximation as

$$I(z) \equiv \exp(z\Psi(t_s)) \int_{-\infty}^{\infty} \exp\left(-z\left\{\Psi(t_s) - \Psi(t)\right\}\right) dt \tag{14.114}$$

and then make the change of coordinate

$$w \equiv \sqrt{\Psi(t_s) - \Psi(t)}. \tag{14.115}$$

The integral becomes

$$I(z) \equiv \exp(z\Psi(t_s))J(z), \qquad J(z) \equiv \int_{-\infty}^{\infty} \exp(-zw^2)\frac{dt}{dw}(w)\,dw. \tag{14.116}$$

The local maximum is now at $w = 0$. Because the integrand is sharply peaked about $w = 0$—more and more sharply peaked as $|z| \to \infty$—it is legitimate to expand the *metric factor* as a power series about $w = 0$.

First, introduce

$$\tau \equiv t - t_s \tag{14.117}$$

so that the stationary point is at the origin in the new coordinate. Expand the steepest descent phase function as

$$\Psi(t) \approx P_0 + \sum_{j=2} P_j \tau^j, \tag{14.118}$$

where the linear term $P_1\tau$ is absent because of the stationary point condition, which is $d\Psi/d\tau(0) = P_1 = 0$. Inverting the function $w(\tau) = \sqrt{\Psi(0) - \Psi(\tau)}$ to obtain $\tau(w)$ as a power series in w, we can then differentiate to obtain $d\tau/dw = dt/dw$ to obtain the metric factor in power series form.

If we define the set a_j as the coefficients of the power series for $\tau(w)$,

$$\tau(w) = \sum_{j=1} a_j w^j \qquad \rightarrow \qquad \frac{d\tau}{dw} = \sum_{k=0} (k+1)a_{k+1}w^k, \tag{14.119}$$

then

$$J(z) \equiv \sum_{k=0} a_{k+1}(k+1) \int_{-\infty}^{\infty} \exp(-zw^2) w^k\,dw. \tag{14.120}$$

When k is odd, the integrand is antisymmetric and therefore the integral is zero. The even integrals are

$$\int_{-\infty}^{\infty} \exp(-zw^2)\, w^{2n}\,dx = \frac{1}{z^{n+1/2}}\Gamma(n+1/2). \tag{14.121}$$

We obtain

$$J(z) \equiv \sum_{n=0} a_{2n+1}(2n+1)\Gamma(n+1/2)\frac{1}{z^{n+1/2}}. \tag{14.122}$$

Table 14.7. *Maple program: Power series expansion in w for the metric factor of the steepest descent method.*

```
# τ ≡ t − t_s; Ψ(τ) is the steepest descent phase function;
  N := 5;
# N odd for convenience;
  Psis := sum(P[j] * tau**j,j = 2..N);
# alternatively, Psis = series(Psi(tau) − Psi(0),N + 1);
# Note that P[0] = Ψ(t_s)
# cancels out in Psis = Ψ(t_s + τ). P[1] is identically zero;
  g := sqrt(−Psis);
# w = g(tau) = sqrt(Psi(ts) − Psi(t));
  tau := sum(a[j]*w**j,j = 1..N); gs := series(g − w,w,N + 1);
# gg[1] := sqrt(−P[2]*a[1]**2) − 1, solved explicitly in the next line;
  a[1] := 1/sqrt(−P[2]);
# Explicit solution of gg[1] = 0;
# Next loop allows better control of Maple's square root;
  for j from 2 to N do gg[j] := coeff(gs,w,j);
  a[j] := solve(gg[j],a[j]); od:
  for n from 0 to ((N − 1)/2) do s[j] :=
  simplify((2*n + 1)*a[2*n + 1]*GAMMA(n + 1/2)/sqrt(Pi)); print(n, s[j]); od:
```

Table 14.8. *Coefficients s_n of z^{-n} in higher-order steepest descent.*

n	s_n
0	$\frac{1}{\sqrt{-P_2}}$
1	$-3/16\,\frac{-5P_3{}^2+4P_4P_2}{(-P_2)^{7/2}}$
2	$-\frac{15}{512}\,\frac{504P_3{}^2P_4P_2-224P_5P_3P_2{}^2-231P_3{}^4+64P_6P_2{}^3-112P_4{}^2P_2{}^2}{(-P_2)^{13/2}}$
3	$-\frac{105}{8192}\{-20592P_3{}^2P_4{}^2P_2{}^2-7293P_3{}^6+512P_8P_2{}^5+2112P_4{}^3P_2{}^3$ $-1152P_5{}^2P_2{}^4-13728P_3{}^3P_5P_2{}^2-2304P_7P_3P_2{}^4+25740P_3{}^4P_4P_2$ $+6336P_3{}^2P_6P_2{}^3-2304P_4P_2{}^4P_6+12672P_4P_2{}^3P_5P_3\}(-P_2)^{-19/2}$

By convention, the series is usually expressed in terms of coefficients s_n defined by (Table 14.8)

$$s_n \equiv a_{2n+1}(2n+1)\Gamma(n+1/2)\frac{1}{\sqrt{\pi}}, \tag{14.123}$$

$$\boxed{I(z) \sim \exp(z\Psi(t_s))\,\tfrac{\sqrt{\pi}}{\sqrt{z}}\,\textstyle\sum_{n=0} s_n z^{-n}}. \tag{14.124}$$

We have skipped the details of inverting $w(\tau) = \sqrt{\Psi(t_s) - [P_0 + P_1\tau + P_2\tau^2 + \cdots]}$ to obtain $\tau(w) = \sum_j a_j w^j$. This is, however, a straightforward exercise of substituting the series for $\tau(w)$ into

$$w - \sqrt{\Psi(t_s) - [P_0 + P_1\tau(w) + P_2\tau(w)^2 + \cdots]} = 0, \tag{14.125}$$

matching powers of w, and solving order by order. The Maple code is Table 14.7.

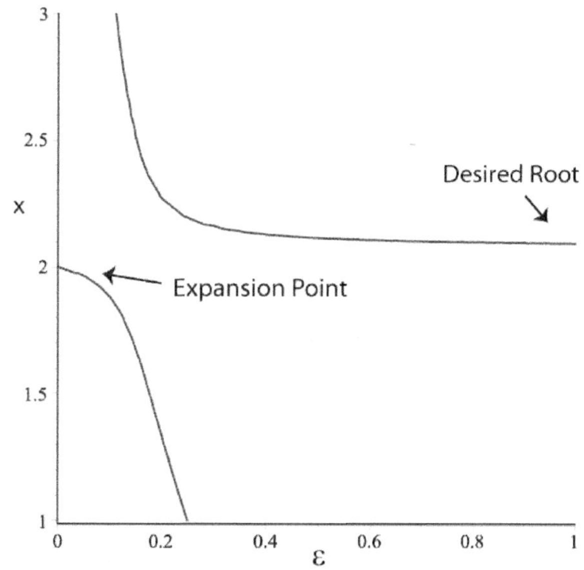

Figure 14.9. *Zero isolines of* $H(x,\epsilon) \equiv \epsilon x^3 - 2x + 4 - 9\epsilon$, *a homotopy for Newton's cubic, showing the upper branch, which is the desired root,* 2.09 *at* $\epsilon = 1$, *and the lower branch, which is tracked by the perturbative expansion about* $(x = 2, \epsilon = 0)$.

14.20 ▪ Case Study: Newton's Cubic

Isaac Newton illustrated his rootfinding methods by computing the one real root of the cubic polynomial[21]

$$f(x) \equiv x^3 - 2x - 5, \qquad x_* = 2.094551482. \tag{14.126}$$

From inspection, the real root is around $x = 2$. A homotopy that has this as the root for $\epsilon = 0$ is

$$H(x,\epsilon) \equiv \epsilon x^3 - 2x + 4 - 9\epsilon. \tag{14.127}$$

One solution branch coincides with the first guess, $x = 2$, at $\epsilon = 0$. One solution branch does go through the exact root of Newton's cubic at $x = 2.094551482$ when $\epsilon = 1$. Unfortunately, Fig. 14.9 shows that these two branches are *different*. Thus, this plausible homotopy is completely useless, showing that one must be careful in creating perturbative homotopies. (Complex-valued starting points for homotopies are much more appealing after an example like this.)

For Newton's cubic, the following alternative homotopy looks plausible because it also has an exact root at $x = 2$ for $\epsilon = 0$:

$$H(x,\epsilon) \equiv \epsilon(x^3 - 2x - 5) + (1-\epsilon)(x-2). \tag{14.128}$$

[21]This example has been used over and over again, illustrating not only various refining techniques, but also the longevity of canonical examples. Augustus de Morgan wrote to William Whewell, January 20, 1861: "The reason I call $x^3 - 2x - 5$ a celebrated equation is because it was the one on which Wallis chanced to exhibit Newton's method when he first published it, in consequence of which *every* numerical solver has felt bound in duty to make it one of his examples. Invent a numerical method, neglect to show how it works on this equation, and you are the pilgrim who does not come in at the little wicket (*vide* J. Bunyan)."

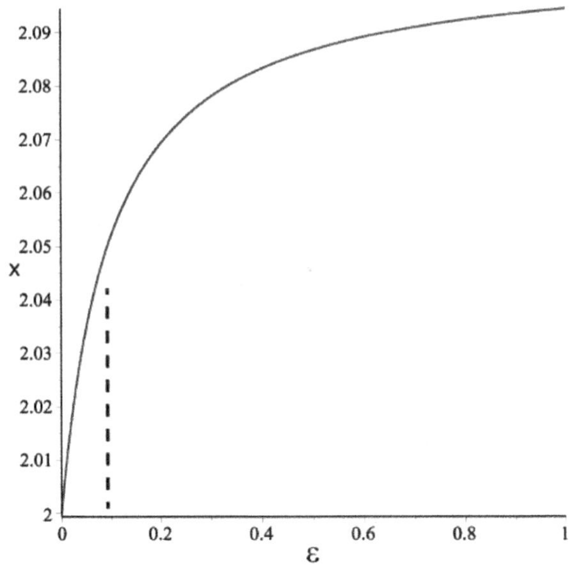

Figure 14.10. *Plot of the root $x(\epsilon)$ that solves $H(x,\epsilon) \equiv \epsilon(x^3 - 2x - 5) + (1-\epsilon)(x-2) = 0$. It is desirous that perturbative homotopies yield roots that vary roughly linearly with ϵ so that the perturbation series converges rapidly even at $\epsilon = 1$, where the original problem is recovered. Here, $x(\epsilon)$ is highly nonlinear, and the ϵ power series converges only for $\epsilon < 0.099$. The dashed vertical line marks the convergence boundary. Fortunately, the Padé approximants converge rapidly anyway.*

Table 14.9. *Padé approximants and errors for the perturbative approximation of a root of Newton's cubic.*

Padé Order	Error	Approx. Root
[1/1]	−0.0054	2.1
[2/2]	0.00030	2.09425
[3/3]	−0.0000168	2.094568
[0/1]	−1.905	4
[1/2]	0.056	2.038
[2/3]	−0.00323	2.0978
exact	0	2.094551482

Although the first guess differs from the true root by only 5%, the perturbation series is

$$x(\epsilon) = 1 + \epsilon - 9\epsilon^2 + 75\epsilon^3 - 568\epsilon^4 + 3753\epsilon^5 - 19329\epsilon^6, \tag{14.129}$$

which clearly diverges at $\epsilon = 1$, the value that recovers Newton's problem. The resultant of H and H_x with elimination of x yields a polynomial in ϵ whose roots are where solution branches of $H(x,\epsilon) = 0$ merge. (Resultants and limit point computations are discussed in section 20.5.) The convergence-limiting double roots of $H(x,\epsilon)$ are at $\epsilon_c = -0.0864 \pm i0.0480$, restricting the convergence of the perturbation series to $|\epsilon| \leq 0.0989$ (Fig. 14.10).

However, the diagonal Padé approximants formed from the perturbation series are very accurate, as cataloged in Table 14.9. (The off-diagonal approximants with numerator one degree less than the denominator converge with equal rapidity but much larger errors at each order, as shown in the bottom three rows of the table.) The linear over linear

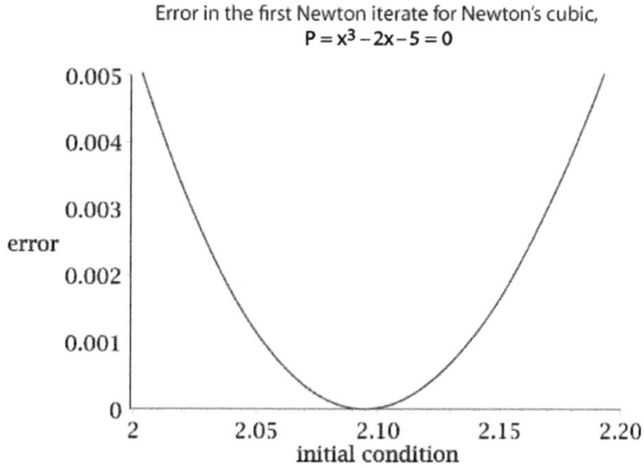

Figure 14.11. *Error of the first Newton's iterate* $x^{(1)}$ *versus the initial condition* $x^{(0)}$. *The first iterate is* $x^{(1)} = x^{(0)} - \{(x^{(0)})^3 - 2x^{(0)} - 5\}/(3[x^{(1)}]^2 - 2)$. *The error is defined as* $x^{(0)} - x_*$.

approximant, $x[1/1](\epsilon) = (2 + 19\epsilon)/(1 + 9\epsilon)$, is 18 times as accurate at $\epsilon = 1$ as the first guess, $x \approx 2$.

Actually, if one can estimate the root to within one-tenth of its true value, Newton's iteration is best. Fig 14.11 shows that for an initialization $x^{(0)} \in [2, 2.2]$, the error of a single Newton's iteration is at most 1/200, and further iterates show the usual digit-doubling convergence (Fig. 14.11).

It is possible to be too clever in hunting zeros.

Never forget that given only a good initialization, boring old Newton's iteration is a tool of great power.

Chapter 15

Singular Perturbation Methods: Fractional Powers, Logarithms, and Exponential Asymptotics

"The devotee of perturbation methods is continually being surprised by the appearance of logarithmic terms where none could reasonably have been anticipated. The recent history of fluid mechanics records a number of well-known investigators who fell victim to the plausible assumptions that their expansion proceeded by powers of the small parameter."
—Milton van Dyke [357, p. 201]

15.1 ▪ Introduction

Perturbative series for roots are not restricted to powers of the small parameter. Useful expansions may also involve fractional powers, logarithms, powers of logarithms, and even more complicated expressions.

When both powers of ϵ and exponentials of reciprocals of ϵ appear, one major complication is that $\exp(-1/\epsilon)$ cannot be usefully represented as a power series. Recall that the coefficients of a power series are proportional to derivatives of the function that is being approximated. The derivatives of $\exp(-1/\epsilon)$ with respect to ϵ are proportional to the exponential multiplied by a polynomial in reciprocal powers of ϵ. However,

$$\lim_{\epsilon \to 0}\left(\epsilon^{-k} \exp(-1/\epsilon)\right) = 0 \qquad \forall |k| < \infty. \tag{15.1}$$

This implies that the power series of a reciprocal exponential has coefficients that are all zeros:

$$\exp(-1/\epsilon) \sim 0 + 0\epsilon + 0\epsilon^2 + 0\epsilon^3 + \cdots. \tag{15.2}$$

In a formal sense, we should sum an infinite number of powers of ϵ before including a correction proportional to $\exp(-1/\epsilon)$ because ϵ^k goes to zero more slowly than the exponential for any k. For any finite ϵ, however, the exponentials will be more important than high, finite powers of ϵ.

A practical way forward is to pretend that

$$\delta \equiv \exp(-1/\epsilon) \tag{15.3}$$

and ϵ are *independent* small parameters, generating a *bivariate* perturbation series.

Series that mix powers of $\log(\epsilon)$ and ϵ are similar; we can convert such expansions to powers and exponentials by defining a new parameter ν such that

$$\nu \equiv -\frac{1}{\log(\epsilon)} \qquad \leftrightarrow \qquad \epsilon = \exp(-1/\nu). \tag{15.4}$$

In this chapter, some of these generalizations are illustrated. The Lambert W-function, described at length in section 13.6, is the inverse of $x\exp(x)$ or, equivalently, the solution $W(\lambda)$ to

$$W\exp(W) = \lambda. \tag{15.5}$$

We shall present three expansions, each appropriate for a different range of λ. The series about $\lambda = 0$ is an expansion in powers of λ. The expansion about the limit point $\lambda = -\exp(-1)$ is a singular series in half-integral powers. The series for large λ is a *double* series in powers of two parameters, which are $1/\log(\lambda)$ and $\log(\log(\lambda))/\log(\lambda)$, as described in section 15.4.

First, though, why do these expansions arise? It is well known that a differential equation like

$$\epsilon u_{zz} + u = 0 \tag{15.6}$$

has a perturbation series in *half-integral* powers of ϵ, even though the coefficients of the differential equation depend only on ϵ itself. This is because the solutions, $\exp(\pm z/\sqrt{\epsilon})$, are functions of the square root of ϵ. Similarly, in cylindrical coordinates, the appearance of a term like $\log(r)$ in the generalized small-r expansion of the solutions often creates terms in $\log(\epsilon)$ in a perturbative expansion.

However, no general predictor for log and power series is known. Van Dyke's rueful comments of half a century ago are still applicable: "In other problems their source is even more obscure.... One can only philosophize that description by fractional powers fails to exhaust the myriad phenomena in the universe, and logarithms are the next simplest function" [357, p. 201].

One particular peril is the *stealth logarithm*. That is to say, it is quite common for the first three or four terms of a series to proceed in powers of ϵ, and then $\log(\epsilon)$ and its powers appear at all higher orders.

Because singular perturbation theory is diverse, this chapter is mostly a collection of illuminating case studies. There are, however, a few general guidelines, which we develop as we proceed and summarize in the final section.

15.2 ▪ Asymptotics, Superasymptotics, and Hyperasymptotics

"Divergent series are the invention of the devil, and it is a shame to base on them any demonstration whatsoever."

—Niels Henrik Abel (1802–1829)

"This series is divergent, therefore we may be able to do something with it."

—Oliver Heaviside (1850–1925)

Singular perturbation series are often asymptotic but divergent. If such a series is truncated at the Nth term for fixed, small ϵ, it is usually found that the error first decreases with N, reaches a minimum at some $N = N_{optimum}(\epsilon)$, and then rises exponentially with N for $N > N_{optimum}$. Abel was a great mathematician, but unduly pessimistic. Modern scientists and applied mathematicians find asymptotic series extremely useful.

The error at $N = N_{optimum}$ is typically an exponential function of $1/N$. This moved Sir Michael Berry to introduce the following neologism.

Definition 15.1 (Superasymptotic). *An* optimally truncated *asymptotic series is a* superasymptotic *approximation. The error is typically an* exponential *function of the* reciprocal *of the perturbation parameter* ϵ [43, 42].

In recent years, as explained in the books [315, 74, 156, 161, 229, 239, 257, 342, 217, 381, 244, 288] and the reviews [75, 289], a rather large variety of tools for improving upon the superasymptotic approximation have been developed.

Definition 15.2 (Hyperasymptotic). *A* hyperasymptotic *approximation in the broad sense is one that achieves higher accuracy than a superasymptotic approximation. A hyperasymptotic approximation in the narrow sense achieves this improvement by adding one or more terms of a* second *asymptotic series, with different scaling assumptions, to the optimal truncation of the original asymptotic expansion* [42]. *(With another rescaling, this process can be iterated by adding terms of a third asymptotic series, and so on.) Hyperasymptotic methods are also called* exponential asymptotics *and* asymptotics beyond all orders.

Sequence and sum acceleration methods are powerful tools for improving upon a superasymptotic approximation, but some authors do not like to label them as "hyperasymptotic" because these methods are very useful for convergent series, too, and also have a history that is much older than the modern surge of interest in hyperasymptotic methods that directly attack the cause of the divergence. The alternative label "exponential asymptotics" is also used because to improve upon the superasymptotic approximation, one must calculate terms that are an *exponential* of $1/\epsilon$. The "beyond all orders" appellation arises because $\exp(-[\text{constant}]/\epsilon)$ decreases to zero faster than any finite *power* of ϵ. Because $\exp(-1/\epsilon)$ cannot be calculated by a power series in ϵ even with a billion terms, effects proportional to such an exponential are completely missed by an ϵ power series and lie beyond all orders in powers of ϵ.

15.3 ▪ A Problem Whose Corrections Lie beyond All Orders

The goal is to solve [212, p. 22]

$$f(x,\epsilon) = 1 + x + \epsilon\,\text{sech}(x/\epsilon) = 0, \qquad \epsilon \ll 1. \tag{15.7}$$

The limiting root as $\epsilon \to 0$ is $x = -1$. All corrections are *exponentially small* in $1/\epsilon$. Nevertheless, we shall successfully discover the *exponential asymptotics*, also known as *hyperasymptotics*, for these roots.

15.3.1 ▪ The Walkaround

Before we expand, we should first look over the problem. Observe that

$$\text{sech}(x/\epsilon) \approx \begin{cases} 1 - \dfrac{1}{2}\epsilon^{2(\alpha-1)} + \cdots, & x \sim O(\epsilon^\alpha),\ \alpha > 1, \\ O(1), & x \sim O(\epsilon), \\ 2\exp(-\epsilon^{\alpha-1}) \ll 1, & |x| \sim O(\epsilon^\alpha),\ \alpha < 1. \end{cases} \tag{15.8}$$

Furthermore, the hyperbolic secant is always positive, and writing the equation as

$$x = -1 - \epsilon\,\text{sech}(x/\epsilon), \tag{15.9}$$

we see that

$$x \leq -1 \qquad \forall \Re(x). \tag{15.10}$$

It follows that for small ϵ,

$$|\epsilon\,\mathrm{sech}(x/\epsilon)| \leq 2\epsilon\,\exp(-|x|/\epsilon) \qquad \forall\, x. \tag{15.11}$$

Therefore we have the following:

1. $x \approx -1 +$ higher-order terms.
2. At least for small ϵ, there is only one root.
3. The residual of the lowest-order term is *exponentially* small in ϵ.

The last suggests that a *power series* in ϵ will be inadequate. Newton's iteration is a useful tool to deduce the form of corrections provided that $df/dx(x)$, the denominator of Newton's correction, is nonzero for $x = x^{(0)}$. For this problem,

$$x^{(n+1)} = x^{(n)} - \frac{f(x^{(n)}}{df/dx(x^{(n)})} \tag{15.12}$$

$$= x^{(n)} - \frac{1 + x^{(n)} + \epsilon\,\mathrm{sech}(x^{(n)}/\epsilon)}{1 - \mathrm{sech}(x^{(n)}/\epsilon)\tanh(x^{(n)}/\epsilon)}. \tag{15.13}$$

However, it is usually best to minimize complexity by approximating df/dx as much as possible. Standard perturbation theory implicitly makes the substitution

$$df/dx(x^{(n)}) \to df/dx(x^{(0)}). \tag{15.14}$$

(In numerical circles, freezing df/dx at its initial value replaces an "honest" Newton's iteration (which updates $df/dx(x^{(n)}) \to df/dx(x^{(n+1)})$ at every n) and yields the "chord" or "Shamanski" iteration; perturbation theory is almost always Shamanski-like rather than Newton-like.)

Here the limit $\epsilon \to 0$ gives $df/dx = 1$, and the iteration simplifies to

$$x^{(n+1)} = x^{(n)} - \left\{(1 + x^{(n)}) + \epsilon\,\mathrm{sech}(x^{(n)}/\epsilon)\right\}. \tag{15.15}$$

The first correction is

$$x^{(1)} = -1 - \epsilon\,\mathrm{sech}(1/\epsilon). \tag{15.16}$$

This can be simplified still further by using $\mathrm{sech}(z) = 2/(\exp(z) + \exp(-z))$:

$$\mathrm{sech}(1/\epsilon) = 2\exp(-1/\epsilon)\frac{1}{1 + \exp(-2/\epsilon)} \tag{15.17}$$

$$\approx 2\exp(-1/\epsilon) - 2\exp(-3/\epsilon) + \cdots. \tag{15.18}$$

Keeping just the lowest exponential term in the hyperbolic secant function gives finally

$$x^{1st} = -1 - 2\epsilon\,\exp(-1/\epsilon). \tag{15.19}$$

This suggests defining a new perturbation parameter

$$\delta \equiv \exp(-1/\epsilon) \tag{15.20}$$

so that

$$x^{1st} = -1 - 2\epsilon\,\delta. \tag{15.21}$$

The next Shamanski iteration, initialized from x^{1st}, is

$$x^{2nd} = -1 - (-1 - 2\epsilon\delta) - \epsilon\,\mathrm{sech}(-1/\epsilon - 2\delta). \tag{15.22}$$

We want to extract the factors of $\exp(-1/\delta)$, but it is not so easy to do this. Substituting $\epsilon = -1/\log(\delta)$ will rewrite the powers of ϵ, too. Maple's "sub" command and its analogs in other algebraic manipulation languages will leave the powers of ϵ alone, but the entanglement of $-1/\epsilon$ with other terms inside the exponentials fools the pattern-matching subroutine, which does nothing.

To avoid such problems, it is helpful to define a new unknown $d(\epsilon)$ such that

$$x = -1 + \epsilon\delta\, d(\epsilon) \tag{15.23}$$

and rewrite the equation as

$$\begin{aligned} f(d,\epsilon,\delta) &= d\epsilon\delta + 2\epsilon\,\exp(-1/\epsilon)\exp(\epsilon\,\delta\, d/\epsilon)\frac{1}{1+\exp(-2/\epsilon)\exp(2\epsilon\,\delta\, d/\epsilon)} = 0 \\ &= \epsilon\,\delta d + 2\epsilon\,\delta\,\exp(\delta\, d)\frac{1}{1+\delta^2\exp(2\delta d)}, \end{aligned} \tag{15.24}$$

where $d = -2 + \cdots$ as $\epsilon \to 0$. Fortuitously, it is now possible to divide out the common factor of $\epsilon\delta$ to obtain

$$\hat{f}(d,\delta) = d + 2\,\exp(\delta\, d)\frac{1}{1+\delta^2\exp(2\delta d)}. \tag{15.25}$$

This can be solved by a *regular* perturbation series in the new perturbation parameter δ:

$$x \approx -1 + \epsilon\delta\left\{-2 + 4\,\delta - 10\,\delta^2 + \frac{80}{3}\,\delta^3 - \frac{206}{3}\,\delta^4 + \frac{756}{5}\,\delta^5 + \cdots\right\}.$$

The coefficients of the series do not exhibit a simple pattern, but the radius of convergence is about $\delta < 1/4$.

The perturbation series can be computed in Maple by the lines in Table 15.1.

Table 15.1. *Maple program to compute the perturbation series for Holmes's problem with beyond all orders corrections.*

```
# initialize "da", array of coefficients of δ-series for d
  da[0] := −2; d := da[0]; maxorder := 10; for n from 1 to maxorder do
# expand F ≡ 2exp(dδ)/(1+δ^2 exp(2dδ)) in δ;
  F := series( 2*exp(delta*d)/(1 + delta**2 * exp(2*d*delta) ),delta,n + 2);
# nth coeff of d is negative of the nth coefficient of F.
  da[n] := − coeff(F,delta,n);
  d := d + delta**n * da[n]; od:
  x := − 1 + epsilon*delta*d;
```

Table 15.2. *Perturbation series for the root of* $1+x+\epsilon\,\mathrm{sech}(x/\epsilon)$.

δ	ϵ	x	5-Term Error	10-Term Error	20-Term Error
0.05	0.334	−1.0304108	5.2×10^{-8}	8.2×10^{-11}	1.2×10^{-17}
0.10	0.434	−1.0729134	0.0000090	3.7×10^{-7}	5.3×10^{-11}
0.15	0.527	−1.1233920	0.00018	0.000050	4.1×10^{-7}
0.20	0.621	−1.1815219	0.0016	0.0017	0.00024
0.25	0.721	−1.2479632	0.0083	0.025	0.033

The numerical success of this expansion is shown in Table 15.2.

An alternative approach suggested by R. M. Corless (private communication) is to write $x=-\epsilon z$:

$$f(z,\epsilon)=1-\epsilon z+2\epsilon\,\exp(-z)\frac{1}{1+\exp(-2z)}=0. \tag{15.26}$$

Since $z\approx 1/\epsilon$, we can make only an error of $O(\exp(-2/\epsilon))$ by writing

$$f(z,\epsilon)\approx 1-\epsilon z^{(0)}+2\epsilon\,\exp(-z^{(0)})+O(\exp(-2/\epsilon)), \tag{15.27}$$

which can be solved exactly in terms of the Lambert W-function, which is extensively discussed in section 13.6:

$$z^{(0)}=\frac{W(-2\exp(1/\epsilon))\epsilon-1}{\epsilon}. \tag{15.28}$$

One can then calculate higher-order corrections, but this is left as an exercise.

15.4 ▪ Log and Log-Log Series for Large x for the Lambert W-Function

The Lambert function $W(x)$ is defined as the solution of

$$W\,\exp(W)=x. \tag{15.29}$$

This arises in many applications and has been intensively studied, as explained in depth in section 13.6. Here, our goal is derive an asymptotic series for $W(x)$ for large x. Note that here to follow conventional notation, W is the root and x is the parameter instead of λ.

By taking the logarithm of both sides of the Lambert equation, it becomes

$$W+\log(W)=\log(x). \tag{15.30}$$

If $x\gg 1$, then W is also large. When $W\gg 1$, W is much bigger than $\log(W)$, implying the first approximation

$$W^{(1)}\approx\log(x). \tag{15.31}$$

Replacing $\log(W)$ by the first approximation gives

$$W^{(2)}\approx\log(x)-\log(W^{(1)})=\log(x)-\log(\log(x)). \tag{15.32}$$

Define a new unknown v by

$$W=W^{(2)}+v=\log(x)-\log(\log(x))+v. \tag{15.33}$$

Then the Lambert equation is transformed to

$$\exp(-v)-1-\frac{1}{\log(x)}v+\frac{\log(\log(x))}{\log(x)}=0, \tag{15.34}$$

where v appears only in the third and higher approximations. Therefore,

$$v<\log(x),\log(\log(x))$$

as $x\to\infty$ so that v is indeed a small correction to the second approximation already found. As $x\to\infty$, $O(\log(\log(x))/\log(x))$ decays. If v is increasing with x or is even bounded by a constant for large x, then $\exp(-v(x))$ and $v(x)/\log(x)$ will be unable to cancel the -1 term in the transformed equation. Therefore, v must be a decreasing function of x. This in turn implies that for large x, $\exp(-v(x))\approx 1-v+(1/2)v^2-(1/6)v^3+\cdots$.

There is a problem, however: the logarithms and logarithms of logarithms imply that the expansion will be a blizzard of logarithms, and a rather confusing one, too, since $\log(\log(x))$ grows with x much more slowly than $\log(x)$. De Bruijn [148] introduced the two-parameter stratagem by defining

$$\sigma=\frac{1}{\log(x)},\qquad \tau\equiv\frac{\log(\log(x))}{\log(x)}. \tag{15.35}$$

The Lambert equation becomes

$$\exp(-v)-1-\sigma v+\tau=0. \tag{15.36}$$

If we *ignore* the connection between the two, i.e., that both are completely determined by x, then the problem can be expanded as an *ordinary* power series in the *two* variables σ and τ.

De Bruijn proved that the series is convergent for sufficiently large $|x|$, and not merely asymptotic, but apparently the numerical value of the radius of convergence is not known.

Comtet [131, pp. 228–229] showed that the coefficients could be expressed analytically in terms of the so-called Stirling cycle numbers.

It is better, however, to use a general strategy to derive the coefficients order by order. First, note that for $x\geq\exp(\exp(1))\approx 15.2$, $\sigma(x)\leq\tau(x)$. This implies that the terms of the double expansion can be ordered by treating σ as no larger than τ (Fig. 15.1). For very large x where $\sigma\ll\tau$, the worst fate that can befall us is that we may retain unnecessary terms to achieve a given accuracy for such large x.

Second, introduce the artificial ordering parameter ϵ such that both σ and τ are assumed to be $O(\epsilon)$. This is equivalent to assuming

$$v=\sum_{j=1}\epsilon^j\sum_{k=0}^{j-1}a_{jk}\tau^{j-k}\sigma^k, \tag{15.37}$$

where the summation in k has been truncated at $j-1$ instead of j because of the a posteriori observation that $a_{jj}=0$ for all j. After this expansion is substituted into the transformed Lambert equation, the residual can be ordered in powers of ϵ. All the coefficients of $O(\epsilon^j)$ are determined by collecting the residual at this order and then collecting powers of τ and setting each term equal to zero. The scheme is very similar to an expansion in a single perturbation parameter even though there are technically two such parameters. The ordering parameter ϵ is then replaced by unity after the coefficients have been computed; the expansion through $O(\epsilon^j)$ then has a smallness of $O(\tau^j)$. A complete Maple program is given in Table 15.3 and contains only 18 statements.

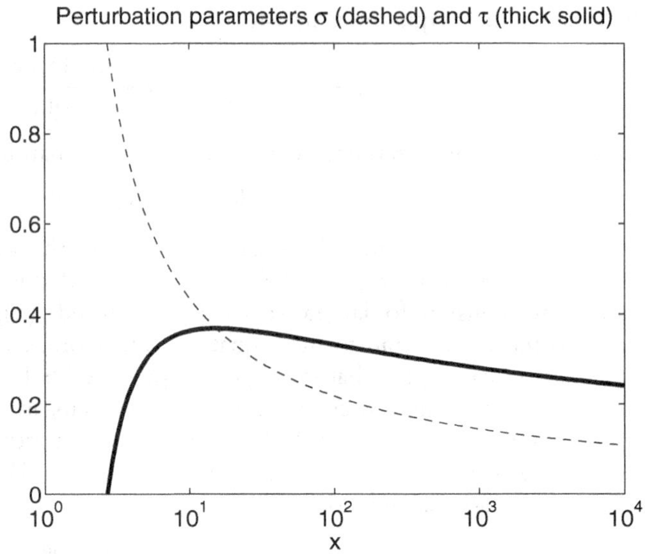

Figure 15.1. *Variation of the perturbation parameters* $\sigma \equiv 1/\log(x)$ *(dashed) and* $\tau \equiv \log(\log(x))/\log(x)$ *(solid) with x. Note that* $\sigma \leq \tau$ *for all* $x \geq \exp(\exp(1)) = 15.154$.

Table 15.3. *Maple program for the log/log-log* ($\sigma - \tau$) *series for the W-function.*

```
  v := 0; jmin := 1; jmax := 6;
# Define v with unknown coefficients "a[j,k]";
  for j from jmin by 1 to jmax do for k from 0 by 1 to (j-1) do
  v := v + epsilon**j * a[j,k] * tau**(j-k) * sigma**k; od: od:
# Compute the residual, then expand as series in dummy parameter ϵ;
  f := exp(-v) - 1 - epsilon*sigma*v + epsilon*tau;
  fs := simplify(series(f,epsilon,jmax + 1));
# "fcoeff" is the coefficient of ϵ^j in the expansion of f;
  for j from jmin by 1 to jmax do fcoeff := collect(coeff(fs,epsilon,j),tau);
  for k from 0 by 1 to (j-1) do eq := coeff(fcoeff,tau,j-k);
# "eq" is the coefficient of τ^(j-k) in "fcoeff";
  a[j,k] := solve(eq,a[j,k]);
# solve "eq = 0" for the coefficient "a[j,k]";
# of the double series in τ and σ
  print(j,k,a[j,k]); od: od:
# print result; terminate loops
```

15.4.1 ▪ The Log/Log-Log Series as a Log and Power Series

In applications of the Lambert function, the argument is often $x = \exp(1/y)$. In terms of y,

$$\sigma = y, \qquad \tau = -y\log(y). \tag{15.38}$$

It follows that the fourth-order $[O(\epsilon^4)]$ term, say, can be written as

$$y^4\left\{\frac{1}{4}\log(y)^4 + \frac{11}{6}\log(y)^3 + 3\log(y)^2 + \log(y)\right\}, \tag{15.39}$$

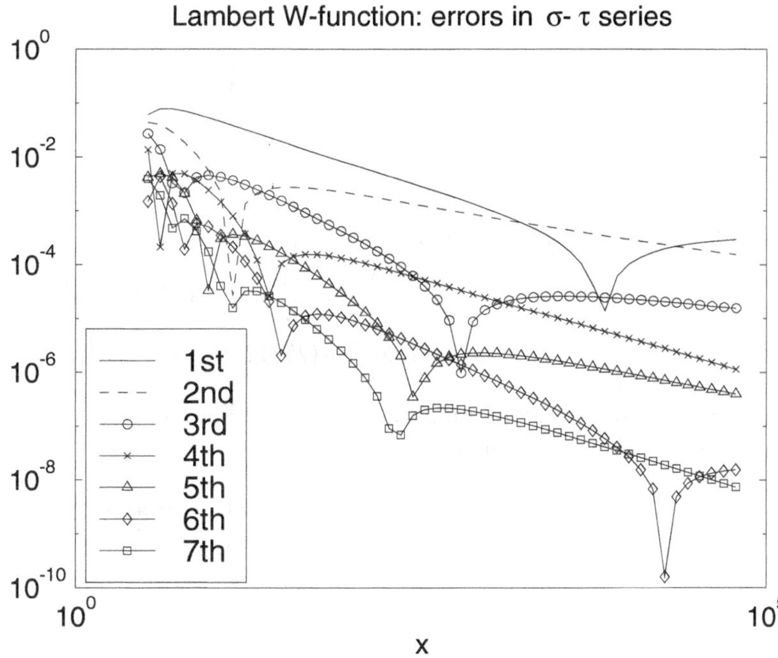

Figure 15.2. *Errors for the first seven orders in the σ-τ perturbation series, where it is assumed that σ and τ are of the same order.*

that is, as the product of a power of the parameter and a weighted sum of various powers of the logarithms. Such power and logarithm expansions are very common and are really just a log and log(log) series in the disguise of a change of parameter.

15.4.2 ▪ Interval-of-Convergence Conjecture: The W-Function Series Converges for $x > \exp(1)$

Fig. 15.2 shows that the error in a given order decreases rapidly with increasing $\log(x)$, but with superimposed oscillations. A similar graph for small x (not shown) shows that there is monotonic divergence with order for $x \in [0, \exp(1)]$ but monotonic convergence for x a little larger than $\exp(1)$. This motivates the conjecture, unproved and apparently not previously stated, that the series (when ordered such that $\sigma \sim O(\tau)$) converges for all $x > \exp(1)$ and diverges for smaller x.

The maximum errors for a given order for all $x > \exp(1)$ are respectively (0.078, 0.043, 0.027, 0.14, 0.0049, 0.0043, 0.0039), which shows that high-order approximations are quite accurate over the whole range $x \in [\exp(1), \infty]$.

15.4.3 ▪ Padé Approximants and Improvement of the Log/Log-Log Series

The simplest strategy to convert the perturbation series into rational approximations is to apply the Padé method in the dummy parameter ϵ. One technical complication is that the coefficients of the numerator and denominator polynomials in the Padé approximants are products of the coefficients of the perturbation series. This yields rather complicated explicit expressions in σ and τ for the approximants.

It is more efficient to write

$$v = \epsilon \left\{a_0 + a_1\epsilon + a_2\epsilon^2 + a_3\epsilon^3 + a_4\epsilon^4\right\}, \tag{15.40}$$

where explicitly

$$a_0 = \tau,\ a_1 = \frac{1}{2}\tau^2 - \tau\sigma,\ a_2 = \frac{1}{3}\tau^3 - \frac{3}{2}\tau^2\sigma + \tau\sigma^2,$$
$$a_3 = \frac{1}{4}\tau^4 - \frac{11}{6}\tau^3\sigma + 3\tau^2\sigma^2 - \tau\sigma^3,$$
$$a_4 = (1/5)\tau^5 - (25/12)\tau^4\sigma + (35/6)\tau^3\sigma^2 - 5\tau^2\sigma^3 + \tau\sigma^4, \tag{15.41}$$

and then substitute these into the forms of the linear over linear and quadratic over quadratic (in ϵ) Padé approximants:

$$v[1/1] = \frac{a_1 a_0 + (a_1 a_1 - a_2 a_0)\epsilon}{a_1 - a_2\epsilon}, \tag{15.42}$$

$$v[2/2] = N_2/D_2, \tag{15.43}$$
$$N_2 = a_0 a_1 a_3 - a_0 a_2^2 + (a_1^2 a_3 - a_1 a_2^2 - a_0 a_1 a_4 + a_0 a_2 a_3)\epsilon$$
$$+ (2a_2 a_1 a_3 - a_2^3 - a_1^2 a_4 - a_0 a_3^2 + a_0 a_2 a_4)\epsilon^2, \tag{15.44}$$
$$D_2 = (a_1 a_3 - a_2^2) + (a_2 a_3 - a_1 a_4)\epsilon + (a_2 a_4 - a_3^2)\epsilon^2. \tag{15.45}$$

Fig. 15.3 shows that the [1/1] approximant is not as good as the perturbation theory whence it came. However, the [2/2] is better although not by a huge amount, and by an amount that seems to disappear asymptotically. For $x < \exp(1)$, the Padé approximants are much better than the perturbation series, but fall apart at $x = 1$. The un-Padé series appears to converge only for $x > \exp(1)$.

It is encouraging that the single, dummy parameter approach is successful both with and without Padé-izing.

15.4.4 ▪ Asymptotic Approximation to the Lower Branch for Small Negative *x*

The problem is

$$-W_{-}\exp(W_{-}) = -x. \tag{15.46}$$

Taking the logarithm of both sides gives

$$\log(-W_{-}) + W_{-} = \log(-x) \tag{15.47}$$

without approximation. If $W_{-} \ll 0$, then $\log(-W_{-}) \ll |W|$. This implies that the lowest-order approximation is

$$W_{-} \approx \log(-x). \tag{15.48}$$

The second approximation is

$$W_{-} \approx \log(-x) - \log(\log(-x)), \qquad |x| \ll 1,\ \operatorname{sgn}(x) < 0. \tag{15.49}$$

One can then continue to higher order in the same manner as for the principal branch.

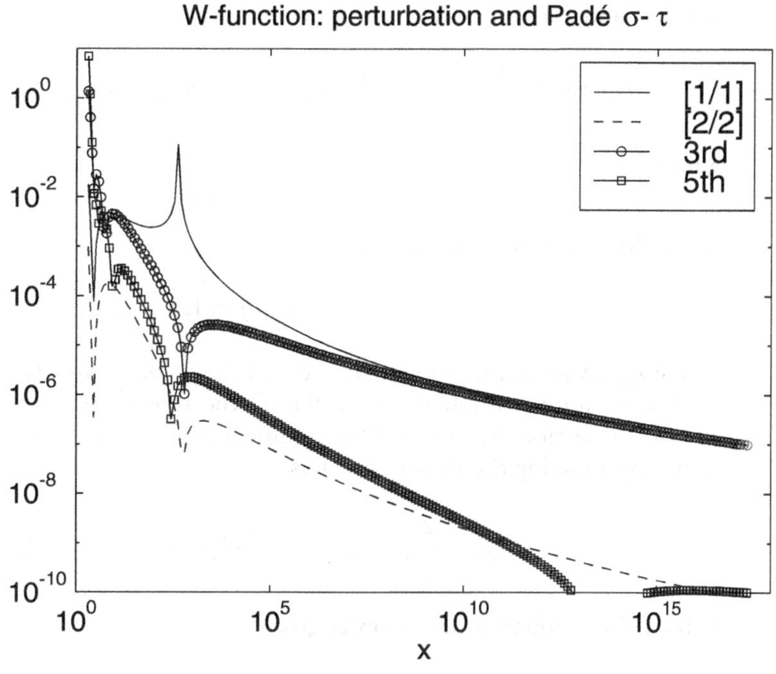

Figure 15.3. *Errors in the σ-τ series. The symbols denote the regular perturbation series of the third and fifth orders. The solid and dashed lines are the $[1/1]$ and $[2/2]$ approximants in ϵ. Note that a common factor of ϵ is factored out before computing the Padé approximants; thus the third-order perturbation solution is necessary and sufficient to compute the $[1/1]$ Padé approximant. The leftmost plotted points are at $x = 2$.*

15.5 ▪ Modon Dispersion Relation

Modons are vortex pairs that are important in geophysical fluid dynamics. The modon dispersion relation [53] is

$$f \equiv J_2(x)\epsilon K_1(\epsilon) + xJ_1(x)K_2(\epsilon). \tag{15.50}$$

The inverse power series for large ϵ was derived in the previous chapter. The small-ϵ perturbation theory is more complicated because it is a log and power series. Logarithms arise in the series for the root x because the small-ϵ series for the K-Bessel functions include logarithms.

In the next section, we shall describe an example in which an *infinite* series of powers of the logarithms forms the first correction to the zeroth-order approximation. In contrast, the log and power series of the modon dispersion relation is an illustration of stealth logarithms: the first logarithm does not appear until fourth order, and $\log^2(\epsilon)$ arises only at eighth order.

The underlying strategy is the same as for deriving the series for the Lambert W-function: pretend that ϵ and $\nu \equiv \log(\epsilon)$ are *independent* variables, and perform a bivariate perturbation series.

Let us write

$$x = x_0 + \epsilon^2 x_2 + \epsilon^4 x_4 + \epsilon^4 \log(\epsilon) x_{41} + \epsilon^6 x_6 + \epsilon^6 \log(\epsilon) x_{61} + \cdots. \tag{15.51}$$

If we expand $\tilde{f}$ as a series in ϵ, we find

$$\tilde{f}(x) = r_{-2}\frac{1}{\epsilon^2} + r_0 + r_2\epsilon^2 + r_{21}\epsilon^2\log(\epsilon) + r_4\epsilon^4 + r_{41}\epsilon^4\log(\epsilon) + \cdots, \tag{15.52}$$

where

$$r_{-2} = 2x_0 J_1(x_0). \tag{15.53}$$

Thus, the lowest-order residual demands that

$$x_0 = j_{1,j}, j = 1, 2, \ldots, \tag{15.54}$$

where $j_{m,n}$ denotes the nth positive zero of the J_m Bessel function.

For notational simplicity, we will omit the arguments of the Bessel functions in the rest of this section, but the hidden argument is always x_0. We shall repeatedly simplify results by applying the Bessel identities

$$J_2 = \frac{2}{x_0}J_1 - J_0, \qquad \frac{d}{dx}J_n = J_{n-1} - (n/x)J_n(x). \tag{15.55}$$

Solving the residuals order by order gives

$$x_2 = \frac{1}{2}\frac{1}{x_0}, \tag{15.56}$$

$$x_{41} = \frac{1}{4}\frac{1}{x_0}, \tag{15.57}$$

$$x_4 = \frac{\gamma - \log(2)}{4x_0} - \frac{5}{8}\frac{1}{x_0^3}, \tag{15.58}$$

$$x_{61} = \frac{1}{8x_0} - \frac{5}{8}\frac{1}{x_0^3}, \tag{15.59}$$

$$x_6 = \frac{\gamma - \log(2) - (1/2)}{8x_0} + \frac{5}{8}\frac{\log(2) - \gamma}{x_0^3} - \frac{1}{24x_0^3} + \frac{21}{16}\frac{1}{x_0^5}, \tag{15.60}$$

where $\gamma = 0.577\ldots$ is Euler's constant. The coefficients of the logarithmic terms are obtained from the corresponding coefficients in the residual, the conditions that $r_{41} = 0$ and $r_{61} = 0$, respectively.

Symbolic algebra packages such as Maple and Mathematica make it easy to keep x_0 as a symbol and thus perturbatively approximate all the infinite number of branches of solutions *simultaneously* (Table 15.4). For the lowest branch of solutions, numerical evaluation of $j_{1,1}$ gives

$$\begin{aligned} x = {} & 3.83171 + 0.13049\epsilon^2 - 0.018674\,\epsilon^4 + 0.065245\,\epsilon^4\log(\epsilon) \\ & - 0.017957\,\epsilon^6 + 0.021513\,\epsilon^6\log(\epsilon) + O\left(\epsilon^8, \log(\epsilon)\epsilon^8, \log^2(\epsilon)\epsilon^8\right). \end{aligned} \tag{15.61}$$

We did not explicitly introduce a second parameter $\nu = -1/\log(\epsilon)$ even though logarithms appear explicitly in the dispersion relation. Instead, we introduced a second parameter *implicitly* by solving the order-by-order residual equations as if the powers of ϵ and factors of $\log(\epsilon)$ were independent parameters. Fig. 15.4 shows the errors.

Table 15.4. *Log-and-power series for modon dispersion relation: Maple program.*

```
# The O(ε^{-2}) residual, 2x_0 J_1(x_0), requires J_1(x_0) = 0 <-> J1 = 0.
# J_2 = -J_0 is a simplification of J_2(x) = (2/x)J_1 - J_0 when J_1 = 0
  Norder := 3; # Order of perturbation series; highest power in x is ε^{2Norder};
  f := BesselJ(2,x)*(epsilon*BesselK(1,epsilon)) + BesselK(2,epsilon)*x*BesselJ(1,x);
  x0 := evalf( BesselJZeros(1,1)) ;
  x := x0 + epsilon*epsilon*x2 + epsilon*epsilon*x21 + epsilon**4 *x4 +
  epsilon**4*log(epsilon)*x41 + epsilon**6 *x6 + epsilon**6 *log(epsilon)*x61;
  fseries := series(f,epsilon,2*Norder + 2);
# expand f as log-and-power series using power series in dummy placeholder ε
# Next, simplify notation, apply Bessel function identities, and log(ε) -> ν;
  fseries := subs(BesselJ(1,x0) = 0,BesselJ(0,x0) = J0,
                   BesselJ(2,x0) = - J0,log(epsilon) = nu,fseries);
  for j from 1 by 1 to Norder do jj := 2*j - 2; jtwo := 2*j; # loop over order
  r||jj := coeff(fseries,epsilon,jj);
  r||jj := collect(r||jj,nu); # pick off ε^{2j-2} part of f(x)
  rnu||jj := coeff(r||jj,nu,1); # pick off log(ε)(≡ ν) part of O(ε^{2j-2}) part of f(x)
  if rnu||jj =0 then x||jtwo||1 := 0 else x||jtwo||1 := solve(rnu||jj,x||jtwo||1) fi;
# solve for log(ε) coefficient
  x||jtwo := solve(r||jj,x||jtwo);
# solve for non-logarithmic coefficient x_{2j} ;
  print(2*j,x||jtwo,x||jtwo||1); od: # print x_{2j} and x_{2j,1} and end loop
  J0:= evalf( BesselJ(0,x0) ); print(x);
```

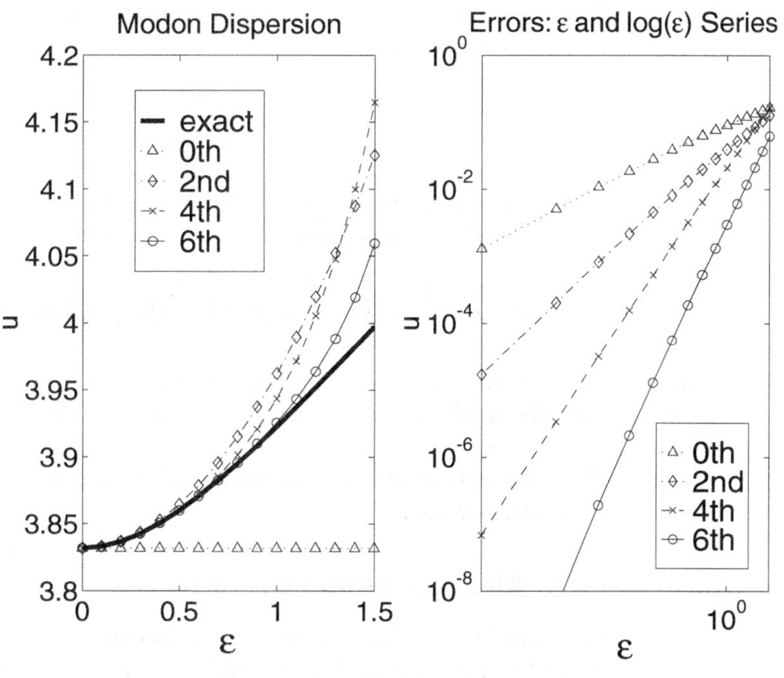

Figure 15.4. *Left: comparisons of the exact first branch of the modon dispersion relation (thick solid curve) with perturbation theory. Right: errors.*

15.6 ▪ The Laplacian Eigenvalue in an Annulus

15.6.1 ▪ Background

An ancient problem is to find the eigenvalues of the Laplace operator with homogeneous Dirichlet boundary conditions:

$$\Delta u + k^2 u = 0 \in \Omega, \qquad u = 0 \in \partial\Omega. \tag{15.62}$$

When the domain Ω is an annulus with outer radius one and inner radius ϵ, the eigenproblem can be solved by separation of variables into modes that are the product of $\cos(m\theta)$ or $\sin(m\theta)$ in the polar angle θ multiplied by a linear combination of Bessel functions of the same integer order m. The condition that the boundary conditions are satisfied can be reduced to the single equation or *dispersion relation*

$$\mathscr{D}(k = x; m) \equiv J_m(x\epsilon)Y_m(x) - J_m(x)Y_m(x\epsilon) = 0, \tag{15.63}$$

where we have replaced k by x to conform the notation here to that in the rest of the book. For illustrative purposes, it will suffice to consider the special case $m = 0$ (radially symmetric modes) and further specialize to the smallest of the countable infinity of roots of this equation for a given m.

In the limit $\epsilon \to 0$ so that the annulus becomes the unit disk, the lowest eigenvalue is given exactly by the first root of $J_0(r)$:

$$x(\epsilon = 0) \approx 2.404825558. \tag{15.64}$$

The complication is that the perturbative approximation in ϵ includes an *infinite* number of powers of the logarithmic parameter

$$\nu \equiv -\frac{1}{\log(\epsilon)}, \tag{15.65}$$

even before the first correction proportional to a power of ϵ.

The origin of the logarithms is that for small argument,

$$Y_0(k\epsilon) \approx \frac{2}{\pi}\log(\epsilon) + \frac{2}{\pi}\log(x) + \frac{2}{\pi}(\log(1/2) + \gamma) + O(\epsilon^2). \tag{15.66}$$

Since the coefficients of the dispersion relation depend on both ϵ and $\log(\epsilon)$, it is hardly surprising that the small-ϵ expansion of the root $x(\epsilon)$ will also involve both powers and logs of ϵ.

Just as for the Lambert W-function, the *key idea is to pretend that there are two independent parameters*, ϵ and ν.

15.6.2 ▪ Corrections to All Logarithmic Orders

It is illuminating to first consider the special case in which all powers of ϵ are ignored. In this limit, $J_0(x\epsilon) \equiv 1$ (neglecting $O(\epsilon^2)$) and $Y_0(x\epsilon)$ is given by (15.66). The dispersion relation to all *logarithmic* orders is then

$$\frac{\pi}{2}\frac{Y_0(x)}{J_0(x)} - \{\log(x) + (\log(1/2) + \gamma)\} = -\frac{1}{\nu}. \tag{15.67}$$

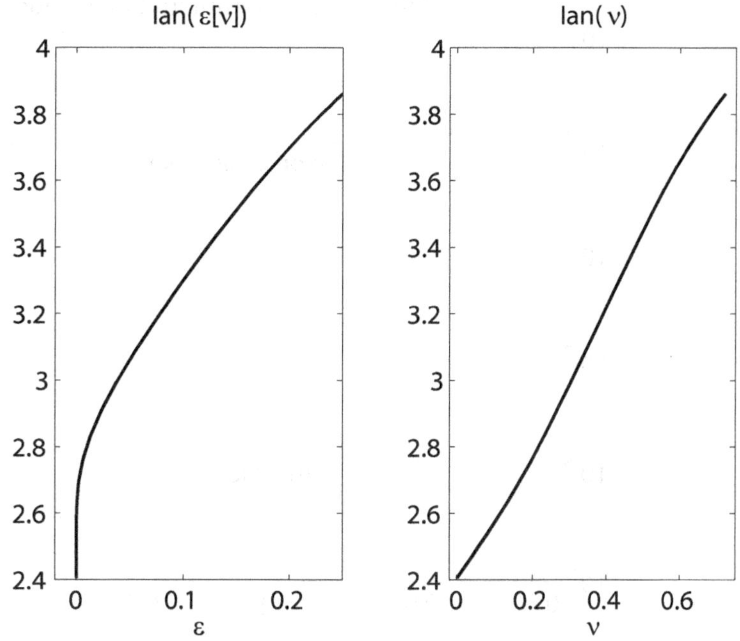

Figure 15.5. *The function $IAN(\nu)$, which incorporates the full infinite series in ν while omitting all terms in ϵ in approximating the root of the Laplacian in the annulus.*

The solution defines a new function of a single variable ν:

$$\mathcal{IAN}(\nu) = x(\nu). \tag{15.68}$$

This new transcendental is worth illustrating (Fig. 15.5) because Ward, Henshaw, and Keller [362] have shown that if the definition of ν is slightly altered to $\nu = -1/\log(d\epsilon)$, then $\mathcal{IAN}(\nu)$ gives the all-logarithmic-orders correction for the eigenvalue of *any* domain with an interior boundary that is $O(\epsilon)$ in size. The only shape dependence, to all logarithmic orders, is the constant d. Ward, Henshaw, and Keller show how to numerically calculate d for various inner boundaries, such as an ellipse or a smoothed star shape.

15.6.3 ▪ The ϵ-ν Expansion for the Annulus

Unless ϵ is very small, however, the $O(\epsilon)$ corrections to x—which do not depend on the shape of the inner boundary—are not negligible. To calculate these, it is necessary to split the dispersion relation into a function of ϵ and ν. The key is to note that without approximation,

$$Y_0(x\epsilon) = \frac{2}{\pi} J_0(x\epsilon)\left\{\log(x/2) - \frac{1}{\nu}\right\} + g(x\epsilon), \tag{15.69}$$

where $J_0(x\epsilon)$ and $g(x\epsilon)$ have power series, convergent for all finite values of their argument, so that the sole dependence of $Y_0(x\epsilon)$ on $\log(\epsilon)$ is that which is explicitly displayed as $1/\nu$. We can then introduce a dummy parameter μ such that

$$\epsilon \to \mu\epsilon, \qquad \nu \to \mu\nu. \tag{15.70}$$

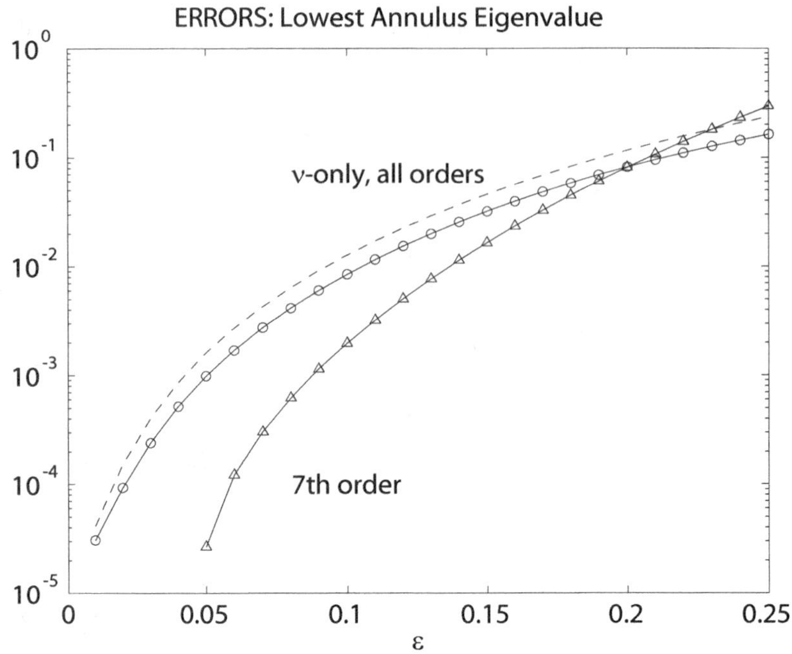

Figure 15.6. *Errors in approximations to the Laplacian in an annulus dispersion mode ($m = 0$, smallest root). The thin dashed curve is the all-orders approximation in the logarithmic variable ν only, i.e., $x \approx IAN(\nu)$. The unlabeled curve with circles adds just the lowest term in ϵ to the all-orders approximation in ν. The lowest curve (triangles) includes all terms up to seventh order in both ϵ and ν.*

Expanding the dispersion relation as a series in the single parameter μ gives the series shown below; μ is replaced by one after it has done its duty of ordering the expansion:

$$\begin{aligned} x = {} & 2.404825558 + 1.542889743\nu + 1.190267317\nu^2 + 0.6309119731\nu^3 \\ & - 0.2571633856\nu^4 + 2.230704578\,\nu^2\,\epsilon^2 - 1.382165285\nu^5 \\ & + 6.304122772\,\nu^3\,\epsilon^2 - 2.364828961\nu^6 + 10.27924388\,\nu^4\,\epsilon^2 + 2.015715534\,\nu^2\,\epsilon^4 \\ & - 2.498642726\nu^7 + 9.435068911\,\nu^5\,\epsilon^2 + 11.50818247\,\nu^3\,\epsilon^4 \text{ [Annulus Relation]}. \end{aligned}$$

The power series for $\mathcal{IAN}(\nu)$ is the result of dropping all the powers of ϵ from the bivariate expansion.

Fig. 15.6 shows that the logarithm-only approximation is accurate to within 0.01 for $\epsilon < 1/10$. However, accuracy is greatly improved for ϵ up to 1/5 by adding in terms proportional to powers of ϵ.

15.7 ▪ Approximation for Large *N* through a Power Series in the Root: Maximum of the Cauchy Factor for Equispaced Polynomial Interpolation

Cauchy showed in the eighteenth century that the error in polynomial interpolation is proportional to the product of all monomials of the form $|x - x_j|$, where the x_j are the interpolation points. It is obviously very important to understand this *Cauchy factor* $\omega(x; x_0, \ldots, x_N)$.

Table 15.5. *Roots of* $\rho(z;N)$.

Degree N	Exact z_*	z_0	z_1	Zeroth Error	z_1 Error
1	1/2	1	0	−100%	100%
2	0.4226497	2/3	0.296	−57%	30.%
3	0.38196601	0.545	0.325	−43%	15.%
4	0.35556732	0.48	0.323	−35.%	9.2%
5	0.336553473	0.438	0.315	−30.1%	6.4%
10	0.2854313370	0.3414	0.2797	−19.6%	2.0%
20	0.2449954044	0.2779	0.24368	−13.5%	0.54%
40	0.2131436496	0.2337	0.213038	−9.6%	0.050%

When the points are evenly spaced with $x_j = -1 + hj, j = 0, \ldots, N$ and $h = 2/N$, it is known that there are two equal maxima on $x \in [-1, 1]$, located symmetrically at $x = \pm x_m$, where $x_m \in [(N-1)h, Nh]$. To estimate the maxima, one must first determine x_m by solving an equation for a root of the first derivative of the Cauchy factor.

It is convenient to introduce the shifted coordinate $y = x+1$ so that the Cauchy factor becomes

$$\omega = \prod_{j=0}^{N} |(y - jh)|. \tag{15.71}$$

One of the maxima is located at the *smallest positive root* in y of the *first derivative* of ω with respect to y.

The second trick is to observe that a root of the first derivative of the *logarithm* of the Cauchy factor is also a root of $d\omega/dy$. It is convenient to find a root of the derivative of the logarithm because taking the logarithm converts the *product* into a *sum*, which is easier to handle. The equation for the root is, after multiplication by y,

$$\sum_{j=0}^{N} 1/(y - jh) = 0. \tag{15.72}$$

The dependence on h can be removed by introducing the new coordinate $z = y/h$. The problem then becomes that of finding the root z_* of

$$\rho(z;N) \equiv 1 + \sum_{j=1}^{N} z/(z - j) = 0. \tag{15.73}$$

It is easy to solve this equation numerically, and selected roots are shown in the second column of Table 15.5. The important point is that as N increases, the root $z_*(N) \to 0$ and is always less than 1/2.

It is not possible to perform a direct expansion in N to obtain an approximation for large N because N is the *summation limit*, which is meaningful only for *integer* N. Furthermore, we shall see below that any such expansion must be a double series in both $\log(N)$ and $1/N$. Nevertheless, we shall eventually obtain a very useful large-N approximation.

The trick is to exploit the smallness of the root z_* for large N by performing a Taylor series expansion of $\rho(z;N)$ in powers of z. This is easy because each factor of $1/(z - j)$ can be expanded in a geometric series:

$$\frac{1}{z-j} = -\frac{1}{j}\frac{1}{1-z/j} = -\frac{1}{j}\sum_{m=0}^{\infty}\frac{1}{j^m}z^m. \tag{15.74}$$

Collecting powers of z gives

$$\rho(z;N) \approx 1 - z\{\gamma + \Psi(N+1)\} - z^2 \left\{ \frac{\pi^2}{6} - \frac{d\Psi}{dN}(N+1) \right\} + O(z^3), \tag{15.75}$$

where Ψ is the usual digamma function, the logarithmic derivative of the gamma function, and $\gamma = 0.57721566\ldots$ is the Euler–Mascheroni constant, and we have used the identities

$$\sum_{j=1}^{N} \frac{1}{j} = \gamma + \Psi(N+1), \qquad \sum_{j=1}^{N} \frac{1}{j^2} = \frac{\pi^2}{6} - \frac{d\Psi}{dN}(N+1). \tag{15.76}$$

Keeping just the linear approximation gives

$$z_0 = \frac{1}{\gamma + \Psi(N+1)} = \frac{1}{\sum_{j=1}^{N} \frac{1}{j}}. \tag{15.77}$$

The third and fifth columns of Table 15.5 show that the approximation is poor for small N, but the error falls below 10% for $N > 40$.

The quadratic approximation can be solved explicitly, of course, but the neglect of cubic and higher terms in z is sensible only if the power series in z is converging rapidly. It is therefore sensible to simplify the quadratic approximation by treating the z^2-term as a small perturbation and solving

$$1 - \frac{1}{z_0} z + \epsilon z^2 = 0, \tag{15.78}$$

where

$$\epsilon \equiv - \left\{ \frac{\pi^2}{6} - \frac{d\Psi}{dN}(N+1) \right\}. \tag{15.79}$$

Solving this to first order in ϵ gives

$$z_1 = z_0 + z_0^3\,\epsilon, \tag{15.80}$$

which is explicitly

$$z_1 = \frac{1}{\gamma + \Psi(N+1)} - \frac{(\pi^2/6) - d\Psi/dN(N+1)}{(\gamma + \Psi(N+1))^3} = \frac{1}{\sum_{j=1}^{N} \frac{1}{j}} - \frac{\sum_{j=1}^{N} \frac{1}{j^2}}{\left(\sum_{j=1}^{N} \frac{1}{j}\right)^3}. \tag{15.81}$$

The first-order approximation has an error that falls very rapidly with N and is smaller than 1 part in 2000 for $N > 40$.

The transcendental functions can be eliminated by using their asymptotic expansions in N, giving

$$z_* \approx z_a \equiv \frac{1}{\gamma + \log(N)} - \frac{\pi^2}{6} \frac{1}{(\gamma + \log(N))^3} + O(1/N). \tag{15.82}$$

One could of course expand the denominators further to obtain a double series in $\log(N)$ and $1/N$, but this would be a bad idea. The expansion of $1/(\gamma + \log(N))$ converges rather slowly unless N is really huge. In contrast, the approximation $z_1(N)$ is accurate to better than three decimal places for all $N > 40$.

15.8 ▪ Logarithms in ϵ from Logarithms in the Coordinate: The Charney Baroclinic Instability Problem of Atmospheric Dynamics

The Charney eigenproblem is solved by a Whittaker function that can be represented as

$$u(x) = W_{r,1/2}(x) = P(x) + Q(x)\log(x), \tag{15.83}$$

where $r > 0$ is a real parameter, P and Q are entire functions, and $x = z - c$, where z is the vertical coordinate with $z = 0$ as the ground. The phase speed c is determined as the negative of the zero x_0 of the function

$$f(x) = x u_x - u = 0. \tag{15.84}$$

The principal branch of instability is for $r \in [0,1]$.

When $r \approx 1$, Branscome [112] have shown the existence of an expansion for the root of $f(x)$ as a series in the small parameter

$$\epsilon \equiv \sqrt{1-r} \tag{15.85}$$

and scale the eigenparameter via

$$x_0 = \epsilon\, \tau. \tag{15.86}$$

If the Whittaker function is replaced in the eigenrelation by $P(x)+Q(x)\log(x)$ and P and Q are then replaced by their truncated power series, then this equation becomes the sum of a polynomial in τ plus the product of $\log(\tau)$ and a polynomial. (Because the functions $P(x)$ and $Q(x)$ are entire functions, the radii of convergence for their power series are infinite.) One can then substitute an expansion in ϵ with symbolic coefficients into the eigenequation, match powers of ϵ, and solve order by order to obtain a series for $c(r)$.

Because the eigenequation is not a polynomial, but rather a *log polynomial*, the expansion of τ (and therefore of $c = -x_0 = -\epsilon\tau$) must contain logarithms, too. (The factor $\log(x)$ in these Whittaker power series becomes $\log(\epsilon\tau) = \log(\epsilon) + \log(\tau)$; the $\log(\epsilon)$ in the residual of the eigenrelation can only be canceled by $\log(\epsilon)$ terms in its solution.) The result is

$$\begin{aligned} c \sim{}& i\sqrt{2}\epsilon + \frac{3}{2}\epsilon^2 + \left\{\pi\frac{\sqrt{2}}{4} - 0.142277482108753i\right\}\epsilon^3 + \left\{\frac{\sqrt{2}}{2}i\right\}\epsilon^3\log(\epsilon) \\ &+ \left\{1.13568388277225 - \frac{3}{4}\pi i\right\}\epsilon^4 + \left\{\frac{3}{2}\right\}\epsilon^4\log(\epsilon) + O(\epsilon^5\log(\epsilon)). \end{aligned} \tag{15.87}$$

Unfortunately, the series is not accurate over the whole interval $r \in [0,1]$, so $c(r)$ must be approximated by other means for $r < 1/2$.

15.9 ▪ Power and Exponential Series in Troesch's Problem

The Troesch ODE nonlinear boundary value problem is a popular testbed for numerical algorithms:

$$u_{xx} - \lambda \sinh(\lambda u) = 0, \qquad u(0) = 0, u(1) = 1. \tag{15.88}$$

Because of the exponential nonlinearity, there is, for $\lambda \sim O(10)$, a narrow boundary layer near the right endpoint where the solution explodes faster than exponentially. This is very challenging to differential equation solvers.

The Troesch problem has an exact analytic solution for all positive values of the parameter. Introducing the parameter $b(\lambda)$ found convenient for analysis by Chin [127],

$$b = \sqrt{1-m} \leq 1/2, \tag{15.89}$$

where m is the elliptic modulus, the solution is

$$u(x;\lambda) = \frac{2}{\lambda}\,\mathrm{arcsinh}\left\{ b\,\mathrm{sc}(\lambda x; m = 1-b^2)\right\}, \tag{15.90}$$

where b must be determined from the parameter λ by solving the transcendental equation

$$f^{Troesch} = b\,\mathrm{sc}(\lambda; m = 1-b^2) - \sinh(\lambda/2) = 0, \tag{15.91}$$

where $\mathrm{sc}(\lambda; m = 1-b^2)$ is one of the canonical Jacobian elliptic functions, the *elliptic tangent function*.

The Troesch rootfinding problem, as we shall dub solving $f^{Troesch}(b;\lambda) = 0$, is very challenging. The term $\sinh(\lambda/2)$ grows exponentially with the parameter λ and therefore is huge even when the parameter is only moderately large. Chin has proved that $b < 1/2$ for all positive λ, and Troesch himself proved that for large λ, $b \sim 4\exp(-\lambda)$. It follows that for large λ,

$$\mathrm{sc}(\lambda; m = 1-b^2) \sim O(\exp([3/2]\lambda)). \tag{15.92}$$

However, the elliptic tangent function is the ratio of the elliptic sine divided by the elliptic cosine, and these functions both have maxima of one. The only way the elliptic tangent can be large is if its argument λ is within $O(\exp(-[3/2]\lambda))$ of the pole of the elliptic tangent function at $K(m = 1-b^2)$, where $K(m)$ is the usual complete elliptic integral. The Troesch function will therefore have its root exponentially close to a pole!

This creates multiple challenges. First, preliminary scouting by graphing f is difficult because a function of the form $[x/(x-\epsilon)]g(x)$, where $g(x)$ is a smooth function with neither zeros nor poles near the origin, will resemble $g(x)$ except very close to the zero/pole pair since $x/(x-\epsilon) \approx 1$ except when $|x| \sim O(\epsilon)$. It is easy, unless f is evaluated with a very fine spacing, to miss a close zero/pole pair.

Newton's iteration is equivalent to constructing the tangent to $f(x)$ at the current iterate; the next iterate is the intersection of this tangent line with the x-axis. If the initial point is high on the infinitely tall peak that is a pole, then the slope of the tangent will be steep and Newton's correction will be very small compared to the actual distance to the root, as illustrated schematically in Fig. 15.7.

For the Troesch problem, the lowest-order approximation to the *zero* is actually an approximation to the *pole*. Perturbation theory as worked out below shows that

$$b_{zero} \sim 4\exp(-\lambda) - 8\exp(-(3/2)\lambda), \tag{15.93}$$

$$b_{pole} = K(m = 1-b^2)^{-1}(\lambda) \qquad [\text{inverse of} K(1-b^2) = \lambda] \tag{15.94}$$

$$\sim 4\exp(-\lambda) + 16(\lambda - 1)\exp(-3\lambda). \tag{15.95}$$

Initialize with $b = 4\exp(-4\lambda)$, the lowest-order approximation to the zero, and the resulting iterates are graphed in Fig. 15.8. The corrections are *initially* tiny; only after iteration 30 does the root begin the precipitous fall of the digit-doubling rate of convergence that is normal for the iteration sufficiently close to the root. Stopping criteria such as a correction smaller than 10^{-11} or an iteration limit of 25 would both prematurely stop

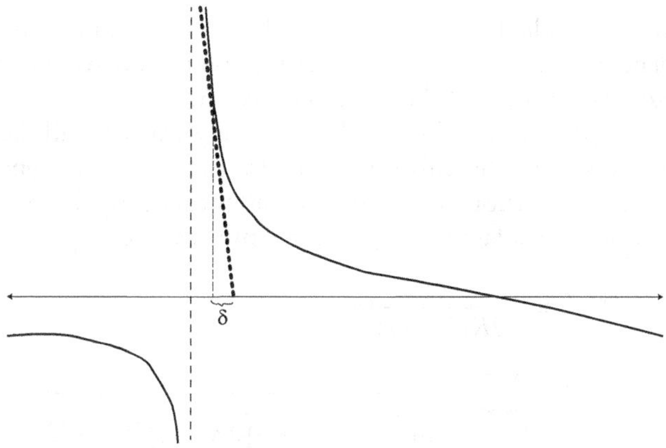

Figure 15.7. *When initialized near a pole whose location is marked by the vertical short-dashed line, the tangent line from that point has an almost vertical slope (thick dotted line). Consequently, the Newton correction δ, which is the difference between the vertical dashed line and the steeply sloping thick dotted line, is small compared to the distance to the zero where the graph of $f(x)$ crosses the x-axis.*

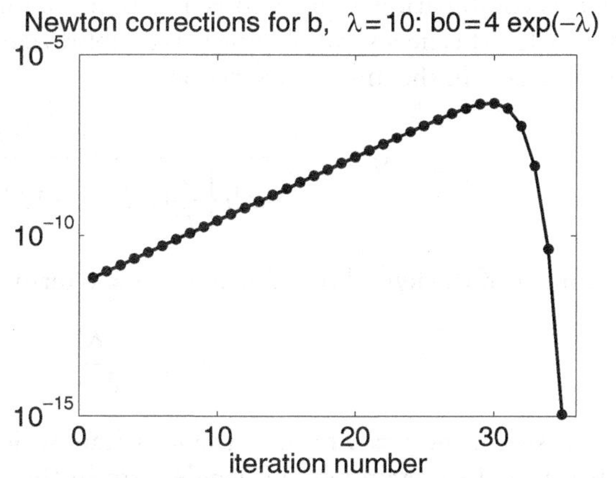

Figure 15.8. *Newton iterates for the Troesch problem for $\lambda = 10$ begun from $b^0 = 4\exp(-\lambda)$.*

the iteration while $\sqrt{(\lambda)}$ is still much closer to the pole than to the zero. Inserting this spurious root into the explicit solution for $u(x)$ would yield a function for which $u(1)$ is huge compared to its true boundary value of 1. This function is thus a morality tale whose moral is: Be careful not to stop Newton's iteration too soon.

Furthermore, the asymptotic series for large values of the parameter will be a mixture of exponentials and powers, the singular perturbation expansion.

Lastly, there is the by no means inconsiderable difficulty of wrestling with elliptic functions and elliptic integrals.

A little-known piece of mathematical folklore is that every Fourier series can be represented as a so-called imbricate series of the form $f(x) = \sum_{m=-\infty}^{\infty} A(x - mP)$, that is to say, all analytic periodic functions can be written as the periodization of a pattern function $A(x)$, as proved in Appendix D. The periodization can be accomplished by placing a graph of the pattern function in a photocopier, making an infinite number of duplicates, spacing these duplicates at integral multiples of the spatial period P, and then adding. The sum is periodic by construction. The Poisson summation theorem shows that the coefficients of the usual Fourier series representation are the evaluation at the integers of the Fourier transform of the pattern function.

The pattern functions are known analytically for all the classical Jacobian elliptic functions. Inserting the imbricate series for the elliptic tangent allows us to write the Troesch problem as, without approximation after canceling the factor of b multiplying the tangent function with a factor of $1/b$ in the imbricate series,

$$f^{Troesch} = -\frac{\pi}{2K(1-m)}\Theta, \tag{15.96}$$

$$\Theta \equiv \sum_{j=-\infty}^{\infty} \frac{1}{\sinh\left\{\frac{\pi}{2K(1-m)}[2K(m)](j-1/2)+\frac{\pi}{2K(1-m)}\lambda\right\}} - \sinh(\lambda/2),$$

where $m=\sqrt{1-b^2}$. It is convenient to define

$$\frac{2K(1-m)}{\pi} \equiv \tilde{K} \tag{15.97}$$

because this combination appears three times in the imbricate series and is equal to one at lowest order; this new symbol allows us to abbreviate $K(m)$ to K.

The jth term in the imbricate series is

$$A_j \equiv -\frac{1}{\tilde{K}} \frac{1}{\sinh\left\{\frac{1}{\tilde{K}}[2K](j-1/2)+\frac{1}{\tilde{K}}z\right\}}. \tag{15.98}$$

The argument of the hyperbolic sine in the $j=0$ term will be denoted by

$$Z \equiv \frac{\lambda - K}{\tilde{K}}. \tag{15.99}$$

When Z is small, the $j=0$ term in the imbricate series is nearly singular and therefore large. It is useful to retain just this single term in the imbricate series, solve for $b(\lambda)$ to lowest order, and then verify that the result is self-consistent at least at low order. Since Z is small, sinh(Z) can be approximated by its Taylor series. Thus,

$$f^{Troesch} \approx -\frac{1}{\tilde{K}}\frac{1}{\sinh\{Z\}} - \sinh(\lambda/2) \tag{15.100}$$

$$-\frac{1}{\tilde{K}}\frac{1}{Z+(1/6)Z^3+(1/120)Z^5+\cdots} - \frac{1}{2}\frac{1}{W} + \frac{1}{2}W, \tag{15.101}$$

where we have introduced the abbreviation

$$W \equiv \exp(-\lambda/2) \qquad \leftrightarrow \qquad \lambda = -2\log(W). \tag{15.102}$$

The lowest-order approximation to $b(\lambda)$ arises from asking: What must b be so that $Z=0$? The pole of $f^{Troesch}$ at $Z=0$ is the solution to $K(m=1-b^2)=\lambda$. Inserting $K(1-b^2) \sim -\log(b/4)+O(b^2)$ and solving $Z=0$ gives

$$b \sim 4\exp(-\lambda). \tag{15.103}$$

This is the zeroth-order approximation to both the pole and the zero of $f^{Troesch}$.

To calculate a higher-order approximation, neglect Z^3 compared to Z and W compared to $1/W$ and write $b = 4W^2 + 4\sigma$, where σ is a small correction:

$$f^{Troesch} \approx -\frac{1}{\tilde{K}}\frac{1}{Z+\cdots} - \frac{1}{2}\frac{1}{W} \tag{15.104}$$

$$\approx -\frac{1}{\tilde{K}}\frac{\tilde{K}}{\lambda - K} - \frac{1}{2}\frac{1}{W} \tag{15.105}$$

$$\approx -\frac{1}{\lambda - K} - \frac{1}{2}\frac{1}{W} \tag{15.106}$$

$$\approx -\frac{1}{\lambda + \log(W^2+\sigma)} - \frac{1}{2}\frac{1}{W} \tag{15.107}$$

$$\approx -\frac{1}{\lambda + \log(W^2[1+\sigma/W^2])} - \frac{1}{2}\frac{1}{W} \tag{15.108}$$

$$\approx -\frac{1}{\lambda + 2\log(W) + \log[1+\sigma/W^2]} - \frac{1}{2}\frac{1}{W} \tag{15.109}$$

$$\approx -\frac{1}{\sigma/W^2 + O(\sigma^2/W^4)} - \frac{1}{2}\frac{1}{W}, \tag{15.110}$$

where we used $\lambda = -2\log(W)$ by the definition of $W(\lambda)$. This gives

$$b \sim 4W^2 - 8W^3 = 4\exp(-\lambda) - 8\exp(-[3/2]\lambda). \tag{15.111}$$

To proceed to higher order, we must expand everything in powers of W in the full imbricate series.

A frequent complication in singular perturbation theory is the need to explicitly partition some transcendental functions. For example, the argument of the hyperbolic sine in the jth copy of the pattern function (and of the pair of exponentials that define the sinh function) is

$$\zeta_j \equiv \frac{1}{\tilde{K}}[2K](j-1/2) + \frac{1}{\tilde{K}}z \tag{15.112}$$

$$= 2\lambda j + [2\frac{1}{\tilde{K}}K - 2\lambda]j + Z \tag{15.113}$$

$$= 2\lambda j + \tilde{\zeta}_j, \tag{15.114}$$

where $\tilde{\zeta}_j \sim O(W) \ll 1$. The tricky part is that the argument of the exponentials is *not small* because of the $2\lambda j$ part. To successfully expand the exponentials, one must write

$$\exp(-\zeta_j) = \exp(-2\lambda j + \tilde{\zeta}_j) \tag{15.115}$$

$$= W^4 \exp(-\tilde{\zeta}_j), \qquad j > 0. \tag{15.116}$$

The remaining exponential does have a tiny, $O(W)$ argument and can be expanded as a W power and log series without difficulty. Then, for $j > 0$,

$$A_j \equiv -\frac{1}{\tilde{K}}\frac{1}{\sinh\left\{\frac{1}{\tilde{K}}[2K(m)](j-1/2) + \frac{1}{\tilde{K}}z\right\}} \tag{15.117}$$

$$= -\frac{2}{\tilde{K}}\exp(-\tilde{\zeta}_j)\frac{W^4}{1 - W^8\exp(-2\tilde{\zeta}_j)}, \tag{15.118}$$

without approximation, where we have used $\sinh(x) = 2\exp(-x)/(1-\exp(-2x))$. The treatment for $j < 0$ is similar.

Up to fourth order in W, all but one term in the imbricate series can be ignored:

$$f^{Troesch} \approx -\frac{1}{\tilde{K}}\frac{1}{\sinh\{Z\}} - \sinh(\lambda/2) \tag{15.119}$$

$$-\frac{1}{\tilde{K}}\frac{1}{Z+(1/6)Z^3+(1/120)Z^5+\cdots} - \frac{1}{2}\frac{1}{W} + \frac{1}{2}W. \tag{15.120}$$

To proceed to higher order, write

$$b \sim 4W^2 + \sum_{j=3}^{\infty}\sum_{k=0}^{\infty} b_{jk}\, W^j \log^k(W). \tag{15.121}$$

Order by order in the series for $f^{Troesch}$,

$$O(1/W): \to b_{30} = -8, \tag{15.122}$$

$$O(1): \to b_{40} = 8, \tag{15.123}$$

$$O(W): \to b_{50} = -8, \tag{15.124}$$

$$O(W^2): \to b_{60} = -8, \tag{15.125}$$

$$O(W^2\log(W)): \to b_{61} = -32, \tag{15.126}$$

$$O(W^3): \to b_{70} = 120, \tag{15.127}$$

$$O(W^3\log(W)): \to b_{71} = 192. \tag{15.128}$$

Rewriting in terms of λ gives

$$\begin{aligned} b \sim\ & 4\exp(-\lambda) - 8\exp(-(3/2)\lambda) + 8\exp(-2\lambda) \\ & -8\exp(-(5/2)\lambda) - 8\exp(-3\lambda) + 16\lambda\exp(-3\lambda) \\ & +120\exp(-(7/2)\lambda) - 96\lambda\exp(-(7/2)\lambda). \end{aligned} \tag{15.129}$$

15.10 ▪ Summary: General Concepts for Singular Perturbation Theory

The appearance of powers of $\log(\epsilon)$, $\log(-\log(\epsilon))$, $1/\log(\epsilon)$, or $\exp(-1/\epsilon)$ need not dishearten us. A mighty battle-ax is our perturbation theory, defeating even problems armored with logarithms and exponentials of a reciprocal. Even when divergent, asymptotic series tell a story and generate useful numbers [39, 273], thank goodness! For problems like the Lambert W-function, the only hope of generating a first guess within the convergence domain of Newton's iteration is to initialize with a couple of terms of the singular perturbation series. Chebyshev-proxy rootfinders (CPRs) are good at bagging zeros at specific values of ϵ, but extrapolating a point cloud of zeros to $\epsilon = 0$ is difficult, especially when the limit is not analytic. A perturbation series is the better way.

Although singular perturbation problems are more individual and less routine than regular perturbation problems, there are some general principles and recurring themes worth review:

1. Newton's iteration is very helpful to deduce the form of corrections, provided that df/dx is nonsingular at $x = x^{(0)}$.

2. Perturbation theory via Newton's iteration should be implemented as the *frozen Jacobian*, *chord*, or *Shamanski* iteration; whatever the name, $df/dx = df/dx(x^{(0)})$ for all iterations.

3. When the goal is an explicit series for the root, a key secondary goal is to make the residual of the current iterate as simple as possible. This requires expanding the residual in terms of the small parameter or parameters and keeping only the lowest term before going on to the next iteration (as well as the Shamanski approximation $df/dx \approx df/dx(x^{(0)})$).

4. The *separate-symbol* strategy is often necessary. (The separate-symbol device is the introduction of new symbols for exponentially small terms or logarithmic terms, followed by the treatment of these *independent* of powers of the small parameter in bivariate Taylor expansions (ϵ, ν), where for example $\nu \equiv -1/\log(\epsilon)$.)

5. The brutally difficult case is when there is an *infinite* series of *slow* terms at a given order of the faster parameter. An example is the need to sum an infinite series of *logarithms* of ϵ before going on to the next *power* of ϵ. As yet, there is no general strategy for this difficulty.

   However, one can truncate a double power series in $\nu = 1/\log(\epsilon)$ and ϵ at finite order up to ν^M and ϵ^N and show consistency at $O(\nu^M)$ for $\epsilon < \epsilon_T$ if

$$\nu^M(\epsilon_T) \geq (\epsilon_T)^N. \tag{15.130}$$

Part VI

Classics, Special Functions, Inverses, and Oracles

Chapter 16

Classic Methods for Solving One Equation in One Unknown

"...the separation of closely bunched multiple roots is generally a cumbersome task which cannot be accomplished without a great deal of computational work."
—Cornelius Lanczos [236, p. 43]

16.1 ▪ Introduction

When there is only a single unknown, or perhaps just two or three, one can use a wide variety of methods that generalize badly or not at all to a large number of unknowns. The first such tool is graphics.

The simplest strategy is MAG: make a graph. Several different types of plots are useful.

When, for a fixed value of the parameter λ, only the real roots of $f(x,\lambda)$ on a certain interval are interesting, the most reliable way to find them all is to make a graph of $f(x)$ with a very fine grid spacing, count the number of intersections with the real axis, and then use Newton's iteration or the secant iteration to refine the approximate roots. This will be dubbed "MAG-1D" since it applies only to a single equation in a single unknown. MAG may miss a pair of roots whose separation is smaller than the interval between plotted points, and it has other vices, cataloged later in this chapter. Nevertheless, there is no denying that when there are only a few unknowns, graphics are a powerful defense against the dreaded missed root.

When the function depends upon a parameter λ, an effective strategy for identifying multiple roots is to make a *contour* plot of $f(x;\lambda)$, graphing the zero isoline only. MAG-Contour has the glorious virtue of displaying multiple, nonintersecting branches without requiring any input from the user except a search range.

Second, it is common in applications that $f(x;\lambda) = g(x) - \lambda$ so that $x(\lambda) = g^{-1}(\lambda)$, the inverse of $g(x)$. One can then plot the root by graphing $\lambda = g(x)$, which is trivial, and simply reversing the roles of the horizontal and vertical axes.

When there are two unknowns or when the goal is *complex* roots of a single unknown, a good strategy is to make a mesh plot or a contour plot of the *negative* of the *logarithm* of the residual norm function,

$$\rho(x_1,x_2) \equiv \frac{1}{2}\left\{f_1(x_1,x_2)^2 + f_2(x_1,x_2)^2\right\}. \tag{16.1}$$

This will be called "MAG-2D." The minus sign in $-\log(\rho)$ turns valleys into peaks, which are easier to see than depressions. The logarithm sharpens and raises the immediate

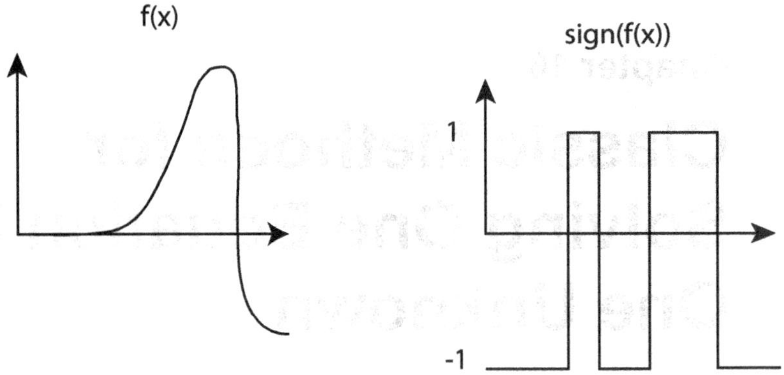

Figure 16.1. *Left: a plot of $f(x)$. The part of the curve that coincides with the axis hides roots that are revealed by plotting the composite function sign($f(x)$) (right).*

neighborhood of the roots. A plot of isosurfaces of the logarithm of the residual norm can produce little bubbles around each trivariate root if the appropriate value of ρ is chosen for the isosurface.

In the rest of this chapter, we shall discuss these and other special tools for a single unknown.

16.2 ▪ MAG-1D: Rootfinding by Plotting for One Unknown: Real Roots on an Interval

16.2.1 ▪ Graphs as Scouts

Plotting $f(x)$ is often the simplest way to identify roots in an interval. An aviation aphorism is: "Instruments are nice, but it's still a good idea to look outside the airplane." Graphs are the advance scouts and Recon Marines of the hunt for zeros.

Unfortunately, some functions may have small-amplitude oscillations that are graphically invisible because of much larger peaks close by, as illustrated on the left in Fig. 16.1. A graph of sign($f(x)$) will alternate between ± 1, crossing the axis at every zero, regardless of the varying amplitude of the maxima and minima of $f(x)$, where the sign function is defined by

$$\text{sign}(z) = \begin{cases} 1, & z > 0, \\ -1, & z < 0. \end{cases} \tag{16.2}$$

However, some $f(x)$ have poles and zeros alternating in x. The composite function sign($f(x)$) will flip polarity at simple *poles* as well as the zeros and not change sign at *double* zeros. A better strategy for such functions is to plot $|f(x)|$ on a logarithmic scale. Zeros appear as downward spikes; poles appear as upward-stretching spikes. Different visual representations are best for different classes of functions.

Unfortunately, sometimes zeros are almost invisible on a graph. A systematic, moderately reliable strategy is the following.

16.2.2 ▪ Sampling

Definition 16.1 (Search by Sampling). *An algorithm for detecting real-valued roots has the following steps:*

1. *Evaluate $f(x)$ at a large set of finely spaced points x_j.*
2. *When $f(x_{j+1})$ is opposite in sign to $f(x_j)$, define an approximate root by*

$$x_{guess} = (1/2)(x_{j+1} + x_j). \tag{16.3}$$

3. *Refine each approximate root x_{guess} by an iterative algorithm, such as Newton's or the secant algorithm.*

A for/next loop in j that compares the signs of $f(x_{j+1})$ and $f(x_j)$ may be replaced by plotting the points as a line graph and visually identifying the points x_{guess} where the curve crosses the x-axis.

The graph is optional, but search and sign detection is not. Monitor resolution limits visual sampling to a few hundred points since even the priciest computer screens are only two thousand pixels across. However, graphs are better than a sign-change-detecting loop at (i) identifying double zeros or close doublets where a set of discrete samples of $f(x)$ may show no sign changes and (ii) identifying sign changes due to simple poles rather than roots.

This method is not useful for finding *complex-valued* roots. However, a mesh plot or contour plot in the complex plane can serve the same function (section 16.3).

There are several difficulties. The loop search relies on the elementary theorem that if f is continuous and $f(a)f(b) < 0$, then there must be at least one root of f on the interval $x \in [a, b]$. One worry is that there might be *three* or *five* or any odd number of roots greater than one in the interval. A second difficulty is that if there are two roots between $x = a$ and $x = b$, then it is likely that $f(a)$ and $f(b)$ are of the *same sign*, and the *pair* of roots will be missed.

Both of these difficulties disappear as the spacing between adjacent zeros goes to zero; since it is often cheap to evaluate a single function in a single unknown at even as many as a million points, the best advice is: Apply MAG with a *very* fine grid, and it will work *most* of the time.

The Chebyshev-proxy method (CPR) described earlier can overcome these difficulties on a rather coarse grid.

When f has the special form

$$f(x;\lambda) \equiv g(x) - \lambda \qquad \leftrightarrow \qquad = g^{-1}(\lambda), \tag{16.4}$$

so that the solution is the inverse of $g(x)$, MAG takes a simple form we may dub "MAG-Reversed Axes": make a plot of $\lambda = g(x)$ and then simply *interchange the horizontal and vertical axes*. This is illustrated with a couple of examples in section 18.3.

16.3 ▪ MAG-Contour: Tracing Entire Branches of Solutions

When the solution to a single equation in a single unknown depends upon a parameter λ, a very good strategy is to make a contour plot of the zero isoline of $f(x,\lambda)$ in the λ-x plane. Fig. 16.2 shows the zero-valued contours of a particular arbitrary example. Often, perturbation theory can give an approximation to the branch that begins at $x = 0$ where the nonlinear equation may be well approximated by a linear equation. Numerical continuation can then trace the branch to large x, where the perturbation theory fails.

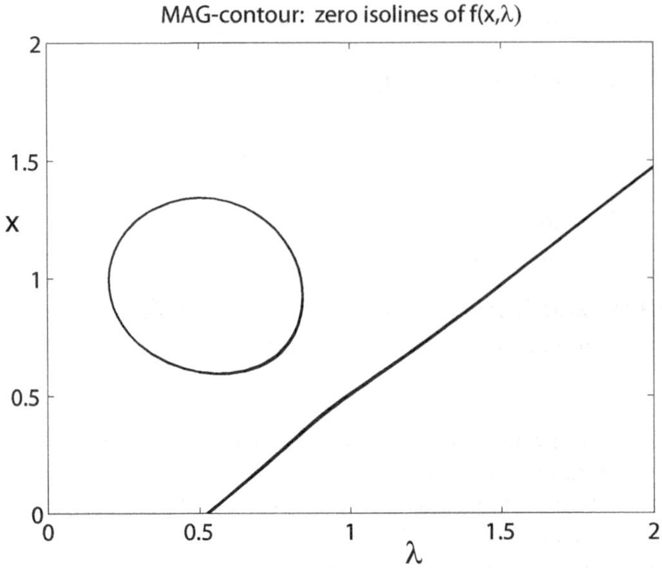

Figure 16.2. *A contour plot of the function $f(x,\lambda)$, showing only the zero-valued isolines. The function is $f(x,\lambda) = -1 + 2\exp(-6(x-1)^2 - 8(\lambda - 1/2)^2) + 2\{1 + erf(-(x-\lambda-1))\}$.*

However, for this example there are two other branches, collectively forming the nearly circular isoline, which do not exist at $x = 0$, but only for finite x. It is usually very difficult to identify such finite amplitude branches through perturbation theory or other analytical tools. However, the contour plot shows all branches. The contours are not highly accurate—most contouring routines use linear interpolation between four nearest neighbors—but one can easily read a couple of values off the graph with sufficient accuracy to initialize Newton's iteration and then continue with pseudoarclength continuation to trace both the upper and lower branches.

A very accurate contour option is the Weyl/marching triangles (MT) algorithm described in Chapter 20.

16.4 ▪ Successive Substitution, Fixed Point Iteration, and Contraction Mapping

When a nonlinear equation can be written as

$$x = g(x), \tag{16.5}$$

then it *may* be possible to solve it by means of the simple iteration

$$x^{(n+1)} = g\left(x^{(n)}\right) \qquad \text{[Successive Substitution]}. \tag{16.6}$$

The power of this algorithm is simplicity: no derivatives are required. If the iteration converges, it converges geometrically fast. Successive substitution can be applied to systems of nonlinear equations, too.

Because of its simplicity, successive substitution is almost universally employed to solve the transcendental equations for trajectories in semi-Lagrangian advection schemes in weather forecasting and climate modeling [76] (one equation in one dimension, two equations in two unknowns in two dimensions, and trivariate systems with three unknowns in three dimensions). For the same reason, successive substitution is popular for many other applications.

Successive substitution is also frequently called the *fixed point iteration* or *application of the contraction mapping principle* in the literature.

However, the method has a weakness. If we write $x^{(n+1)} \equiv x_* + \delta^{(n+1)}$ and similarly for $x^{(n)}$, then very close to the root, we can expand about x_*. Canceling the common factors from $x_* + \delta^{(n+1)} = g(x_*) + dg/dx(x_*)\delta^{(n)} +$ higher-order terms gives

$$\delta^{(n+1)} \approx \frac{dg}{dx}(x_*)\delta^{(n)}. \tag{16.7}$$

It follows that the iteration will *diverge* if $|dg/dx(x_*)| > 1$; if the derivative of $g(x)$ at the root is smaller than one in magnitude, then the iteration will converge geometrically in the vicinity of the root, with the error in each iterate smaller than its predecessor by a factor of $|dg/dx(x_*)|$.

Any equation of the form $f(x) = 0$ can be trivially rearranged into the form required for successive substitution by defining $g(x)$ as

$$g(x) \equiv f(x) + x \qquad \longleftrightarrow \qquad f(x) = 0. \tag{16.8}$$

A problem of the form $x = g(x)$ is referred to as a *fixed point problem* and the root is called the *fixed point*.

The catch is that the first derivative of g at the root may be larger than one in magnitude, in which case the root is a *repellor* in the language of dynamical systems theory. The reason that successive substitution has been particularly popular for a single equation in a single unknown is that the same equation $f(x) = 0$ can be written as a fixed point problem in more than one way, and the freedom to rearrange terms is especially wide for a single equation.

Fletcher [171, p. 122] gives the example of

$$f(x) = 1 + x - \exp(2x), \tag{16.9}$$

which can be written as

$$x = \exp(2x) - 1 \tag{16.10}$$

or as

$$x = (1/2)\log(1 + x). \tag{16.11}$$

Successive substitution diverges rapidly for the first fixed point problem, but when rewritten as the second form, successive substitution converges, and each iterate is only half as large as its predecessor.

A crucial step in obtaining the second fixed point problem is to take the logarithm of both sides of the equation. Such functional transformations are usually not feasible for systems of equations. Nevertheless, for one or many equations, a little algebraic rearrangement will often turn failure into success.

Underrelaxation is a general strategy for making successive substitution work in one dimension. Suppose that $df/dx(x_*) = F$. Suppose further that we generalize the relationship between the original problem, $f(x) = 0$, and the function that appears in the

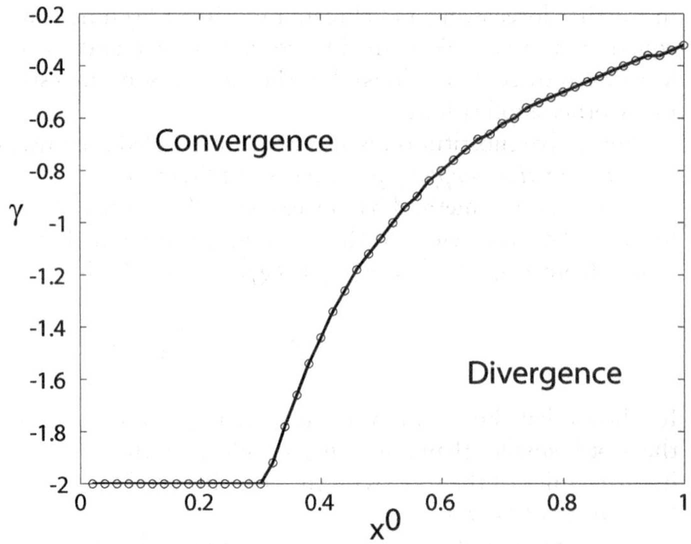

Figure 16.3. *The heavy solid curve is the boundary between convergence and divergence for the solution of $f(x) = 1 + x - \exp(2x)$ by underrelaxed successive substitution in the form $x^{(n+1)} = x^{(n)} + \gamma f(x^{(n)})$. The exact root is $x = 0$. Here x^0 denotes the first guess, while γ is the underrelaxation parameter. The highest rate of convergence is obtained for $\gamma = -1$ for this example, but the iteration converges for this γ only when the initial guess is less than 1/2. For larger γ, it is possible to obtain convergence from a first guess that is twice as bad.*

fixed point problem by writing

$$g(x) \equiv x + \gamma f(x) \quad \Rightarrow \quad \frac{dg}{dx}(x_*) = 1 + \gamma F, \tag{16.12}$$

where γ is a constant. It follows that we can always force convergence by choosing

$$\text{sign}(\gamma) = -\text{sign}\left(\frac{dg}{dx}(x_*)\right) \quad \text{and} \quad |\gamma| < 2/F. \tag{16.13}$$

The optimum convergence is achieved for

$$\gamma = -1/F. \tag{16.14}$$

It is important to warn that our analysis is *local*, valid only near the root. If $x^{(0)}$ is far from the zero x_*, $|dg/dx|$ may be larger than one and the iteration will diverge even though $|dg/dx(x_*)| < 1$. Thus, successive substitution is usually a *local* method; only if $|dg/dx| < 1$ *everywhere* is the iteration *global*, convergent from any starting point.

As often happens, there is a tradeoff between the rate of convergence and tolerance of a poor first guess. For Fletcher's problem, as illustrated in Fig. 16.3, the optimum γ, which is -1 for this example, yields convergence only with a rather good first guess. For smaller $|\gamma|$, the radius of convergence in the first guess x_0 increases significantly.

Fig. 16.4 shows error versus iteration number for four different values of γ for a different example. There is again no convergence for $\gamma > 0$.

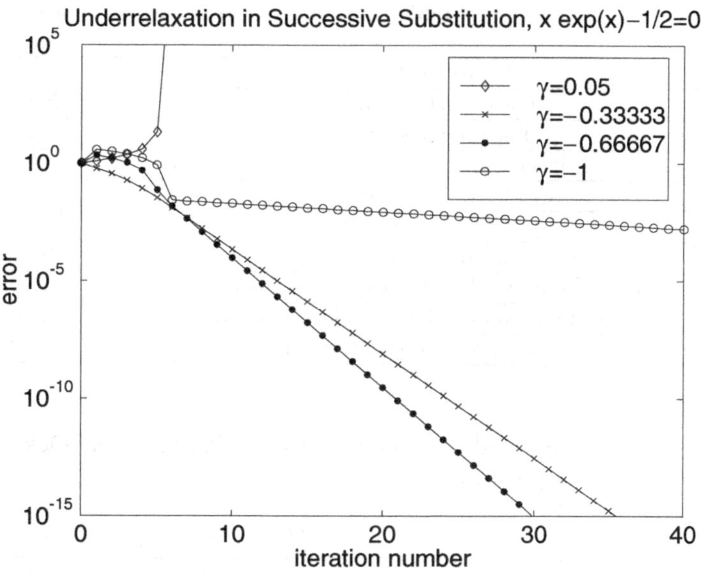

Figure 16.4. *Successive substitution for the equation $f(x) = x\exp(x) - 1/2 = 0$, which was converted to the fixed point problem $x = g(x)$, where $g(x) = x + \gamma f(x) \Leftrightarrow x^{n+1} = x^n + \gamma f(x^n)$. The graph shows the error versus iteration for four different values of the underrelaxation parameter γ, beginning with a first guess $x_0 = x_* + 1$. The iteration diverges for all positive γ. With $\gamma = -1$, the iteration converges, but only slowly. When $\gamma \in [-2/3, -1/3]$, convergence is quite rapid.*

16.5 ▪ Inverse Interpolation Methods

Alternatively, one can approximate the *inverse* function $x(f)$ by a polynomial or rational function of f, $U(f)$. Since the inverse function isn't known analytically—if it were known, the root would be $x_* = x(0)$ and we would not need a numerical algorithm—inverse proxies are always found by *interpolation*. The strategy is to pick some values x_k close to the current iterate (usually, the x_k are chosen to be the last $(n+1)$ iterates), evaluate $f_k = f(x_k)$, compute the polynomial $X(f_k)$, and then take the next iterate as $x^{(n+1)} = U(0)$.

For example, linear inverse interpolation through the last two iterates gives

$$X(f) \equiv x^{(n-1)} + \frac{x^{(n)} - x^{(n-1)}}{f(x^n) - f(x^{(n-1)})}(f - f(x^{(n-1)})). \tag{16.15}$$

This strategy is called variously *regula falsi* and the *method of false position.* From this proxy, $x^{(n+1)} \approx X(0)$, i.e.,

$$x^{(n+1)} \equiv x^{(n-1)} - f(x^{(n-1)})\frac{x^{(n)} - x^{(n-1)}}{f(x^n) - f(x^{(n-1)})}. \tag{16.16}$$

One important variant is to replace $x^{(n-1)}$ as the second interpolation point by $x^{(0)}$. If the sign of $f(x^{(1)})$ is opposite to that of $f(x^{(0)})$, this variation ensures that the root remains bracketed by the iterates, guaranteeing convergence to a root.

Table 16.1. *Solve the proxy methods in one unknown.*

Approximation	Name
linear Taylor series	Newton [Newton–Raphson] iteration
linear interpolant	secant iteration
inverse linear interpolant	regula falsi
quadratic Taylor series	Cauchy's method
quadratic interpolant	Muller's method
inverse quadratic interpolant	Brent's algorithm[21]
[1/1] (linear-over-linear) Padé approximant	Halley's scheme
[2/2] (quadratic-over-quadratic) Padé approximant	Shafer's method [320]
Chebyshev polynomial interpolant	Boyd [96]

Brent's method, also known as the van Wijngaarden–Deker–Brent scheme, uses quadratic inverse interpolation:

$$\begin{aligned} X_{Brent}(f) = & \frac{[f - f(x^{(n-1)})][f - f(x^{(n-2)})]}{[f(x^{(n)}) - f(x^{(n-1)})][f(x^{(n)}) - f(x^{(n-2)})]} f(x^{(n)}) \\ & + \frac{[f - f(x^{(n)})][f - f(x^{(n-2)})]}{[f(x^{(n-1)}) - f(x^{(n)})][f(x^{(n-1)}) - f(x^{(n-2)})]} f(x^{(n-1)}) \\ & + \frac{[f - f(x^{(n)})][f - f(x^{(n-1)})]}{[f(x^{(n-2)}) - f(x^{(n)})][f(x^{(n-2)}) - f(x^{(n-1)})]} f(x^{(n-2)}). \end{aligned} \tag{16.17}$$

In Brent's implementation, this quadratic inverse interpolation is not applied blindly but is combined with bisection ideas to improve reliability.

These various proxy and inverse interpolation strategies are summarized in Table 16.1.

16.6 ▪ An Arbitrary-Order Approximation of an Isolated Root as a Ratio of Two Chebyshev Quadratures

Suppose that $f(x)$ has only a single root on $x \in [a, b]$. Let ρ denote the (unknown!) location of the root. The integral

$$I \equiv \int_a^b \frac{x - \rho}{f(x)} \frac{1}{\sqrt{1 - x^2}} dx \tag{16.18}$$

can be integrated with exponential accuracy by a Chebyshev quadrature because the zero in $f(x)$ cancels the zero in the numerator. There are, however, two different Chebyshev polynomial-based quadrature schemes in widespread use. The *Gauss-Chebyshev* N-point method employs the zeros of $T_N(x)$ as quadrature points. The *Chebyshev-Lobatto* grid is the union of the endpoints, $x = \pm 1$, plus the roots of dT_{N-2}/dx [76, 63]. Because both choices of quadrature/interpolation points yield an accuracy proportional to $\exp(-\mu N)$ for some $\mu > 0$, it follows that the results of Gauss-Chebyshev and Chebyshev-Lobatto quadrature must differ by an amount exponentially small in N,

$$I_{Gauss}(N) = I_{Lobatto} + O(\exp(-\mu N)), \tag{16.19}$$

[22] As usually implemented, Brent's method also bounds the corrections and shifts to bisection as needed.

for some positive constant μ (neglecting powers of N and other algebraic factors of N multiplying the exponential), where

$$I_{Gauss}(N) \equiv \sum_{j=1}^{N} w_j^G \frac{(x_j^G - \rho)}{f(x_j^G)}, \tag{16.20}$$

and similarly for the Lobatto quadrature, where the w_j are the quadrature weights. Neglecting the exponential error terms, we can rearrange the terms in the equation $I_{Gauss} = I_{Lobatto}$ to obtain

$$\rho \left\{ \sum_{j=1}^{N} \left(\frac{w_j^G}{f(x_j^G)} - \frac{w_j^L}{f(x_j^L)} \right) \right\} = \left\{ \sum_{j=1}^{N} \left(x_j^G \frac{w_j^G}{f(x_j^G)} - x_j^L \frac{w_j^L}{f(x_j^L)} \right) \right\}. \tag{16.21}$$

Dividing by the sum on the left gives an explicit formula for the root ρ as a ratio of weighted values of $f(x)$. For Chebyshev quadrature with the usual Chebyshev weight function in the integral (16.18), the quadrature weights w_j are all identical except for the two endpoints, and the quadrature points for both grids combined are the images of an evenly spaced grid under the cosine mapping. This greatly simplifies the final approximation to the root to

$$\rho = \left\{ \sum_{j=0}^{2N}{}'' (-1)^j \frac{x_j}{f(x_j)} \right\} \Bigg/ \left\{ \sum_{j=0}^{2N}{}'' (-1)^j \frac{1}{f(x_j)} \right\}, \tag{16.22}$$

where the double prime on the sums denotes that the first and last terms in both the numerator and denominator should be multiplied by (1/2), and

$$x_j = \frac{1}{2} \left\{ (a+b) - (a-b) \cos\left(\frac{j\pi}{2N} \right) \right\}. \tag{16.23}$$

Fig. 16.5 shows that for a typical example, the convergence of the approximation with N is indeed geometric, that is, proportional to $\exp(-\mu N)$ for some $\mu > 0$.

16.7 ▪ Phase-Plane Analysis

When the coefficients of a set of equations depend on a parameter, continuation in that parameter is both a powerful way to initialize the Newton–Raphson iteration and a means to trace the entire branch of solutions. When the branch has a fold or limit point, it is necessary in general to use pseudoarclength or tangent vector continuation to successfully round the fold.

When there is but a single equation in a single unknown, one can use a simpler procedure. The idea is to introduce a new parameter T such that both the root $x(\lambda)$ and the parameter λ are *single-valued* functions of T. This is called *phase-plane analysis* (McLeod and Baart [264]). The idea is to replace the Davidenko equation, in which the marching variable is one of the parameters, by a pair of ODEs in which the marching variable is a dummy parameter introduced to parameterize what otherwise would be multivalued curves.

For a single, scalar unknown, the problem is

$$f(x, \lambda) = 0. \tag{16.24}$$

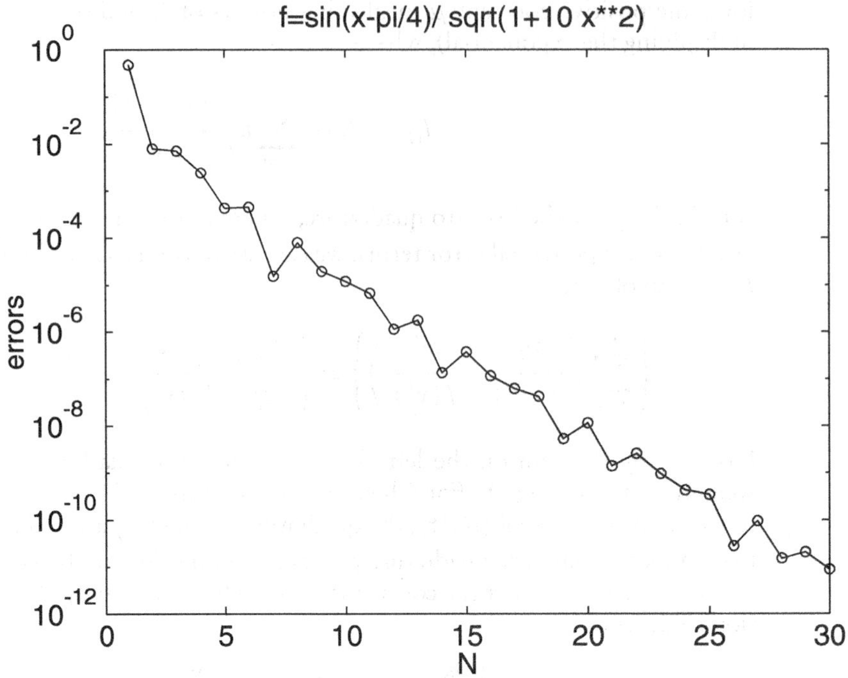

Figure 16.5. *Absolute value of the absolute error in approximating the root nearest the origin of the function* $\sin(x-\pi/4)/(1+10\,x^2)^{1/2}$ *using the quadrature interval* $[-1,1]$ *in Ioakimidis's noniterative Chebyshev rootfinding algorithm. The number of evaluations of* $f(x)$ *is* $2N+1$, *where* N *is the order of the method.*

If both x and λ are functions of a parameter T, then obviously

$$\frac{df}{dT} = \frac{\partial f}{\partial x}\frac{\partial x}{\partial T} + \frac{\partial f}{\partial \lambda}\frac{\partial \lambda}{\partial T} = 0, \tag{16.25}$$

since $f(x[T], \lambda[T]) \equiv 0$ for all T. If we now choose the relationships between (x, λ) and T to be defined by the pair of ODEs

$$\frac{d\lambda}{dT} = \frac{\partial f}{\partial x}, \qquad \frac{dx}{dT} = -\frac{\partial f}{\partial \lambda} \qquad \text{[Phase-Plane Equations]}, \tag{16.26}$$

then substitution into (16.25) shows that the constraint $df/dT \equiv 0$ is satisfied. If we choose initial conditions such that $f(x[T=0], \lambda[T=0]) = 0$, then $f = 0$ everywhere along the curve in the x-λ plane parameterized by T.

The good news about the phase-plane equations is that, unlike the Davidenko equation and other continuation-in-λ schemes, it does *not fail* at *limit points* where $\partial f/\partial x = 0$. The bad news is that the phase-plane equations do fail at a transcritical *bifurcation* point where *both* $\partial f/\partial x$ and $\partial f/\partial \lambda$ are zero. In the language of ODE theory, a bifurcation point is a *stationary point* where the solutions of the ODEs come to rest, making it impossible (at least in theory) to track the branch beyond the bifurcation point. In practice, if one integrated the phase-plane equations numerically with a fixed minimum stepsize, one could march beyond the stationary point and thus continue to trace the branch. However, special tricks would be needed to switch to the other branch to trace it, too.

The phase-plane equations are particularly useful for cubic and quartic equations because the phase-plane system can be solved *analytically* (sections 11.10 and 12.5). This makes it rather easy to graph the complete solution surface for these polynomial equations even though this surface contains folds.

The phase-plane analysis is similar to tangent vector continuation when the latter is applied to one equation in one unknown. The difference is that the right-hand sides of the tangent vector continuation equations are the same as for the phase-plane analysis, but divided by $\sqrt{\left(\frac{\partial f}{\partial x}\right)^2 + \left(\frac{\partial f}{\partial \lambda}\right)^2}$. This normalizes the parameter T to be arclength. However, this normalization is entirely optional.

Chapter 17

Special Algorithms for Special Functions

"The zeros of special functions appear in a great number of applications in mathematics, physics, and engineering, from the computation of Gauss quadrature rules in the case of orthogonal polynomials to many applications in which boundary value problems for second order ordinary differential equations arise."

—Amparo Gil, Javier Segura, and Nico M. Temme on p. 191 of their book *Numerical Methods for Special Functions* [195]

17.1 ▪ In Praise of Narrowness: Stenarithms

In an earlier chapter, we quoted Tai Tsun Wu's aphorism, paraphrased here as: "The more general a method, the less useful." Wu, whose career at Harvard as student, Junior Fellow, and faculty is now in its 60th year, meant that special tricks are underrated. General-purpose methods are great, but for special problems and special functions, one can usually devise special algorithms of narrow applicability that are superior to computational procedures with a large range.

Gil, Segura, and Temme, who devote an entire lengthy chapter of their book on special functions to zerofinding algorithms, agree with Wu [195]. Indeed, during his talk at the SIAM Annual Meeting in 2013 (with both coauthors in attendance), Nico Temme advised, "Don't generalize too much. Keep it simple." He went on to explain that Airy functions, for example, can be expressed as Bessel functions of orders $\pm 1/3$. It is almost always better, however, to write a routine just to evaluate $\mathrm{Ai}(x)$ and $\mathrm{Bi}(x)$.

The Wu/Gil–Segura–Temme precepts motivate the following.

Definition 17.1 (Stenarithm). *A numerical algorithm that is useful only for a* single *function or a* single, narrow class *of functions (from Greek $\sigma\tau\epsilon\nu o\sigma$, "narrow," and "algorithm.") The opposite is* euryarithm.

The neologism is not important, but the philosophy of *narrow algorithms* is very significant.

Our treatment of special function root calculations will be brief here for two reasons. The first is that Chapter 7 of [195] is comprehensive and excellent.

The other reason for brevity is that general-purpose methods actually work very well for special functions. The zeros of classical orthogonal polynomials and J-type Bessel functions are all real, simple, and well separated. Perturbation theory gives good approximations for the larger roots; the Chebyshev-proxy algorithms have not the slightest difficulty with the smaller zeros.

Thus, the casual user need make no distinction between special functions and everything else. However, special function libraries make heavy use of special algorithms, not merely for computing zeros, but also for calculating the special functions themselves. Elliptic functions have rapidly convergent Fourier series with explicit coefficients, but the AGM iteration is faster still. Bessel functions have rapidly convergent power series, but three-term recurrences simultaneously compute Bessel functions of different orders in a single for/next loop. And so on for many other special functions, as cataloged by Gil, Segura, and Temme and many earlier authors whom they cite. A library-writer is not a casual user, however, nor is the architect of a big time-marching code that needs to compute Bessel zeros a billion times per timestep. So a brief overview of special rootfinders is warranted.

17.2 ▪ What Is a Special Function?

The narrow definition is: A function is *special* if it appears in the *NIST Digital Library of Mathematical Functions* [284] or the five-volume *Bateman Manuscript Project* [162]. We shall refer to these as *canonical* special functions.

A broader definition is: A function is special if someone has written a paper about its power series, asymptotic expansions, singularities, and so on. The Blasius function is an important similarity solution in fluid mechanics, defined as the solution to

$$2f_{xxx} + f f_{xx} = 0, \qquad f(0) = f_x(0) = 0, \qquad f_x(\infty) = 1. \tag{17.1}$$

It is noncanonical, but definitely special, in that its power series, asymptotic expansions, singularity location and type, dilation group invariances, and rational Chebyshev series accurate to 15 decimal places have been published over the past century [86]. Similarly, the balanced oscillator gain function defined by

$$k(z) \equiv \frac{2}{z}\int_0^1 \tanh(9z)\sin(2\pi t)\sin(2\pi t)dt \tag{17.2}$$

is not a canonical special function either, but Boyd and Visser have derived power series and asymptotic approximations to the inverse, a blend of these that is a never-failing Newton's initialization (NFNI) [108] plus a pair of Chebyshev series that approximate the inverse $z(k)$ to eight decimal places over its entire range. The significance of noncanonical special functions is that for these functions, perturbation series and valuable formulas and insights are only a few clicks away.

George Carrier, who was a rare double electee to both the National Academy of Science and the National Academy of Engineering, advised his students that there were three ways to learn:

Precept 2 (Carrier's Three Modes of Information Acquisition). 1. *Run up and down the hallway, asking, until you find a human source. [BEST]*

2. *Scour the library and look it up in a book or journal.*
3. *Work it out yourself. [WORST]*

Repeating previously published research yourself is the best, not worst, in terms of deeply learning a subject, but it is the worst in terms of time expenditure. Albert Migliori's *bon mot*: "Six months in the lab can save you a day in the library" is equally true if "in the lab" is replaced by "deriving special methods" [260].

Valuable background, such as NFNIs for iterative rootfinders, can be found in the literature for many noncanonical special functions.

17.3 ▪ Themes in Special Function Rootfinding

First, we shall summarize these themes and then elaborate on each in turn.

1. Perturbation methods have a high success rate.
2. Much asymptotic information is readily available at the *NIST Digital Library of Mathematical Functions.*
3. It is often possible to identify or invent an initialization or set of initializations such that Newton's iteration and/or contraction mapping *always converge.*
4. An always-convergent Newton's iteration or contraction mapping is often the sole algorithm needed for a given special function.
5. Inverse functions are important, and series reversion is a common strategy to find the inverses.
6. If a function satisfies a three-term recurrence relation (TTRR), its roots can be computed through the eigenvalues of a tridiagonal matrix.

The special functions are perturbatively special in two different ways. First, it is often possible to calculate power series or Chebyshev or inverse power coefficients in *closed form* or by a recurrence with explicit coefficients. Second, a vast assortment of perturbation series, often including series for zeros, are cataloged in the *NIST Digital Library of Mathematical Functions* and similar reference books [284, 1, 162]. However, Chapters 14 and 15 here show that perturbation series are a *general* tool, not at all restricted to Bessel functions, hypergeometric functions, and the "NIST Digital Library Canon" of special functions.

Truncated perturbation series give only approximate answers, but often these approximations lie safely within the convergence interval of Newton's iteration. Table 17.1 is an example. The zeros of the Airy function have an asymptotic expansion that becomes increasingly accurate as the index k increases. However, using just the zeroth-order term in the asymptotic series (third column) is an initialization that is accurate to 0.018 even for the zero of smallest magnitude, $k = 1$. The third Newton correction for $k = 1$ is only -6.7×10^{-17}, and the errors slowly become even smaller as the root index k increases; higher-precision arithmetic was required to compute this correction accurately. Errors decrease slowly with increasing k. In other words, the result of two Newton's iterations, initialized with the leading order of the asymptotic series, is accurate to full machine precision! Despite its great simplicity and low expense, CPR is unnecessary here. Every component of CPR except Newton-polishing can be replaced by

$$x_{k,asy} \sim -\{(3/8)\pi(4k-1)\}^{2/3}\left\{1+O(k^{-2})\right\}. \tag{17.3}$$

As an alternative to Newton's that does not require df/dx, the iteration

$$x^{(n+1)} = g(x^{(n)}) \tag{17.4}$$

Table 17.1. *Airy roots: asymptotic initialization of Newton's iteration.*

k	Exact x_k	$-((3/8)\pi(4k-1))^{2/3}$	$\delta^{(1)}$	$\delta^{(2)}$	$\delta^{(3)}$
1	−2.3381	−2.3203	−0.0179	0.0000044	-6.7×10^{-17}
2	−4.0879	−4.0818	−0.0061	3.2×10^{-7}	-4.3×10^{-20}
3	−5.5206	−5.5172	−0.0034	7.2×10^{-8}	-6.9×10^{-22}
4	−6.7867	−6.7844	−0.0022	2.6×10^{-8}	-3.9×10^{-23}

is so popular that it has three names: *successive substitution*, *fixed point iteration*, and *contraction Mapping*. (Any equation of the form $f(x) = 0$ can be rearranged into a fixed point problem by defining $g(x) = f(x) + x$.) As explained in section 16.4, this iteration converges geometrically, rather than the digit-doubling rate of Newton's iteration, but contraction mapping will diverge, even very, very close to the fixed point, if $|dg/dx| > 1$ at the fixed point. It is often very difficult to predict when this convergence condition is satisfied.

For special functions, however, the vast array of identities in the handbooks often makes it possible to prove convergence. Gil, Segura, and Temme [195] do a couple of nice examples in their section 7.6 under the appellation of "global fixed point" methods.

Special function libraries omit the inverses for many important special functions. By another clause of the arithmurgical Murphy's law, your application is likely to need the inverse of one of those f-but-not-f^{-1} functions—hence Boyd's recent article on the inverse of the incomplete elliptic integral of the second kind [93]. We shall say no more of inverse functions here because this is the theme of the next chapter.

The zeros of all orthogonal polynomials can be found as eigenvalues of tridiagonal matrices whose elements are simple functions of the coefficients in the recurrence. Legendre spectral elements and Legendre spectral methods for solving PDEs require the zeros of $P_N(x)$ for various integers N, which are the essential interpolation points and quadrature abscissas. The widely used MATLAB library of Reddy and Weideman [365] calculates the needed roots through

```
 function r=legroots(N); n=[1:N-1];  d=n./sqrt(4*n.^2-1);

J=diag(d,1)+diag(d,-1);    r = sort(eig(sparse(J)));
```

(The comment lines have been omitted because the purpose in presenting the code is to emphasize its brevity!) The eigenvalues of a tridiagonal matrix method is a stenarithm, applicable only to zeros of orthogonal polynomials, but it is very fast, reliable, and easily programmed, and it is the recommended choice for its narrow range of applicability.

Gil, Segura, and Temme [195] show that the zeros of almost any set of functions that satisfy TTRRs can also be obtained from tridiagonal matrix eigenvalues. Bessel functions (of all varieties) are perhaps the most important nonpolynomial case, but Gil et al. [195] also do conical functions.

17.4 ▪ The Virtues and Failings of Special Methods for Special Functions

The authors of library software should usually employ special methods, such as Newton's iteration or contraction mapping with an always-convergent initialization. Stenarithms usually execute much faster than general-purpose rootfinders of the same reliability.

Casual users should generally use the Chebyshev-proxy rootfinder (CPR), perturbation series, and other general-purpose techniques. Remember that

Weeks of mathematical reading and computer programming can save milliseconds on a laptop.

The exceptions are (i) when the computation of $f(x) = 0$ must be done billions of times as part of a larger computation and (ii) when the application demands not only the roots of $f(x) = 0$ but also a thorough understanding of the function's properties. An always-convergent Newton's iteration initialization is a blessing not be rejected, if found effortlessly. An extended research program to find such an initialization is rarely worth the bother.

Chapter 18

Inverse Functions of One Unknown

"The purpose of this small work is to make available to students a reasonably complete self-contained treatment of inverse functions, a subject which is discussed in most calculus books with distressing brevity. The result, frequently, is inadequate understanding of the concept."

—William K. Smith, from the preface to his book *Inverse Functions* (1966) [335]

18.1 ▪ Introduction

When the solution to a single equation in a single unknown is the inverse of a function $g(x)$, that is, when the equation to be solved is

$$f(x;\lambda) \equiv g(x) - \lambda \quad \Leftrightarrow \quad x(\lambda) = g^{-1}(\lambda), \tag{18.1}$$

a great many simplifications and new strategies are possible, as summarized in Table 18.1.

In particular, much can be learned simply by examining the function to be inverted, $g(x)$. The values of k different real-valued branches of the inverse $g^{-1}(\lambda)$ are merely the places where the straight line $x \equiv \lambda$ intersects the graph of $g(x)$.

18.2 ▪ Polynomialization of Inverse Functions

18.2.1 ▪ Overview

Definition 18.1 (Parametric Interval Polynomialization (PIP)). *The conversion of a transcendental equation depending upon a parameter λ, $f(x,\lambda) = 0$, into the polynomial equation*

$$f_N(x,\lambda) = 0, \tag{18.2}$$

where the approximation is accurate on some finite interval $\lambda \in [a,b]$, is known as parametric interval polynomialization.

The adaptive Chebyshev-proxy algorithm of course replaces the transcendental equation by a polynomial equation, but generally this is true only at a single *point* in the parameter λ. When the unknown and parameter are mixed in intricate ways such as in $f = \exp(\lambda x)/(\lambda + x)$, one must compute a fresh Chebyshev expansion for each value of λ.

Table 18.1. *Relationships between properties of $g(x)$ and properties of its inverse $g^{-1}(\lambda)$.*

Properties of the Inverse	Corresponding Property of $g(x)$
domain of inverse, $\lambda \in [\lambda_{min}, \lambda_{max}]$	$\lambda_{min} = \min(g(x))$ or $= \max(g(x))$ or $= \lim_{x\to\pm\infty} g(x)$ $\lambda_{max} = \min(g(x))$ or $= \max(g(x))$ or $= \lim_{x\to\pm\infty} g(x)$
limit points of the inverse, $dx/d\lambda(\lambda_{limit}) = \infty$	λ_{limit} = max. or min. of $g(x)$
number of real-valued branches	$1 + N_{max/min}$, where $N_{max/min}$ is the number of local or global maxima or minima
domain of branch k, $\lambda \in [\lambda^k_{min}, \lambda^k_{max}]$	$\lambda_{min} = k$th minimum of $g(x)$, $\lambda_{max} = k$th maximum of $g(x)$

The key phrase in the definition is therefore "accurate on a finite interval" in the sense that we shall use it here. Parametric interval polynomialization (PIP) means that a single expansion suffices to compute roots at n_g points on an interval in λ, whereas the Chebyshev-proxy algorithm would compute n_g different Chebyshev series.

Inverse functions are easily polynomialized because the part that is dependent upon the unknown and the part that is dependent upon the parameter are *separate.*

Let the goal be a good approximation to the inverse on the interval $\lambda \in [\lambda_{min}, \lambda_{max}]$. Let

$$a \equiv g^{-1}(\lambda_{min}), \qquad b \equiv g^{-1}(\lambda_{max}). \tag{18.3}$$

Then the inverse function is polynomialized by calculating

$$g(x) \approx g_N \equiv \sum_{n=0}^{N} a_n T_n\left(\frac{2x-(b+a)}{b-a}\right), \qquad x \in [a,b]. \tag{18.4}$$

The inverse function $x(\lambda) = g^{-1}(\lambda)$ can be calculated for any $\lambda \in [\lambda_{min}, \lambda_{max}]$ by solving the polynomial equation

$$g_N(x) = \lambda. \tag{18.5}$$

18.2.2 ▪ Example: Inverse of the Complete Elliptic Integral of the First Kind for Modulus Less than One-Half

The goal is to solve

$$K(m) = \lambda, \qquad \lambda \in [\pi/2, \infty], \tag{18.6}$$

where m is the usual elliptic modulus and $K(m)$ is the usual complete elliptic integral of the first kind. Complications include a semi-infinite parametric range and the logarithmic singularity of the integral for m near one. In this section, we avoid these problems by polynomializing only for the restricted interval $x \in [0, 1/2]$, corresponding to $\lambda \in [\pi/2, K(1/2) = 1.8540746775]$. The harder, singular case of treating the entire real parameter space is deferred to [99].

It is convenient to replace m by the new unknown x, where $m = 1 - x$. The Chebyshev approximation to $g_N(x) \equiv K(m = 1 - x)$ on the restricted domain is

$$g_N = \sum_{n=0}^{N} a_n T_n(4x - 1), \qquad x \in [0, 1/2] \tag{18.7}$$

$$\approx 1.698886 + 0.140031\,(4x-1) + 0.013338\,T_2(4x-1) + 0.001579\,T_3(4x-1) + 0.000207\,T_4(4x-1) + 0.000029\,T_5(4x-1) + 0.000004\,T_6(4x-1). \tag{18.8}$$

Direct calculation shows that $N = 19$ is accurate to machine precision. The inverse can then be calculated by solving the polynomial equation $g_N(x) - \lambda = 0$ using the Chebyshev companion matrix algorithm described earlier.

Of course, there are other options, including NFNI for a simple fixed point iteration. Define the auxiliary parameter

$$\epsilon \equiv \frac{\lambda - \pi/2}{K(1/2) - \pi/2}. \tag{18.9}$$

Then

$$x^{(0)} \equiv 0.72136\epsilon - 0.29270\epsilon^2 + 0.0909101\epsilon^3 - 0.02445996\epsilon^4. \tag{18.10}$$

Numerically, the following initialization/iteration combination never failed for $\lambda \in [\pi/2, K(1/2)]$:

$$x^{(0)} \equiv 0.72136\epsilon - 0.29270\epsilon^2 + 0.0909101\epsilon^3 - 0.02445996\epsilon^4,$$
$$x^{(n+1)} = x^{(n)} - (K(x^{(n)}) - \lambda)/0.566), \tag{18.11}$$

where the quasi-Newton's iteration was adopted to avoid computation of derivatives of the elliptic integral. This NFNI was derived using regular perturbation theory (Chapter 14).

18.3 ▪ Graphing the Inverse of a Single-Valued Function: MAG-Reversed-Axes

When the inverse function $x(\lambda)$ is defined as the inverse of a single-valued function $g(x)$, i.e., as the solution to $\lambda - g(x) = 0$ for some $g(x)$ that is known in simple explicit form, it is trivial to graph the inverse. One merely plots $g(x)$ on the interesting range and then interchanges the x- and λ-axes. This strategy is a special case of the MAG (make-a-graph) strategy discussed in the previous chapter.

Figs. 18.1 and 18.3 are typical.

In discrete form, one chooses a set of points x_j in the *unknown* and then evaluates

$$\lambda_j = g(x_j). \tag{18.12}$$

For example, to graphically determine the arcsine function, one can evaluate

$$\lambda_j = \sin(x_j). \tag{18.13}$$

The plot of the points (λ_j, x_j)—with parameter λ as the *horizontal* axis—is then a plot of the arcsine function, $x = \arcsin(\lambda)$.

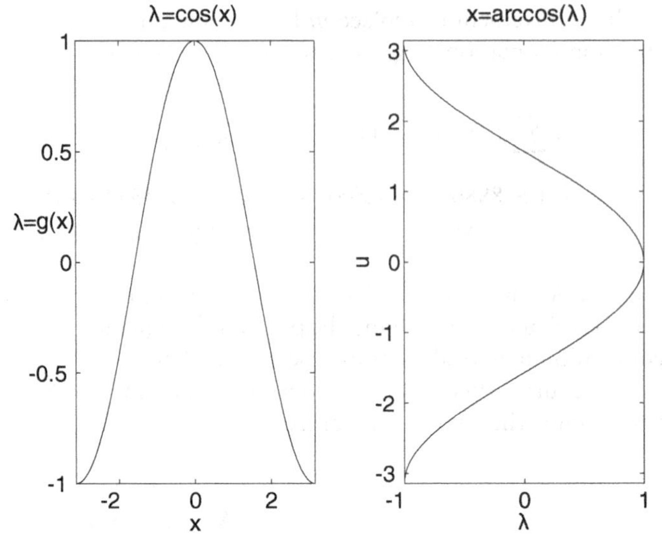

Figure 18.1. *The cosine function (left) and its inverse (right). The two plots are identical except the second was made from the first merely by reversing the axes.*

A more elaborate example is furnished by the solitary waves of the Camassa–Holm equation. Camassa, Holm, and Hyman have shown that the traveling waves of this PDE satisfy an ODE that can be formally integrated twice and then solved implicitly by the method of separation of variables [120]. The word "implicit" means that the solitary wave $u(X)$ is defined as the solution of the algebraic equation, depending upon the spatial coordinate X as a parameter,

$$X = -\left\{ \log(1+v) - \log(v-1) - \rho \log\left(\frac{v-\rho}{v+\rho}\right)\right\}, \tag{18.14}$$

where c is the phase speed, $\rho = \sqrt{c/(c-2x)}$, $v = \sqrt{(c-u)/(c-2x-u)}$, and k is an additional parameter. Solving for $u(X)$ is obviously rather a mess; Camassa, Holm, and Hyman note that for small x, the perturbation series in x involves logarithms in x as well as powers. However, because the inverse function $X(u)$ is given *explicitly* — it is only $u(X)$ that is implicit—it is trivial to graph the solitary wave by specifying a range of points in u, evaluating the corresponding X, and then connecting the points as shown in Fig. 18.2.

One obvious complication with MAG-Reversed-Axes is that the desired range in λ is known while the range in u is unknown. However, it is known for the Camassa–Holm solitary waves that $u \in [0, 1-2x]$. In other situations, one may have to experiment by guessing a large interval in x and then omitting from the plot all values of the parameter λ that are complex valued or otherwise unphysical.

This tactic of evaluating the inverse can also be turned into a numerical assault. To evaluate $x(\lambda)$ for an arbitrary choice of λ, one may first compute an array of pairs (λ_j, x_j) as in drawing the graph. One may then search and interpolate in this array of numbers—computer scientists call this *table look-up*—to generate an approximation to $x(\lambda)$. If this approximation is insufficiently accurate, it can be easily refined by Newton's iteration, successive substitution, or whatever.

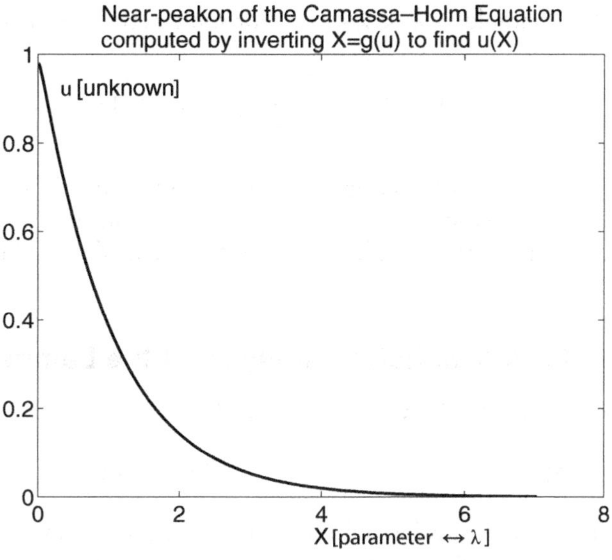

Figure 18.2. *This graph of $u(X)$ was generated by evaluating the inverse function, $X = g(u)$, which is known explicitly, and then plotting the points with X as the horizontal axis and u as the vertical axis. In the implicit solution, $X = g(u)$, the parameter is X (λ elsewhere in the book) and u (x elsewhere) is the unknown. The solitary waves depend on two other parameters, κ and c, which here are equal to $1/50$ and one. When $\kappa = 0$, the function $u(X)$ has a discontinuous slope at $X = 0$ for all c, a so-called peakon. The slope discontinuity of a* corner wave *has a bad effect on numerical methods for solving differential equations, so the implicit, algebraic solution is very useful for Camassa–Holm solitary waves. The soliton is symmetric about $X = 0$, so it is sufficient to graph it (and solve the algebraic equation) only for positive X.*

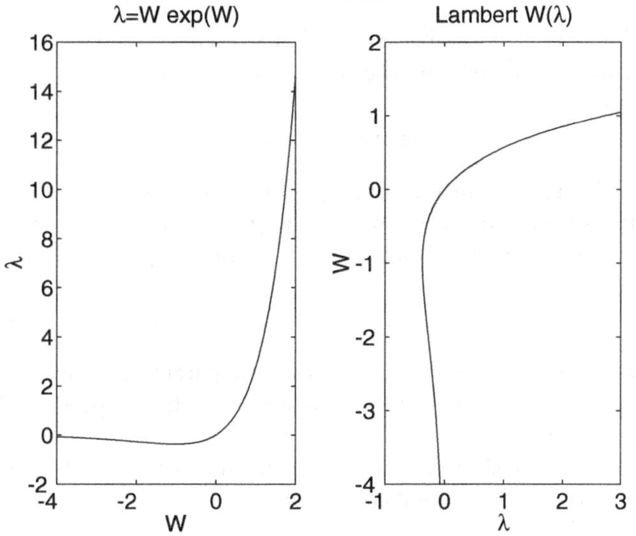

Figure 18.3. *The Lambert W-function is defined as the inverse of the function $g(W) = W\exp(W)$. The left graph is a plot of $g(W)$, which is easy to make because $g(W)$ is a single-valued function with an explicit analytical form. The plot on the right is of the transcendental $W(\lambda) \equiv g^{-}(\lambda)$ for which no simple analytical form is known. The two plots are identical except the second was made from the first merely by reversing the axes.*

For example, if $\lambda_j < \lambda < \lambda_{j+1}$ for some j, then linear interpolation gives

$$x(\lambda) \approx \frac{1}{\lambda_{j+1} - \lambda_j} \left\{ \left(x_{j+1} - x_j \right) \lambda + x_j \lambda_{j+1} - x_{j+1} \lambda_j \right\}. \tag{18.15}$$

Solving this for a better approximation to the root is just the secant iteration. Higher-order interpolation gives a more accurate initialization, but if the look-up table is sufficiently dense, the secant iteration, perhaps refined by additional secant iterations, is quite satisfactory.

18.3.1 ▪ Example: A Qualitative Analysis of the Lambert W-Function

The W-function $W(\lambda)$ is the inverse of

$$g(x) \equiv x \exp(x) \qquad \longleftrightarrow \qquad W(\lambda) = \text{root of} f(x, \lambda) = x \exp(x) - \lambda. \tag{18.16}$$

First, note that this has a finite minimum $\lambda_{min} = \min_{x \in [-\infty, \infty]} = -\exp(-1)$ but a maximum only at infinity. Therefore, $W(\lambda)$ is real valued only when the domain of λ (range of $g(x)$) is

$$\lambda \in [-\exp(-1), \infty]. \tag{18.17}$$

Second, there is only a single finite minimum or maximum of $g(x)$, that is, $x_{min} = -\exp(-1)$, where $g(x_{min}) = -1$. Therefore, there are two real-valued branches, which meet at the limit point $(-\exp(-1), -1)$.

18.4 ▪ Inverse Functions: Perturbation Methods and Lagrange's Method for Reversion of Series

The general perturbation methods discussed in Chapter 14 can be specialized to perturbatively determine the inverse of a function $g(x)$. This is equivalent to finding the roots of

$$f(x; \lambda) \equiv g(x) - \lambda \qquad \Leftrightarrow \qquad x(\lambda) = g^{-1}(\lambda). \tag{18.18}$$

There are multiple strategies for perturbative inversion because one can make different assumptions about where the perturbation parameter appears. In this section, we assume that $g(x)$ is a *near-linear* function, regardless of whether x is large or small, and write

$$g(x) = x + \epsilon p(x), \qquad \epsilon \ll 1, \tag{18.19}$$

where $p(x)$ is arbitrary and ϵ is a small parameter. Lagrange gave a formula for determining the coefficients of the expansion of $x(\lambda; \epsilon)$ in powers of ϵ, which is called Lagrange's reversion of series.

Lagrange showed that for equations written in this particular form, it is possible to write down the general coefficient:

$$\begin{aligned} g^{-1}(\lambda) = x(\lambda) &= \lambda + \sum_{j=1}^{\infty} \frac{\epsilon^j}{j!} \frac{d^{j-1}\left([p(\lambda)]^j \right)}{d\lambda^{j-1}} \qquad (18.20) \\ &= \lambda + \epsilon p(\lambda) + \epsilon^2 p(\lambda) \frac{dp(\lambda)}{d\lambda} + \epsilon^3 \left\{ p(\lambda) \left(\frac{dp(\lambda)}{d\lambda} \right)^2 + \frac{1}{2} [p(\lambda)]^2 \frac{d^2 p(\lambda)}{d\lambda^2} \right\} + \cdots . \end{aligned}$$

Another form of the Lagrange expansion, expressed in terms of $g(x)$ (and without the explicit appearance of the perturbation parameter ϵ defined above), is [1, p. 14]

$$x = x_0 + \frac{\lambda - \lambda_0}{dg/dx(x_0)} + \sum_{j=2}^{\infty} \frac{(\lambda - \lambda_0)^j}{j!} \left\{ \frac{d^{j-1}}{dx^{j-1}} \left(\frac{x - x_0}{g(x) - \lambda_0} \right)^j \right\} \Bigg|_{x=x_0} \tag{18.21}$$

for some arbitrarily chosen x_0, where $\lambda_0 \equiv g(x_0)$.

18.5 ▪ Inverse Functions: Power Series in *x* and λ

Lagrange's formula is powerful because it is only a series in powers of $\lambda - \lambda_0$; it does not require a power series of $g(x)$ in powers of $(x - x_0)$. However, this is also a weakness. When $|\lambda - \lambda_0|$ is small, (18.21) implies that $x - x_0 \approx (\lambda - \lambda_0)/dg/dx(x_0)$, which means that $x - x_0$ is also small. It follows that when we expand in powers of λ, we shall usually be in a region of the x-λ plane where $g(x)$ has a convergent power series.

A useful alternative approach is therefore to analyze what happens when $g(x)$ is explicitly, rather than implicitly, expanded in a series of powers of $(x - x_0)$. Like the Lagrange formula, this requires that $g(x)$ be analytic in a neighborhood of x_0. It excludes functions like

$$g(x) = \tanh(x) \exp(-1/x^{1/3}) \tag{18.22}$$

when x_0 is zero because then $g(x)$ is not analytic at the expansion point. The dependence of x on λ for this example is a complicated mixture of logarithms and logarithms of logarithms, and $x(\lambda)$ cannot be expressed as a simple power series.

In this section, we shall assume that $dg/dx(0) \neq 0$; when the dependence of $g(x)$ on x is parabolic rather than linear in the neighborhood of x_0, we are expanding about a limit point, which requires a special treatment given in the next section.

18.5.1 ▪ First Case: *g*(*x*) Is Linear Near the Expansion Point

Define

$$\lambda_0 \equiv g(x_0), \tag{18.23}$$

$$\lambda = g(x) = \lambda_0 + g_1(x - x_0) + g_2(x - x_0)^2 + g_3(x - x_0)^3 + \cdots, \qquad |x| \ll 1. \tag{18.24}$$

It is convenient, both to simplify notation and for conceptual understanding, to define

$$\tilde{\lambda} \equiv \lambda - \lambda_0, \qquad \tilde{x} \equiv x - x_0, \qquad \tilde{g}(\tilde{x}) \equiv g(\tilde{x}) - \lambda_0. \tag{18.25}$$

We can then substitute

$$\tilde{x} = \frac{1}{g_1} \tilde{\lambda} + a_2 \tilde{\lambda}^2 + \cdots \tag{18.26}$$

into

$$\tilde{\lambda} = \tilde{g}(x) = g_1 \tilde{x} + g_2 \tilde{x}^2 + g_3 \tilde{x}^3 + \cdots, \qquad |\tilde{x}| \ll 1. \tag{18.27}$$

The complete Maple code (with the tildes omitted for programming simplicity) is so trivial we shall give it as Table 18.2.

Note that a_j depends only upon the first j power series coefficients of $g(x)$ since the $\tilde{x}^{j+1} \sim O(\tilde{\lambda}^{j+1})$ because $\tilde{x}$ and $\tilde{\lambda}$ are of the same order of smallness. We can therefore

Table 18.2. *Maple program: Perturbation series for the inverse function away from a limit point.*

```
# User must replace the next line with his/her own choice;
# of g(x) and N, and supply the choice of expansion point x0.;
g := sin(x) + cos(x) - 1 - x*x/2;     x0 := 0;  lambda0 := subs(x = x0,g);  N := 6;
x := xtilde + x0;
gtilde := g - lambda0;
 xtilde := sum(a[j]*lambdatilde**j,j = 1..N);
gs := series(gtilde - lambdatilde,lambdatilde,N + 2);
for j from 1 to N do
gg[j] := coeff(gs,lambdatilde,j);
a[j] := solve(gg[j],a[j]); od:                xtilde;
residualseries := series (gtilde - lambdatilde, lambdatilde, N + 2);
```

write symbolically

$$x = x_0 + \frac{1}{g_1}(\lambda-\lambda_0) - \frac{g_2}{g_1^3}(\lambda-\lambda_0)^2 + \frac{2g_2^2 - g_3 g_1}{g_1^5}(\lambda-\lambda_0)^3 + \frac{5g_3 g_2 g_1 - 5g_2^3 - g_4 g_1^2}{g_1^7}(\lambda-\lambda_0)^4$$
$$+ \frac{14g_2^4 - 21g_3 g_2^2 g_1 + 6g_4 g_2 g_1^2 + 3g_3^2 g_1^2 - g_5 g_1^3}{g_1^9}(\lambda-\lambda_0)^5 + \cdots. \quad (18.28)$$

This is in fact identical to the Lagrange reversion series (18.21). The only difference is that the output of tricky repeated applications of l'Hôpital's rule to evaluate the repeated derivatives, followed by taking the limit $x \to x_0$, has been rendered explicitly in (18.28). Lagrange's reversion generalizes easily to arbitrary functions of functions, not merely the inverse, but for the inverse, a humble matching of powers is much more efficient.

The Maple **"solve"** command is sufficiently intelligent to respond to the solution of an equation involving a power series by returning another power series. Thus, Robert Corless pointed out that Table 18.2 can be shortened to **restart; g:= sin(x)+cos(x)-1 - x*x/2; fgs:=series(g,x=Pi/4,13); xx:= solve(y=fgs,x)**. However, this abbreviated code is not easily implemented in a different computer language.

18.6 ▪ Expansion about a Limit Point

If $dg/dx(x_0) = 0$, then x_0 is a limit point where the inverse function $g^{-1}(\lambda)$ has infinite slope. We will identify the limit value of x by x_{lim} and define

$$\lambda_{lim} \equiv g(x_{lim}), \quad (18.29)$$

$$\tilde{\lambda} \equiv \lambda - \lambda_{lim}, \qquad \tilde{x} \equiv x - x_{lim}. \quad (18.30)$$

The problem is to invert the equation for $\tilde{x}(\tilde{\lambda})$:

$$\tilde{\lambda} = g_2\tilde{x}^2 + g_3\tilde{x}^3 + g_4\tilde{x}^4 + \cdots, \qquad |\tilde{\lambda}| \ll 1. \quad (18.31)$$

It is still possible to expand the inverse as a series, but the powers are now fractional powers of $\tilde{\lambda}^{1/2}$ and the expansion is a Puiseux series. Furthermore, the lowest-order approximation is two valued, with each branch generating a separate Puiseux series:

$$\tilde{x} \approx \pm\sqrt{\tilde{\lambda}/g_2}. \quad (18.32)$$

Table 18.3. *Maple program: Expansion of the inverse function about a limit point.*

```
# User must replace the next line with his/her own choice;
# of g(x) and N, and supply the limit point xlim;
  g := cos(x); xlim := 0; lambdalim := 1
  x := xtilde + xlim; gtilde := g - lambdalim;   N := 6;
# Solves gtilde(xtilde(epsilon)) - lambdatilde = 0; xtilde(lambdatilde) = inverse(gtilde);
# lowest order is g2*epsilon**2 - lambdatilde = 0; We use this to define;
# the auxiliary perturbaiion parameter epsilon in terms of lambdatilde;
   xtilde := sum(a[j]*epsilon**j,j = 1..N);   a[1] := 1;
  gs := series(gtilde,epsilon,N + 2); g2 := coeff(gs,epsilon,2); gs := gs - g2*epsilon**2;
  for j from 2 to N do gg[j] := coeff(gs,epsilon,j + 1); a[j] := solve(gg[j],a[j]); od:
  epsilon := sqrt(lambdatilde/g2); x;
  xtilde2 := sum( (-1)**jj * a[jj]*epsilon**jj,jj = 1..N); # second branch;
```

Note that real-valued branches are possible only when $\tilde{\lambda}/g_2 \geq 0$, that is, the real branches occur for

$$\begin{aligned} &\lambda \geq \lambda_{limit}, \qquad g_2 > 0, \\ &\lambda \leq \lambda_{limit}, \qquad g_2 < 0. \end{aligned} \tag{18.33}$$

It is convenient to define the new parameter

$$\epsilon \equiv \sqrt{\tilde{\lambda}/g_2} \tag{18.34}$$

because now the Puiseux series in half-integral powers of $\tilde{\lambda}$ become ordinary power series in ϵ.

Each of the two solutions of the lowest-order approximation (18.32) generates a separate power series in ϵ that is a series in λ. The program in Table 18.2 needs to be slightly modified, as shown in Table 18.3. The latter code calculates only one of the two branches; the second branch is obtained by flipping the signs of the odd coefficients of ϵ, as in the last line of the table.

Note that limit point expansions were discussed in greater generality in section 14.13.

18.7 ▪ Robust Numerical Inversion

A reliable but expensive method is to expand $g(x)$ as a Chebyshev series or several such series on nonoverlapping subintervals that span the desired domain and then apply polynomial rootfinders. The only disadvantage is expense, and this is not an issue unless one needs to evaluate $g^{-1}(\lambda)$ for a billion values of λ.

A major reason for applying a reliable but costly method is that simple, one-at-a-time iterations are *not* reliable unless one has trustworthy *oracles* that can guide these local algorithms so that no roots are missed (Chapter 19). Inverse functions are special in that it is possible to perform a preliminary analysis to deduce information that is usually unavailable for a general $f(x)$ as illustrated in Fig. 18.4.

The first step is to compute all the zeros $r_j^{(1)}$ of the first derivative of $g(x)$. Next, calculate the second derivative of $g(x)$ at each of the first derivative roots. Then

$$g_{xx}(r_j^{(1)}) < 0 \Rightarrow \text{local maximum}, \tag{18.35}$$

$$g_{xx}(r_j^{(1)}) = 0 \Rightarrow \text{double root of first derivative}, \tag{18.36}$$

$$g_{xx}(r_j^{(1)}) > 0 \Rightarrow \text{local minimum}. \tag{18.37}$$

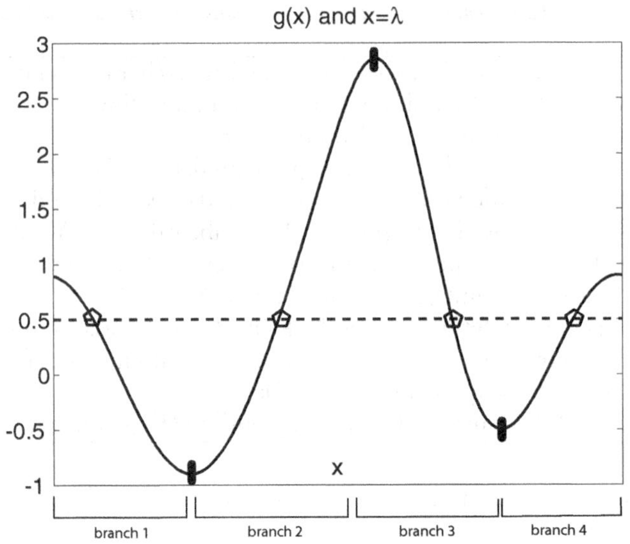

Figure 18.4. *Schematic of a function $g(x)$. The values of its inverse at an argument of λ are the points where $x=\lambda$ (dashed) intersects the graph of $g(x)$ (solid). These points, marked by pentagons, are the roots of $f(x,\lambda)=g(x)-\lambda$. Different branches are bounded by maxima and minima of $g(x)$, which are marked by thick vertical bars.*

Suppose for simplicity that $g(x)$ has only maxima and minima. Each branch of the inverse is then bounded by one maximum and one minimum of $g(x)$ in the sense that the kth branch exists only when

$$\lambda \in [\lambda_{min}^{(k)}, \lambda_{max}^{(k)}] \qquad \text{or} \qquad \lambda \in [\lambda_{\max}^{(\mathrm{k})}, \lambda_{\min}^{(\mathrm{k})}]. \tag{18.38}$$

Furthermore, there is one and only one solution to the equation $f(x,\lambda) \equiv g(x)-\lambda$ when λ is in the indicated range, and the value of that root, $x_{*,k}$, must lie in the range

$$x_{*,k} \in [x_{min,k}, x_{max,k}]. \tag{18.39}$$

The bisection method will reliably find one root on an interval $x \in [a,b]$ provided that $f(a)f(b)<0$. Its curse is that it will miss additional roots if present on the interval. When the bisection method is applied to compute the kth branch of the inverse, we are guaranteed that there will be one and only one root on the specified interval. Thus, bisection can be applied with complete reliability to find the branches of the inverse of the function provided that we do the preliminary analysis of calculating the maxima and minima of the function being inverted, $g(x)$. One caveat: if a branch has a semi-infinite interval, which is common, then one must derive asymptotic approximations or the like to impose a finite bound on the zero-containing interval for a given parameter value λ, and then and only then can bisection be applied to compute $g^{-1}(\lambda)$ for that branch.

18.8 ▪ Illustrations

The Bratu problem, $w_{yy}+\lambda\exp(w)=0, w(\pm 1)=0,\ \lambda \geq 0$, has the exact solution $w(y;\lambda)=\log\left(x^2\operatorname{sech}^2(x\sqrt{\lambda}y)\right)$ provided that

$$\lambda = 2\operatorname{arccosh}^2(x)/x^2. \tag{18.40}$$

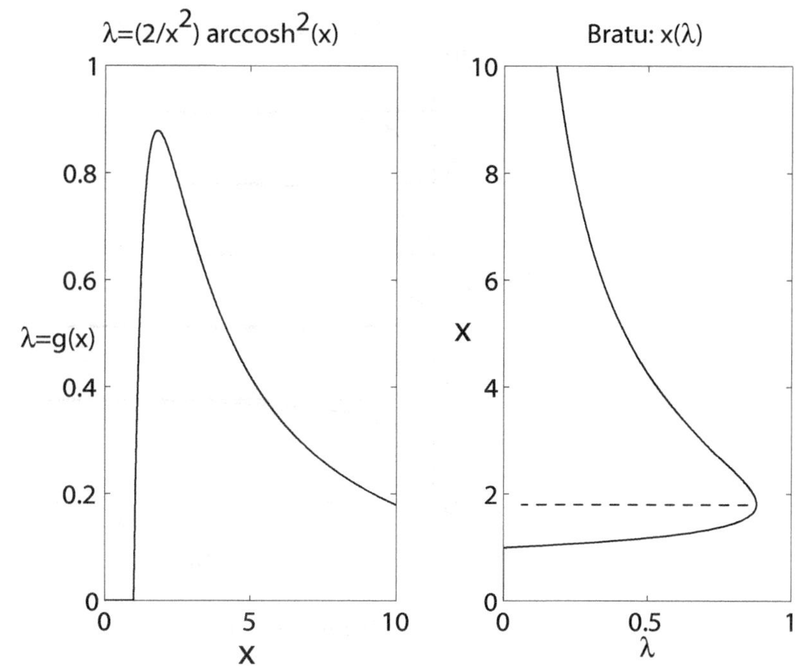

Figure 18.5. *Eigenrelation for the one-dimensional planar Bratu problem. The dashed line separates the upper and lower branches at $x = x_{lim}$.*

The graph of $\lambda(x)$ shows a single maximum at $\lambda_{limit} = 0.87845767978129030l5$ at $x_{max} = 1.8101705806989772753$. Thus, there are two real-valued branches and both span the same range $\lambda \in [0, \lambda_{limit}]$. The function and its inverse are shown in Fig. 18.5.

Because the eigenparameter λ is meaningful only when positive, the range of the lower branch is restricted to $x \geq 1$; this lower limit must be treated as a boundary even though it is not a minimum or maximum of $\lambda(x)$. The branches are

$$x \in [1.81017058, \infty] \text{ [upper]}, \quad x \in [1, 1.81017058] \text{ [lower]}. \tag{18.41}$$

To generate a perturbative approximation near the limit point, it is helpful to define

$$g(x) \equiv 2\text{arccosh}^2(x)/x^2. \tag{18.42}$$

The limit point is not at $x = 0$, but rather at $x_{lim} = 1.8101705806989772753$. Furthermore, $\lambda_{lim} = 0.8784576788$. The perturbation program in Table 18.3 yields

$$x \approx 1.810170581 \pm 1.609882406\sqrt{-\tilde{\lambda}} + 1.507563549\tilde{\lambda} \pm 1.456445498(-\tilde{\lambda})^{3/2} + 1.435317944\tilde{\lambda}^2 \pm 1.433667611(-\tilde{\lambda})^{5/2} + 1.445853195\tilde{\lambda}^3, \tag{18.43}$$

where the expansion variable is

$$\tilde{\lambda} \equiv \lambda - \lambda_{lim} = \lambda - 0.8784576788 \tag{18.44}$$

and where the upper signs give the upper branch and the same series, taking the lower signs, gives the lower branch.

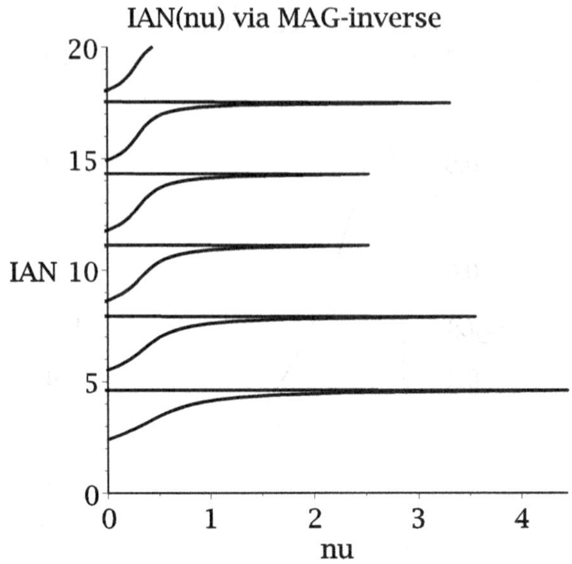

Figure 18.6. $\mathscr{IAN}(\nu)$.

To Newton-polish the perturbative approximation, Newton's iteration is

$$x^{n+1} = x^n - \frac{2\operatorname{arccosh}^2(x^n)/(x^n)^2}{4\left(\operatorname{arccosh}(x^n)/(x^n)^2\right)\left(1/\sqrt{(x^n)^2-1} - \operatorname{arccosh}(x^n)/x^n\right)}, \quad (18.45)$$

where the superscript n denotes the iteration level.

18.8.1 ▪ $\mathscr{IAN}(\nu)$

In section 15.6.2, a function $\mathscr{IAN}(\nu) = x(\nu)$ was defined implicitly as the inverse of $g(x)$, where

$$\nu = g(x) \equiv -\left\{ \frac{1}{(\pi/2)Y_0(x)/J_0(x) - \{\log(x) + (\log(1/2)+\gamma)\}} \right\}. \quad (18.46)$$

The first step is apply MAG-Reversed-Axes: we evaluate $\nu = g(x)$ for many discrete values of x but then make the ν-axis horizontal and the x-axis vertical. Fig. 18.6 shows that the function qualitatively resembles the arctangent. There is an infinite number of branches. In applications, $\nu \in [0, \infty]$ is relevant.

When $x = j_{0,m}$, the mth root of J_0, the denominator of $f(x)$ has a pole, implying that $\nu = 0$. Thus, the lower limit of the mth branch of $\mathscr{IAN}(\nu)$ is $\mathscr{IAN}(0) = j_{0,m}$, the mth root of $J_0(x)$.

The upper limit is the mth root $k_{0,m}$ of

$$f_0(x) \equiv (\pi/2)Y_0(x) - \{\log(x) + (\log(1/2)+\gamma)\}J_0(x). \quad (18.47)$$

Therefore, the first two steps in computing the mth branch are to compute its limits, $x \in [j_{0,m}, k_{0,m}]$. The lowest branch, for example, is $x \in [2.40482555, 4.69152764]$.

The Chebyshev-proxy rootfinder (CPR) has the weakness that in general a new Chebyshev series is required for each value of the parameter. Here, however, we can write

$$f_0(x) + \frac{1}{\nu} f_1(x) = 0, \tag{18.48}$$

$$f_0 \equiv (\pi/2) Y_0(x) - \{\log(x) + (\log(1/2) + \gamma)\} J_0(x), \tag{18.49}$$

$$f_1(x) \equiv J_0(x), \tag{18.50}$$

where both f_0 and f_1 are *independent* of the parameter ν. Thus, a Chebyshev expansion of these two completely polynomializes the problem.

Defining the argument $z \equiv 0.874622x - 3.103313$ so that

$$z \in [-1, 1] \longleftrightarrow x \in [2.40482555, 4.69152764],$$

the quadratic polynomialization is

$$0.5508 - 0.4158 T_1(z(x)) - 0.1547 T_2(z(x)) \\ + \frac{1}{\nu}(-0.2582 - 0.1355 T_1(z(x)) + 0.1251 T_2(z(x))) = 0. \tag{18.51}$$

The relative error in the root $x_{quad}(\nu)$ to this quadratic polynomial, $|(x_* - x_{quad})/x_*|$, is less than or equal to 0.0077 for the first branch for all $\nu \in [0, \infty]$. The maximum relative errors in fourth-degree and sixth-degree Chebyshev approximations are 0.00046 and 6.0×10^{-7}.

With just three interpolation points, the quadratic interpolant is easy to write down explicitly. Introduce the abbreviations $a = j_{0,m}$ and $b = k_{0,m}$ as the limits of the mth branch. Then $x \in [a, b]$ is transformed to and from the Chebyshev argument $z \in [-1, 1]$ by

$$x = ((b-a)/2) z + (b+a)/2 \qquad \longleftrightarrow \qquad z = (2x - (b+a))/(b-a). \tag{18.52}$$

The interpolant is

$$f^{quad}(z) = Az^2 + Bz + C, \tag{18.53}$$

$$A = (1/2)\, g_0 - g_1 + (1/2) g_2, \tag{18.54}$$

$$B = -(1/2)\, g_0 + (1/2)\, g_2, \tag{18.55}$$

$$C = g_1, \tag{18.56}$$

$$g_0 = f(s), \qquad g_1 = f([b+a]/2), \qquad g_2 = f(a). \tag{18.57}$$

As always, only the root with $z_* \in [-1, 1]$ is relevant.

Numerical experiments showed the quadratic interpolant is an NFNI for all branches up to $m = 10000$ and all $\nu \in [0, \infty]$.

Chapter 19

Oracles: Theorems and Algorithms for Determining the Existence, Nonexistence, and Number of Zeros

"What is the elementary particle of understanding? It's the 'clariton.' The problem is that there are also anticlaritons that come tomorrow and annihilate the one you had today. We have lots of them in this subject, claritons followed by anticlaritons."
—Sir Michael Berry, FRS [303, p. 177]

19.1 ▪ Rootfinding Oracles: Predicting the Number of Roots in a Region or Interval

Definition 19.1 (Rootfinding Oracle). *A theorem or algorithm is said to be an* oracle *when it calculates (or bounds) the number of roots within a given interval or specified region of the complex plane* [219], *or provides similar qualitative information about zeros. An oracle that predicts the roots within an interval is an* interval oracle.

For a general $f(x)$, interval oracles include

1. the derivative-bound root-free oracle,
2. Bolzano's criterion (Theorem 9.15),
3. Bolzano's criterion with one-signed derivative,
4. the Kronecker–Picard integral, and
5. the contour integral oracle.

A contour integral provides an oracle for a region in the complex plane, furnishing the heart of the Delves–Lyness algorithm, explained in Chapter 5. The first four on the list will be discussed in this chapter.

For the special case that $f(x)$ is a polynomial, there are additional oracles, including Descartes's rule of signs, Sturm's criterion, and the Fourier–Budan rule.

Definition 19.2 (Isolated Root). *A root is said to be* isolated *when a rootfinding oracle predicts that there is only a single root within a given region or interval.*

When a zero is isolated, a method derived directly from Bolzano's oracle is very popular and reliable.

Table 19.1. *Bisection: Matlab program.*

```
function root = bisect(f,intervalbounds);
% Finds a root of f(x) on the interval x ∈ [intervalbounds(1) intervalbounds(2)]
itermax = 50; % maximum number of iterations
epsilon = 1.E–12; % desired error tolerance
x1 = intervalbounds(1); x2 = intervalbounds(2);
ff = feval(f,x1); fmiddle = feval(f,x2);
if (ff*fmiddle >= 0), disp('Oops! Root not bracketed in interval'); pause; end
if ff < 0, root = x1; deltax = x2 – x1; else root = x2; deltax = x1 – x2; end
for j = 1:itermax
deltax = deltax*0.5; xmiddle = root + deltax; fmiddle = feval(f,xmiddle);
if(fmiddle <= 0), root = xmiddle; end % if
if ((abs(deltax) < epsilon) | (fmiddle == 0)), break; end % if
end % j loop
```

19.2 ▪ Bisection and the Bolzano Oracle

Bolzano's oracle is the basis for a geometrically converging algorithm. The oracle is that if $f(a)f(b) < 0$, then there is at least 1 root in $x \in [a,b]$. Evaluate $f(x)$ at the midpoint of the interval. If $f(a)f(a+(b-a)/2) < 0$, then there is at least 1 root in the left half of the original interval; otherwise, there is at least 1 root in the right half of the original interval. Repeat these steps, replacing $[a,b]$ by $[a+(b-a)/2,b]$ or $[a,a+(b-a)/2]$ as appropriate. With each iteration, a root is isolated within an interval that is only half the size of its predecessor. After 30 to 40 iterations, the root will be isolated within an interval whose size is equal to machine precision.

Bisection has two strengths. The first is inevitability: if one can find an interval such that $f(x)$ has opposite sign at the 2 endpoints, then bisection is guaranteed to converge to a root. The second is that bisection is very easy to program. Table 19.1 is a MATLAB implementation in just 12 lines.

Bisection has serious flaws, too. Bolzano's oracle cannot detect multiple roots or roots of even order. For example, the function $f(x) \equiv x^2$ has a double root at the origin, but because $f(x) \geq 0$ everywhere, the oracle's rootfinding condition, $\text{sign}(f(a)) \neq \text{sign}(f(b))$, is never satisfied. Bisection *fails* for *double zeros* and higher-order zeros of even multiplicity.

Even if all roots are simple, bisection will find only a single root even if there is more than one zero on $x \in [a,b]$. A missed root, either simple or higher order, is a constant peril with bisection. However, it is sometimes possible to prove that an interval contains only a single simple root; in that case, bisection is completely safe and reliable, as discussed below.

19.3 ▪ Bolzano's Oracle with One-Signed Derivative: Bisection Made Trustworthy

A common numerical strategy for solving differential equations is to use a change of coordinate, $x = g(y)$, where the grid is uniform in y but highly nonuniform in the physical coordinate x. These coordinate changes are always constructed so that the derivative is nonnegative everywhere on the interval because such mappings are singular and multivalued otherwise. To calculate the grid points in the physical coordinate requires repeatedly solving

$$f(x;y) \equiv g(x) - y, \tag{19.1}$$

where $y = y_j$, each of the uniformly spaced grid points in the computational coordinate. However, all reasonable mappings (to avoid multivaluedness on the domain) have $dg/dx = df/dx$ *positive* everywhere on the domain of interest, $x \in [a,b]$.

This is important because when df/dx is zero free on $x \in [a,b]$, Bolzano's oracle and bisection can both be strengthened. The proof invokes Rolle's theorem, which is useful in its own right.

19.3.1 ▪ Rolle's Theorem and Its Implications

Theorem 19.3 (Rolle's Theorem). *Between two consecutive zeros of a real analytic function $f(x)$, the first derivative df/dx must vanish an* odd *number of times and hence* at least once *(Householder [213, p. 6]).*

Although we shall not repeat the textbook proof, Fig. 19.1 is almost as good.

When there is just a single root of the derivative between a pair of zeros of f, the Newton flow will converge to one or the other of the roots when started at any point on the interval between the roots except the zero of df/dx, which is a stationary point of the Newton flow.

However, when there are *three* zeros of the derivative between a pair of roots, one of those roots of df/dx will be a local *minimum* of f^2. The Newton flow, a strongly underrelaxed Newton's iteration, and a steepest descent iteration will all march downward to stall at the local minimum when the initial point is anywhere on the interval bounded by the two local maxima of $f^2(x)$, as shown by the dashed horizontal line in the right panel of Fig. 19.1.

19.3.2 ▪ Bolzano Oracle with One-Signed First Derivative

Theorem 19.4 (Bolzano Oracle with One-Signed First Derivative). *Suppose that the first derivative (slope) of $f(x)$ is provably one signed everywhere on an interval $x \in [a,b]$:*

$$\left|\frac{df}{dx}\right| > 0 \qquad \forall x \in [a,b]. \tag{19.2}$$

Then

1. *$f(x)$ has at most one root on the interval, and this root must be simple.*
2. *If $f(a)f(b) < 0$, then there is precisely one simple root in $x \in [a,b]$.*
3. *If $f(a)f(b) > 0$, then there are no zeros on $x \in [a,b]$.*

Proof: A double root by definition is a point where $df/dx = 0$, so all roots on the interval must be simple. The condition obviously implies that df/dx is zero free on the interval. Rolle's theorem requires that df/dx vanish at least once between any two roots and thus allows a maximum of one root. The Bolzano oracle, if it is also satisfied, allows a minimum of one root, removing all uncertainty. ■

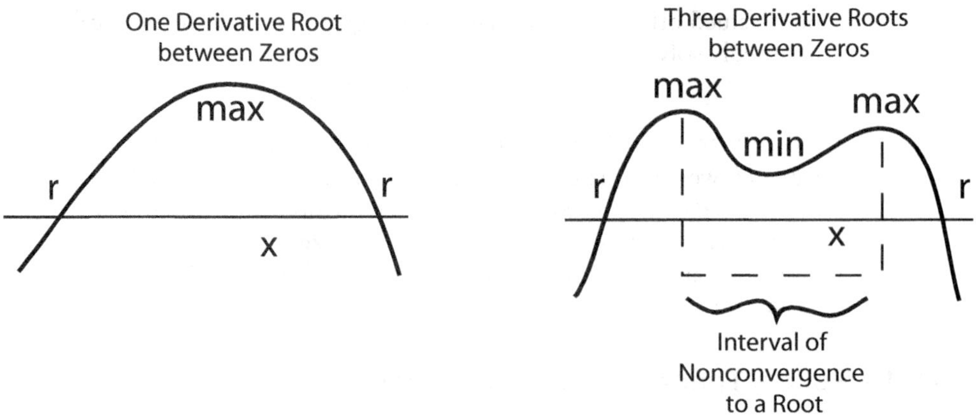

Figure 19.1. *Schematic of Rolle's theorem: Between any two real roots of a real function must lie an* odd *number of roots of the derivative of that function. Here, the roots of f are marked with "r," while the simple zeros of the first derivative of f, which are also local maxima or minima of f, are marked with "max" or "min." If initialized anywhere in the middle subinterval in the right graph, a steepest descent algorithm, which strives to find a zero of $f(x)$ by minimizing $f^2(x)$, would converge only to the local minimum of $f^2(x)$ at the bottom of the valley.*

Corollary 19.5 (Bisection Convergence). *When $f(a)f(b) < 0$ and df/dx is one signed on $x \in [a,b]$, bisection always converges and converges to all the roots (one) on the interval $x \in [a,b]$.*

19.4 ▪ Curvature Oracles

Theorem 19.6 (One-Signed Curvature Oracle). *Suppose that the second derivative (curvature) of a twice-differentiable $f(x)$ is provably one signed everywhere on an interval $x \in [a,b]$:*

$$\left|\frac{d^2 f}{dx^2}\right| > 0 \qquad \forall x \in [a,b]. \tag{19.3}$$

Then

1. *There are at most two simple roots or a single double root on the interval.*
2. *If $f(a)f(b) < 0$, then there is precisely one simple root in $x \in [a,b]$.*

Proof: Positive definiteness of the second derivative implies that the first derivative has at most one root on the interval by the first proposition of Theorem 19.4, applied to df/dx instead of f. However, Rolle's theorem applied to the derivative asserts that d^2f/dx^2 must vanish once between any two real roots of the first derivative. Therefore, df/dx can have at most *one* root on $[a,b]$. Rolle's theorem applied to f then allows at most two roots on the interval. If there is a double root x_*, then this is the one point on the interval where $df/dx = 0$. In the vicinity of a double root, $f(x)$ is same signed both slightly to the left and slightly to the right of x_*. In order to have a second root, $|f|$ will have to rise to a local maximum and then descend to zero again. This requires a second zero of df/dx on the interval, which we have already seen is excluded. Therefore, if there is a double root on the interval, there cannot be another zero of f on the interval. The Bolzano oracle by itself tells us only that there is an odd number of roots on the interval, but the second derivative one-signed property allows at most two roots. The only odd number less than or equal to two is one. Neither oracle by itself excludes a single double root. However, we have already noted that in the vicinity of a double zero, f has the same sign on both sides of the double root. It is possible for $f(a)f(b)<0$ only if there is at least one more simple root on the interval, and we have already seen that there can be only one root when both conditions are true. ■

19.5 ▪ Kronecker–Picard Oracle: The Number of Simple Roots on an Interval

In the late nineteenth century, building on earlier work of Kronecker, E. Picard showed the following.

Theorem 19.7 (Kronecker–Picard Integral Oracle: The Number of Real Roots on an Interval). *The number of roots on a real interval is $n_r(a,b)$, where*

$$n_r(a,b)=-\frac{\gamma}{\pi}\int_a^b \frac{f(u)f''(u)-f'(u)^2}{f^2+\gamma^2(f')^2}\,du+\frac{1}{\pi}\arctan\left(\gamma\frac{f'(b)}{f(b)}\right)-\frac{1}{\pi}\left(\gamma\frac{f'(a)}{f(a)}\right), \quad (19.4)$$

where γ is a small, positive constant. Picard showed that this expression is independent *of γ, so, in theory, any reasonable γ may be used.*

Fig. 19.2 shows how this oracle succeeds in practice for four different $f(x)$ when the integral is evaluated by numerical quadrature and the interval is chosen to be $a=-1$, $b=1$. When γ is tiny, the integrand has singularities just off the real axis, which creates large quadrature errors. These are visible as the curving portions of the graphs on the left-hand side of the diagram where γ is too small for the quadrature scheme (with a fixed number of quadrature points) to handle.

The bottom curve shows that multiple roots are reported the same as simple roots; x^2 has a single root of multiplicity two on $x\in[-1,1]$, and $n_r(-1,1)=1$. The good news is that the correct number of roots is reported; a rootfinding algorithm that stops after finding one root on the interval would terminate correctly. The bad news is that a double root is often numerically morphed into a pair of simples, separated by $O(\sqrt{\text{machine epsilon}})$, in violation of the oracle! Most iteration schemes converge much more slowly for double roots and root doublets than for simple zeros, and a truly optimal oracle would warn us.

The thick-dashed curve, second from the bottom, shows the numerically calculated value of the Kronecker–Picard integral for a function with two distinct but closely spaced roots. When $\gamma\in[0.0075,0.03]$, $n_r\approx 2$, the correct answer. For larger γ, however, the numerically computed n_r drifts downward. Computations with a larger number of grid points show that this downward drift, like the upward curl for $\gamma<0.0075$, is a quadrature error.

This and other experiments not illustrated show that one needs an increasingly accurate quadrature scheme to correctly count a pair of roots as the roots become closer and closer together. If the roots are separated by only 0.0004 instead of 0.004 as in the figure, then $n_r\approx 1$ (with the same 300 quadrature points). Thus, closely spaced roots are identified as a single root, where "closely spaced" is inversely proportional to the number of grid points of a (nonadaptive) quadrature scheme.

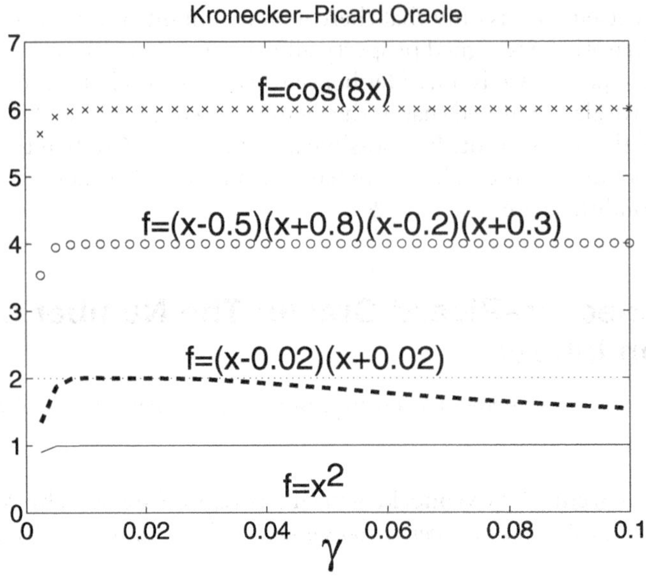

Figure 19.2. *Values of the Kronecker-Picard integral for the interval $x \in [-1, 1]$ for four different functions $f(x)$ as computed by Chebyshev quadrature with* 300 *abscissas.*

It follows that the Kronecker–Picard integral is a good but imperfect oracle. However, some advice is helpful:

1. Pick a moderately small γ, or better yet, choose several small γ and compare results. Both very small and very large γ give large quadrature errors.

2. Use a large number of grid points in a uniform quadrature, or else use an adaptive quadrature. The latter will place many points close to the near-singularities of the integrand that occur when $\gamma \ll 1$.

3. Note that multiple roots are *not* counted according to their multiplicity by the Kronecker–Picard integral.

4. A pair of very closely spaced roots will be reported as a single root unless the quadrature scheme is highly adaptive or uses a huge number of quadrature points.

Hoenders and Slump, Kavvadias and Vrahatis, and Kavvadias, Makri, and Vrahatis discuss the Kronecker–Picard integral in detail [211, 210, 220, 219]. Hoenders and Slump show that by computing additional integrals, one can identify multiple roots and their multiplicity [210].

19.6 ▪ Regional Oracles: Delves–Lyness Complex Plane Algorithm

19.6.1 ▪ Computing the Number of Roots

If (and only if) $f(z)$ is analytic in a region Ω of the complex z-plane, then the number of roots within the region is given exactly by a contour integral, which is therefore a *regional* rootfinding oracle).

Theorem 19.8 (Contour Integral for the Number of Roots within the Contour). *If a function $f(x)$ is analytic everywhere within a region Ω of the complex plane including its boundary $\partial\Omega$, then the number n_r of roots in the region is given exactly by*

$$n_r = \frac{1}{2\pi i} \oint \frac{f_z(z)}{f(z)} dz, \tag{19.5}$$

where roots are counted according to their multiplicity. That is, for example, when there is either a pair of simple roots or a single double root within Ω, then the contour integral evaluates to $n_r = 2$.

The practical difficulty is that it is often difficult to prove that $f(z)$ is analytic within a region. It is important to note that a simple *pole* within the contour of integration will contribute to the integral the same as a zero. Blind application of this oracle, with no knowledge of analyticity in Ω, is a really bad idea. This oracle *lies* if $f(z)$ has singularities within Ω.

Adaptive interpolation along a disk or ellipse can detect such interior-to-Ω singularities. Subdivision of the original domain into pieces can then identify regions of analyticity and other regions containing poles. This and the mechanics of contour integration are discussed in detail in Chapter 5.

19.7 ▪ Fourier Derivative Bounds on an Ordinary Polynomial

19.7.1 ▪ Equivalence of Polynomial and Trigonometric Cosine Polynomial Rootfinding Problems

It is much easier to bound derivatives of trigonometric polynomials than ordinary polynomials. It is for this reason that the following is important.

Theorem 19.9 (Polynomial/Fourier Cosine Equivalence). *The problem of finding the real roots x_j on $x \in [-1,1]$ of a polynomial $f_N(x)$ of degree N is equivalent to computing the zeros t_j on $t \in [0,\pi]$ of a trigonometric cosine polynomial*

$$g_N(t) \equiv f_N(\cos(t)) \tag{19.6}$$

$$= \sum_{n=0}^{N} a_n \cos(nt) \tag{19.7}$$

with the same coefficients as the Chebyshev series of $f_N(x)$,

$$f_N(x) = \sum_{n=0}^{N} a_n \cos(n)(x), \tag{19.8}$$

and the roots are related by

$$x_j = \cos(t_j) \qquad \forall j. \tag{19.9}$$

Proof: This is an immediate consequence of Chebyshev's identity, $T_n(\cos(t)) = \cos(nt)$. ■

19.8 ▪ Derivative Bounds on a Trigonometric Polynomial

If $f(x)$ is a polynomial, $f_N(x)$, then its derivatives can be bounded. Unfortunately,

$$|dT_n(x)| \leq n^2, \tag{19.10}$$

which escalates quadratically with degree. If we use the theorem of the previous section to convert the polynomial to the equivalent cosine polynomial, then much better bounds are possible.

Theorem 19.10 (Fourier Derivative Bounds). *Let $g_N(t)$ denote the trigonometric cosine polynomial*

$$g_N(t) = \sum_{n=0}^{N} a_n \cos(nt). \tag{19.11}$$

Then

$$\left| \frac{d g_N}{dt}(t) \right| \leq \sum_{n=1}^{N} n |a_n|, \tag{19.12}$$

$$\left| \frac{d^2 g_N}{dt^2}(t) \right| \leq \sum_{n=1}^{N} n^2 |a_n|. \tag{19.13}$$

Proof: The kth derivative of $\cos(nt)$ is bounded for all real t by n^k by explicitly taking the derivative and recalling that the sine and the cosine are bounded by one. Applying this bound to each individual term and recalling that $|a_n \cos(nt)| \leq |a_n||\cos(nt)|$ proves the theorem. ■

19.9 ▪ Bounds on Derivatives of Polynomials

19.9.1 ▪ The Derivative-Bound Zero-Free-Interval Oracle

Theorem 19.11 (Derivative-Bound Oracle). *Let a be a point. Suppose that for a given grid interval h we can prove*

$$\max_{x \in [a-h, a+h]} \left| \frac{df}{dx}(x) \right| < h f(a), \tag{19.14}$$

where $f(x)$ is continuously differentiable on the interval but otherwise arbitrary.

Then f has no zeros on $x \in [a-h, a+h]$.

Proof: The proof is given in Fig. 19.3 and its caption. ■

When $f(x)$ is a polynomial, or is *approximated* by a polynomial, then a precise bound on df/dx is given by the following.

Theorem 19.12 (Markoff's Derivative Bound). *If $P(x)$ is a polynomial of degree N whose absolute value on $x \in [-1,1]$ is bounded by one,*

$$|P(x)| \leq 1 \,\forall x \in [-1,1], \tag{19.15}$$

then

$$|dP/dx| \leq N^2 \,\forall\, x \in [-1,1]. \tag{19.16}$$

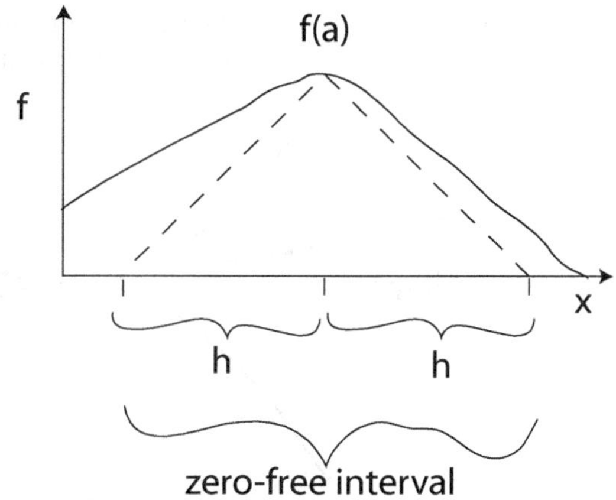

Figure 19.3. *Derivative-bound root-free-interval oracle. If a is the center of an interval of width $2h$ and one can bound the derivative f_x on the interval such that* $\max|f_x| \leq h f(a)$*, then the interval is provably free of zeros. The dashed lines show straight lines with slopes* $\pm\max|f_x|$*; if these bounding lines cannot cross zero when descending from $f(a)$, then neither can $f(x)$.*

This result is the best possible, with equality when $P = T_N(x)$. Under the same conditions,

$$\left|\frac{d^k P}{dx^k}\right| \leq \prod_{j=0}^{k-1} \frac{N^2 - j^2}{2j-1}, \tag{19.17}$$

which is again the best possible, with the upper bound attained only when $P = T_N$ [349].

If $g_N(t)$ is a trigonometric polynomial of degree N and

$$|g_N(t)| \leq 1 \;\forall\, t \in [-\pi, \pi], \tag{19.18}$$

then

$$\left|\frac{d^k g_N(t)}{dt^k}\right| \leq N^k \;\forall t \in [-\pi, \pi]. \tag{19.19}$$

19.9.2 ▪ Bounded Curvature Zero-Free Theorems

Suppose that we have computed a bound C on the curvature (second derivative) of $f(x)$, perhaps by using the theorems of the previous two sections if $f(x)$ is a polynomial or a cosine polynomial. Note that we obtain tighter bounds if we convert a polynomial $f(x)$ to a cosine polynomial as in the previous section; the kth derivative of $T_n(x)$ is $O(n^{2k})$, but the t-derivatives of $\cos(nt) \equiv T_n(\cos(t))$ are bounded by n^k.

Let us examine a small interval $x \in [a, a+h]$. If $f(a)f(a+h) < 0$, then by the Bolzano oracle there is at least one root on the interval.

The more subtle case is when $f(a)$ and $f(a+h)$ have the same sign. Roots are still possible, but there must be an *even* number of roots on the interval. If the number of roots is not zero when $f(a)f(a+h) > 0$, then Rolle's theorem asserts that there must be at least one point x_m on the interval where $df/dx(x_m) = 0$.

We can then prove the following lemma.

Lemma 19.13 (Curvature-Bounded Movement from a Stationary Point). *Let x_m be a stationary point of $f(x)$, that is, a point where $df/dx(x_m) = 0$. Suppose that the second*

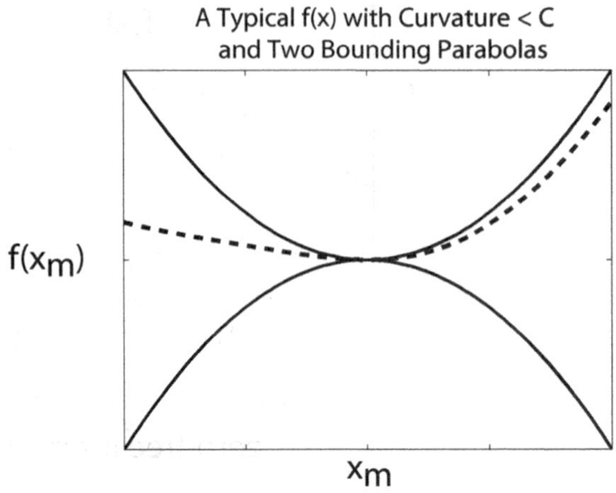

Figure 19.4. *If an analytic function $f(x)$ has the properties that (i) $df/dx = 0$ at $x = x_m$ and (ii) $|d^2f/dx^2(x)| \leq C$ on some interval including $x = x_m$, then on that interval $f(x)$ (dashed) is confined within the two parabolas $P^{\pm} = f(x_m) \pm (C/2)(x - x_m)^2$.*

derivative of f is bounded in magnitude by C on some interval including the stationary point. Then everywhere within the interval where the bound holds,

$$|f(x) - f(x_m)| \leq \frac{1}{2} C |x - x_m|^2. \tag{19.20}$$

Equivalently, let $P(x; C, x_m)$ denote the parabola through $f(x_m)$ with curvature C, and let P^- denote a similar parabola but with negative curvature, $-C$. The lemma asserts that the graph of $f(x)$ is bounded by the parabolas shown in Fig. 19.4.

Proof: By elementary calculus,

$$f_x(x) = \int_{x_m}^{x} f_{xx}(y)\, dy. \tag{19.21}$$

If we have a bound on the magnitude of the second derivative C, it follows that

$$|f_x(x)| \leq C |x - x_m|, \tag{19.22}$$

where we have invoked the elementary calculus bound $\int_a^b g(x)dx \leq (b-a) \max_{x \in [a,b]} |g(x)|$ for any $g(x)$. A classic calculus identity is

$$f(x) - f(x_m) = \int_{x_m}^{x} f_x(y)\, dy. \tag{19.23}$$

If we have a bound on the first derivative of the form of (19.22), then, again invoking the bound on the integral,

$$|f(x) - f(x_m)| \leq \int_{x_m}^{x} C |y - x_m|\, dy \tag{19.24}$$

$$\leq \frac{1}{2} C |x - x_m|^2 \tag{19.25}$$

by explicit integration of a linear function of x. ■

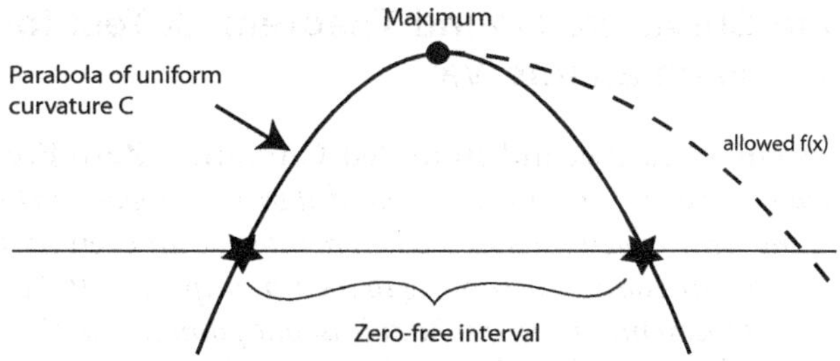

Figure 19.5. *The solid curve is the parabola of maximum curvature that passes through the maximum of $f(x)$. If $f(x)$ (dashed) has a curvature of smaller magnitude than the parabola, its graph will bend less dramatically toward the axis, and it will cross the axis beyond the interval bounded by the roots of the parabola (stars). It follows that the open interval between the roots of the parabola is provably free of zeros of $f(x)$.*

There are two ways to exploit this lemma, described in turn in the next two subsections. The first uses only knowledge of $f(x)$ at the stationary point. The second employs values of $f(x)$ at two other points bracketing the stationary point.

19.9.3 ▪ The First Bounded Curvature Theorem

Theorem 19.14 (First Bounded Curvature Zero-Free Interval). *Suppose that the second derivative (curvature) is bounded by C. Let x_m denote a maximum or minimum of $f(x)$. Then the interval*

$$x \in \left(x_m - \sqrt{\frac{2|f(x_m)|}{C}},\, x_m + \sqrt{\frac{2|f(x_m)|}{C}} \right) \tag{19.26}$$

is free of zeros.

Proof: At a minimum or a maximum, the slope of $f(x)$ is zero. In order to cross the $f = 0$ axis, the function must curve. If the curvature is very gentle, then $f(x)$ will cross the axis only at a great distance from the stationary point x_m. The greater the curvature, the more rapidly $f(x)$ can develop a steep slope that will take it back across the axis. Given the existence of a bound on the curvature, the best $f(x)$ can do is to curve at the maximum curvature. This shape is the parabola

$$P = f(x_m) - \frac{1}{2}C(x - x_m)^2, \tag{19.27}$$

which has roots at

$$x_{*,parab} = x_m \pm \sqrt{\frac{2|f(x_m)|}{C}}. \tag{19.28}$$

If a function curves less than this parabola, its roots will be farther from x_m than the roots of the parabola. See Fig. 19.5. ■

19.9.4 ▪ Second Curvature Bound Theorem: A Test for the Absence of Zeros on an Interval

Theorem 19.15 (Second Bounded Curvature Zero-Free Interval). *Suppose that the second derivative (curvature) is bounded by C. Suppose further that $f(a)$ and $f(a+h)$ have the same sign, which implies that there must be an* even *number of roots on $x \in [a, a+h]$, when roots are counted according to their multiplicity. If the number of roots is not zero, then by Rolle's theorem, there must be at least one point x_m on the interval where $df/dx(x_m) = 0$.*
Then the interval $x \in [a, a+h]$ is free of zeros if

$$\min|f(a)|, |f(a+h)|) \geq \frac{C}{8} h^2. \tag{19.29}$$

Proof: If there are pairs of roots on the interval, then x_m can be chosen so that sign($f(x_m)$) is opposite to that of $f(a)$ and $f(a+h)$. Then $|f(a) - f(x_m)| \geq |f(a)|$ and similarly at the other endpoint. The lemma above bounds $|f(a) - f(x_m)|$. Thus, we must have simultaneously

$$|f(a)| < |f(a) - f(x_m)| \leq \frac{1}{2} C |a - x_m|^2, \tag{19.30}$$

$$|f(a+h)| < |f(a+h) - f(x_m)| \leq \frac{1}{2} C |a + h - x_m|^2, \tag{19.31}$$

where the right inequalities follow from the previous theorem. These can be written as

$$\min(|f(a)|, |f(a+h)|) \leq \frac{1}{2} C |a - x_m|^2, \tag{19.32}$$

$$\min(|f(a)|, |f(a+h)|) \leq \frac{1}{2} C |a + h - x_m|^2. \tag{19.33}$$

Because the length of the interval is h, it must be true that the smaller of $|a - x_m|$ and $|a + h - x_m|$ must be less than or equal to half the width of the interval, $h/2$. It follows that for $f(x)$ to cross the $f = 0$ axis,

$$\min(|f(a)|, |f(a+h)|) \leq \min\left(\frac{1}{2} C |a - x_m|^2, \frac{1}{2} C |a + h - x_m|^2\right) \tag{19.34}$$

demands

$$\min(|f(a)|, |f(a+h)|) \leq \frac{C}{8} h^2. \tag{19.35}$$

If this inequality is not satisfied, then we have a contradiction and the assumption that there are zeros on the interval is false. ■

19.10 ▪ Field–Morgan Oracle for Quadratic Polynomials

If $f(a)f(b) > 0$, the Bolzano oracle is agnostic; it is impossible for general $f(x)$ to determine whether an interval $[a, b]$ contains a zero or not. Field and Morgan observed that if the function is a quadratic polynomial, one can be much more precise.

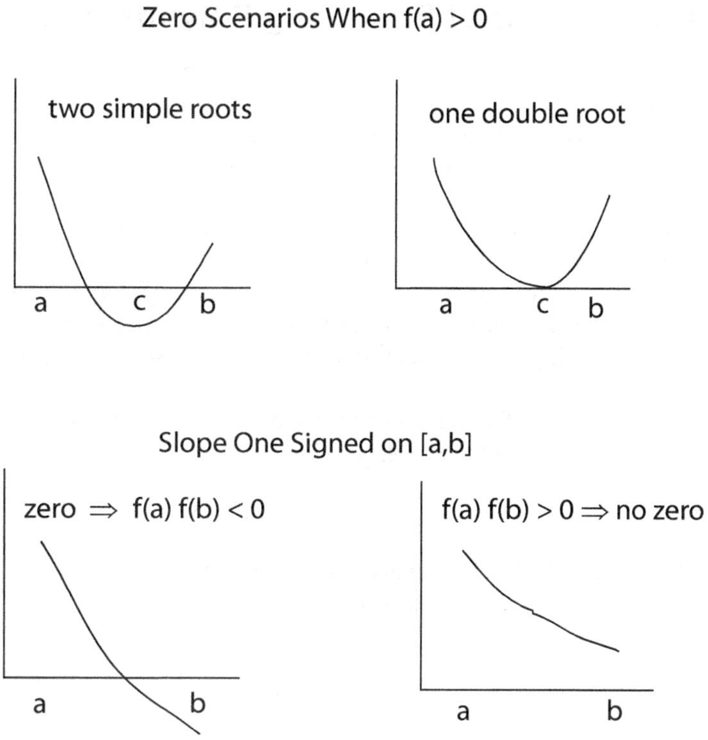

Figure 19.6. *Schematic of univariate quadratic polynomial oracle.*

Theorem 19.16 (Univariate Quadratic Polynomial/Field–Morgan Oracle). *If $f(x)$ is a quadratic polynomial, $f(x) = Ax^2 + Bx + C$, and $f(a)f(b) > 0$, then f has a zero on the interval $x \in [a, b]$ if and only if* (i) *the unique zero of df/dx, call it $c = -C/(2A)$, is on the interval and* (ii) $f(a)f(c) \leq 0$ [170].

Proof: If $f(a)$ and $f(b)$ are of the same sign, then zeros between these endpoints are possible only if $f(x)$ crosses the axis, levels off at $x = c$ where by definition $df/dx = 0$, and then crosses the axis a second time. If the derivative root is outside $[a, b]$, then df/dx cannot change sign on $x \in [a, b]$. If $f(x)$ crosses the axis with a slope everywhere one signed on the interval, then $f(a)f(b) < 0$, contradicting an assumption of the theorem. Fig. 19.6 illustrates the possibilities. ■

The multidimensional generalization is immediate because the gradient vector must be zero at a maximum or minimum of the function, and for a quadratic polynomial, the components of its gradient vector are always *linear* polynomials. There is then a unique point where the gradient is zero.

Theorem 19.17 (Multivariate Quadratic Polynomial/Field–Morgan Oracle). *Suppose that $f(\vec{x})$ has the same sign at all vertices of a simplex in d dimensions. Suppose further that $f(\vec{x})$ is a quadratic polynomial. Let $\vec{c}$ denote the unique point where the gradient of the function is zero. The function has a zero isosurface within the simplex or box if and only if* (i) *$\vec{c}$ lies within the simplex or box and* (ii) $f(\vec{c})f(\vec{a}) \leq 0$, *where $\vec{a}$ is any of the vertices of the simplex or box* [170].

19.11 ▪ Applications of Enclosing Polynomials

Because the cost of finding the roots of a companion matrix grows as the *cube* of the matrix, it is obviously desirable to identify situations when truncating a polynomial or trigonometric polynomial of degree N to one of lower degree yields reliable information about the roots. Is it possible, then, to compute the roots of a low-degree trigonometric polynomial $f_M(x)$ and then safely use local, iterative methods like Newton's iteration to obtain the zeros of a higher-degree truncation $f_N(x)$ where $N > M$? The theorems below provide an affirmative answer.

These theorems use the concept of *enclosing polynomials*, as defined below. Since these enclosure ideas are very general and not restricted to polynomials or trigonometric polynomials per se, we shall state and prove these theorems in general form.

Definition 19.18 (Enclosing Functions). *Two functions $\mathcal{B}^{\pm}(x)$ are said to* enclose *a function $f(x)$ on an interval $x \in [a,b]$ if*

$$\mathcal{B}^{-}(x) \leq f(x) \leq \mathcal{B}^{+}(x). \tag{19.36}$$

Definition 19.19 (Enclosure Tolerance). *If $\mathcal{B}^{\pm}(x)$ enclose a function $f(x)$ on an interval $x \in [a,b]$, then their* enclosure tolerance *is*

$$\epsilon \equiv \frac{1}{2} \max_{x\in[a,b]} |\mathcal{B}^{+}(x) - \mathcal{B}^{-}(x)|. \tag{19.37}$$

We shall show below that it is in fact easy to determine ϵ when the enclosing functions are polynomials, truncated from higher-degree polynomials or infinite spectral series.

The enclosing functions are useful if and only if it is somehow easier to compute their roots than those of $f(x)$. Let us define $\{r_j^{\pm}\}$ to denote the set of the roots of $\mathcal{B}^{\pm}$ on $[a,b]$ plus the endpoints a and b. Let $I_k^{\pm}$ denote the subintervals that extend from one of these points to the next point to its right; each subinterval is thus free of a root of $\mathcal{B}^{\pm}$, but either endpoint may be a root. It is then easy to determine subintervals where the enclosing functions are positive or negative: if $\mathcal{B}^{-}$ is positive at the midpoint of I_j^{-} for some j, then $\mathcal{B}^{-}$ will be positive everywhere on that subinterval since the interval has been constructed to have roots only at its endpoints. This makes the following theorem useful.

Theorem 19.20 (Zero-Free Intervals from Enclosing Functions). *Let $\mathcal{B}^{\pm}(x)$ be enclosing functions for $f(x)$ on an interval $x \in [a,b]$. Then every subinterval in $[a,b]$ where*

$$\mathcal{B}^{-}(x) > 0 \tag{19.38}$$

and also every subinterval where

$$\mathcal{B}^{+}(x) < 0 \tag{19.39}$$

is free of zeros of the enclosed function $f(t)$.

Proof: The definition of enclosing function requires that on all subintervals where $\mathcal{B}^{-}(x) > 0$, $f(t)$ must be larger still and therefore positive also. Similarly, $f(t)$ must be negative wherever its upper bound $\mathcal{B}^{+}(t)$ is negative, too. ■

Enclosing functions can also identify intervals that must contain zeros, as expressed by the following theorem.

Theorem 19.21 (Enclosing Functions: Bolzano Zero-Containing Intervals). *Let $\mathcal{B}^{\pm}(x)$ be enclosing functions for $f(x)$ on an interval $t \in [a,b]$. Let I be the interval $t \in [\alpha, \beta]$, which is contained within $[a,b]$. Suppose that either*

$$B^{-}(\alpha) > 0 \qquad and \qquad B^{+}(\beta) < 0 \tag{19.40}$$

or

$$B^{+}(\alpha) < 0 \qquad and \qquad B^{-}(\beta) > 0. \tag{19.41}$$

Then the function $f(t)$ has at least one zero *on the interval $x \in [\alpha, \beta]$.*

Proof: If B^- is positive at $t = \alpha$, then $f(\alpha)$ must be positive, too, since $f(\alpha) \geq B^-(\alpha)$. Similarly, if B^+ is negative at $x = \beta$, then, since $f(x) \leq B^+(x)$, $f(\beta) \leq 0$. The usual Bolzano oracle then implies that $f(x)$ must have at least one root on the interval. ∎

This theorem is a generalization of Bolzano's theorem, which predicts at least one zero on an interval when $f(t)$ changes sign between the endpoints of the interval. The enclosing functions Bolzano oracle is imperfect because $f(t)$ may have more than one zero.

Another deficiency of this theorem is that α and β are chosen at random. It would be desirable to more systematically divide the interval $[a,b]$ into zero-free subintervals and zero-containing subintervals. A theorem similar to Bolzano's, but identifying the largest subinterval that can contain one or more zeros, is given by the following.

Theorem 19.22 (Enclosing Functions: Largest Interval with a Root). *Let $\mathcal{B}^{\pm}(x)$ be enclosing functions for $f(x)$ on an interval $t \in [a,b]$. Suppose that I is a subinterval of $[a,b]$ whose endpoints are a simple zero of $\mathcal{B}^+$ and a simple zero of $\mathcal{B}^-$ such that*

$$\mathcal{B}^{-} \leq 0, \qquad x \in I, \tag{19.42}$$

and

$$\mathcal{B}^{+} \geq 0, \qquad x \in I. \tag{19.43}$$

Then there must be at least one zero of $f(x)$ on the interval I.

Proof: Suppose that the simple zero of $\mathcal{B}^-$ is at the left endpoint α of the interval I, as illustrated in Fig. 19.7. Then to the left of α, both $\mathcal{B}^-$ and $\mathcal{B}^+$ are positive and $f(\alpha) \geq 0$. Similarly, at the right endpoint of the interval, which we shall dub β, both $\mathcal{B}^-$ and $\mathcal{B}^+$ are negative and therefore $f(\beta) \leq 0$. It follows by Bolzano's theorem that $f(x)$ must have at least one zero on the interval since it has either changed sign on the interval or is zero at one or both endpoints. The same argument applies if the simple zero of $\mathcal{B}^-$ is at the right endpoint while that of $\mathcal{B}^+$ is the left endpoint of I. ∎

Unfortunately, the figure shows that it is not possible to replace the vague assertion "at least one zero" by something more precise without imposing additional conditions. In the figure, for example, the bounding functions each have only a single root but $f(x)$ has *three*. It is possible for $f(x)$ to have two extra roots and still remain within ϵ of the bounding functions if the local maximum and minimum within the interval are very shallow, that is, less than ϵ in magnitude.

We can exclude such shallow minima and obtain a more precise theorem by introducing additional conditions.

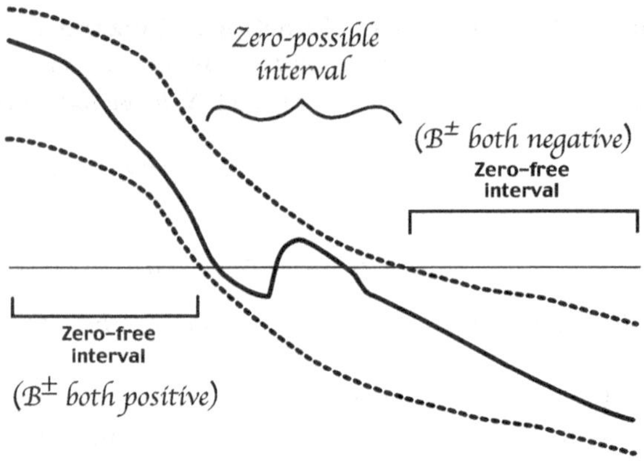

Figure 19.7. *Enclosing functions (dotted lines) for $f(x)$ (solid). Note that although the two bounding polynomials have only a* single *zero each on the interval, the enclosed function $f(x)$ has* three.

Theorem 19.23 (Enclosing Functions: Interval Containing Precisely One Zero). *Let $\mathcal{B}^{\pm}(x)$ be enclosing functions for $f(x)$. Suppose that I is a subinterval of $[a,b]$ whose endpoints are a simple zero of $\mathcal{B}^{+}$ and a simple zero of $\mathcal{B}^{-}$ such that*

$$\mathcal{B}^{-} \leq 0, \qquad x \in I, \tag{19.44}$$

and

$$\mathcal{B}^{+} \geq 0, \qquad x \in I. \tag{19.45}$$

Then the interval I contains precisely one *zero of $f(x)$ if any conditions are satisfied that guarantee that I is a zero-free interval for the* first derivative *of $f(t)$.*

Proof: From the previous theorem, we know that there is at least one zero on I. If df/dx cannot change sign, then $f(x)$ cannot descend through the real axis, reach a minimum, and then rise through a second root because this requires a change of the first derivative of f. ■

A wide variety of conditions are possible to show that the first derivative is zero free on an interval. The theorem below collects two variations on earlier theorems, modified to apply to df/dt rather than $f(t)$ itself.

Theorem 19.24 (Zero-Free First Derivative). 1. *Let $\mathcal{B}\mathcal{B}^{\pm}(x)$ be enclosing functions for $df/dx(x)$ on an interval $x \in [a,b]$. Then $df/dx(x)$ is free of derivatives on I if either or both of the following conditions are satisfied everywhere on the interval I:*

$$\mathcal{B}\mathcal{B}^{-}(x) > 0, \qquad \mathcal{B}\mathcal{B}^{+}(x) < 0. \tag{19.46}$$

2. *If f is a periodic function with Fourier coefficients a_n and b_n, B_2 is defined by*

$$B_2 \equiv \sum_{j=1}^{\infty} |j^2 a_j| + \sum_{j=1}^{\infty} |j^2 b_j|, \tag{19.47}$$

and τ denotes the midpoint of the interval $[a,b]$, then the first derivative of $f(x)$ has no zeros on $x \in [a,b]$ if

$$\left| \frac{df}{dx}(\tau) \right| \geq B_2 (b-a)/2. \tag{19.48}$$

Proof: The propositions are merely restatements of Theorem 19.22 and Theorem 4.9 with the substitution of df/dx for f. ∎

We have restated these two conditions on the first derivative to make a couple of points. The first is that it is not necessary to compute the roots of a second set of enclosing functions; because the zero-possible intervals are very narrow for a function with well-separated simple roots, other criteria such as that depending on a bound for the second derivative (as well as other conditions not stated here) can be used to show that the first derivative is free of zeros, thus tightening up the Bolzano oracle to exclude multiple roots on the interval.

19.12 ▪ Polynomial Enclosing Functions

It is easy to construct enclosing functions that are polynomials, assuming that the Chebyshev or Fourier coefficients of $f(x)$ can be computed up to some finite order by using the following theorem.

Theorem 19.25 (Trigonometric Polynomial Enclosing Function). *Let $f(t)$ be a function with a Fourier series, either finite or infinite,*

$$f(t) \equiv \sum_{j=0}^{\infty} a_j \cos(jt) + \sum_{j=1}^{\infty} b_j \sin(jt). \tag{19.49}$$

Define the N-term truncation of the Fourier series to be

$$f_N(t) \equiv \sum_{j=0}^{N} a_j \cos(jt) + \sum_{j=1}^{M} b_j \sin(jt) \tag{19.50}$$

and the positive constant ϵ by

$$\epsilon \equiv \sum_{j=N+1}^{\infty} |a_j| + |b_j|. \tag{19.51}$$

Then the trigonometric polynomials of degree N

$$\mathcal{B}^{\pm}(t) \equiv \pm\epsilon + f_N(t) \tag{19.52}$$

are enclosing functions for $f(t)$.

Proof: It has long been known [76] that the error in a truncated Fourier series $f_N(t)$ is bounded, usually rather tightly, by the sum of the absolute values of all neglected coefficients, which here defines ϵ. Common to both $\mathcal{B}^+$ and $\mathcal{B}^-$ is the Fourier truncation of $f(t)$ of degree N. Since this truncation $f_N(t)$ cannot deviate from $f(t)$ by more than ϵ, it follows that the polynomials defined by taking the Fourier truncation and either adding or subtracting ϵ must be an upper and lower bound on $f(t)$, respectively. ∎

Example: The elliptic function $\mathrm{Dn}(x;p)$, differing from the usual dn by a multiplicative constant, has the Fourier series

$$\mathrm{Dn}(x;p) = \frac{1}{2} + \sum_{n=1}^{\infty} 2\frac{p^n}{1+p^{2n}} \cos(nx), \tag{19.53}$$

where $0 < p < 1$ is a constant. The error in truncating at the Nth term is bounded by

$$\epsilon(N) = \sum_{n=N+1}^{\infty} 2\frac{p^n}{1+p^{2n}} \leq \sum_{n=N+1}^{\infty} 2p^n = 2\frac{p^{N+1}}{1-p}. \tag{19.54}$$

Then the following functions are enclosing functions of degree N:

$$\mathcal{B}^{\pm}(x;p,N) = \pm 2\frac{p^{N+1}}{1-p} + \frac{1}{2} + \sum_{n=1}^{N} 2\frac{p^n}{1+p^{2n}}\cos(nx). \tag{19.55}$$

It follows that $\mathrm{Dn}(x;p)$ has no real zeros if we can demonstrate that $\mathcal{B}^{-}(x;N,p)$ is zero free for any N. For $p = 1/10$, for example,

$$\mathcal{B}^{-}(x;p=1/10,1) = \left\{-\frac{1}{45} + \frac{1}{2}\right\} + \frac{20}{101}\cos(x). \tag{19.56}$$

Since the constant is bigger than the sum of the absolute values of the higher coefficients (here just a single coefficient), it follows from Theorem 4.8 that $\mathcal{B}^{-}$ has no roots on the interval. Similarly, because the constant in $\mathcal{B}^{-}$ is positive, it follows that $\mathcal{B}^{-} > 0$ for all real x, and therefore $\mathrm{Dn}(x; p = 1/10)$ must similarly always be positive.

With a little more algebra, one can extend this argument to all $p \in [0, 0.2]$. By using enclosing polynomials of higher degree N, one can extend the argument to larger values of p.

The equivalent theorem for ordinary polynomials is virtually identical because of the Chebyshev polynomial/Fourier equivalence principle.

Theorem 19.26 (Polynomial Enclosing Function). *Let $f(x)$ be a function with the Chebyshev series*

$$f(t) \equiv \sum_{j=0}^{\infty} a_j T_j(x). \tag{19.57}$$

Define the Nth-degree truncation of the Chebyshev series to be

$$f_N(x) \equiv \sum_{j=0}^{N} a_j T_j(x) \tag{19.58}$$

and the positive constant ϵ by

$$\epsilon \equiv \sum_{j=N+1}^{\infty} |a_j|. \tag{19.59}$$

Then the polynomials of degree N,

$$\mathcal{B}^{\pm}(x) \equiv \pm\epsilon + f_N(x), \tag{19.60}$$

are enclosing functions for $f(x)$.

If $f(x)$ is a function with only simple zeros on the interval and if the enclosure tolerance ϵ is small, then the zero-possible intervals will be very small. The subintervals that flunk the zero-free test will be bounded by a root of $\mathcal{B}^{-}$ and the nearest root of $\mathcal{B}^{+}$. Now $\mathcal{B}^{\pm}(x) = \pm\epsilon + f_N(x)$. Let x_* denote a root of $f_N(x)$ and let S denote $dg/dx(x_*)$.

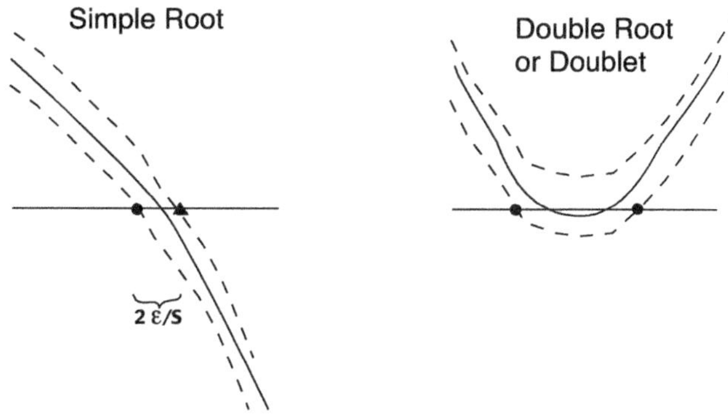

Figure 19.8. *Left: enclosing functions (dashed) at a simple root of $f(x)$ (solid); the roots of the lower and upper bounding functions, $\mathscr{B}^-$ and $\mathscr{B}^+$, are marked by the disk and triangle, respectively. Right: enclosing functions at a double root, or a pair of closely spaced roots (doublet) of $f(x)$. The two roots of $f(x)$ are flanked by two roots of the lower bounding function if $f(x)$ is concave upward at the double root/doublet. If $f(x)$ is concave downward, then the doublet will be enclosed by a pair of roots of the other enclosing function $\mathscr{B}^+$.*

Then Newton's iteration, which becomes exact in the limit $\epsilon \to 0$, predicts that roots of the enclosing functions are at

$$x^{\pm} \approx x_* \pm \epsilon/S. \tag{19.61}$$

Thus, for simple roots with enclosing polynomials constructed as above, the zero-possible intervals will have the small width of $2\epsilon/|S|$ at simple roots of $f(x)$, as shown in Fig. 19.8.

Recall that if $f(x)$ has two coincident roots, i.e., a double root, or a pair of simple roots that are very close together, then the double or root pair is a root doublet. This terminology, borrowed from atomic spectroscopy, is convenient because even a tiny perturbation of $f(t)$, either analytical or numerical, can split a double root into a pair of closely spaced simple roots, so the double root/close pair situations are *numerically* indistinguishable, at least without special tactics.

In contrast to a simple root that is well separated from neighboring roots, a doublet of $f(x)$ is flanked by roots of only one of the enclosing functions. Indeed, we may regard the presence of two neighboring roots of, say, $\mathscr{B}^-$ without an intervening root of the upper enclosing function $\mathscr{B}^+$ to be the definition of a doublet/double root situation. If $f(x) \sim \delta + C(x - x_d)^2$ near the double root or doublet, where $|\delta| \ll 1$ and $C > 0$, then it is easy to see that the two roots of $\mathscr{B}^-$ will be approximately at

$$x_{\pm}^{-} \approx x_d \pm \sqrt{\epsilon/C}. \tag{19.62}$$

The zero-possible interval will be still be small if ϵ is small.

Part VII

Bivariate Systems

Chapter 20

Two Equations in Two Unknowns

"Keep writing the same paper until you are sure someone is getting your message."
—Kirk Bryan (1929–) Advice to H. W. Menard while Menard was a graduate student [272, p. 113]

20.1 ▪ Two Unknowns: An Introduction

The story of one equation in one unknown is incomplete without a discussion of higher-dimensional systems for at least two reasons. The first is that some of the concepts and algorithms from the univariate case extend to N equations in N unknowns, and some don't. The second reason is that bifurcation points (x_{bif}, λ_{bif}) are the solutions to a system of two equations in two unknowns:

$$f(x_{bif}, \lambda_{bif}) = 0 \qquad \text{and} \qquad f_x(x_{bif}, \lambda_{bif}) = 0, \tag{20.1}$$

where the parameter λ is now one of the unknowns; the parameter value λ_{bif} is where a double root occurs, and also the limit point (fold) of the two solution branches that merge into the double zero. Triple roots occur in systems with two parameters and generate systems of three equations in three unknowns, the value of the unknown and the two parameter values at the triple point.

The Chebyshev-proxy method generalizes directly to two unknowns, as shown in Fig. 20.1. Adaptive bivariate Chebyshev interpolation using a tensor product basis with a tensor product grid replaces a transcendental system by a polynomial system or, with subdivision, by a set of polynomial systems. Once each polynomial system has been solved, the roots can be improved by Newton's iterations with the original transcendental functions.

The polynomial system solver is surrounded by a black box in the diagram because ideally it *is* a black box: a combination of well-conditioned algorithms that will yield the polynomial roots with no user input except the coefficients of the polynomials in Chebyshev form. Good bivariate polynomial rootfinders will be described below, but these are not yet at black-box maturity.

Each of the competing tactics by itself possesses a *fail set*, a combination of circumstances where the idea breaks. The most sensible approach is to combine multiple algorithms with only weakly overlapping fail sets so that the grand strategy is unsuccessful only for extremely improbable situations.

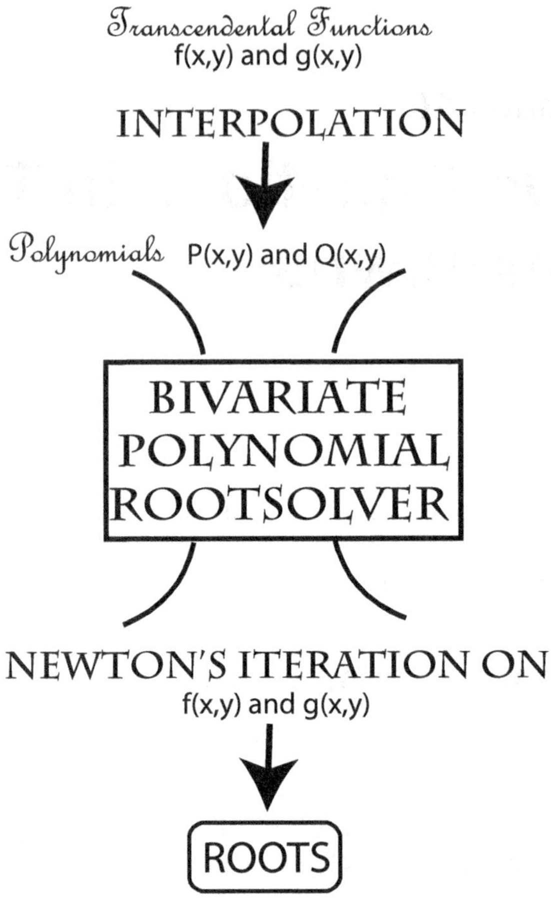

Figure 20.1. *The Chebyshev-proxy method in two dimensions. A system of two transcendental equations* $\{f(x,y), g(x,y)\}$ *is converted to a bivariate polynomial system* $\{P(x,y), G(x,y)\}$*. This is solved by a bivariate polynomial system solver. The zeros are then refined by Newton's iterations to obtain final approximations whose accuracy is limited only by the precision to which* $f(x,y)$ *and* $g(x,y)$ *can be evaluated.*

The roots of a pair of equations are the intersections of the zero contours (isolines) of each individual equation. It follows that the *contouring problem* is closely related to bivariate rootfinding, as amplified below.

The first half of the chapter explicates algorithms that are indifferent to whether $f(x,y)$ and $g(x,y)$ are polynomials or not—graphical schemes and local iterations such as Newton's method. In the second part, the analysis is restricted to polynomial systems only. Thanks to the magic of Chebyshev interpolation, this is really no restriction at all.

20.2 ▪ Rootfinding by Plotting in Two Unknowns: MAG-Intersecting-Contours and MAG-Residual-Norm

When there are only two unknowns, the roots can be graphically determined in two different ways. The first is the superimposed contour plot method: make a contour plot of the zero isolines of $f(x,y)$, i.e., the contour lines where $f(x,y) = 0$, and superimpose

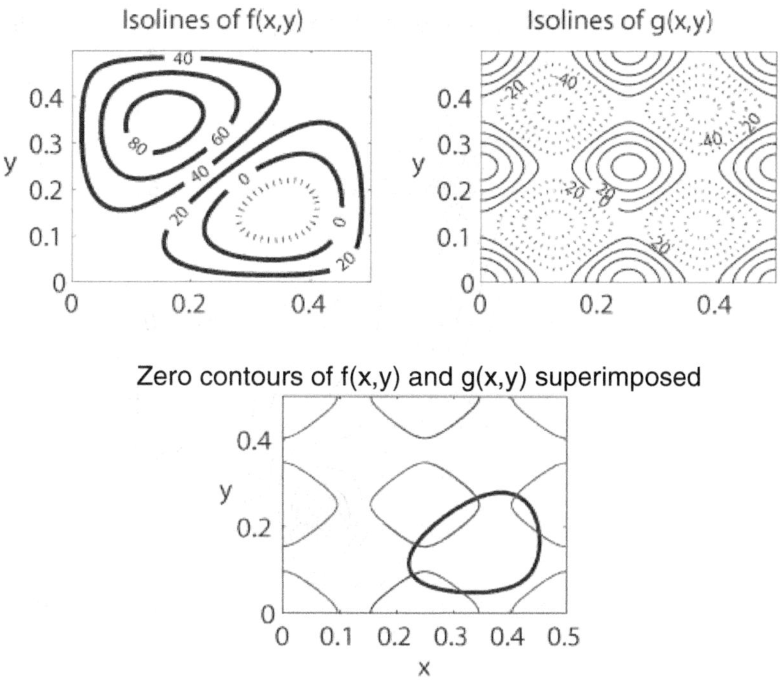

Figure 20.2. *The lower panel shows the zero contours for both of the functions* $f(x,y) = 40\{\sin(4\pi x)\sin(2\pi y) - \sin(2\pi x)\sin(4\pi y)\} + 30$ *and* $g(x,y) = 40\{\cos(8\pi x) + \cos(8\pi y)\} - 10$. *When the zero contours are superimposed, their intersections approximate roots of the system* $f(x,y) = g(x,y) = 0$. *Contour plots of* $f(x,y)$ *and* $g(x,y)$ *separately are shown as the two upper panels (example from Booth [47]).*

them on the curves where $g(x,y) = 0$; the intersections are the roots. Fig. 20.2 is an illustration.

The second graphical strategy is the logarithm of the residual function method. Surface mesh plots or contour plots of the residual norm,

$$\rho(x,y) \equiv (1/2)\{f^2 + g^2\}, \tag{20.2}$$

also isolate roots because these are the points where the nonnegative function $\rho(x,y)$ dips to zero.

To make the roots stand out more clearly, it is better to contour the *logarithm* of the residual function. As illustrated in Fig. 20.3, the residual norm has *zero gradient* at a minimum. Taking the logarithm exaggerates the relief of the bottom of the valley.

Even with this trick, however, it may be difficult to distinguish root valleys from shallower valleys, free of a root, that do not dip all the way to zero, as also illustrated.

A mesh plot is another alternative, as illustrated in Fig. 20.4. It is better to plot the *negative* of the logarithm so that the roots appear as tall, narrow peaks. However, it is again difficult to distinguish roots from near-roots. In contrast, both the root and the near-root are easily visible in the zero isoline plot for this same example, Fig. 20.5.

The intersecting zero isolines method is therefore recommended.

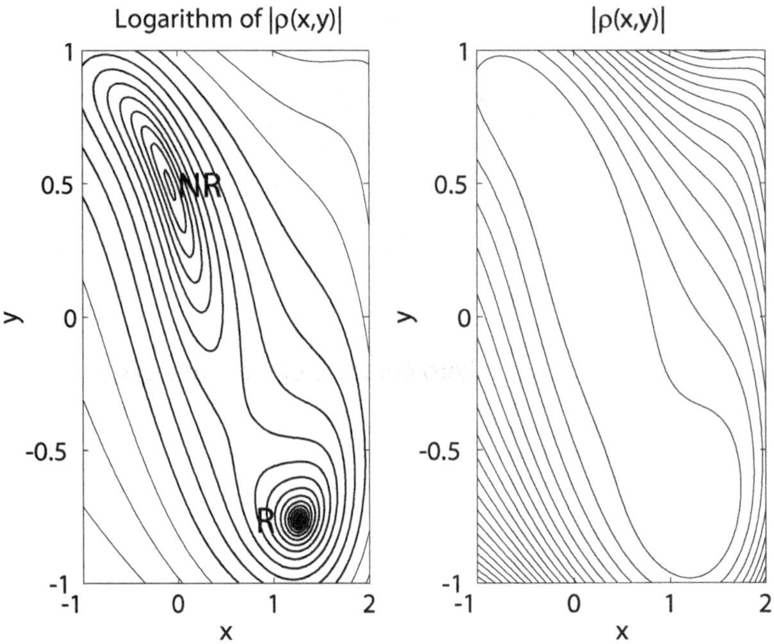

Figure 20.3. *Contour plot of the* logarithm *of the residual norm* ρ *(left) and of the residual norm itself (right) for a two-dimensional function. Each plot displays isolines of* 20 *evenly spaced values whose range spans that of the plotted function. Both the root and nonroot minima appear only as very broad valleys in the plot on the right. However, it is easy to visually identify the minima of the* logarithm *of the residual. Further, the root (marked by "R") is easily distinguished, because the contours are much more concentrated, from the nonroot local residual minimum ("NR").*

Rootfinding by plotting is discussed in greater length in sections 16.2 and 16.3. Log-contour plots are equally effective at finding *complex-valued* roots in a single unknown.

20.2.1 ▪ Disadvantages

The two main flaws of these graphical methods are (i) crude accuracy and (ii) poor automation. The first drawback is obvious: it is impossible to estimate any quantity from the graph with high precision because of the finite thickness of printed curves and pixels. However, it is straightforward to refine graphical estimates of roots by Newton's iteration.

What we mean by poor automation is that contouring software will automatically draw the contour lines but does not compute the intersections of isolines. It is trivial to approximately determine the intersections by eye, but writing an automatic procedure may not be trivial and requires detective work to understand the way the underlying contouring software represents contour lines and how this representation can be extracted, usually by hacking the contouring software.

In contrast, it is straightforward to determine the smallest element of any matrix. This in particular includes the matrix that stores values of the residual function. However, if there are several roots on the plotted domain, then we must automatically identify the *local* minima associated with each root. This is not complicated; each interior point $(x_{i,j}, y_{i,j})$ where $\rho(x,y)$ is smaller than its nearest neighbors, i.e.,

$$\rho(x_{i,j}, y_{i,j}) \leq \rho(x_{i\pm1,j\pm1}, y_{i\pm1,j\pm1}), \tag{20.3}$$

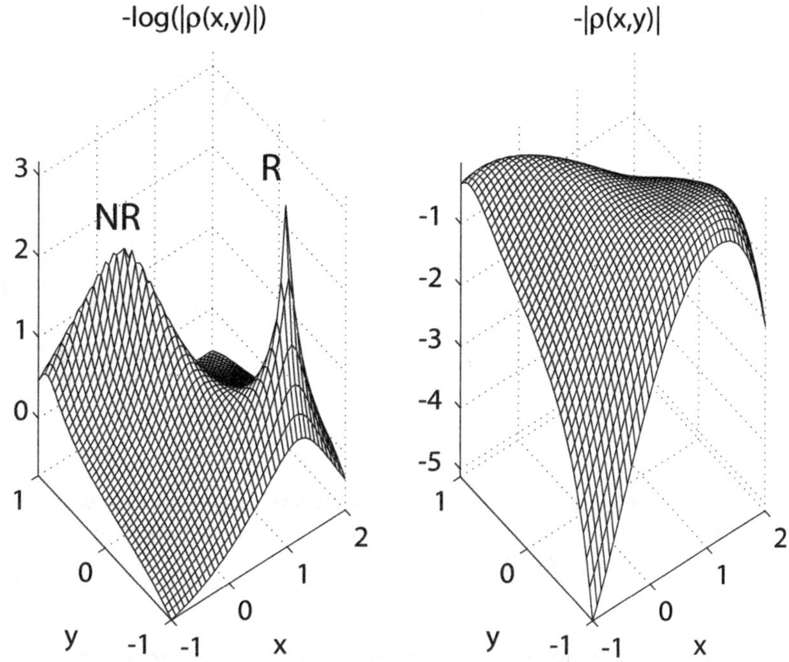

Figure 20.4. *Same as previous figure except that* log(ρ) *(right) and* ρ *(left) are shown as surface mesh plots where* $\rho(x,y) \equiv (1/2)\{f(x,y)^2 + g(x,y)^2\}$. *The very tall spike at the root ("R") and the shorter, broader nonroot minimum ("NR") are obvious in the plot of the* logarithm *(left panel), but so flat as to be almost invisible in the surface of* ρ *itself (right).*

is a local minimum. At the boundaries, we restrict the comparisons with neighbors to include only those that are actually on the computed array of points. As illustrated above, however, the residual function will often have local minima that are not roots. One slow but reliable procedure is to use each local minimum as a first guess for a Newton's iteration. Those iteration sequences that converge to a point on the domain are accepted as true roots; divergent (or redundant) sequence limits are discarded.

This procedure of computing a large array of samples of the residual function, followed by computation of local minima, and finally a set of Newton's iterations, begun from each local minimum, is both automated and fairly reliable. However, it is easy to miss roots unless the number of samples is very large.

20.3 ▪ Bivariate Newton's Iteration

Like its one-dimensional counterpart, the bivariate Newton's method replaces $f(x,y)$ and $g(x,y)$ by proxies that are linear Taylor series and then solves the proxy. The Taylor series are

$$f(x,y) \approx f(x^{(n)},y^{(n)}) + \frac{\partial f}{\partial x}(x^{(n)},y^{(n)})(x - x^{(n)}) + \frac{\partial f}{\partial y}(x^{(n)},y^{(n)})(y - y^{(n)}) + HOT,$$
$$g(x,y) = g(x^{(n)},y^{(n)}) + \frac{\partial g}{\partial x}(x^{(n)},y^{(n)})(x - x^{(n)}) + \frac{\partial g}{\partial y}(x^{(n)},y^{(n)})(y - y^{(n)}) + HOT,$$

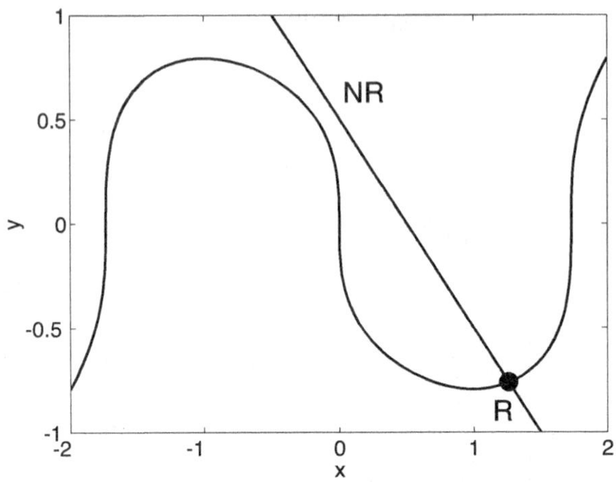

Figure 20.5. *Same example as the previous figure except that the zero isolines of both $f(x,y)$ and $g(x,y)$ are plotted on the same graph. The root is marked by the black disk and letter "R"; the near-root is marked with "NR."*

where *HOT* is an abbreviation for the neglected higher-order terms. Using f_x as an abbreviation for $\partial f/\partial x$ evaluated at $(x^{(n)}, y^{(n)})$, and denoting the Newton correction vector by $\vec{\delta}$, we must solve the system

$$\begin{vmatrix} f_x & f_y \\ g_x & g_y \end{vmatrix} \begin{vmatrix} \delta_1 \\ \delta_2 \end{vmatrix} = \begin{vmatrix} -f \\ -g \end{vmatrix} \tag{20.4}$$

with the explicit solution

$$\vec{\delta} = \begin{bmatrix} -\dfrac{-f_y\, g + f\, g_y}{f_x\, g_y - g_x\, f_y} \\ \dfrac{-f_x\, g + g_x\, f}{f_x\, g_y - g_x\, f_y} \end{bmatrix}. \tag{20.5}$$

The iteration is then

$$\vec{x}^{(n+1)} = \vec{x}^{(n)} + \vec{\delta} \tag{20.6}$$

until the correction is negligibly small.

With a poor initialization, Newton's iteration may diverge, but when the current iterate is sufficiently close to the zero, the number of correct digits roughly doubles at each iteration until reaching a *saturation* level controlled by floating point roundoff errors.

Because it will not converge unless the initialization $(x^{(0)}, y^{(0)})$ is very close to a root, Newton's iteration is unreliable for the "opening game," to borrow chess jargon. It is terrific, however, because of its digit-doubling rate of convergence, for *refining* and *polishing* approximate zeros found by other algorithms (endgame).

20.4 ▪ Bivariate Polynomials: Background

20.4.1 ▪ Numerical Algebraic Geometry

Several different families of algorithms are in widespread use to solve general systems of n polynomial equations in n unknowns:

1. Groebner bases;
2. resultants, multiresultants, Bezoutians, and other elimination-based strategies;
3. divide-and-conquer subdivision algorithms;
4. all-roots homotopies.

The reason that all flourish is that all work most of the time, and also all have significant fail sets.

We shall concentrate on two representative schemes. One is a new subdivision method chosen because it has close affinities to CPR, can be described entirely with undergraduate mathematics, and can be applied to transcendental systems as well as polynomials. The other is a resultant/Chebyshev–Bezout scheme.

Homotopy methods for a pair of equations will not be discussed because this follows the same guiding principle as for homotopy for a single equation in a single unknown. However, homotopy is far from trivial when the number of unknowns is large [337, 36, 155, 279].

20.4.2 ▪ Bezout's Theorem

Definition 20.1 (Total Degree). *The* total degree *of a monomial $x^j y^k$ is $d = j + k$.*

The total degree of a polynomial is the maximum of the total degrees of the monomials that compose it.

Theorem 20.2 (Bivariate Bezout). *Let n_r denote the number of common zeros of two polynomials in two unknowns of total degrees M and N, counting complex roots and also enumerating zeros according to their multiplicity. Then*

$$n_r \leq MN, \tag{20.7}$$

where the upper bound MN is the "Bezout number." In projective geometry, the inequality becomes an equality [137].

The theorem sometimes gives only an *upper bound* because of so-called solutions at infinity. These do not occur for *generic* bivariate polynomials, that is to say, if the coefficients of the polynomials are chosen randomly, the probability of having fewer than MN roots is zero. (The more precise and technical statement is that the set of polynomials with roots at infinity is, in the parameter space of polynomial coefficients, a set of measure zero. However, in the real world of robotic arm movements and other engineering and physics problems, nongeneric polynomials with solutions at infinity are in fact very common.)

Definition 20.3 (Solutions at Infinity). *Let P and Q denote polynomials of total degree M and N, respectively. Let P_{hom} and Q_{hom} denote the sum of monomials in P and Q whose total degree is exactly equal to M and N. The roots of the system $P_{hom} = 0$ and $Q_{hom} = 0$ are* *the* solutions at infinity *of the system* $P = 0, Q = 0$. *If roots at infinity exist, the system is said to be "deficient."*

Homotopy methods are bedeviled by solutions at infinity since, without special precautions, the homotopy software will actually track these roots to infinity, or a finite approximation thereof.

In applications such as robot kinematics and chemical equilibrium systems, the polynomial systems are often sparse, with special properties that homotopy methods can exploit to reduce costs by orders of magnitude. Unfortunately, the bivariate polynomial systems that are Chebyshev polynomial interpolants of transcendental functions do not have such sparseness properties, as far as currently known.

20.4.3 ▪ Groebner Bases

In infinite precision arithmetic with unlimited computer speed, the general way to solve a system of n polynomial equations in N unknowns is by the calculation of a Groebner basis with lexicographic ordering. This algorithm is the nonlinear generalization of Gaussian elimination. The system is *triangularized* in the sense that the first element of the Groebner basis is a *univariate* polynomial. Its roots provide all possible x_1-values for the N-dimensional vectors that are the roots of the system. The second element of the Groebner basis is *bivariate*, but plugging in each x_1-value in turn yields a univariate polynomial that can be solved to generate all possible x_2-values that appear in system root vectors paired with that x_1. The third Groebner basis function is trivariate, but plugging in the known values of x_1 and x_2 gives a univariate polynomial in x_3 only (actually, a *set* of such polynomials, one for each pair of (x_1, x_2)). And so it goes.

We shall not elaborate or describe reducible systems and other exceptions because in finite precision and finite computational resources, Groebner bases are unreliable except for the important special case of polynomials with rational coefficients in exact arithmetic, as available in computer algebra systems.

20.5 ▪ Resultants

"Previous robust algorithms [for intersecting algebraic curves] have employed resultants.... This method works very well for curves of low degree (and is the method of choice for low degrees), but in the author's experience it can be numerically unreliable for curves with degrees above five."

—T. W. Sederberg [313, p. 547]

More than a century and a half ago, J. J. Sylvester invented a general procedure for obtaining a univariate polynomial from a pair of bivariate polynomials that have the property that all y-values of the system roots are zeros of the polynomial in y only. The univariate polynomial is said to be the x-resultant of the pair of polynomials whence it came.

Sylvester's key idea was the observation that at a common root of P and Q, these polynomials are zero so that one can multiply them by arbitrary powers of x, and the product will still be zero. Let $P(x,y)$ and $Q(x,y)$ be of degree M and N in x, respectively. Then this multiplication-by-x trick creates the system

$$\begin{aligned} x^p\, P(x,y) &= 0, \qquad p = 0, 1, \ldots, N-1, \\ x^q\, Q(x,y) &= 0, \qquad q = 0, 1, \ldots, M-1. \end{aligned} \tag{20.8}$$

By multiplying out the powers of x in this system of $(M+N)$ equations, one finds that the highest power of x that appears is x^{M+N-1}.

Sylvester then made an imaginative leap. The powers of x are all connected because as soon as x is specified, all the powers of x are immediately determined. If we pretend that these powers are independent quantities, then the system (20.8) can be regarded as a system of $(M+N)$ *linear* equations in the $(M+N)$ unknowns $1, x, x^2, \ldots, x^{M+N-1}$. The $(M+N)\times(M+N)$ matrix of this system will be dubbed the *Sylvester matrix*. Because this matrix problem is homogeneous, it has a solution if and only if the determinant of the Sylvester matrix is zero. This determinant is the *resultant* of the two polynomials $P(x,y)$ and $Q(x,y)$ with respect to x.

If the polynomials are written

$$P(x,y)=\sum_{j=0}^{M} p_j(y)x^{M-j}, \qquad Q(x,y)=\sum_{j=0}^{N} q_j(y)x^{M-j}, \tag{20.9}$$

then

$$\text{Resultant}(P,Q;y)=\det\begin{vmatrix} p_0(y) & p_1(y) & \cdots & p_M(y) & 0 & 0 & \cdots & 0 \\ 0 & p_0(y) & p_1(y) & \cdots & p_M(y) & 0 & \cdots & 0 \\ 0 & 0 & p_0(y) & p_1(y) & \cdots & p_M(y) & 0 & \cdots \\ & & & \cdots & & & & \\ 0 & 0 & \cdots & 0 & p_0(y) & p_1(y) & \cdots & p_M(y) \\ q_0(y) & q_1(y) & \cdots & q_N(y) & 0 & 0 & \cdots & 0 \\ 0 & q_0(y) & q_1(y) & \cdots & q_N(y) & 0 & \cdots & 0 \\ & & & \cdots & & & & \\ 0 & 0 & q_0(y) & q_1(y) & \cdots & q_N(y) & 0 & \cdots \end{vmatrix}. \tag{20.10}$$

It is a univariate polynomial in y of degree MN.

For example, the Maple statements **f:=(1/3)*x**3−x−(4/3)*y**3; g:=x+y−1/2; h:=resultant(f,g,x);** generate the resultant polynomial $h=(5/3)y^3-(1/2)y^2-(3/4)y+11/24$. The Maple command **y:=fsolve(h,y)** followed by the back substitution and solution **x:=fsolve(g)** gives the sole real root of the system $f(x,y)=0, g(x,y)=0$ as $x=1.2627, y=-0.7627$.

Three warnings are needed. First, the resultant method is known by various other names, including eliminants, Sylvester's dialytic method, and the Euclidean algorithm for polynomials. Second, the resultant usually has roots that are not zeros of the original polynomial, so each root must be *tested*. (The test is to substitute the candidate root (x_*, y_*) into the system; for true roots, the norm of the residual, $\sqrt{f(x_*,y_*)^2+g^2(x_*,y_*)}$, will be tiny.) Third, in the words of Allgower, Georg, and Miranda [9, p. 831], "the calculation of the determinant of the resultant [i.e., Sylvester matrix] is an unstable operation [in floating point arithmetic]."

On the plus side, the resultant operator can be applied pairwise to reduce a system of several equations in several unknowns to a single polynomial in a single unknown.

Resultants are particularly effective when the polynomial coefficients are integers or rational numbers, and symbolic manipulation is applied because then the roundoff difficulty evaporates and the resultant is completely trustworthy. Even when the coefficients are floating point numbers, reliability can be restored by using arbitrary precision arithmetic, a built-in feature of Maple and Mathematica and available through the VPA (Variable Precision Arithmetic) Toolbox in MATLAB.

Even in ordinary MATLAB/IEEE double precision with 16 digits of accuracy, resultants are useful for *low degree*, as noted by Sederberg above.

20.5.1 ▪ The Block Companion Matrix

An equivalent way of describing the floating point folly of resultants (of medium or higher degree) is that computing determinants is usually a bad idea for any large matrix if there is an alternative. Any eigenvalue problem can reduced to computing the zeros of the characteristic polynomial of the matrix; this polynomial is the determinant. We have argued earlier that the opposite direction is much more stable: it is better to solve the companion matrix eigenvalue problem to find the zeros of the associated polynomial.

When the elements of the Sylvester matrix are polynomials of higher than linear degree in y, the zeros of the Sylvester determinant are the eigenvalues of a *nonlinear* eigenvalue problem in the terminology of Peters and Wilkinson [293] and Ruhe [300] in the sense that the matrix elements are nonlinear functions of y, which is the eigenparameter. When the nonlinearity is a polynomial of degree $\overline{N}$, then the matrix can be trivially inflated to a conventional, *linear* matrix eigenvalue whose dimension is $\overline{N}$ times the size of the original.

Theorem 20.4 (Block Companion Matrix). *Suppose $\vec{\vec{S}}$ is a matrix polynomial of degree $\overline{N}$ in λ where each matrix $\vec{\vec{A}}_r$ is of dimension $\overline{M}\times\overline{M}$:*

$$\vec{\vec{S}} \equiv \sum_{n=0}^{\overline{N}} \vec{\vec{A}}_n \lambda^n \tag{20.11}$$

Then the zeros of the determinant of $\vec{\vec{S}}$ are the eigenvalues of the generalized $\overline{MN}\times\overline{MN}$ linear eigenvalue problem

$$\vec{\vec{C}} \equiv \begin{vmatrix} \vec{\vec{0}} & \vec{\vec{I}} & \vec{\vec{0}} & \vec{\vec{0}} & \cdots \\ \vec{\vec{0}} & \vec{\vec{0}} & \vec{\vec{I}} & \vec{\vec{0}} & \cdots \\ \vdots & \vdots & \vdots & \vdots & \vdots \\ \vec{\vec{0}} & \vec{\vec{0}} & \cdots & \vec{\vec{I}} & \vec{\vec{0}} \\ -\vec{\vec{A}}_0 & -\vec{\vec{A}}_1 & \cdots & -\vec{\vec{A}}_{N-2} & -\vec{\vec{A}}_{N-1} \end{vmatrix} \begin{vmatrix} \vec{x}_1 \\ \vec{x}_2 \\ \vdots \\ \vec{x}_{N-1} \\ \vec{x}_N \end{vmatrix} = \lambda \begin{vmatrix} \vec{\vec{I}} & \vec{\vec{0}} & \vec{\vec{0}} & \vec{\vec{0}} & \cdots \\ \vec{\vec{0}} & \vec{\vec{I}} & \vec{\vec{0}} & \vec{\vec{0}} & \cdots \\ \vdots & \vdots & \vdots & \vdots & \vdots \\ \vec{\vec{0}} & \vec{\vec{0}} & \cdots & \vec{\vec{I}} & \vec{\vec{0}} \\ \vec{\vec{0}} & \vec{\vec{0}} & \cdots & \vec{\vec{0}} & \vec{\vec{A}}_N \end{vmatrix} \begin{vmatrix} \vec{x}_1 \\ \vec{x}_2 \\ \vdots \\ \vec{x}_{N-1} \\ \vec{x}_N \end{vmatrix} = \lambda. \tag{20.12}$$

If the matrix $\vec{\vec{A}}_N$ is invertible, then the generalized eigenproblem may be replaced by

$$\vec{\vec{C}} \equiv \begin{vmatrix} \vec{\vec{0}} & \vec{\vec{I}} & \vec{\vec{0}} & \vec{\vec{0}} & \cdots \\ \vec{\vec{0}} & \vec{\vec{0}} & \vec{\vec{I}} & \vec{\vec{0}} & \cdots \\ \vdots & \vdots & \vdots & \vdots & \vdots \\ \vec{\vec{0}} & \vec{\vec{0}} & \cdots & \vec{\vec{I}} & \vec{\vec{0}} \\ \vec{\vec{B}}_0 & \vec{\vec{B}}_1 & \cdots & \vec{\vec{B}}_{N-2} & \vec{\vec{B}}_{N-1} \end{vmatrix} \begin{vmatrix} \vec{x}_1 \\ \vec{x}_2 \\ \vdots \\ \vec{x}_{N-1} \\ \vec{x}_N \end{vmatrix} = \lambda \begin{vmatrix} \vec{x}_1 \\ \vec{x}_2 \\ \vdots \\ \vec{x}_{N-1} \\ \vec{x}_N \end{vmatrix}, \tag{20.13}$$

where

$$\vec{\vec{B}}_j \equiv -\left(\vec{\vec{A}}_4\right)^{-1} \vec{\vec{A}}_j, \qquad j = 0, 1, \ldots, (\overline{N}-1). \tag{20.14}$$

Proof: The generalized eigenvalue problem is equivalent to

$$\vec{\vec{I}}\vec{x}_{j+1} = \lambda \vec{\vec{I}}\vec{x}_j, j = 1,2,\ldots,(\overline{N}-1), \tag{20.15}$$

$$-\sum_{n=0}^{\overline{N}-1} \vec{\vec{A}}_n = \lambda \vec{\vec{A}}_{\overline{N}} \vec{x}_{\overline{N}}. \tag{20.16}$$

The first $(\overline{N}-1)$ equations yield

$$\vec{x}_j = \frac{1}{\lambda^{\overline{N}-j}} \vec{x}_{\overline{N}}, j = 1,2,\ldots,(\overline{N}-1). \tag{20.17}$$

The last equation becomes, after multiplication by $-\lambda^{\overline{N}-1}$,

$$\left\{ \sum_{n=0}^{\overline{N}-1} \vec{\vec{A}}_n \lambda^n + \lambda^{\overline{N}} \vec{\vec{A}}_{\overline{N}} \right\} \vec{x}_{\overline{N}} = \vec{0}. \tag{20.18}$$

This homogeneous equation can be satisfied only if the matrix $\vec{\vec{S}}$ has a zero determinant. Whenever λ is an eigenvalue, it must be a root of the determinant of the matrix polynomial and vice versa. ■

Both propositions were stated without proof or reference by Peters and Wilkinson [293] in 1970 as their (8.1) and (8.3).

Manocha and Demmel [249, 250] pioneered the application of block companion matrices to compute the intersections of polynomial or rational curves and to solve bivariate polynomial systems in general.

The theorem and the Manocha–Demmel article employ the monomial basis, which is even more ill conditioned for high-degree matrix polynomials than for ordinary polynomials of the same size. However, Maroulas and Barnett constructed the equivalent Chebyshev (and general orthogonal polynomial) block companion matrices (labeled "confederate matrices" in section 7 of [252]).

The Sylvester/generalized eigenproblem strategy seems to be never failing, which is high praise indeed, but there are two speedups. First, a standard eigenproblem can be solved by the QR iteration, but the generalized two-matrix eigenproblem must be solved by the QZ algorithm, and: "Generally it is slower by a factor of 2.5 to 3 as compared to the QR algorithm" (Manocha and Demmel [249, p. 87]). Thus, the second, single matrix eigenproblem in the theorem is preferable. If $\vec{\vec{A}}_N$ is not invertible, but $\vec{\vec{A}}_0$ is, one can multiply the matrix polynomial by λ^N and reinterpret the result as a polynomial in $\daleth \equiv 1/\lambda$ to recover a standard eigenproblem, as suggested by Peters and Wilkinson [293].

Second, the Sylvester matrix can be replaced by the Bezout matrix. For simplicity, assume the polynomials have equal degree, $M = N$. The Bezout matrix in x for the bivariate polynomials P and Q has elements B_{ij} that are the coefficients in the expansion

$$\frac{P(x,y)Q(z,y) - P(z,y)Q(x,y)}{x-z} = \sum_{i=1}^{N}\sum_{j=1}^{N} B_{ij} x^{i-1} z^{j-1}. \tag{20.19}$$

(Note that this definition implies that the Bezout matrix is *always symmetric*, i.e., $B_{ij} = B_{ji}$.) If the polynomials are written

$$P(x) = \sum_{n=0}^{N} p_n x^n, \qquad Q(x) = \sum_{n=0}^{N} q_n x^n, \tag{20.20}$$

and defining the auxiliary matrix with elements

$$m_{ij} \equiv \min(i, N+1-j), \tag{20.21}$$

then

$$B_{ij} = \sum_{k=1}^{m_{ij}} p_{j+k-1} q_{i-k} - p_{i-k} q_{j+k-1}. \tag{20.22}$$

When the polynomials are of equal degree N, the Bezout matrix has only half the dimension of the Sylvester matrix, but the formula for the elements is more complicated.

The Bezout matrix is a matrix polynomial of the same degree in y as the Sylvester matrix and can similarly be replaced by a block matrix eigenproblem. However, the dimension is as small as half that of the block matrices generated by the Sylvester matrix. Because the QR and QZ costs grow as the *cube* of the dimension, the Bezout matrix reduces the linear algebra burden by a factor of eight if the degree in y is no larger than for the Sylvester matrix. (Often the Bezout degree is higher, as discussed more below.) Thus, the Bezout eigenproblem approach could be cheaper than a Sylvester-generalized eigenproblem by as much as a factor of twenty-five!

As a concrete example, suppose the goal is to find the intersections of the ellipse P and parabola Q where the curves are the zero isolines of the polynomials

$$\begin{aligned} P(x,y) &\equiv -3.727 - 0.3333\,y + 0.3255\,x + 0.9536\,x^2 + 0.1500\,xy + 0.9536\,y^2, \\ Q(x,y) &= 0.5000 + 5.667\,y - 0.1429\,x - 2.646\,x^2 + 0.4500\,xy - 0.2000\,y^2. \end{aligned}$$

These curves are shown in the top panel of Fig. 20.6.

The Sylvester matrix is the 4×4 matrix

$$\begin{bmatrix} -3.727 - 0.3333\,y + 0.9536\,y^2 & 0.3255 + 0.1500\,y & 0.9536 & 0.0 \\ 0.0 & -3.727 - 0.3333\,y + 0.9536\,y^2 & 0.3255 + 0.1500\,y & 0.9536 \\ 0.5000 + 5.667\,y - 0.2000\,y^2 & -0.1429 + 0.4500\,y & -2.646 & 0.0 \\ 0.0 & 0.5000 + 5.667\,y - 0.2000\,y^2 & -0.1429 + 0.4500\,y & -2.646 \end{bmatrix}.$$

The matrices of the generalized eigenproblem are 8×8:

$$\begin{bmatrix} 0.0 & 0.0 & 0.0 & 0.0 & 1.0 & 0.0 & 0.0 & 0.0 \\ 0.0 & 0.0 & 0.0 & 0.0 & 0.0 & 1.0 & 0.0 & 0.0 \\ 0.0 & 0.0 & 0.0 & 0.0 & 0.0 & 0.0 & 1.0 & 0.0 \\ 0.0 & 0.0 & 0.0 & 0.0 & 0.0 & 0.0 & 0.0 & 1.0 \\ 3.727 & -0.3255 & -0.9536 & 0.0 & 0.3333 & -0.1500 & 0.0 & 0.0 \\ 0.0 & 3.727 & -0.3255 & -0.9536 & 0.0 & 0.3333 & -0.1500 & 0.0 \\ -0.5000 & 0.1429 & 2.646 & 0.0 & -5.667 & -0.4500 & 0.0 & 0.0 \\ 0.0 & -0.5000 & 0.1429 & 2.646 & 0.0 & -5.667 & -0.4500 & 0.0 \end{bmatrix},$$

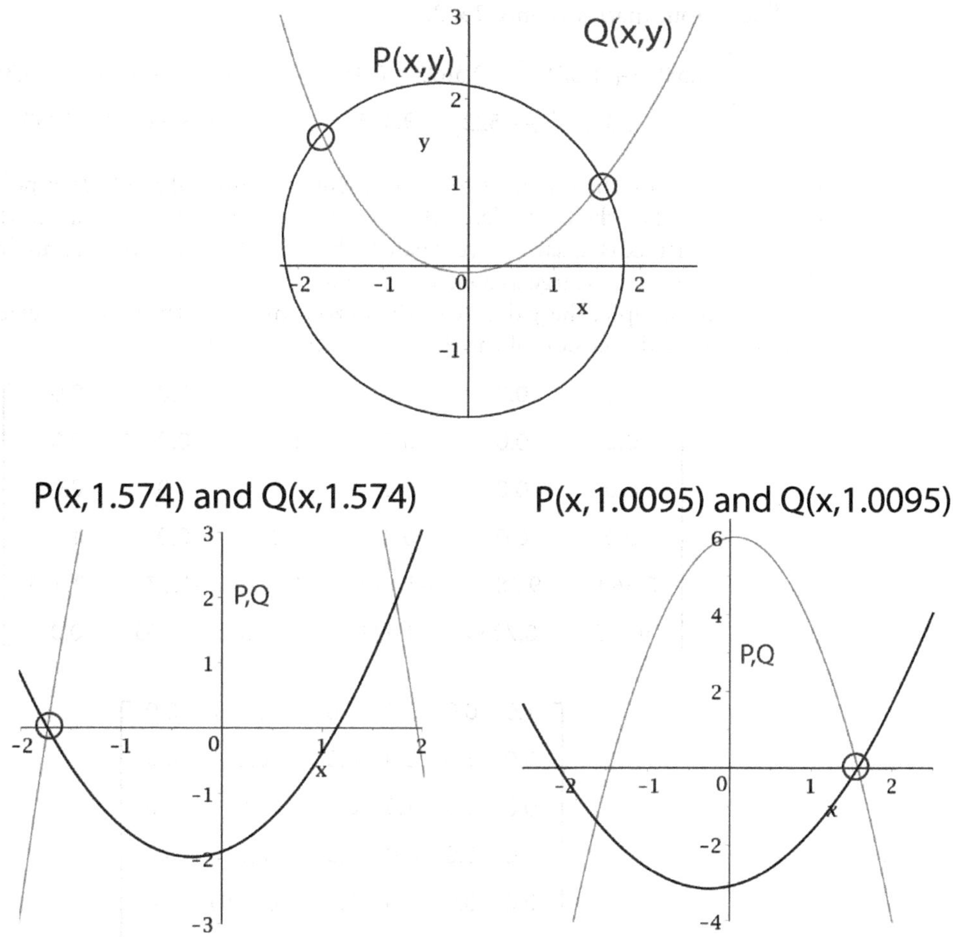

Figure 20.6. *Top: the ellipse (black) $P(x,y)=0$ and parabola (gray) $Q(x,y)=0$; their two real intersections are circled. Bottom: the univariate polynomials that result when one of the eigenvalues is substituted into P and Q. Left: $y=1.574$. Right: $y=1.0095$. Only zeros of* both *P and Q (for the same y) are zeros of the system (circled).*

$$\begin{bmatrix} 1.0 & 0.0 & 0.0 & 0.0 & 0.0 & 0.0 & 0.0 & 0.0 \\ 0.0 & 1.0 & 0.0 & 0.0 & 0.0 & 0.0 & 0.0 & 0.0 \\ 0.0 & 0.0 & 1.0 & 0.0 & 0.0 & 0.0 & 0.0 & 0.0 \\ 0.0 & 0.0 & 0.0 & 1.0 & 0.0 & 0.0 & 0.0 & 0.0 \\ 0.0 & 0.0 & 0.0 & 0.0 & 0.9536 & 0.0 & 0.0 & 0.0 \\ 0.0 & 0.0 & 0.0 & 0.0 & 0.0 & 0.9536 & 0.0 & 0.0 \\ 0.0 & 0.0 & 0.0 & 0.0 & -0.2000 & 0.0 & 0.0 & 0.0 \\ 0.0 & 0.0 & 0.0 & 0.0 & 0.0 & -0.20000 & 0.0 & 0.0 \end{bmatrix}.$$

The Bezout matrix is only 2×2:

$$\begin{bmatrix} 3.549\,y + 1.071\,y^2 - 0.3698 - 0.4591\,y^3 & 2.333\,y^2 + 4.522\,y - 9.385 \\ 2.333\,y^2 + 4.522\,y - 9.385 & 0.8260\,y + 0.7250 \end{bmatrix}.$$

However, because the formulas for its elements are sums of *products* of polynomial coefficients, instead of the coefficients themselves as in the Sylvester matrix, the y-degree of the Bezout matrix is usually *larger* than N, but less than $2N$, so there are indeed savings, though usually not as large as a factor of eight.

For this example, the y-degree of the Bezout matrix is three, so the eigenmatrices are three times the dimension of the Bezout matrix, or 6×6:

$$\begin{bmatrix} 0.0 & 0.0 & 1.0 & 0.0 & 0.0 & 0.0 \\ 0.0 & 0.0 & 0.0 & 1.0 & 0.0 & 0.0 \\ 0.0 & 0.0 & 0.0 & 0.0 & 1.0 & 0.0 \\ 0.0 & 0.0 & 0.0 & 0.0 & 0.0 & 1.0 \\ 0.3698 & 9.385 & -3.549 & -4.522 & -1.071 & -2.333 \\ 9.385 & -0.7250 & -4.522 & -0.8260 & -2.333 & 0.0 \end{bmatrix}, \tag{20.23}$$

$$\begin{bmatrix} 1.0 & 0.0 & 0.0 & 0.0 & 0.0 & 0.0 \\ 0.0 & 1.0 & 0.0 & 0.0 & 0.0 & 0.0 \\ 0.0 & 0.0 & 1.0 & 0.0 & 0.0 & 0.0 \\ 0.0 & 0.0 & 0.0 & 1.0 & 0.0 & 0.0 \\ 0.0 & 0.0 & 0.0 & 0.0 & -0.4591 & 0.0 \\ 0.0 & 0.0 & 0.0 & 0.0 & 0.0 & 0.0 \end{bmatrix}.$$

Both the 6×6 and 8×8 eigenproblems have only four eigenvalues:

$$\begin{bmatrix} 1.57401 \\ 1.00952 \\ -3.05672 + 0.462418\,i \\ -3.05672 - 0.462418\,i \end{bmatrix}. \tag{20.24}$$

Bezout's theorem asserts that a pair of quadratic equations may have at most four solutions; the Sylvester and Bezout matrices have found all the roots.

The remaining step is to compute the x-component of each root. The procedure is to substitute each eigenvalue into P and Q, thus reducing each to a *univariate* polynomial whose roots are easily computed by the ordinary companion matrix discussed in Chapter 2. If both roots of P for each eigenvalue y_j were true system roots, we would end up with eight solutions for a system that cannot have more than four. It follows that we must apply Newton's iteration and accept only roots that are only slightly adjusted thusly, or compute the zeros of both $P(x, y_j)$ and $Q(x, y_j)$ and accept only zeros that are common to both. A visualization of zeros common to both and zeros not in common is provided by the two lower graphs in Fig. 20.6.

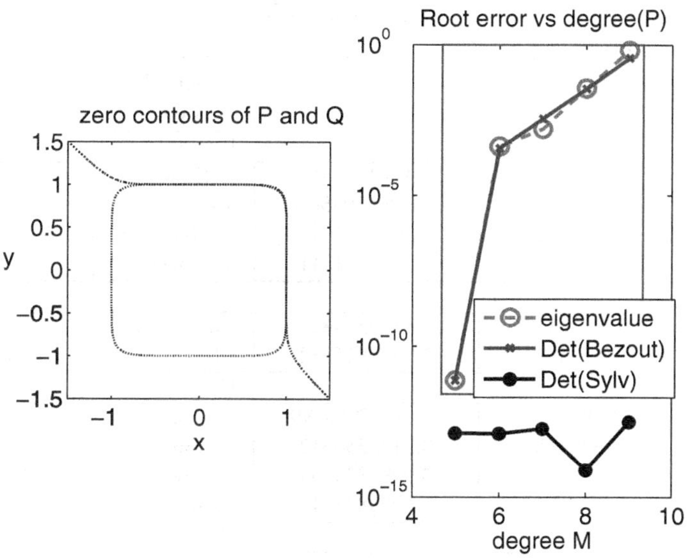

Figure 20.7. *Left: the zero contours of Zeng's squircle intersection problem when the degrees of the polynomials are nine and eight. The squircle is the square with rounded corners, defined as the zero isoline of the eighth-degree polynomial Q. The ninth-degree algebraic curve $P = 0$ parallels two sides of the squircle before curving away from the upper left and lower right corners. The curves for other degrees M are similar. Right: the largest absolute error in any of the roots is plotted versus $M = \text{degree}(P)$ as calculated through the eigenvalues of the block companion matrix (bottom curve, solid disks) and through the determinants of the Sylvester matrix (x's) and Bezout matrix (open circles).*

20.5.2 ▪ The Chebyshev–Bezout Matrix

The analysis earlier in this section is restricted to the monomial basis, but it is well known, as quantified by Walter Gautschi, that the condition number of this grows exponentially fast with degree as $(1+\sqrt{2})^N$ [185, 100]. For small degree, the monomial basis is fine, but for moderate and large N, one needs Chebyshev equivalents of the Sylvester, Bezout, and block companion matrices.

Stephen Barnett and his student John Maroulas developed block Sylvester and Bezout matrices for general orthogonal polynomials in an extensive series of articles around 1980 but did no actual rootfinding [254, 23, 24, 28]. Nakatsukasa, Noferini, and Townsend greatly extended this work and developed practical strategies for bivariate rootfinding including adaptive Chebyshev interpolation, automatic subdivision, dynamic range detection, and Bezout matrix regularization [283, 350].

20.6 ▪ Block Companion Matrix Eigenvalues versus Resultants: Numerical Examples

The first family of problems is borrowed from Zeng [384]:

$$P = x^M + y^M - 0.9999, \tag{20.25}$$

$$Q = x^{M-1} + y^{M-1} - 1.000001. \tag{20.26}$$

The left panel of Fig. 20.7 shows that although the zeros are simple, the problem is difficult because the two algebraic curves are almost parallel on two sides of the squircle.

Table 20.1. *Intersections of squircles: errors prior to Newton-polishing.*

j	x_j	y_j	Eigenvalue Error	det(Bezoutian) Error	det(Sylvester) Error
			$M = 5$		
1	0.9999814041	−0.9317861331e−1	0.4524e−13	0.9445e−13	0.9445e−13
2	0.9999782993	0.9679965955e−1	0.13428e−12	0.3751e−13	0.3751e−13
3	0.9679965955e−1	0.9999782993	0.600e−13	0.68337e−11	0.68337e−11
4	−0.9317861331e−1	0.9999814041	0.1260e−12	0.73522e−11	0.73522e−11
			$M = 6$		
1	0.9999807832	0.1575528427	0.129e−12	0.106e−5	0.106e−5
2	0.1575528427	0.9999807832	0.923e−13	0.424e−3	0.374e−3
			$M = 7$		
1	0.9999878719	−0.2047903899	0.	0.00095	0.00295
2	0.9999823968	0.2177556512	0.186e−12	0.00161	0.00345
3	0.2177556512	0.9999823968	0.139e−12	0.143e−4	0.183e−4
4	−0.2047903899	0.9999878719	0.109e−12	0.880e−5	0.128e−4
			$M = 8$		
1	0.9662967816	−0.8365189533	0.77e−14	0.0357	0.0291
2	−0.8365189533	0.9662967816	0.27e−14	0.0319	0.0340
			$M = 9$, DIGITS = 16		
1	0.9999912720	−0.3028802992	0.2189e−12	0.338	0.274
2	0.9999843101	0.3256616194	0.3075e−12	0.651	0.354
3	0.3256616194	0.9999843101	0.965e−13	0.2291867005e−3	0.156e−4
4	−0.3028802992	0.9999912720	0.1529e−12	0.2222248230e−3	0.862e−5
			$M = 9$, DIGITS = 96		
1	0.9999912720	−0.3028802992	0.821e−93	0.331e−74	0.119e−74
2	0.9999843101	0.3256616194	0.2969e−92	0.326e−74	0.121e−74
3	0.3256616194	0.9999843101	0.3898e−92	0.347e−50	0.105e−50
4	−0.3028802992	0.9999912720	0.2281e−92	0.964e−49	0.292e−49

Nevertheless, Table 20.1 and the right graph in the figure show that the block companion matrix is very accurate, and the error does not grow with degree at least up to ninth degree. In contrast, the error of the determinantal methods grows exponentially with degree, falling apart completely, failing to return even the correct number of intersections, for $M > 9$, which is why the errors for degree 10 and higher are not shown.

Insofar as *accuracy* is concerned, the differences between the Sylvester and Bezout matrices are insignificant. The Bezout matrix has triumphed in popularity only because it is a matrix of lower dimension than the Sylvester matrix and therefore cheaper to manipulate.

The last entry of the table used very high precision arithmetic to show that the errors of the determinantal methods are entirely due to roundoff error; in 96-digit precision, the zeros of the determinants of the Sylvester and Bezout approximate the intersections of the plane algebraic curves to 60 digits. This would be impressive except the block companion matrix eigenvalues are accurate root approximations to an additional 30 digits!

For polynomial degrees less than 10, switching from the monomial basis to the Chebyshev basis produces little improvement. For larger degrees, "Chebyshevization" is essential. Nakatsukasa, Noferini, and Townsend describe the very good Chebyshev–Bezout block companion matrix bivariate rootsolver now incorporated into *Chebfun2* [283, 351].

20.7 ▪ Marching Triangles

20.7.1 ▪ Introduction and Summary

Marching triangles (MT) is an elementary and ancient strategy for computing contour lines and zeros of bivariate systems. MT is very simple and not restricted to polynomials. It also has, alas, serious limitations, but most of these can be fixed by combining it with Weyl bisection.

Define the goal to be the calculation of the common zeros of the two functions $f(x,y)$ and $g(x,y)$. MT is as follows:

1. Sample both functions on a logically rectangular array of points (x_{jk}, y_{jk}).

2. Organize the data into a list of triangles. For each triangle, catalog the coordinates of the vertices of the triangle along with the value of each function at the vertices.

3. Process each triangle independently of every other triangle. First, determine if at least one sample of $f(x,y)$ at a vertex is negative and also at least one vertex sample is positive. If the function is of the same sign at all three vertices of a given triangle, then the piecewise bivariate linear interpolant on the triangle will have no zeros on the triangle by virtue of the theorem stated below. Exclude that triangle from any further processing. If a triangle has at least one vertex of each sign, then pass it on to the next stage of processing.

4. Calculate the bivariate linear interpolant for both functions on each surviving triangle.

5. Solve the system of two linear equations in two unknowns to determine where the zero contours of the polynomials intersect.

6. After the intersection point has been determined, convert it into barycentric coordinates on the triangle. A theorem states that a point is inside the triangle if and only if all three of the barycentric coordinates of the point are nonnegative.

7. If the intersection is on the interior or boundary of the triangle, then accept that intersection as a legitimate approximate root of the system of equations $f(x,y=0)$ and $g(x,y)=0$.

8. Apply Newton's iteration to refine the roots to near machine precision.

The necessary formulas are given in Appendix C.

20.7.2 ▪ The Defects of Marching Triangles

It is possible for contour lines to self-intersect even when $f(x,y)$ is a polynomial of degree as small as three. However, the contours of a bivariate linear polynomial are always a single line segment; it is not possible for a linear approximation to self-intersect. The polynomial of lowest degree that can describe self-intersections is a *quadratic* polynomial, and only in the degenerate case that the conic section is the product of the two linear polynomials that individually describe intersecting lines. Thus, the MT contouring algorithm, which draws the contours of the bivariate linear interpolant as a proxy for $f(x,y)$, must inevitably fail in a triangle where two contours of f or two contours of g intersect.

In addition, an isola, which is a contour that consists of a single point, cannot be described by the linear polynomial either. Furthermore, a closed contour confined within a circle of small diameter ϵ may hide entirely within a triangle with an incircle of diameter ϵ. Such a closed contour includes an isola as the limiting case $\epsilon \to 0$. A quadratic polynomial, which implicitly defines a conic section (ellipse, hyperbola, or parabola), is the approximation of lowest degree that can describe a closed contour even crudely.

A closed contour that spans many triangles will be approximated by a polygon, but the polygonal approximation will not be very accurate unless the number of triangles that contains the contour is large.

In summary, the following are described poorly in the MT algorithm:

1. Islets, which we define as closed contours that lie entirely within the interior of a single triangle, not touching any of the three sides of the triangle. Such small contours are always possible even if the values of the function at all three vertices of the triangle are of the same sign.

2. Multiple zero contours within the triangle. Bivariate linear interpolation must invariably fail because a linear polynomial in two variables is zero only along a single line.

3. Self-intersections of a contour, which is a special case of 2.

Last but not least, linear interpolation is very inaccurate unless the triangles are *very* small.

20.8 ▪ Two-Dimensional Weyl Subdivision

20.8.1 ▪ One-Dimensional Rootfinding through Subdivision with Lipschitz Zero-Free Testing

In 1924, Herman Weyl proposed a remarkably simple global algorithm for finding all roots on an interval of a univariate function. He supposed that one could prove a Lipschitz condition of the form

$$|f(x) - f(y)| \leq L|x - y|, \tag{20.27}$$

where L is called the *Lipschitz constant.* He further supposed that one could at least estimate the Lipschitz constant. Suppose then that the search interval is subdivided into subintervals, each of a small length h. The Lipschitz concept makes it possible to prove that many of these subintervals are free of zeros. The key observation is the following.

Theorem 20.5 (Lipschitz Zero-Free Interval). *If the function satisfies a Lipschitz condition on the interval*

$$|f(x) - f(y)| \leq L|x - y|, \tag{20.28}$$

then an interval $x \in [a, a+h]$ is provably free of zeros if

$$\min(|f(a)|, |f(a+h)|) \geq hL/2. \tag{20.29}$$

Proof: Suppose that y is a root of $f(x)$. It can be at most a distance $h/2$ from one of the endpoints of the interval. It follows that $h/2$ is the largest possible value of $|x - y|$ on the subinterval formed by the root and the nearer of the points $x = a$ and $x = a + h$. The Lipschitz condition on this subinterval of length at most $h/2$ is still true if the right side of the Lipschitz inequality, $L|x - y|$, is replaced by its upper bound on this interval, $Lh/2$. Recalling that by our choice of y, $f(y) = 0$, the Lipschitz condition is then

$$|f(x)| \leq Lh/2. \tag{20.30}$$

This is a necessary condition for a zero on the interval. If the values of $f(a)$, $f(a+h)$, h, and L are inconsistent with the inequality, then $f(x)$ cannot equal zero on the interval. ■

An important fact is that the constraint becomes more and more severe as the interval size h decreases. Weyl therefore proposed to subdivide the initial interval, which we can take as $x \in [-1, 1]$ without loss of generality. At each level, the Lipschitz test is applied to each subdomain still in play. Those subintervals that are provably zero free are *discarded*. Those subdomains of size h that flunk the test, and perhaps contain zeros, are *bisected*, the parent spawning two "children," each of length $h/2$. At level K, there are 2^K subintervals of size $h = 2^{1-K}$ if no subintervals have yet been excluded. The whole point is that eventually subdomains will be proved zero free, and the subintervals still active will become for large K small clusters of very narrow subintervals, clumped around each true zero of $f(x)$. The output of the Weyl algorithm is a set of *intervals*, each containing a root; the centroid of each narrow cluster of intervals can be fed to Newton's iteration to refine each root-containing interval to a single number, accurate to the limits of machine precision.

Weyl's strategy can be applied in any number of dimensions. Kavvadias, Vrahatis, and Makri [220, 219]; Smiley and Chun [334]; Sederberg and Nishita [314]; Barton and Juettler [34]; and Schultz [310] have developed similar, multivariate subdivision strategies for reliable transcendental rootfinding.

Such *exclusion methods* have blossomed into the subfields of *interval arithmetic* and *reliable computing*. Moore, Kearfott, and Cloud's book is a good introduction to the former [275], while Rump's long review is the same for reliable computing [301]. Software includes Rump's MATLAB package INTLAB and the Alias C++ library developed by the COPRIN project, led by Jean-Pierre Merlet.

The Weyl subdivision algorithm does not require a very accurate estimate for L in the sense that the algorithm will converge even if L is way too large. However, a loose bound requires a lot more work because many subdomains will be in play at each stage when a tighter bound would prove that many are zero free.

20.8.2 ▪ Improved Zero-Free Tests for Polynomials

When $f(x)$ is a polynomial, one easy improvement is to use tighter bounds derived from bounds on derivatives of $f(x)$.

In section 19.9.4, the following was proved as Theorem 19.15.

Theorem 20.6 (Second Bounded Curvature Zero-Free Interval). *Suppose that the second derivative (curvature) is bounded by C. Suppose further that $f(a)$ and $f(a+h)$ have the same sign. Then the interval $x \in [a, a+h]$ is free of zeros if*

$$\min(|f(a)|, |f(a+h)|) \geq \frac{C}{8} h^2. \tag{20.31}$$

This bound is greatly preferable to the Lipschitz bound because the curvature diminishes as the *square* of the interval size, instead of being proportional to h. The curvature

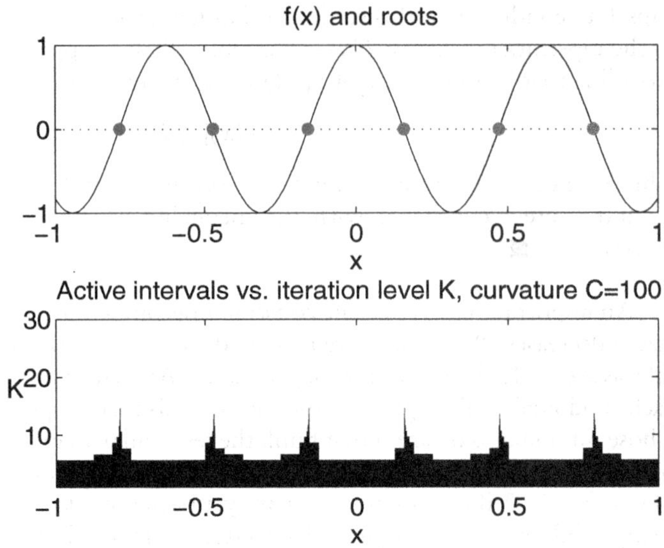

Figure 20.8. *One-dimensional Weyl bisection. Upper plot: the target function,* $f(x) = \cos(10x)$*; its six roots on* $[-1, 1]$ *are marked by small disks. Lower panel: the black squares show the intervals active ("suspect") at each level K. For the first five levels, the curvature bound is unable to exclude any intervals; the number of subintervals is* $(1, 2, 4, 8, 16)$*. Then the number of active intervals drops to* 14 *and then descends to* 12 *intervals, remaining there for the rest of the process. (For levels* $K > 15$*, the remaining suspect intervals are too narrow to be visible on the plot.) Each root is bracketed by a pair of adjacent subintervals, whose centroids can be averaged to give the best approximation to the zero. (However, Newton-polishing is highly recommended.)*

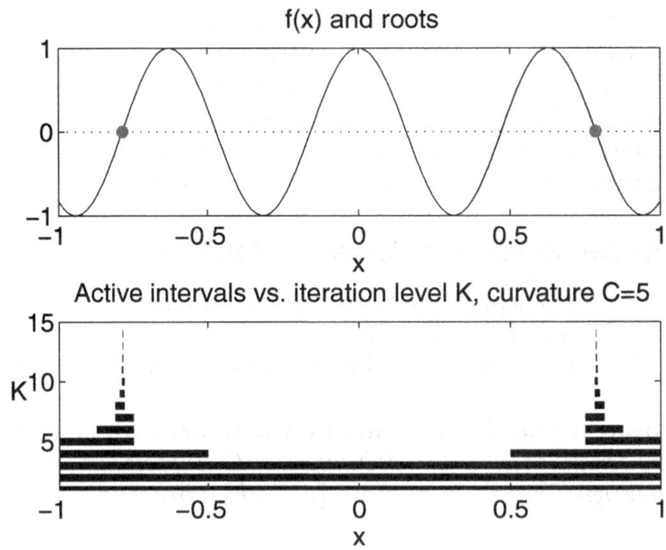

Figure 20.9. *When the curvature is too small—here* $C = 5$ *instead of the true maximum of* 100 *for* $f = \cos(10x)$*—Weyl bisection converges to only some of the zeros.*

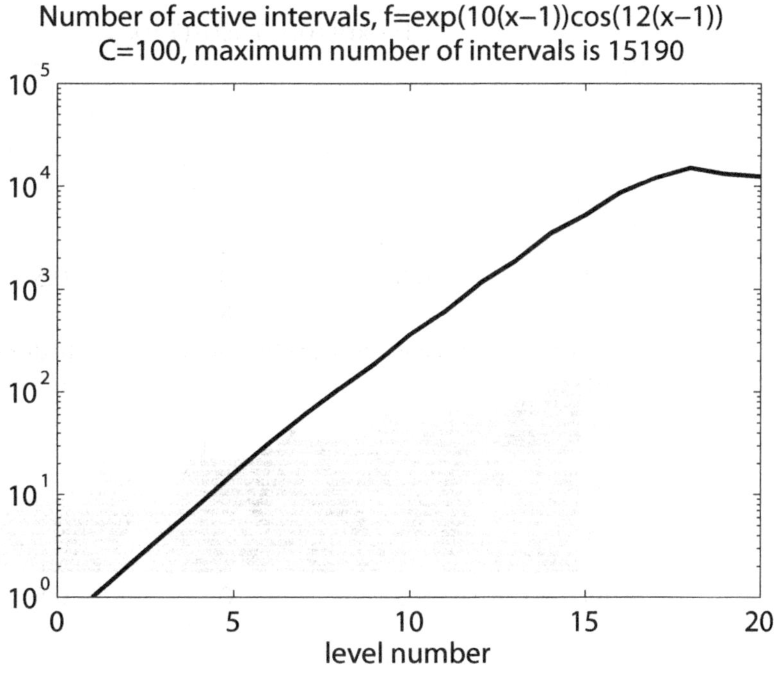

Figure 20.10. *The number of active intervals in Weyl's subdivision scheme for* $f(x) \equiv \exp(10[x-1])\cos(12[x-1])$*; this number peaks at more than* 15000 *suspects, even with a curvature condition.*

bound tightens more rapidly as h decreases, allowing us to exclude more subintervals at each level, reducing cost.

The curvature can be bounded (in any number of dimensions) by expanding f as a Chebyshev series, recalling that $T_n(\cos(t)) = \cos(nt)$, and then exploiting Theorem 19.10. Figs. 20.8 and 20.9 illustrate the algorithms and the role of c.

20.8.3 ▪ Vulnerability of Weyl Bisection to Dynamic Range Problems

The great weakness of Weyl bisection is that it is very vulnerable to large dynamic range. For example,

$$f(x) \equiv \exp(10[x-1])\cos(12[x-1]) \tag{20.32}$$

is one at the right endpoint, but oscillates between peaks and valleys with amplitude $O(\exp(-20)) \approx 2 \times 10^{-9}$ near $x = -1$. If the Lipschitz constant is $O(1)$, then subintervals near the *peaks* near $x = 1$ will not be excluded until the subinterval size has shrunk to $O(10^{-9})$, and something like a billion subintervals will still be active because no subintervals can be excluded near $x = -1$. In contrast, the Chebyshev-proxy rootfinder (CPR) will yield all 8 roots on $[-1, 1]$ to a maximum absolute error of 2×10^{-11} from an expansion with just 41 terms.

Curvature bounds are more powerful than Lipschitz conditions: for this example, the number of possible zero-containing intervals maxes out—with the $O(h^2)$ inequality—at 15,190 (Fig. 20.10). This is a vast reduction from a billion! Fig. 20.11 shows that although all roots on the interval are found, convergence is highly nonuniform: roots near $x = 1$

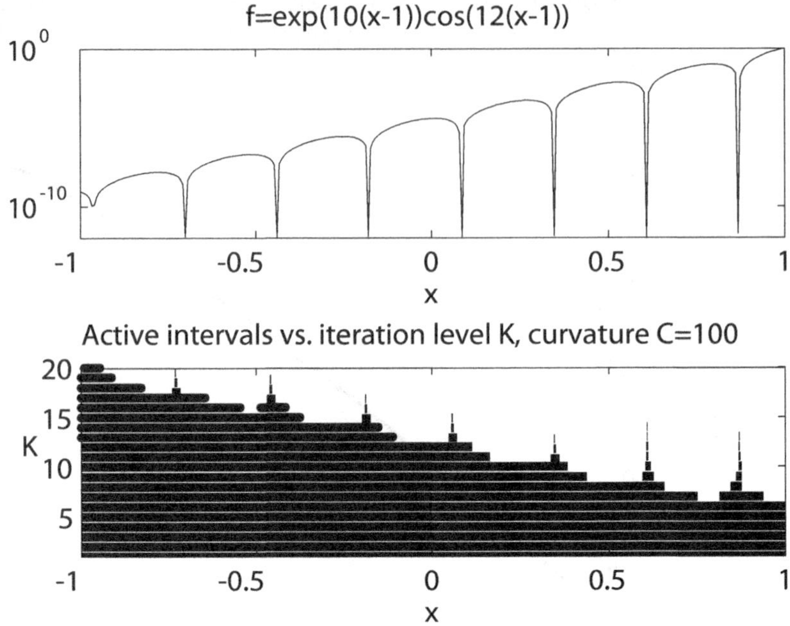

Figure 20.11. *Same case as the previous figure, but showing the absolute value of $f(x)$ on a logarithmic scale in the upper panel. The lower graph shows the intervals still active versus the iteration number.*

where f is $O(1)$ converge rapidly, while roots on the far left near $x = -1$ converge much more slowly since there, f is oscillating between tiny peaks and valleys.

Another example with a large dynamic range is a version of the Wilkinson polynomial of the first kind,

$$f(x) = x \prod_{j=1}^{5} \left(1 - \frac{100x^2}{j^2}\right). \tag{20.33}$$

The slow convergence when bisection begins on the interval $x \in [-1, 1]$ is illustrated in Fig. 20.12.

Subdivision enormously accelerates convergence: when the interval is shortened to $x \in [-1/2, 1/2]$, the maximum curvature $C = \max|f_{xx}|$ is reduced from 247 to only 0.0031. With $C = 0.004$, the maximum number of suspect intervals is only 19 instead of the over 5000 required when the starting domain was $x \in [-1, 1]$, as illustrated in Fig. 20.13.

In general, a high-order root or a cluster of closely spaced simple roots, as in the Wilkinson polynomial, is troublesome.

Typically, the zero-free Lipschitz test will not begin to omit subintervals in a part of the domain where $f(x) \sim O(\epsilon)$ until the subinterval size h has shrunk to $O(\epsilon/L)$, leaving $O(L/\epsilon)$ subintervals still in play. This can easily lead to out-of-memory crashes. Not without reason, Barton commented: "Subdivision solvers are frequently used to achieve this goal; however, the subdivision process is expensive, and a vast number of subdivisions is to be expected, especially for higher-dimensional systems" [33, p. 1870].

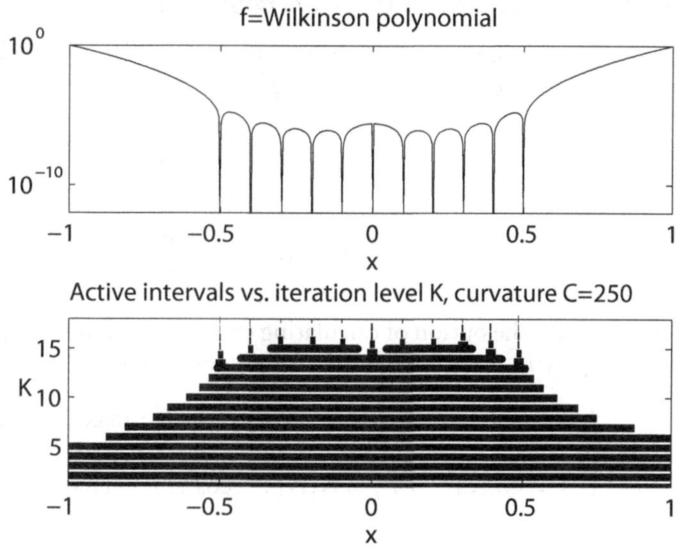

Figure 20.12. *Wilkinson polynomial: the maximum of the second derivative is 247. The number of intervals peaks at 5132 and then slowly decreases.*

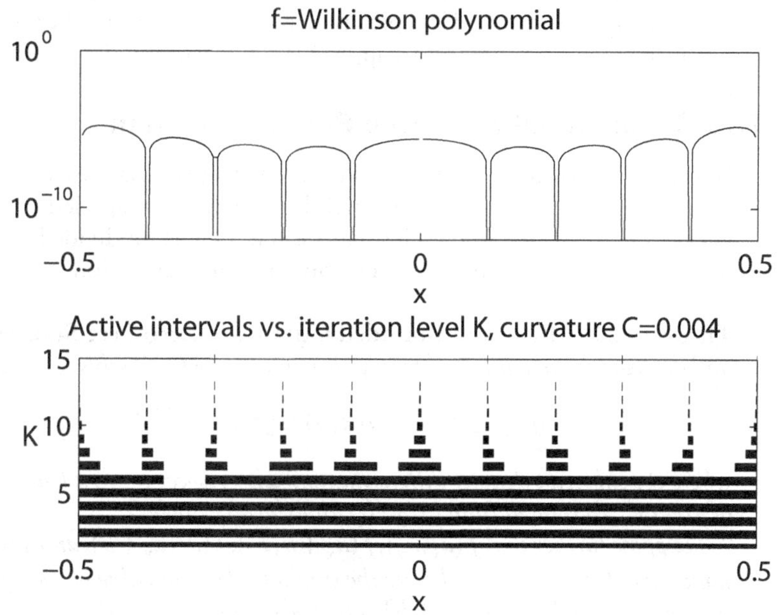

Figure 20.13. *Weyl bisection for the Wilkinson polynomial when the initial domain is halved. No more than 19 subintervals are active at any time.*

20.8.4 • Fourier Enhancement of Weyl Bisection

In one dimension, the first step is to replace the original $f(x)$ by a truncated Chebyshev interpolant. Chebyshev's identity implies

$$f(x) = \sum_{n=0}^{N} a_n T_n(x), \tag{20.34}$$

$$\hat{f}(t) = f(\cos(t)) = \sum_{n=0}^{N} a_n \cos(nt). \tag{20.35}$$

Thus, we have the option of calculating curvature bounds and performing Weyl bisection in either x or t. A couple of examples are instructive:

$$x^N \Rightarrow \begin{cases} f = x^N \to f_{xx} = N(N-1)x^{N-2} : \max_{x\in[-1,1]} |f_{xx}| = N(N-1), \\ f = \cos(t)^N \to f_{tt} = N(N[1-\cos^2(t)]-1)\cos(t)^{N-2} : \max_{t\in[0,\pi]} |f_{tt}| = N; \end{cases}$$

$$T_N \Rightarrow \begin{cases} f = T_N \to f_{xx} = T_{N,xx} : \max_{x\in[-1,1]} |f_{xx}| = N^2(N^2-1)/3, \\ f = \cos(Nt) \to f_{tt} = -N^2\cos(Nt) : \max_{t\in[0,\pi]} |f_{tt}| = N^2. \end{cases}$$

The interval in t is larger by $\pi/2$, implying that for the same number of subdivisions, h^2 will be larger by $\pi^2/4 \approx 2.5$. Even so, because of the (much larger) factors proportional to N, the curvature bound is typically so much smaller using the trigonometric coordinate t that the preferred choice is to apply Weyl bisection to the *cosine* polynomial.

20.8.5 • Two-Dimensional Zero-Free Square Theorem

Bisection in two dimensions is very similar to univariate exclusionary bisection. Tests for the possibility of zeros in a small domain are now applied to a square (or rectangle) instead of an interval, and each square still in play is subdivided into four squares instead of an interval divided into two. The bivariate zero-exclusion test is the following.

Theorem 20.7 (Bivariate Bounded Curvature Zero-Free Square). *Suppose that the second derivative (curvature) of $f(x,y)$ is bounded by C. By this, we mean that*

$$|f_{x'x'}| \leq C \qquad \forall (x,y) \in [a_x, a_x+h] \otimes [a_y, a_y+h] \; \forall \theta, \tag{20.36}$$

where the subscript denotes differentiation with respect to x', a Cartesian coordinate in a set of axes (x', y') rotated from (x,y) by an arbitrary angle θ.

Suppose further that $f(a_x,a_y), f(a_x+h,a_y), f(a_x,a_y+h)$, and $f(a_x+h,a_y+h)$ have the same sign. Let $f_{corner,min}$ denote the smallest absolute value of these four corner values. The square $[a_x,a_x+h] \otimes [a_y,a_y+h]$ is free of zeros of $f(x,y)$ if

$$f_{corner,min} > \frac{C}{4} h^2. \tag{20.37}$$

Proof: If the corner values satisfy the weaker inequality

$$f_{corner,min} \geq \frac{C}{8} h^2, \tag{20.38}$$

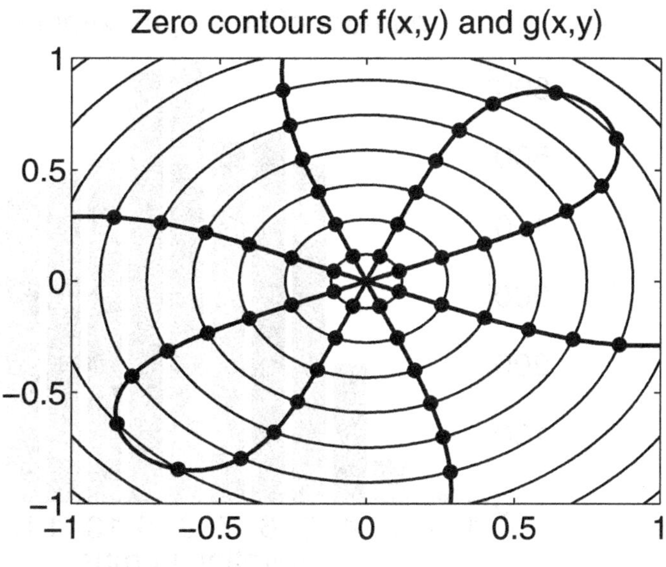

Figure 20.14. *The zero contours of $f(x,y) = r^4\{\cos(4\theta)+0.7r^4\sin(2\theta)\}$ are the thick curves; the thin, concentric isolines are the zero contours of $g(x,y)=J_0(20r)$, where $\sqrt{x^2+y^2}$ is the usual radial polar coordinate and θ is the angular coordinate. There are 52 roots, all simple, where the contours intersect; these zeros of the system are marked with small black disks.*

then the univariate theorem just above proves that the four sides of the square are free of zeros. If there is a zero contour inside the square and if $f(x,y)$ is everywhere positive on the sides of the square, it follows that there must be a local minimum of $f(x,y)$ on the interior of the square. From this point (x_m,y_m), $f(x,y)$ must increase by at least $f_{corner,min}$ along all four lines from the minimum to the corners. The weakest inequality results when the minimum is in the center of the square and the distance to each corner is $h/\sqrt{2}$. Along these diagonals, one can apply the one-dimensional reasoning of the proof of the previous theorem to show that it is possible for f to rise to the specified corner values only if $f_{corner,min} \le \frac{C}{4}h^2$. Otherwise, we have a contradiction that proves the theorem. (If f is negative at all four corners, the same reasoning applies with all signs reversed.) ■

20.8.6 ▪ Numerical Examples

The first example is, converting Cartesian coordinates to polar through $\theta = \arctan2(y,x)$, where arctan2 denotes the two-argument form of the arc tangent function as in MATLAB, and $r=\sqrt{x^2+y^2}$,

$$f(x,y) = r^4\left\{\cos(4\theta)+0.7r^4\sin(2\theta)\right\}, \tag{20.39}$$

$$g(x,y) = J_0(20r). \tag{20.40}$$

The zero contours of these functions are illustrated in Fig. 20.14. There are 52 simple roots on $x\in[-1,1]\otimes y\in[-1,1]$. The Weyl subdivision has no trouble finding all of them; after only 10 subdivision levels, the maximum error norm for any of the roots $(\|(x_{exact}-x_{approx},y_{exact}-y_{approx})\|_2)$ was only 0.0099. The maximum number of active squares peaks at 640 on the eighth step and then slowly declines as shown in Fig. 20.15.

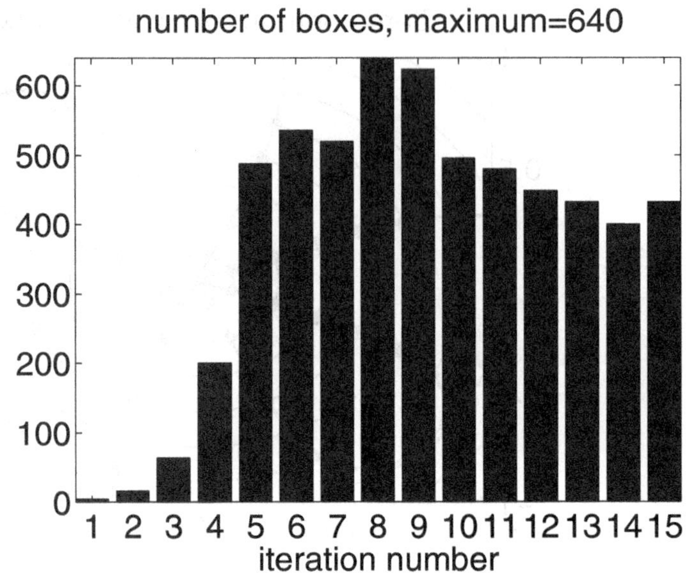

Figure 20.15. *Same case as previous figure, but showing the number of active squares versus the iteration level.*

The second example has only four roots, but is much more challenging (Fig. 20.16):

$$f(x,y) = (x^2+y^2)^2 - x^3 + 3xy^2, \tag{20.41}$$
$$g(x,y) = x^4 + 2x^2y^2 + y^4 - 1/2x^3\sqrt{2} + 3/2x^2\sqrt{2}y + 3/2x\sqrt{2}y^2 - 1/2y^3\sqrt{2}.$$

The equation $f(x,y)=0$ defines a trifolium; $g(x,y)$ is the same curve, rotated by an angle of $-\pi/8$. Three roots are simple, but the root at the origin is sixth order. Consequently, the high-order root is visualized by a large number of squares even at rather late stages in the iteration. The root at the origin ensures that f is very flat near the center of the disk. This creates a dynamic range problem and a huge increase in the number of active squares, as shown in Fig. 20.17.

20.8.7 ▪ The Weyl Bivariate Bisection Endgame

One limitation of the Weyl algorithm, as shown in Fig. 20.18, is that after many iterations, each isolated root is not represented by a point, i.e., a pair of numbers, but rather is a *cluster* of boxes, only one of which actually contains the root.

Thus, the repeated iterations of test and subdivide must be supplemented by an endgame in which approximations to the root are computed. We shall describe three endgames:

1. cluster identification,
2. MT, and
3. Newton's iteration with pruning.

The cluster identification endgame is to identify the nonoverlapping clusters of boxes around each zero and then return the centroid or center of mass of each cluster as an approximation to the root within it. The identification of clusters is straightforward but

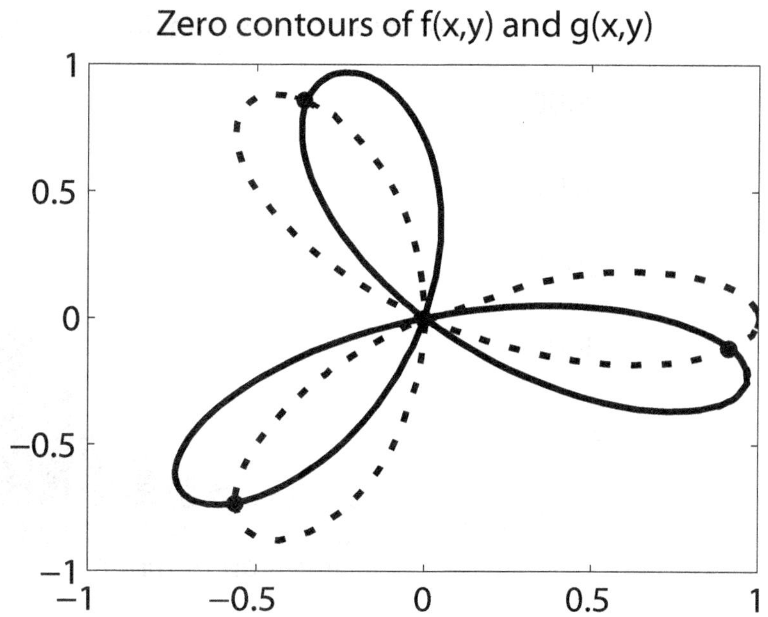

Figure 20.16. *The zero contour of two trifolium curves, identical but for a rotation.*

requires some fussy bookkeeping. First, create a matrix in which each row contains critical information about each box still in play. The first two columns identify the coordinates of the lower left-hand corner of the box. The third column is an integer that is the number of the cluster to which the box will be assigned; this is initialized to zero. The fourth column is either one or zero and flags whether there has been a search for the nearest neighbors of that box.

Next, pick an arbitrary square that is still in play and assign it to cluster one. Then search for its nearest neighbors by comparing the distance from the centroid of that box to the centroid of all the other boxes. A nearest neighbor is one defined as a box whose centroid is within $1.5h$ of the target box. When all the active nearest neighbors have been found, each is assigned to cluster one also and the search column for that box is changed from zero to one. One then does a similar nearest neighbor search for each of the boxes that have been assigned to cluster one. This process is repeated until all active boxes that are connected to the starting point of the cluster by an unbroken chain of nearest neighbors are found. When no further squares are added to the cluster, the cluster counter is incremented and an arbitrary square that is active but not in the first cluster becomes the seed for the second cluster. This process continues until all active squares are grouped into clusters. A numerical value for the approximate zero is computed by averaging the centroids of all the boxes in the cluster.

This procedure is rather slow, as described, but faster nearest neighbor searches have been developed and can be used to greatly accelerate this cluster identification endgame. The strategy will fail only if the Weyl bisection is stopped prematurely so that some clusters contain more than one root.

The second option is to switch to MT. When the boxes are very small, piecewise linear interpolation on the two triangles in a box will be very accurate. Boxes can then be discarded if the values of $f(x,y)$ are the same at all four corners of the box or if the values of $g(x,y)$ are the same at all four corners. However, as shown by the schematic, these

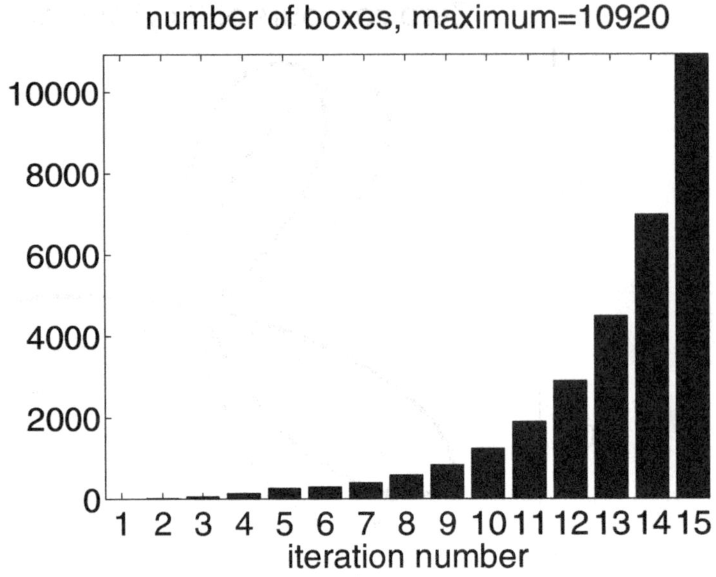

Figure 20.17. *Intersecting trifolia: the number of active squares versus the iteration number has risen to over* 10,000 *by iteration* 15, *and the number is still climbing.*

Figure 20.18. *Exclusionary tests are usually inadequate to completely isolate roots. The curves are the zero isolines of f and g, respectively. Three squares and four triangles contain zero contours of both functions and thus must be left in play by any exclusionary (no-zero-in-this-subdomain) test. There is, however, only one root to the system $f = g = 0$ (marked by the triangle). The Weyl algorithm (and the exclusionary tests in MT) require an endgame to return a single point instead of a cluster of boxes or triangles.*

tests will not reduce the collection of active boxes to one box per zero, but instead will usually yield a small cluster of boxes enclosing each zero. The final step of MT, which is to apply linear interpolation to actually compute the zeros, must be applied here also.

MT is much cheaper as the endgame of the Weyl bisection than when applied ab initio because it is applied not to millions of tiny squares but to only a few hundred or a few thousand.

The third option is to skip both cluster identification and MT, and proceed directly to the final stage of the endgame, *Newton-and-prune*. That is, refine each root by Newton's iteration and then prune the list of duplicates, discarding $(k-1)$ of a group of k roots that lie within some tiny, user-chosen tolerance δ_{prune} such as $\delta_{prune} = 10^{-10}$.

Extensive experiments in [94] show that cluster identification, although very useful for visualizing and debugging, is very inaccurate and slow compared to MT as an endgame. However, MT, because it employs only piecewise linear interpolation, is not very accurate; Newton-polishing of the MT roots is the only route to high accuracy. When a multiple root is misrepresented by MT as a cluster of closely spaced simple roots, Newton's iteration will converge to the same multiple zero from several different approximate roots.

It follows that after Weyl quadtree subdivision has descended to some user-chosen level K, the best middle game is MT followed by Newton-and-prune.

20.8.8 ▪ Subdivision

The best remedy for the dynamic range problem is to subdivide the original domain. In areas where $f(x,y)$ is tiny, computing a new, local Chebyshev series will yield new upper bounds on the curvature that may be many orders of magnitude smaller than those for the global domain. It is important that C be as close as possible to the true maximum curvature, or the Weyl algorithm will partition boxes into smaller and smaller boxes without ever excluding any.

20.9 ▪ Fourier-Enhanced Weyl Exclusionary Subdivision Marching Triangles Bivariate Polynomial Rootfinder: Summary

This fast and robust bivariate rootsolver consists of the following steps:

1. A pair of transcendental functions (f,g) are converted to a polynomial system (P,Q) by adaptive bivariate Chebyshev interpolation. (The domain is subdivided as needed to keep the degree less than or equal to some user-chosen maximum, typically 16.)

2. A global bound on the curvature of each polynomial is computed from its Chebyshev coefficients.

3. Weyl quadtree subdivision is then applied to the domain of each polynomial pair: quartering each remaining square suspected of harboring a zero contour, zero-testing each child, discarding squares that are provably zero free, and repeating to some user-chosen level $\mathfrak{K}$.

4. MT is applied to each tiny square that is still suspected to possibly harbor a zero contour for both polynomials.

5. Newton-and-prune is applied.

This combination of algorithms has been thoroughly explained in this book and is closest in strategy to the univariate CPR scheme. However, other successful subdivision schemes can be found in [334, 281, 322, 149, 183].

20.10 ▪ Homotopy/Continuation for Polynomial Systems

Polynomial systems can also be solved directly by using continuation methods. Bezout's theorem gives an upper bound to the number of roots of a polynomial system.

Given the number of roots, one can then create a simple solvable system of the same order and track each root from the artificial system to the true one. In general, the roots of the artificial system should be *complex valued* so that there is an infinitesimal probability of roots colliding as the continuation parameter τ is varied.

This continuation strategy is a direct method in the sense that no a priori information about the location of the roots is needed. Morgan [278] and Morgan and Sommese [276] describe robust algorithms and software to solve general systems of polynomial equations under the name of the *globally convergent homotopy*.

The great weakness of this method is that the number of roots ν_r escalates rapidly with N, the number of unknowns. Even a quadratic system may have more than a thousand roots for $N = 10$ unknowns, and more than a million zeros for $N = 20$ unknowns! Finding all the roots is simply impractical unless N is small. (One saving grace is that each root can be tracked independently of all the others so that continuation is "embarrassingly parallel" and runs very efficiently on multiprocessor computers.)

However, if one is selective and computes only some of the roots, one is necessarily employing a local method whose success is inevitably tied to a priori knowledge of which roots are important, that is, of where to search in the space of unknowns.

The Bertini system developed by Sommese, Wampler, and many collaborators is a successful and well-maintained software package for homotopy continuation [36, 337].

20.11 ▪ Elimination Methods

Resultant and other elimination strategies are still popular [45, 384]. Nakatsukasa, Noferini, and Townsend [283, 351] have developed a particularly good one. Their scheme takes the Chebyshev form as input and never converts the bivariate polynomials at any stage into the more familiar (but numerically ill-conditioned) monomial form. To minimize the notorious ill-conditioning of resultants, they use the Chebyshev version of a Bezoutian, a matrix with properties similar to the resultant, but with only half the dimension (albeit more complicated formulas for the matrix elements). The usual fast but unstable evaluation of the determinant of the resultant by repeated polynomial division with remainder and the slower but only weakly unstable evaluation of the determinant by Gaussian elimination are replaced by more complicated and slower but better-conditioned matrix procedures. The Chebyshev–Bezoutian method has done well in test problems and has been adopted as the bivariate rootfinder in the two-dimensional *Chebfun2* open source system [351].

20.12 ▪ Summary

Two equations in two unknowns is still an active research frontier. Our personal preference is for the Fourier-enhanced/Weyl exclusionary subdivision/MT combination. However, the Nakatsukasa, Noferini, and Townsend Chebyshev–Bezoutian method is very good [283]. Both are "CPR2D": two-dimensional generalizations of CPR. The Bertini package developed by Bates, Hauenstein, Sommese, and Wampler is excellent for homotopy continuation [36, 337].

It will take some years for the dust to settle. It is comforting that now at least there are some good options for hunting bivariate roots.

Part VIII

Challenges

Chapter 21
Past and Future

Logarithmic Law of Arithmurgy: *Insight grows logarithmically with the number of floating point operations.*

Extended Moore's Law: *Computational power and scientific data, as measured in operations and bytes, both grow exponentially with time.*

Corollary: Temporal Linearity of Progress: *Knowledge, as opposed to mere facts or data, grows linearly with time.*

—J. P. Boyd [87]

Zymurgy is fermentation-working; arithmurgy is number-working. Both beer-making and computational science and engineering are crafts that empower and intoxicate. The Chebyshev-proxy rootfinder (CPR) deserves a champagne toast and the bivariate solvers a round from the tap, but the ancient field of zero-hunting has life yet.

My wish list includes the following:

1. Spectrally accurate parameterizations of solutions $x(\lambda)$ to $f(x;\lambda)=0$ accommodating both open and closed branches, singular endpoints, and bifurcations.
   - Spectrally-accurate contour parameterization and implicitization.
2. CPR in higher dimensions:
   - Major improvements in multidimensional convergence theory.
3. Complete Chebyshevization of algebraic geometry.
4. Black-box software.

The knowledgeable reader will surely object that contours and solution branches are already available. Are not contouring routines included in MATLAB, Maple, and most scientific libraries? Indeed, the Weyl/Marching Triangles (MT) described in this very book is a contour calculator.

Yes, but there is a difference between a thing done, and a thing done well. MT parameterizes contours as a collection of unrelated line segments. With additional steps, these line segments can be organized into branches. But there is a sense in which no applied mathematical task is well understood until we know how to do it with spectral accuracy. By this standard, contour-tracing and parameterization fall short. It is easy to trace a contour with a Davidenko predictor and a Newton corrector, but the points on the contour are not the usual Chebyshev grid points in the parameter. High-order contour-fitting is wiggly and far from optimum.

The bivariate solver in *Chebfun2* and the Fourier-enhanced/Weyl Exclusionary Subdivision/MT combination show that extending CPR to two equations in two unknowns is almost finished. Extending to more unknowns, alas, rapidly runs on the rocks due to the "curse of dimensionality": a tensor product grid of N points in each of d dimensions contains a total of N^d nodes, which exhausts the capacity of even the quantum computers of starship *Enterprise* unless both N and d are small. However, Smolyak *sparse* grids have made computations feasible in 10 or 20 or even 100 dimensions [321]. A "CPR–Smolyak" might considerably expand the range of Chebyshev proxies.

One complication is that sparse grids require sparse coefficients. Multidimensional Fourier and Chebyshev coefficients often cluster around the coordinate axes like ancient sailing ships hugging the coasts. This allows the sparse *hyperbolic cross* truncation; error estimates can be proved in so-called Korobev spaces. It is, however, trivial to construct multidimensional functions whose coefficient isosurfaces are spheres in the d-dimensional space of coefficient degrees. My review of asymptotic spectral coefficients is a list of useful theorems accompanied by a litany of enumerated ignorance [89]. When applying a hyperbolic cross truncation, zero-hunters will not sleep easily until some of these gaps are filled.

Jack Good and especially Stephen Barnett began the recasting of nonlinear algebra from the monomial basis to orthogonal polynomial bases. Hans Stetter's book is a considerable advance in some directions. Rob Corless and collaborators have found companion matrices for the Bernstein, Lagrange, and Newton bases. With apologies to those whose contributions cannot be squeezed into a paragraph, there is still much to be done, especially for the bivariate case and beyond. Stetter's book contains not a word about block companion matrices; there are brilliant papers by many hands, but no synoptic perspective.

The CPR code in *Chebfun* is a black box; its bivariate counterpart in *Chebfun2* is a black box in beta. Point clouds of roots are not parameterized branches, however, and mindless trust in the subroutine is a little dangerous with two unknowns or parameters, and the perils rise steeply with dimension.

Moore's law, the exponential growth in execution speed and memory, will help, but only a little. The Hewlett-Packard personal computer I shared with the entire Earth and Planetary Physics group at Harvard in 1973 boasted an 8-kilobyte memory, a 1-line display, a cassette tape recorder as the only long-term storage, and an execution speed of 50 floating point operations per second. We all thought it was marvelous and signed up eagerly for 1-hour blocks in front of it, supplemented by the occasional all-nighter. My new laptop has 8 gigabytes of memory, a flash drive in lieu of magnetic storage, and has reached 45 gigaflops while chewing up a large matrix. Sadly, as noted in the "laws" above, a million-fold increase in memory and a billion-fold increase in speed have not made me, or anyone else, a million times wiser.

Indeed, last week I found the original manuscript of my first paper on CPR, written as a postdoc at the National Center for Atmospheric Research in the summer of 1976. The internal reviewer hated it, but my boss allowed me to submit it anyway. I mailed my article to *IMA Journal*, full of hope, because Leslie Fox was the editor, and he more than anyone else had developed Chebyshev technology in the sixties and written a fine book. He and his reviewers hated it also. And so I put the paper aside for 17 years to pursue my day job of atmospheric and oceanic dynamics.

And yet it was all there. The dynamic range problem and its cure by subdivision, slightly complex roots, etc. All but the Chebyshev companion matrix, known to Jack Good in 1961, unknown to me for another quarter century.

Gigaflops are good, but moral courage is more important. I finally found mine in middle age when I came upon my old, unloved manuscript and decided that it still had merit [71]. I hope the reader will profit by my bad example.

Publish all your dreams! The challenges listed here are probably unfundable, unimpressive to a tenure committee, and not part of any current fad. But on your deathbed you won't value your life by the number of your NSF grants, and being elected a fellow of some society won't make you live a day longer. But rootfinding is as practical as potatoes and broccoli. And it has been such fun!

Appendix A
Companion Matrices

"Mathematics is not a careful march down a well-cleared highway, but a journey into a strange wilderness, where the explorers often get lost. Rigour should be a signal to the historian that the maps have been made, and the real explorers have gone elsewhere."
—W. S. Anglin [13]

A.1 ▪ Software

Most of the companion matrices cataloged here are available in Maple through the "**CompanionMatrix**" command. However, the formulas are so simple that the reader can translate them into a few lines of code in any desired language. The code to compute the Chebyshev and Legendre matrices is given explicitly as tables below.

A.2 ▪ Why Eigenvalues Are Roots: A Derivation of the One-Dimensional Companion Matrix for General Orthogonal Polynomials

Let $\phi_j(x), j = 0, 1, \ldots$ denote a family of orthogonal polynomials. Any polynomial of an arbitrary degree N can be represented exactly as the finite sum of the first $N+1$ basis functions. In particular, if we seek the roots of a polynomial $f_N(x)$, it can be written as

$$f_N(x) = \sum_{j=0}^{N} a_j \phi_j(x) \tag{A.1}$$

for some coefficients a_j that could be found by imposing $N+1$ interpolation conditions, among other ways.

If we multiply each basis function $\phi_j(x)$ by x, the result is obviously a polynomial of degree $j+1$. This in turn can be reexpanded in terms of the basis functions to yield

$$x\phi_j(x) = \sum_{k=0}^{j+1} H_{j+1,k+1} \phi_k(x) \tag{A.2}$$

for some coefficients $H_{j+1,k+1}$.

Define a vector $\vec{\Phi}$ whose N elements are the basis functions. The expansions of the first N polynomials $x\phi_j(x)$ can be organized into an inhomogeneous matrix equation,

$$\vec{\vec{H}}\vec{\Phi} = x\vec{\Phi} - H_{N,N+1}\phi_N(x)\vec{e}_N, \tag{A.3}$$

where the double arrow labels the matrix whose elements are the $H_{j,k}$, single arrows denote vectors, and $\vec{e}_N$ denotes the unit vector whose elements are all zeros except for a one in row N. The rightmost term arises because the product $x\phi_{N-1}$ is a polynomial of degree N and thus requires $\phi_N(x)$ for its representation, but this is not one of the N basis functions in the vector $\vec{\Phi}$, which includes only the basis functions up to $\phi_{N-1}(x)$.

We can remove the extra term in ϕ_N by adding $qf_N(x)$ to and subtracting it from the last equation of the matrix system, which becomes

$$\sum_{k=0}^{N-1} H_{N,k+1}\phi_k(x) = x\phi_{N-1}(x) - H_{N,N+1}\phi_N(x) + q\left\{a_N\phi_N(x) + \sum_{j=0}^{N-1} a_j\phi_j(x)\right\} - qf_N(x). \tag{A.4}$$

If we choose $q = H_{N,N+1}/a_N$, the terms in ϕ_N will *cancel*, and the terms involving a_j for j up to $(N-1)$ can be absorbed into the matrix by modifying the elements of the lowest row. The only rub is the leftover term $-qf_N(x)$. If, however, x is not a general value but happens to be one of the *roots* x_r of $f_N(x)$, then $-qf_N(x_r)$ is zero and the matrix problem becomes a classic matrix eigenvalue problem with the parameter x_r as the eigenvalue. We have thus proved the following proposition.

Theorem A.1 (Generalized Companion Matrix). *Let $\phi_j(x)$ be a set of polynomial basis functions such that the subscript j is the degree. Then the roots x_j of the degree-N polynomial*

$$f_N(x) = \sum_{n=0}^{N} a_n\phi_n(x) \tag{A.5}$$

are the eigenvalues of

$$\vec{\vec{M}}\vec{\phi} = x\vec{\phi}, \tag{A.6}$$

where the elements of $\vec{\vec{M}}$ are

$$M_{jk} = H_{jk}, \qquad j = 1,2,\ldots,N-1, \tag{A.7}$$

$$M_{Nk} = H_{Nk} - H_{N,N+1}\frac{a_{k-1}}{a_N}, \qquad k = 1,2,\ldots,N, \tag{A.8}$$

where

$$x\phi_j(x) = \sum_{k=0}^{j+1} H_{j+1,k+1}\phi_k(x) \tag{A.9}$$

for some coefficients $H_{j+1,k+1}$.

Although not proved here, it can be shown that the N eigenvalues always give the N roots if both eigenvalues and roots are counted according to their multiplicity; a double root thus appears twice in the list of eigenvalues.

The argument presented here is very similar to that of Barnett [22] and Day and Romero [146]. The proof used almost nothing about the basis functions except that the

$x\phi_j$ are polynomials of degree at most $N-1$ for $j \leq N-2$ and $x\phi_{N-1}$ is a polynomial of at most degree N.

Corless and Litt [136] and also Calvetti, Reischel, and Sgallari independently devised a companion matrix for the Newton basis (divided differences). Corless has devised a related Lagrange basis [134, 133, 136] and demonstrated its usefulness [134]. Independently, Corless [134] and Joab Winkler [376, 377, 378] invented companion matrices for the Bernstein–Bezier basis.

A.3 ▪ Frobenius Companion Matrix (Monomial Basis)

Let the polynomial be $f_N(x) = \sum_{j=0}^{N} a_j x^j$. Its roots are the eigenvalues of the $N \times N$ matrix whose nonzero elements are the following:

$$F_{j-1,j} = 1, \qquad j, = 2, \ldots, N, \tag{A.10}$$

$$F_{N,j} = -a_{j-1}/a_N, \qquad j = 1, 2, \ldots, N. \tag{A.11}$$

When $N = 4$, the Frobenius companion matrix is explicitly

$$\begin{vmatrix} 0 & 1 & 0 & 0 \\ 0 & 0 & 1 & 0 \\ 0 & 0 & 0 & 1 \\ (-1)\frac{a_0}{a_4} & (-1)\frac{a_1}{a_4} & (-1)\frac{a_2}{a_4} & (-1)\frac{a_3}{a_4} \end{vmatrix}. \tag{A.12}$$

A.4 ▪ The Chebyshev Companion Matrix

Theorem A.2 (Chebyshev Companion Matrix). *Let $f_N(x)$ denote a polynomial of degree N written in* Chebyshev form *as*

$$f_N(x) = \sum_{j=0}^{N} a_j \, T_j(x). \tag{A.13}$$

Then all roots of $f_N(x)$, both on and off the canonical expansion interval $x \in [-1,1]$, are eigenvalues of the $N \times N$ matrix $\vec{\vec{A}}$ whose elements are

$$A_{jk} = \begin{cases} \delta_{2,k}, & j = 1, \quad k = 1, 2, \ldots, N, \\ \frac{1}{2}\left\{\delta_{j,k+1} + \delta_{j,k-1}\right\}, & j = 2, \ldots, (N-1), \; k = 1, 2, \ldots, N, \\ (-1)\dfrac{a_{k-1}}{2a_N} + (1/2)\delta_{k,N-1}, & j = N, \; k = 1, 2, \ldots, N, \end{cases} \tag{A.14}$$

where δ_{jk} is the usual Kronecker delta function such that $\delta_{jk} = 0$ if $j \neq k$ while $\delta_{jj} = 1$ for all j [338, 339, 200, 22, 198, 344, 146].

For a quintic polynomial, for example, the Chebyshev companion matrix is

$$\begin{vmatrix} 0 & 1 & 0 & 0 & 0 \\ (1/2) & 0 & (1/2) & 0 & 0 \\ 0 & (1/2) & 0 & (1/2) & 0 \\ 0 & 0 & (1/2) & 0 & (1/2) \\ (-1)\dfrac{a_0}{2a_5} & (-1)\dfrac{a_1}{2a_5} & (-1)\dfrac{a_2}{2a_5} & (-1)\dfrac{a_3}{2a_5} + (1/2) & (-1)\dfrac{a_4}{2a_5} \end{vmatrix}. \tag{A.15}$$

Table A.1. *MATLAB function to compute the Chebyshev companion matrix.*

```
function CM = ChebyshevCompanionMatrix(acoeff);
% Input: "acoeff", a vector whose elements are the (N + 1) Chebyshev
% polynomial coefficients of a polynomial f_N(x) of degree N.
% Output: CM is the N x N Chebyshev companion matrix.
% Its eigenvalues are the N roots of f_N(x).
% Post-processing (i) The roots are then xi = eig(CM).
% (ii) zeros outside the real interval ξ ∈ [−1, 1] should be discarded
% (iii) If the Chebyshev expansion is on x ∈ [A, B], then the eigenvalues
% ξ_j ∈ [−1, 1] should be converted by the affine mapping
% x ≡ (B − A)/2ξ + (B + A)/2 back to the original coordinate.
  N = length(acoeff) − 1; CM = zeros(N,N); CM(1,2) = 1;
  for j = 2:(N − 1) CM(j,j − 1) = 0.5; CM(j,j + 1) = 0.5; end;
  for j = 1:N, CM(N,j) = − acoeff(j)/ (2* acoeff(N + 1)); end
  CM(N,N − 1) = CM(N,N − 1) + 1/2;
```

With **acoeff(1:(N+1)** as the vector of Chebyshev coefficients, the sole input, the Chebyshev companion matrix can be computed in MATLAB in just four lines, as given (with more comments than executable lines) in Table A.1.

A.5 ▪ Rational Chebyshev TB, TL, and TM

The rational Chebyshev functions are images of the usual Chebyshev polynomials under a change of coordinate. It follows that the companion matrices for these, and for all basis sets that are images of Chebyshev polynomials under a mapping, are *identical* to the Chebyshev polynomial companion matrix. However, one must use the appropriate formula to map the computed roots in x to roots of the original problem in y.

TB: For $y \in [-\infty, \infty]$,

$$TB_n(y) \equiv T_n(x) \equiv \cos(nt), \tag{A.16}$$

where the coordinates are related via

$$y = \frac{Lx}{\sqrt{1-x^2}}, \qquad x = \frac{y}{\sqrt{L^2+y^2}}, \tag{A.17}$$

$$y = L\cot(t), \qquad t = \operatorname{arccot}(y/L), \tag{A.18}$$

TL: For $y \in [y_0, \infty]$,

$$TL_n(y) \equiv T_n(x) \equiv \cos(nt), \tag{A.19}$$

$$y = y_0 + \frac{L(1+x)}{1-x} \qquad \longleftrightarrow \qquad x = \frac{y-L-y_0}{y+L-y_0}, \tag{A.20}$$

$$y = y_0 + L\cot^2\left(\frac{t}{2}\right) \qquad \longleftrightarrow \qquad t = 2\operatorname{arccot}\left(\sqrt{\frac{y-y_0}{L}}\right). \tag{A.21}$$

Boyd, Rangan, and Bucksbaum suggested the alternative rational basis [106]

$$TM_j(r;L) \equiv \cos(jx(r)), \qquad r = L\cot(x/2)\cos(x/2), \qquad x \in [0,\pi] \text{ [TM Map]}. \tag{A.22}$$

A.6 ▪ Legendre Companion Matrix

Theorem A.3 (Legendre Companion Matrix). *Let $f_N(x)$ denote a polynomial of degree N written in* Legendre form *as*

$$f_N(x) = \sum_{j=0}^{N} s_j \, P_j(x). \tag{A.23}$$

Then all roots of $f_N(x)$, both on and off the canonical expansion interval $x \in [-1,1]$, are eigenvalues of the $N \times N$ matrix $\vec{\vec{P}}$ whose elements are

$$P_{jk} = \begin{cases} \delta_{2,k}, & j=1, \quad k=1,2,\ldots,N, \\ \frac{j-1}{2j-1}\delta_{j,k+1} + \frac{j}{2j-1}\delta_{j,k-1}, & j=2,\ldots,(N-1), \; k=1,\ldots,N, \\ (-1)\frac{N}{2N-1}\frac{s_{j-1}}{s_N} + \frac{N-1}{2N-1}\delta_{k,N-1}, & j=N, \quad k=1,\ldots,N, \end{cases} \tag{A.24}$$

where δ_{jk} is the Kronecker delta function such that $\delta_{jk}=0$ if $j \neq k$ while $\delta_{jj}=1$ for all j.

For a quintic polynomial, for example, the Legendre companion matrix is

$$\begin{vmatrix} 0 & 1 & 0 & 0 & 0 \\ (1/3) & 0 & (2/3) & 0 & 0 \\ 0 & (2/5) & 0 & (3/5) & 0 \\ 0 & 0 & (3/7) & 0 & (4/7) \\ (-1)\frac{5}{9}\frac{s_0}{s_5} & (-1)\frac{5}{9}\frac{s_1}{s_5} & (-1)\frac{5}{9}\frac{s_2}{s_5} & (-1)\frac{5}{9}\frac{s_3}{s_5}+(4/9) & (-1)\frac{5}{9}\frac{s_4}{s_5} \end{vmatrix}. \tag{A.25}$$

The complete MATLAB code to generate the companion matrix is given as Table A.2.

Table A.2. *MATLAB code for the Legendre polynomial companion matrix.*

```
% Input: coefficients s(1),...,s(N+1) for truncated Legendre series of degree N
  P=zeros(N,N); P(1,2)=1; P(N,1)=-( s(1)/ s(N+1) ) * (N/(2*N-1) );
  for j=2:(N-1), P(j,j-1)=(j-1)/(2*j-1); P(j,j+1)=j/(2*j-1);
  P(N,j)=-( s(j)/ s(N+1)) * (N/(2*N-1) ); end
  P(N,N-1)=P(N,N-1)+(N-1)/(2*N-1); P(N,N)=-(N*s(N)/(s(N+1)*(2*N-1)));
  rootsfromeig=eig(P);
% compute Legendre roots
```

A.7 ▪ Gegenbauer Polynomials and Spherical Harmonics

The Gegenbauer polynomials, also known as the ultraspherical polynomials, depend on an order parameter m, which is the superscript on the functions. The polynomials include the Chebyshev and Legendre polynomials as the special cases $m=0$ and $m=1/2$, respectively. The Gegenbauer polynomials for half-integer m are, modulo a nonnegative factor, the latitudinal parts of spherical harmonics. A multiplicity of notations and normalizations are employed. We shall specify the Gegenbauer polynomials uniquely, as in Appendix A of [76], by their recurrence

$$C_0^{(m)} \equiv 1, \qquad C_1^{(m)}(x) \equiv 2\,m\,x, \tag{A.26}$$
$$(n+1)\,C_{n+1}^{(m)}(x) = 2(n+m)x\,C_n^{(m)}(x) - (n+2m-1)\,C_{n-1}^{(m)}(x).$$

Theorem A.4. *The roots of a series of Gegenbauer polynomials,*

$$f_N(x) = \sum_{j=0}^{N} s_j\, C_j^{(m)}(x), \tag{A.27}$$

are the eigenvalues of the matrix whose elements are

$$G_{1k} = \delta_{2k}(1/[2m]), \qquad j = 1, \tag{A.28}$$

$$G_{jk} = \frac{(j-1+2m-1)}{2(j-1+m)}\delta_{j-1,k} + \frac{j}{2(j-1+m)}\delta_{j+1,k},\; j = 2,\ldots,(N-1), \tag{A.29}$$

$$G_{Nk} = -\frac{s_{k-1}\,\frac{N}{2(N-1+m)}}{s_N} + \delta_{N-1,k}\,\frac{(N-2+2m)}{2(N-1+m)}, \qquad j = N. \tag{A.30}$$

For a polynomial of degree four in Gegenbauer form, we have

$$\begin{vmatrix} 0 & 1/[2m] & 0 & 0 \\ \frac{2m}{2(1+m)} & 0 & \frac{2}{2(1+m)} & 0 \\ 0 & \frac{2m+1}{2(2+m)} & 0 & \frac{3}{2(2+m)} \\ \frac{-4}{2(3+m)}\frac{s_0}{s_4} & \frac{-4}{2(3+m)}\frac{s_1}{s_4} & \left[\frac{-4}{2(3+m)}\frac{s_2}{s_4} + \frac{(2m+2)}{2(3+m)}\right] & \frac{-4}{2(3+m)}\frac{s_3}{s_4} \end{vmatrix}. \tag{A.31}$$

A.8 ▪ Hermite Polynomials and Hermite Functions

Theorem A.5 (Hermite Companion Matrix). *Let $f_N(x)$ denote a polynomial of degree N written in* Hermite form *as*

$$f_N(x) = \sum_{j=0}^{N} h_j\, H_j(x) \tag{A.32}$$

or a function that is a truncated series of Hermite functions,

$$f_N(x) = \sum_{j=0}^{N} h_j\, \psi_j(x), \tag{A.33}$$

where $\psi_j(x) = \exp(-[1/2]x^2)H_j(x)$. (Note that the Gaussian factor that is part of the definition of the Hermite functions has no effect on the zeros of the sum of the series.) Then all roots of $f_N(x)$, both on and off the canonical expansion interval $x \in [-\infty, \infty]$, are eigenvalues of the $N \times N$ matrix $\vec{\vec{H}}$ whose elements are

$$H_{jk} = \begin{cases} (1/2)\delta_{2,k}, & j = 1, \quad k = 1,2,\ldots,N, \\ (j-1)\delta_{j,k+1} + \frac{1}{2}\delta_{j,k-1}, & j = 2,\ldots,(N-1), \\ (-1)\frac{1}{2}\frac{h_{j-1}}{h_N} + (N-1)\delta_{k,N-1}, & j = N, \end{cases} \tag{A.34}$$

where δ_{jk} is the usual Kronecker delta function.

For a quintic polynomial, the Hermite companion matrix is

$$\begin{vmatrix} 0 & (1/2) & 0 & 0 & 0 \\ 1 & 0 & (1/2) & 0 & 0 \\ 0 & 2 & 0 & (1/2) & 0 \\ 0 & 0 & 3 & 0 & (1/2) \\ (-1)\frac{h_0}{2h_5} & (-1)\frac{h_1}{2h_5} & (-1)\frac{h_2}{2h_5} & (-1)\frac{h_3}{2h_5}+4 & (-1)\frac{h_4}{2h_5} \end{vmatrix}. \tag{A.35}$$

A.9 ▪ General Orthogonal Polynomials

Theorem A.6 (General Polynomial Companion Matrix). *If a set of polynomials $\psi_n(x)$ satisfy a three-term recurrence relation (TTRR) such that*

$$x\psi_n = B(n)\psi_{n+1} + D(n)\psi_n + E(n)\psi_{n-1}, \tag{A.36}$$

then the roots of the polynomial

$$f_N(x) \equiv \sum_{j=0}^{N} s_j \, \psi_j(x) \tag{A.37}$$

are the eigenvalues of the companion matrix with elements

$$Q_{jk} = \begin{cases} D(0)\delta_{1k} + B(0)\delta_{1,k-1}, & j=1, \\ E(j-1)\delta_{j,k+1} + D(j-1)\delta_{jk} + B(j-1)\,\delta_{j,k-1}, & j=2,\ldots,N-1, \\ -B(N-1)s_{k-1}/s_N + E(N-1)\delta_{N-1,k} + D(N-1)\delta_{Nk}, & j=N, \end{cases} \tag{A.38}$$

where δ_{jk} is the Kronecker delta function.

For a quartic polynomial, the companion matrix is

$$\begin{vmatrix} D(0) & B(0) & 0 & 0 \\ E(1) & D(1) & B(1) & 0 \\ 0 & E(2) & D(2) & B(2) \\ \frac{-B(3)\,s_0}{s_4} & \frac{-B(3)\,s_1}{s_4} & \frac{-B(3)\,s_2}{s_4} + E(3) & \frac{-B(3)\,s_3}{s_4} + D(3) \end{vmatrix}. \tag{A.39}$$

A.10 ▪ Fourier Companion Matrices

A.10.1 ▪ Complex Companion Matrix (CCM)

The transformation

$$z = \exp(it) \tag{A.40}$$

converts a *trigonometric* polynomial $f_N(t)$ with $2N+1$ terms into an *ordinary* polynomial $h(z)$ of degree $2N$, as independently discovered several times [366, 78, 37]. The *associated polynomial*, $h(z[t]) \equiv \exp(iNt) f_N(t)$, is

$$h(z) = \frac{1}{2}\sum_{k=0}^{2N} h_k\, z^k \equiv z^N f(t[z]), \tag{A.41}$$

where

$$h_j = \begin{cases} a_{N-j} + i\, b_{N-j}, & j=0,1,\ldots,(N-1), \\ 2a_0, & j=N, \\ a_{j-N} - i\, b_{j-N}, & j=N+1,N+2,\ldots,(2N). \end{cases} \tag{A.42}$$

From this transformation comes the following.

Theorem A.7 (Complex Companion Matrix). *Define the trigonometric polynomial*

$$f_N(t) \equiv \sum_{j=0}^{N} a_j \cos(jt) + \sum_{j=1}^{N} b_j \sin(jt). \tag{A.43}$$

The matrix elements B_{jk} of the Frobenius matrix for a trigonometric polynomial of general degree N (and therefore $(2N+1)$ terms) are

$$B_{jk} = \begin{cases} \delta_{j,k-1}, & j = 1,2,\ldots,(2N-1), \\ (-1)\dfrac{h_{k-1}}{a_N - i b_N}, & j = 2N, \end{cases} \tag{A.44}$$

where δ_{jk} is the usual Kronecker delta function such that $\delta_{jk} = 0$ if $j \neq k$ while $\delta_{jj} = 1$ for all j and k, and the h_j are defined by (A.42).

The roots t_k of $f_N(t)$ are the negative of $\sqrt{-1}$ times the logarithm *of the matrix eigenvalues z_k:*

$$t_{k,m} \equiv arg(z_k) + 2\pi m - i \log(|z_k|), \qquad k = 1,2,\ldots,2N,\ m = integer. \tag{A.45}$$

In particular, the real-valued roots of $f_N(t)$ for real $t \in [-\pi,\pi]$ are the angles of the roots of $h(z)$ on the unit circle. Equivalently, each real-valued root t_k of $f(t)$ on $t \in (-\pi,\pi]$ is connected to a root z_k of the associated polynomial through $t_k = arg(z_k)\ \forall\, k$ such that $|z_k| = 1$. Here $arg(z)$ is the usual complex argument function such that, for $z = |z|\exp(i\theta)$, $arg(z) = \theta$ [83].

For $N = 2$, the Fourier–Frobenius matrix is explicitly

$$\begin{vmatrix} 0 & 1 & 0 & 0 \\ 0 & 0 & 1 & 0 \\ 0 & 0 & 0 & 1 \\ (-1)\dfrac{a_2 + i b_2}{a_2 - i b_2} & (-1)\dfrac{a_1 + i b_1}{a_2 - i b_2} & (-1)\dfrac{2a_0}{a_2 - i b_2} & (-1)\dfrac{a_1 - i b_1}{a_2 - i b_2} \end{vmatrix}. \tag{A.46}$$

A.10.2 ▪ The Fourier Division Companion Matrix (FDCM)

This alternative Fourier companion matrix has real-valued elements, but, even so, the CCM is recommended for most situations. The eigenvalues are equal to *double* $\cos(t)$, so one must halve the eigenvalues before taking the inverse arccosine. Otherwise, the postprocessing is similar to the previous section.

Using the auxiliary variables

$$w_1 = -1/2\,\frac{-a_N b_{N-1} + b_N a_{N-1}}{\left(a_N{}^2 + b_N{}^2\right) a_N}, \tag{A.47}$$

$$w_2 = \frac{b_N}{a_N{}^2 + b_N{}^2}, \tag{A.48}$$

$$w_3 = \frac{a_N}{a_N{}^2 + b_N{}^2}, \tag{A.49}$$

the elements of the companion matrix $\vec{\vec{M}}$ are

$$M_{1,j} = 2\delta_{j,2}, \tag{A.50}$$

$$M_{n,j} = \delta_{j,n-1} + \delta_{j,n+1}, \qquad n = 2,\ldots,(N-1), \tag{A.51}$$

$$M_{N+1,j} = \delta_{j,N+2}, \tag{A.52}$$

$$M_{n,j} = \delta_{j,n-1} + \delta_{j,n+1}, \qquad n = (N+2),\ldots,(2N-1), \tag{A.53}$$

$$M_{N,j} = \delta_{j,N-1} - \frac{a_{j-1}}{a_N}, \qquad j = 1, \ldots, N, \tag{A.54}$$

$$M_{N,j} = -\frac{b_{j-N}}{a_N}, \qquad j = N+1, \ldots, 2N, \tag{A.55}$$

$$\begin{aligned}
M_{2N,1} &= -2w_1 a_0 - w_2 a_1 - w_3 b_1, \\
M_{2N,2} &= -2w_1 a_1 - w_2(2a_0 + a_2) + w_3(-b_2), \\
M_{2N,N+1} &= -2w_1 b_1 - w_2(b_2) + w_3(a_2 - 2a_0), \\
M_{2N,2N-1} &= -2w_1 b_{N-1} - w_2(b_{N-2} + b_N) + w_3(a_N - a_{N-2}) + 1, \\
M_{2N,2N} &= -2w_1 b_N - w_2(b_{N-1}) + w_3(-a_{N-1}), \\
M_{2N,n} &= -2w_1 a_{n-1} - w_2(a_{n-2} + a_n) + w_3(b_{n-2} - b_n), \qquad n = 3, \ldots, N, \\
M_{2N,n+N} &= -2w_1 b_n - w_2(b_{n-1} + b_{n+1}) + w_3(a_{n+1} - a_{n-1}), \qquad n = 2, \ldots, (N-2).
\end{aligned} \tag{A.56}$$

Appendix B
Chebyshev Interpolation and Quadrature

"Sciens et prudens manum misi in ignem. (Knowingly and carefully I put my hand in the fire.)"

—Marie-Joseph Lagrange, O. P. (1855–1938), quoting St. Jerome [274]

B.1 ▪ Mapping $x \in [A, B]$ to $\xi \in [-1, 1]$

$$x \equiv \frac{B-A}{2}\xi + \frac{B+A}{2}, \tag{B.1}$$

$$\xi \equiv \frac{2}{B-A}\left(x - \frac{B+A}{2}\right) \tag{B.2}$$

$$= \frac{2x-A-B}{B-A}. \tag{B.3}$$

B.2 ▪ Chebyshev Interpolation of a Function $f(x)$

Our goal is to compute a Chebyshev series, including terms up to and including T_N, on the interval $x \in [A,B]$.

Step 1: Create the interpolation points (Lobatto grid):

$$x_k \equiv \frac{B-A}{2}\cos\left(\pi\frac{k}{N}\right) + \frac{B+A}{2}, \qquad k = 0,1,2,\ldots,N. \tag{B.4}$$

Step 2: Compute the elements of the $(N+1)\times(N+1)$ interpolation matrix.

Define $p_j = 2$ if $j = 0$ or $j = N$ and $p_j = 1, j \in [1, N-1]$. Then the elements of the interpolation matrix are

$$\mathcal{I}_{jk} = \frac{2}{p_j\, p_k\, N}\cos\left(j\pi\frac{k}{N}\right). \tag{B.5}$$

Step 3: Compute the grid point values of $f(x)$, the function to be approximated:

$$f_k \equiv f(x_k), \qquad k = 0,1,\ldots,N. \tag{B.6}$$

Table B.1. *Maple code for Clenshaw's recurrence to sum a Chebyshev series at a point x.*

```
ClenshawSumChebSeries := proc(a,N,A,B,x) local b0, b1,b2,b3,fpoly,j,xi;
# This sums a Chebyshev series at a point x on the interval x ∈ [A,B];
# The elements of "a", a[1],...,a[N+1] are the Chebyshev coefficients of f(x).;
# f_N(x) = Σ_{n=0}^N a[n+1]T_n(x)
  xi := ( 2*x - (B+A))/(B-A); b1 := 0; b2 := 0;
  for j from 1 to (N+1)do b0 := evalf( 2 *xi * b1 - b2 + a[N+2-j ] );
  b3 := b2; b2 := b1; b1 := b0; od:
  fpoly := evalf( (1/2) *(b0 - b3 ) + (1/2)*a[1]) ; end proc;
```

Step 4: Compute the coefficients through a vector–matrix multiply:

$$a_j = \sum_{k=0}^{N} \mathcal{I}_{jk} f_k, \qquad j = 0, 1, 2, \ldots, N. \tag{B.7}$$

The approximation is

$$f \approx \sum_{j=0}^{N} a_j T_j\left(\frac{2x-(B+A)}{B-A}\right) = \sum_{j=0}^{N} a_j \cos\left\{ j \arccos\left(\frac{2x-(B+A)}{B-A}\right)\right\}. \tag{B.8}$$

B.2.1 ▪ Summing the Chebyshev Approximation

The truncated Chebyshev series

$$f_N \equiv \sum_{j=0}^{N} a_j T_j\left(\frac{2x-(B+A)}{B-A}\right) = \sum_{j=0}^{N} a_j \cos\left\{ j \arccos\left(\frac{2x-(B+A)}{B-A}\right)\right\} \tag{B.9}$$

can be written without approximation in terms of trigonometric and inverse trigonometric functions as

$$f_N = \sum_{j=0}^{N} a_j \cos\left\{ j \arccos\left(\frac{2x-(B+A)}{B-A}\right)\right\}. \tag{B.10}$$

However, it is faster to use the following recurrence relation.

First, initialize by defining

$$\xi = \frac{2x-(B+A)}{B-A}, \qquad b_1 = 0, \qquad b_2 = 0. \tag{B.11}$$

Second, apply the loop N times:

$$b_0 = 2\xi\, b_1 - b_2 + a_{N+1-j}, j = 1, 2, \ldots, N, \qquad b_2 = b_1, \qquad b_1 = b_0. \tag{B.12}$$

Then

$$f_N = (b_0 - b_3)/2 + a_0/2. \tag{B.13}$$

Tables B.1 and B.2 provide Maple and MATLAB implementations.

Fox and Parker give an error analysis of Clenshaw's recurrence on pp. 57–58 of [176]. They write, "From this we deduce that our recurrent process is perfectly stable, the local rounding errors are not amplified, and the upper bound to the error cannot exceed the arithmetical sum of the individual local rounding errors."

Table B.2. *MATLAB code for Clenshaw's recurrence to sum a Chebyshev series at a point x.*

```
function fpoly = ChebsumClenshaw(a,A,B,x);
% sums a Chebyshev series
% at a point x in [A,B] where the coefficients
% a(1), a(2),...,a(N) are the Chebyshev coefficients of f(x).
  xi = ( 2*x - (B + A))/(B - A);b0 = 0*xi; b1 = b0; b2 = b0; b3 = b0; N = length(a);
  for j = 1:N, b0 = 2 *xi .* b1 - b2 + a(N + 1 - j );
  b3 = b2; b2 = b1; b1 = b0; end
  fpoly = (1/2) *(b0 - b3 ) + 0.5*a(1);
```

B.2.2 ▪ Computing the Chebyshev Coefficients of Derivatives

Let

$$\frac{df}{dx} = \sum_{k}^{N} a_k^{(1)} T_k(x) \tag{B.14}$$

so that the superscript (1) denotes the coefficients of the first derivative. These may be computed from the Chebyshev coefficients of the $f(x)$ itself by the recurrence relation (in descending order)

$$S = \frac{2}{B-A}, \tag{B.15}$$

$$a_N^{(1)} = 0, \tag{B.16}$$

$$a_{N-1}^{(1)} = 2SNa_N, \tag{B.17}$$

$$a_{k-1}^{(1)} = a_{k+1}^{(1)} + 2Ska_k, \qquad k = N-1, N-2, N-3, \ldots, 2,$$

$$a_0^{(1)} = a_2^{(1)}/2 + Sa_1. \tag{B.18}$$

Tables B.3 and B.4 are Maple and MATLAB implementations.

Table B.3. *Maple code for the recurrence to compute the Chebyshev coefficients of the first derivative df/dx from those of $f(x)$.*

```
# Compute Chebyshev coefficients for df/dx from those of f itself.;
# a[N + 1] is the coefficient of T(N,xi) where xi = ( 2*x - (B + A))/(B - A);
# The expansion interval is x ∈ [A,B];
# Input: integer "N", the degree of f(x), plus a vector "a" of length (N + 1);
# which stores the Chebyshev coefficients of f(x), plus the endpoints (A, B).;
# Output: vector "ax" of length (N + 1) with Chebyshev coefficients of df/dx.;
  S := 2/(B - A): # S scales the derivatives to the length of the interval [A,B].;
  ax[N + 1] := 0; ax[N] := evalf(2*N*a[N + 1]);
  for k from (N - 1) by (-1) to 2 do ax[k] := evalf( ax[k + 2] + 2*k*a[k + 1]); od:
  ax[1] := evalf( ax[3]/2 + a[2]);
# Replace df/dxi by df/dx = S df/dxi where xi ∈ [-1,1];
# where S = dxi/dx = 2/(B - A);
  for j from 1 to N do ax[j] := evalf( S*ax[j]); od:      return ax;      end proc;
```

Table B.4. *MATLAB code for the recurrence to compute the Chebyshev coefficients of the first derivative df/dx from those of $f(x)$.*

```
% Compute coefficients of the Chebyshev series for du/dx from those of u itself.
% a(N+1) is the coefficient of T(N,xi) where xi=(2*x-(B+A))/(B-A)
% The expansion interval is x ∈ [A,B]
  function ax=ChebDerivCofromChebco(a,N,A,B);
% Input: N, and (N+1)-dimensional vector of Chebyshev coefficients "a"
% endpoints of expansion interval [A, B].
% Output: vector "ax" of length (N+1) with Chebyshev coefficients of df/dx
% where the coefficients of f(x) are the input vector "a"
  S=2/(B-A); % scales the derivatives to the length of the interval [A,B].
  ax(N+1)=0; ax(N)=2*N*a(N+1);
  for k=(N-1):-1:2, ax(k)=ax(k+2)+2*k*a(k+1); end
  ax(1)=ax(3)/2+a(2); % convert df/dxi (xi ∈ [-1, 1]) to df/dx
% by multiplication by dxi/dx=S=2/(B-A);
  ax=S*ax;
```

B.2.3 ▪ Avoiding Endpoint Overflow: Chebyshev–Gauss Interpolation

If the target $f(x)$ has an apparent singularity at an endpoint, or is not singular, or is defined by a formula containing $1/x$, overflow errors can be avoided by using the Gauss/interior grid. Theory shows that this is just as accurate as the Lobatto grid described earlier in this section [76, 176]. Adaptive interpolation with doubling of N at each pass is still feasible, but no points of the N-point Gauss coincide with points of the $2N$-point grid, which makes adaptation more costly.

It is important to note that the switch in grids removes only overflow/divide-by-zero problems. Genuine endpoint singularities can and must be handled by the strategies described in section 2.13 and in the author's book [76].

Step 1: Create the interpolation points (roots grid):

$$x_k \equiv \frac{B-A}{2}\cos\left(\pi\frac{2k+1}{2N+2}\right)+\frac{B+A}{2}, \qquad k=0,1,2,\ldots,N. \tag{B.19}$$

Step 2: Compute the elements of the $(N+1)\times(N+1)$ interpolation matrix.

Define $c_j = 2$ if $j=0$ and $c_j = 1\,\forall j>0$. Then the elements of the interpolation matrix are

$$\mathcal{I}_{jk}^{roots} = \frac{2}{c_j(N+1)}\cos\left(j\pi\frac{2k+1}{2N+2}\right). \tag{B.20}$$

Step 3: Compute the grid point values of $f(x)$, the function to be approximated:

$$f_k \equiv f(x_k), \qquad k=0,1,\ldots,N. \tag{B.21}$$

Step 4: Compute the coefficients through a vector–matrix multiply:

$$a_j = \sum_{k=0}^{N}\mathcal{I}_{jk}^{roots} f_k, \qquad j=0,1,2,\ldots,N. \tag{B.22}$$

The approximation is

$$f \approx \sum_{j=0}^{N} a_j T_j\left(\frac{2x-(B+A)}{B-A}\right) = \sum_{j=0}^{N} a_j \cos\left\{j\arccos\left(\frac{2x-(B+A)}{B-A}\right)\right\}. \tag{B.23}$$

Repeat the four steps with different N until the differences between interpolants of different N are smaller than a user-chosen tolerance.

B.3 ▪ Rational Chebyshev Interpolation on a Semi-infinite Domain (TL_j)

Our goal is to approximate a function $f(x)$ by a rational Chebyshev series, including terms up to and including TL_N, on the interval $y \in [0, \infty]$.

Step 1: Choose a map parameter L. ($L = 2$ is a good default choice; some experimentation may be needed; the optimum L is that which yields the fastest-converging Chebyshev coefficients.)

Step 2: Create the interpolation points:

$$y_k \equiv L\cot^2\left(\pi\frac{2k+1}{4N+4}\right), \qquad k = 0, 1, 2, \ldots, N. \tag{B.24}$$

Step 3: Compute the elements of the $(N+1) \times (N+1)$ interpolation matrix.

Define $c_j = 2$ and $c_j = 1, j \in [1, N-1]$. Then the elements of the interpolation matrix are

$$\mathcal{I}_{jk} = \frac{2}{c_j\,(N+1)} \cos\left(j\pi\frac{2k+1}{2N+2}\right), j = 0, 1, \ldots, N, \qquad k = 0, 1, \ldots, N. \tag{B.25}$$

Step 4: Compute the grid point values of $f(x)$, the function to be approximated:

$$f_k \equiv f(y_k), \qquad k = 0, 1, \ldots, N. \tag{B.26}$$

Step 5: Compute the coefficients through a vector–matrix multiply:

$$a_j = \sum_{k=0}^{N} \mathcal{I}_{jk} f_k, \qquad j = 0, 1, 2, \ldots, N. \tag{B.27}$$

The approximation is

$$f \approx \sum_{j=0}^{N} a_j\, TL_j(y; L) = \sum_{j=0}^{N} a_j \cos\left\{j\, 2\operatorname{arccot}\left(\sqrt{y/L}\right)\right\}. \tag{B.28}$$

B.4 ▪ Rational Chebyshev Interpolation on an Infinite Domain (TB_j)

Our goal is to approximate a function $f(x)$ by a rational Chebyshev series, including terms up to and including TB_N, on the interval $y \in [0, \infty]$.

Step 1: Choose a map parameter L. Some experimentation may be needed; the optimum L is that which yields the fastest-converging Chebyshev coefficients.

Step 2: Create the interpolation points:

$$y_k \equiv L\cot\left(\pi\frac{2k+1}{2N+2}\right), \qquad k = 0, 1, 2, \ldots, N. \tag{B.29}$$

Step 3: Compute the elements of the $(N+1) \times (N+1)$ interpolation matrix.

Define $c_j = 2$ and $c_j = 1, j \in [1, N-1]$. Then the elements of the interpolation matrix are

$$\mathscr{I}_{jk} = \frac{2}{c_j\,(N+1)} \cos\left(j\pi\frac{2k+1}{2N+2}\right), j = 0, 1, \ldots, N, \qquad k = 0, 1, \ldots, N. \tag{B.30}$$

Step 4: Compute the grid point values of $f(x)$, the function to be approximated:

$$f_k \equiv f(y_k), \qquad k = 0, 1, \ldots, N. \tag{B.31}$$

Step 5: Compute the coefficients through a vector–matrix multiply:

$$a_j = \sum_{k=0}^{N} \mathscr{I}_{jk} f_k, \qquad j = 0, 1, 2, \ldots, N. \tag{B.32}$$

The approximation is

$$f \approx \sum_{j=0}^{N} a_j\; TB_j(y; L) = \sum_{j=0}^{N} a_j\; \cos\{j\, \mathrm{arccot}(y/L)\}. \tag{B.33}$$

B.5 ▪ Clenshaw–Curtis Adaptive Chebyshev Quadrature

The *quadrature problem* is to evaluate an integral such as

$$I \equiv \int_a^b f(x)\omega(x)dx \tag{B.34}$$

by an approximation of the form

$$I \approx \sum_{j=1}^{N} w_j\, f(x_j). \tag{B.35}$$

Here, $\omega(x) \geq 0$ is a user-chosen weight function and $f(x)$ is a smooth but otherwise arbitrary function. The w_j are *quadrature weights*. Both quadrature weights and abscissas are *independent* of $f(x)$ and depend only on the interval $[a, b]$ and the weight function $\omega(x)$.

Clenshaw and Curtis noted that the Chebyshev–Lobatto grid of $(N+2)$ points [129],

$$x_j = \cos\left(\frac{\pi j}{N+1}\right), \qquad j = 0, 1, \ldots, (N+1), \tag{B.36}$$

has the property that when $N+1$ is doubled, all points on the old grid are also part of the new, higher-resolution grid. This allows inexpensive computation of the same integral for multiple resolutions because it is only necessary to evaluate $f(x)$ on the finest grid, which provides all the evaluations of f needed for the coarser grids, too.

Automatic adaptivity also requires an estimate for the error E_N.The simplest and most conservative is

$$E_N^{estimate} \equiv |I_N - I_{N/2}|, \tag{B.37}$$

where I_N and $I_{N/2}$ denote the $(N+2)$-point and $[(N+1)/2+1]$-point approximations to the integral I. In words, when $N+1$ is doubled, the error is almost certainly less than the difference between two successive estimates of the integral, provided that this difference is small. Gentleman, who discusses and compares several estimates [192], says, "Considerable experience with the subroutine CQUAD [191], however, indicates that this simple estimate [Eq. (B.37)] appears unusually realistic here [for quadrature]....

The naive estimate also has the advantage that when the integral is being computed to essentially full-word accuracy, a reasonable indication of the representational rounding error in the answer is frequently given. The alternative error estimates ... ignore this—often to one's embarrassment."

Because the transformed integrals have integrands that are *antisymmetric* with respect to $t = 0$ for the finite and semi-infinite intervals, Boyd [63] expanded the product of f with the metric factor in terms of *sine* cardinal functions. Thus, the endpoints are *omitted* for these two cases so that each uses only the N interior points of the $(N+2)$-point Lobatto grid.

Alternatively, one may expand only f as a Chebyshev or Fourier series and then evaluate integrals of the products of the cardinal functions with the metric factor, which is $\sin(t)$ for the finite interval. This requires two additional evaluations of f but raises the degree of f for which the approximation is exact only by one. The quadrature weights are given by Fraser and Wilson as their Eq. (4.3) [177].

Clenshaw and Curtis [129] and Gentleman prefer to compute the Chebyshev series for $f(x)$ first by a Fast Fourier Transform (FFT) and then integrate term by term through a recurrence formula [129, 192, 193, 191]. This is an $O(N \log_2(N))$ procedure, whereas the cost of evaluating the weights through Boyd's method is $O(N^2)$. Since these weights need only be computed once for a given N, however, we recommend Boyd's procedure.

B.5.1 ▪ Mechanics

Boyd has given a general treatment for Curtis–Clenshaw quadrature that embraces the infinite and semi-infinite intervals as well as $x \in [-1,1]$ [63]. The first step is to transform the interval in x to the trigonometric coordinate t. The second step is to approximate the integral by an evenly spaced quadrature formula on the interval $t \in [0,\pi]$. The quadrature weights are simply the integrals of the trigonometric cardinal functions with the metric factor that results from the change of coordinates.

B.5.2 ▪ Quadrature on a Finite Interval with a Nonperiodic Integrand: Chebyshev Polynomials

The transformation of the integral is

$$x = \cos(t), \tag{B.38}$$

$$I = \int_{-1}^{1} f(x)dx = \int_{0}^{\pi} f(\cos(t))\sin(t)dt. \tag{B.39}$$

The quadrature approximation is

$$I_N \equiv \sum_{j=1}^{N} w_j f(\cos(t_j)), \tag{B.40}$$

$$t_j \equiv \pi j/(N+1), \qquad j = 1,2,\dots,N, \tag{B.41}$$

$$w_j \equiv \sin(t_j)\frac{2}{N+1}\sum_{m=1}^{N} \sin(m t_j)[1-\cos(m\pi)]/m. \tag{B.42}$$

B.5.3 ▪ Periodic Integrand: Trapezoidal Rule

If the integrand $f(x)$ is *periodic* with period 2π, the trapezoidal rule is *spectrally accurate* (the trapezoidal rule is only second order for *nonperiodic* integrands):

$$I \equiv \int_0^{2\pi} f(x)dx \approx \sum_{j=1}^{N} w_j f(x_j), \tag{B.43}$$

where

$$x_j = 2\pi \frac{(j-1)}{N}, \qquad j = 1,2,\ldots,N, \tag{B.44}$$

$$w_j = \frac{2\pi}{N} \qquad \text{for all } j. \tag{B.45}$$

Appendix C

Marching Triangles

"An algorithm which is conveniently and extensively analyzed and implemented is no substitute for common sense, but seems often to serve as such. There are few final answers of any generality in a model-building activity such as this, at best only partial, tentative, indicative answers ..."

—Donald G. M. Anderson in the introduction to mimeographed Harvard University lecture notes on numerical analysis (1970)

This appendix discusses the essential but elementary mechanics and properties of bivariate linear interpolation, barycentric coordinates, and contouring piecewise linear functions.

C.1 ▪ Bivariate Linear Interpolation

Theorem C.1 (Bivariate Linear Polynomial Interpolation Formulas). *Define a triangle by its three vertices, $(x_1,y_1),(x_2,y_2),(x_3,y_3)$. Let the values of a function $f(x,y)$ at its vertices be*

$$f_1 = f(x_1,y_1), f_2 = f(x_2,y_2), f_3 = f(x_3,y_3). \tag{C.1}$$

Then the linear polynomial interpolating $f(x,y)$ through the vertices is

$$P(x,y) = ax + by + c, \tag{C.2}$$

where

$$a = (-y_2 f_3 + y_1 f_3 - y_1 f_2 - f_1 y_3 + f_1 y_2 + f_2 y_3)/d, \tag{C.3}$$
$$b = -(x_2 f_1 + x_1 f_3 - x_3 f_1 - x_1 f_2 - x_2 f_3 + x_3 f_2)/d, \tag{C.4}$$
$$c = (y_2 x_1 f_3 - f_2 x_1 y_3 - y_1 x_2 f_3 + y_1 x_3 f_2 + f_1 x_2 y_3 - y_2 x_3 f_1)/d, \tag{C.5}$$

and

$$d = (-x_1 y_3 + x_2 y_3 - x_2 y_1 - x_3 y_2 + x_3 y_1 + x_1 y_2). \tag{C.6}$$

C.2 ▪ Properties of Bivariate Linear Interpolation

Parameterize a line segment as

$$x = \gamma t + \upsilon, \qquad y = \delta t + \chi, \; t \in [t_1, t_2], \tag{C.7}$$

where the Greek letters are all constants. Substitute this parameterization into the equation of the bivariate linear interpolant:

$$P(x,y) = ax + by + c \quad \Rightarrow \quad P = a[\gamma t + \upsilon] + b[\delta + \chi] + c[a\gamma + b\delta]t + \upsilon + \chi + c. \tag{C.8}$$

This shows that along an arbitrary line segment in the (x,y)-plane, the interpolant varies linearly with the parameter. It further implies that along the line segment,

$$P \in [P(t_1), P(t_2)]. \tag{C.9}$$

In other words, the value of the interpolant is constrained to live between its values at the endpoints of the line segment.

Theorem C.2 (Bivariate Polynomial Vertex Value Zero-Free Test on a Triangle). *Let the values of a function $f(x,y)$ at its vertices be*

$$f_1 = f(x_1, y_1), f_2 = f(x_2, y_2), f_3 = f(x_3, y_3). \tag{C.10}$$

These are also the vertex values of the interpolant that fits these vertex values. If these vertex values are all positive or all negative, then the bivariate linear interpolant to the vertex values of $f(x,y)$ cannot have a zero or a zero contour inside the triangle or on its boundaries.

Proof: Along each of the sides of the triangle, the bivariate linear interpolant is simply a one-dimensional linear interpolant; it is the straight line drawn between its values at the two vertices that are the ends of a given side of the triangle. It follows that if f at both vertices bounding a given triangle side is positive, then the bivariate linear interpolant will be positive everywhere along that side of the triangle also. It follows that if all three vertex values are positive, then the interpolant will be positive on all three sides of the triangle. If we now draw a line from an arbitrary point on one side of the triangle to an arbitrary point on one of the other sides, the interpolant will again be a one-dimensional linear function along that line. Because the function is linear, if it is positive at both of the endpoints of the line segment, then it must be positive everywhere on the line segment connecting those two points. If we imagine covering the area of the triangle with an ever-finer collection of such lines drawn at arbitrary angles, it becomes clear that it is impossible for the linear interpolant to be negative or even zero anywhere within the triangle or its boundaries. If all the vertex values are negative, then the same argument applies with merely a reversal of all the signs. ■

Theorem C.3 (Different Signs of $f(x,y)$ at Vertices of a Triangle Implies a Zero Contour in the Triangle). *If a function $f(x,y)$ is positive at one vertex of a triangle and negative at one or more of the other vertices of the same triangle, then the function has at least one zero contour on that triangle.*

Proof: Along the side connecting a vertex where the function is positive with a vertex where the function is negative, one can apply the usual univariate mean value theorem to conclude that the function must take every value between its extremes at the vertices somewhere on the side of the triangle. Thus, the function has a zero along the side of the triangle. If the function is smooth, then its contours will be continuous and must extend into the interior of the triangle also. ■

C.3 ▪ Barycentric Coordinates

These off-beat coordinates provide an easy test for inclusion within the triangle: A point is in the triangle if and only if $0 \leq \lambda_j \leq 1$ for all three j.

The coordinates are

$$\lambda_1 = \frac{(y_2 - y_3)(x - x_3) + (x_3 - x_2)(y - y_3)}{\Delta}, \tag{C.11}$$

$$\lambda_2 = \frac{((y_3 - y_1)(x - x_3) + (x_1 - x_3)(y - y_3))}{\Delta}, \tag{C.12}$$

$$\lambda_3 = 1 - \lambda_1 - \lambda_2, \tag{C.13}$$

where

$$\Delta = (x_1 - x_3)(y_2 - y_3) - (x_2 - x_3)(y_1 - y_3). \tag{C.14}$$

C.4 ▪ Contouring of Bivariate Linear Polynomials

Theorem C.4 (Line through Two Points). *The line through the points (x_1, y_1) and (x_2, y_2) is*

$$\Lambda = (y_2 - y_1)x - (x_2 - x_1)y + (x_2 y_1 - x_1 y_2). \tag{C.15}$$

Theorem C.5 (Line through Triangle Vertices). *The lines through the three vertices of a triangle are*

$$L_{12} = (y_2 - y_1)x - (x_2 - x_1)y + (x_2 y_1 - x_1 y_2), \tag{C.16}$$

$$L_{23} = (y_3 - y_2)x - (x_3 - x_2)y + (x_3 y_2 - x_2 y_3), \tag{C.17}$$

$$L_{31} = (y_1 - y_3)x - (x_1 - x_3)y + (x_1 y_3 - x_3 y_1), \tag{C.18}$$

Theorem C.6 (Intersection of Two Lines). *The point (x, y) where two lines, given implicitly by*

$$ax + by + c = 0, \tag{C.19}$$

$$a'x + b'y + c' = 0, \tag{C.20}$$

intersect are the solution of the matrix problem

$$\begin{vmatrix} a & b \\ a' & b' \end{vmatrix} \begin{vmatrix} x \\ y \end{vmatrix} = \begin{vmatrix} -c \\ -c' \end{vmatrix}. \tag{C.21}$$

To contour a linear, bivariate polynomial $Q(x, y)$ at isoline value $\mathfrak{c}$, we first determine if $\mathfrak{c} \in [\min(f_1, f_2, f_3), \max(f_1, f_2, f_3)]$. This condition is necessary so that the contour intersects the triangle; if this condition is not satisfied, then we stop.

Next, compute intersections of the straight line $Q(x,y) - c$ with each of the three lines L_{jk} connecting pairs of vertices of the triangle. Next, determine which of these intersections lies between a pair of vertices of the triangle. These intersections with the side of the triangle determine the endpoints of the line segment that is the desired contour.

Degenerate cases are possible. A contour line could, for example, coincide with one of the three sides of the triangle. However, such a line does not require special treatment.

Appendix D

Imbricate-Fourier Series and the Poisson Summation Theorem

Suppose that $G(x)$ is a function that decays as fast as $1/x^2$ or faster as $x \to \infty$. It is always possible to *periodize* or *imbricate* such a function so as to generate a periodic function $f(x)$. Imagine that a mathematical photocopier generates an infinite number of identical copies of $G(x)$. Place one copy centered at $x = 0$, two more at $x = \pm P$, and another two at $x = \pm 2P$ and continue until one copy of $G(x)$ is centered on $x = \pm m$ for all integers m. Define $f(x)$ as the sum of all these clones:

$$f(x) \equiv \sum_{m=-\infty}^{\infty} G(x - mP). \tag{D.1}$$

The function $G(x)$ is the *pattern function* for the series. The function so constructed is obviously periodic with period P, that is, $f(x + P) = f(x)$ for all x because increasing x by P merely shifts the infinite chain of identical copies. It is also trivial to prove that the imbricate series converges uniformly; if $|G(x)| \leq C$ for $x \in [-1, 1]$ and $|G(x)| < C/x^2$ for all real $|x| > 1$, then each term of the imbricate series at a given x is bounded by a constant times a term of the convergent series $\Sigma = \sum_{m=1}^{\infty} 1/m^2$, which is known to converge to the finite sum $\pi^2/6$. But can one say more?

Remarkably, the Poisson Summation Theorem asserts that the Fourier coefficients of $f(x)$ are the values at integral n of $g(n)$, the Fourier transform of the pattern function $G(x)$ of the imbricate series of $f(x)$.

Theorem D.1 (Poisson Summation Theorem). *Any periodic function with period 2π has the two series representations*

$$f(x) = \sum_{n=-\infty}^{\infty} g(n) \exp(inx) = \sum_{m=-\infty}^{\infty} G([x - m2\pi]), \tag{D.2}$$

where $g(n)$ is the Fourier transform of $G(k)$:

$$g(n) = \frac{1}{2\pi} \int_{-\infty}^{\infty} G(x) \exp(-ixn)\, dx, \tag{D.3}$$

$$G(x) = \int_{-\infty}^{\infty} g(n) \exp(ixn)\, dn. \tag{D.4}$$

Proof: Observe that by the change of variable $x - 2\pi m = y$,

$$\frac{1}{2\pi}\int_{-\pi}^{\pi} G(x-2\pi m)\,\exp(-inx)\,dx = \frac{1}{2\pi}\int_{2\pi m-\pi}^{2\pi m+\pi} G(y)\,\exp(-iky)\,dy, \qquad \text{(D.5)}$$

where we have used $\exp(-in[y+2\pi m]) = \exp(-iny)$. The usual Fourier coefficient integral is then

$$g(n) = \frac{1}{2\pi}\int_{-\pi}^{\pi} f(x)\,\exp(-ixn)\,dx \qquad \text{(D.6)}$$

$$= \frac{1}{2\pi}\int_{-\pi}^{\pi}\left(\sum_{m=-\infty}^{\infty} G([x-m2\pi])\right)\exp(-ixn)\,dx \qquad \text{(D.7)}$$

$$= \frac{1}{2\pi}\sum_{m=-\infty}^{\infty}\int_{-\pi}^{\pi} G([x-m2\pi])\exp(-ixn)\,dx \qquad \text{(D.8)}$$

$$= \frac{1}{2\pi}\sum_{m=-\infty}^{\infty}\int_{2\pi m-\pi}^{2\pi m+\pi} G(x)\exp(-ixn)\,dx \qquad \text{(D.9)}$$

$$= \frac{1}{2\pi}\int_{-\infty}^{\infty} G(x)\exp(-ixn)\,dx. \qquad \text{(D.10)}$$

■

Imbricate series are useful in the theory of traveling waves and solitons [128, 56], wavelets [140], elliptic and hyperelliptic functions [59], asymptotic series of spheroidal wave functions and spherical harmonics, convergence theory and coefficient analysis for radial basis functions [109], series acceleration [256], and a wide variety of other applications [66, 103, 67, 72]. *Imbricate* means to overlap in a regular order like roof tiles; it was used thus in architecture, zoology, medicine, sedimentology, and geology before being appropriated for mathematics some decades ago.

Glossary

"A man of fire-new words."
— W. Shakespeare, *Love's Labour Lost*

ADELPHIC POINT: A cardinal function by definition is zero at all of the interpolation points but one. If x_j is the exceptional point, then by convention the cardinal function is assigned the ordinal j, and x_j is the adelphic point for the cardinal function $C_j(x)$.

AFFINE VARIETY: The set of all simultaneous solutions $\vec{u}_*$ of a set of equations in N variables u_j,

$$\begin{aligned} f_1(u_1, u_2, \ldots, u_n) &= 0, \\ f_2(u_1, u_2, \ldots, u_n) &= 0, \\ \ldots &= 0, \\ f_M(u_1, u_2, \ldots, u_n) &= 0, \end{aligned}$$

is said to be the *affine variety* of the system of equations. Note that the number of equations M need not equal the number of unknowns; if $M < N$, the affine variety is not a set of discrete points, but rather a set of curves or surfaces [137, p. 18].

ALGEBRAIC FUNCTION: Any function that can be defined implicitly as the solution $u(z)$ of $f(u(z), z) = 0$, where f is a polynomial of finite degree in both its arguments.

ALGEBRAIC SOLUTION: An explicit, analytic solution that requires only a *finite* number of operations, where the allowed steps are restricted to addition, multiplication, subtraction, division, and root extractions (square roots, cube roots, etc.). If an equation has an algebraic solution, then it is said to be *solvable by radicals*.

ARITHMURGY: Synonym for "number-crunching" (from the Greek $\alpha\rho\iota\theta\mu o\sigma$, "number," and -$\epsilon\rho\gamma o\sigma$, "working").

ARMIJO LINE SEARCH: See NEWTON–ARMIJO ITERATION.

ASYMPTOTIC: A power series is *asymptotic* to a function $f(\epsilon)$ if, for *fixed N* and *sufficiently small* ϵ, $|f(\epsilon) - \sum_{j=0}^{N} a_j \epsilon^j| \ll \epsilon^N$ [39].

BASIS FUNCTIONS: The members of a basis set. Examples of basis functions are the Chebyshev polynomials and the Hermite functions.

BASIS SET: The collection of functions used to approximate the solution of a differential equation or an arbitrary transcendental function $f(x)$. The Fourier functions $\{1, \cos(nx), \sin(nx) \text{ for } n = 1, 2, \ldots\}$ and the Chebyshev polynomials $\{T_n(x), n = 0, 1, \ldots\}$ are two examples of basis sets.

BEZOUT NUMBER: This is an upper bound on the number of isolated zeros of a polynomial system; $d = \prod_{j=1}^{N} d_j$, where N is the number of polynomial unknowns and equations and d_j is the degree of each equation.

BIFURCATION POINT: A point where two solution branches $x(\lambda)$ cross or merge.

BRANCH-CROSSING POINT: A point where two independent solution branches intersect. The branch-crossing point is common to both branches. The determinant of the Jacobian matrix (df/dx for one unknown) is zero at a branch-crossing point and is a *bifurcation point*. Also called a *transcritical bifurcation point*.

DAVIDENKO EQUATION: This is the ordinary differential equation (ODE) or system of equations $d\vec{x}/d\lambda = -\vec{\vec{J}}^{-1} d\vec{f}/d\lambda$ where λ is the continuation parameter and $\vec{\vec{J}}^{-1}$ denotes the inverse of the Jacobian matrix. The solution of this ODE system will yield a root curve $\vec{x}(\lambda)$ if the initial condition, $\vec{x}(0)$, is a solution of $\vec{f}(\vec{x}, \lambda) = \vec{0}$ for $\lambda = 0$. Usually employed as a "predictor" with Newton's method applied as the "corrector" to refine the root at each desired value of λ. For a single unknown, $dx/d\lambda = -\partial f/\partial\lambda/\partial f/\partial x$.

DEFICIENT POLYNOMIAL SYSTEM: A polynomial system for which the number of roots is less than the *bezout number*.

DEGREE: For a polynomial in one variable x, the degree is the largest exponent of a monomial in the polynomial. For a multivariate polynomial in variables $x_j, j = 1, \ldots, N$, the exponent of a monomial term $x_1^{d_1} x_2^{d_2} \ldots x_N^{d_N}$ is the sum of the exponents of the variables, $d = \sum_{j=1}^{N} d_j$; the degree of the polynomial is the maximum of d over all terms in the polynomial.

DISCRIMINANT OF A POLYNOMIAL: If $f(x;\lambda)$ is a polynomial in x, optionally containing one or more parameters λ, then its discriminant Disc(λ) is the *resultant* of $f(x)$ and df/dx. The polynomial $f(x;\lambda)$ has only single roots if and only if its discriminant Disc(λ) is nonzero. In the space of the parameter λ (or the K-dimensional parameter space of $\lambda_1, \lambda_2, \ldots, \lambda_K$ if there are K parameters), the points where the discriminant is zero are the points where roots merge to form multiple zeros. Two real roots may transition to a complex-conjugate pair at a zero of the discriminant.

DOUBLE ROOT: A point x_d is a double root of $f(x)$ if $f(x_d) = 0$ and $df/dx(x_d) = 0$ and $df^2/dx^2(x_d) \neq 0$.

DOUBLET: A pair of closely spaced simple roots. (Borrowed from spectroscopy, where it is applied to spectral lines that are slightly split by a perturbation; mathematically, these are a double root, perturbed into slightly different simple roots, of the characteristic polynomial of the Galerkin matrix for the Schrödinger equation in the Heisenberg matrix mechanics formulation of quantum mechanics.)

DRAINAGE BASIN: The largest region in the solution space of x that can be reached by moving "uphill" from a minimum of the residual norm function ρ.

ENDGAME: Rootfinding very close to a root. The term is borrowed from chess. Newton's iteration without underrelaxation is usually superior to all competing strategies in the endgame but may diverge when the current iterate is too far from the root.

ENVELOPE OF THE COEFFICIENTS: A smooth, monotonically decreasing curve that is a tight bound on oscillatory Chebyshev or Fourier coefficients $\{a_n\}$ in the sense that the absolute value of the coefficients is arbitrarily close to the envelope infinitely often as $n \to \infty$. (Borrowed from wave theory, where the "envelope of a wave packet" has the identical meaning.)

EURYARITHM: A numerical method of broad applicability, the opposite of a *stenarithm*.

EXPLICIT SOLUTION: An analytical, explicit expression "in which the solution is obtained as a general formula, so that nothing remains but to substitute numerical values in the formula; as for example the solution of the quadratic equation $x^2 + 2bx + c = 0$ by the formula $x = -b \pm \sqrt{b^2 - c}$" [369, p. 78]. Synonym for *literal solution*.

EXPONENTIAL CONVERGENCE: The property of a spectral series where the error decreases faster than any finite inverse power of N as N, the number of terms in the truncated series, increases. Typically, the series coefficients decrease as $O(\exp[-pn^r])$ for some positive constants p and r, which is the reason for the adjective "exponential." A synonym for *infinite order convergence*.

FFT: Abbreviation for *fast Fourier transform*.

FOLD POINT: Synonym for *limit point*.

GEOMETRIC CONVERGENCE: The property of a series whose coefficients a_n decrease as

$$a_n \sim A(n)p^n \longleftrightarrow A(n)\exp(-n\,|\log(p)|), \qquad n \to \infty, \qquad |p| < 1,$$

where $A(n)$ denotes a function that varies algebraically with n (such as a power of n). The reason for the name is that the terms of a geometrically convergent series can always be bounded by those of a geometric series, that is, by the terms of the power series expansion of $\alpha/(\beta + x)$ for some constants α and β. (All convergent power series have geometric convergence. All Chebyshev series for functions that have no singularities on $x \in [-1, 1]$ (including the endpoints) also have geometric convergence.)

GLOBAL: An adjective describing an algorithm that (i) is able to find roots without requiring an initialization or (ii) can compute multiple roots from a single, continuous process that requires only a single starting value or (iii) can climb out of valleys of the residual norm to escape local minima and then converge to a global minimum, which is a root.

GROEBNER BASIS: If $\vec{f}$ denotes a set of N *polynomial* equations in $\vec{x}$, a set of N polynomials, computed by a systematic algorithm due to B. Buchberger, whose roots are the *same* as those of the original system and that has certain algebraic-geometry properties that make the new basis more useful than the original system. In particular, if N is equal to the number of variables, then one member of the Grœbner basis is a *univariate* polynomial, another is a function of only two variables, and in general there is a basis function that depends on exactly j variables from the vector $\vec{x}$. Thus, finding a Grœbner basis for N polynomial equations in N unknowns is analogous to rearranging a matrix equation into an equation with a triangular matrix through Gaussian elimination. The Grœbner functions are said to form a basis because they completely describe the *ideal* generated by $\vec{f}$ (in algebraic geometry).

HOMOTOPY METHOD: A continuation method in which an artificial parameter λ is added to a system of nonlinear equations to define an inflated problem such that (i) $\lambda = 0$ is analytically solvable and (ii) $\lambda = 1$ is the same as the original problem. The inflated system of equations is dubbed a *homotopy* of the original problem.

IDEAL (Generated by a Set of Functions): Let $f_1, \ldots, f_s$ denote a set of functions in the commutative ring of polynomials of N variables, $k[x_1, \ldots, x_N]$. The *ideal* generated by these functions is

$$\langle f_1, \ldots, f_s \rangle = \{p_1 f_1 + \ldots p_s f_s\},$$

where the p_i are arbitrary polynomials that are also in the commutative ring $k[x_1, \ldots, x_N]$. The motive for this concept of algebraic geometry is that just as linear equations, encapsulated in a matrix, can be transformed by row and column operations without changing the solutions of the equations, similarly a set of polynomial equations can be manipulated by taking linear combinations of the polynomials where now the multipliers in these combinations need not be restricted to numbers but can instead be arbitrary polynomials p_i. All sets of functions in an ideal have the *same roots* or, in algebraic geometry jargon, the same *affine variety*. The reason for the "commutative ring" jargon is that if the coefficients of the polynomials f_j are restricted to integers or rational numbers or real numbers, then all possible functions in the ideal are subject to these same restrictions, and all algorithms, such as finding a Grœbner basis for the ideal, can be performed in this same restricted class of functions.

IMBRICATE SERIES: A representation of a *spatially periodic* function that is the superposition of an infinite number of evenly spaced identical copies of a *pattern* function $A(x)$. All periodic functions have imbricate series in addition to their Fourier expansions, and often the imbricate series converge faster. Imbricate series may be generalized to an arbitrary number of dimensions.

INFINITE ORDER CONVERGENCE: The property of a spectral series whereby the error decreases faster than any finite inverse power of N as N, the number of terms in the truncated series, increases. A synonym for *exponential convergence.*

INTERPOLANT: An approximation $f_{N-1}(x)$ whose free parameters or coefficients are chosen by the requirement that

$$f(x_i) = f_{N-1}(x_i), \qquad i = 1, \ldots, N,$$

at a set of N grid points. The process of computing such an approximation is *interpolation.*

LIMIT POINT: A point where a solution $x(\lambda)$ of a real-valued nonlinear equation curves back so that there are two solutions for λ on one side of $\lambda = \lambda_{\text{limit}}$ and no real solutions on the other side of the limit point. As the limit point is approached, $dx/d\lambda \to \infty$. Special methods (*pseudoarclength continuation* or *globally convergent homotopy*) are needed to turn the corner and march from the lower branch through the limit point onto the upper branch or vice versa. Synonyms are *fold point*, *turning point*, and *saddle-node bifurcation*.

LITERAL SOLUTION: An analytical, explicit expression "in which the solution is obtained as a general formula, so that nothing remains but to substitute numerical values in the formula; as for example the solution of the quadratic equation $x^2 + 2bx + c = 0$ by the formula $x = -b \pm \sqrt{b^2 - c}$" [369, p. 78]. Synonym for *explicit solution*.

LOCAL: An adjective that is applied to an algorithm that requires a good first approximation and is effective only when the first guess is sufficiently close to the root.

LYAPUNOV-SCHMIDT REDUCTION: A technique in which a system of nonlinear equations in many unknowns is reduced to one or two equations. The reduction identifies a small number of modes that are eigenfunctions with zero eigenvalue of the Jacobian matrix and projects the system into a "nice" but large subsystem that is solved for the nonsingular modes with the amplitudes of the singular modes as parameters; the result is projected onto the singular modes to yield a difficult nonlinear system with only a small number of unknowns—perhaps a single equation in a single unknown.

MACHINE EPSILON: The difference between 1.0 and the nearest number representable as a machine precision number; equivalently, an upper bound on the relative error due to rounding in floating point arithmetic. Often abbreviated "macheps" or ϵ_{mach}.

MATLAB/IEEE PRECISION: A machine epsilon equal to $2.2 \times 10^{-16} = 2^{-52}$. This is the default in MATLAB; it is also double precision in the IEEE 754 convention adopted by most hardware and software used for scientific computations.

MONIC POLYNOMIAL: A polynomial such that the coefficient of the highest power is one.

MONOMIAL: A product of powers of the unknowns, such as $x^j y^k$.

MULTIPLE ROOT: A root x_* is of *multiplicity k* if $f(x_*)/(x - x_*)^{k-1}$ is zero. By convention, a root is a *single root*, *simple root*, or a *root of multiplicity one* if $f(x_*) = 0$ but $f(x_*)/(x - x_*) \neq 0$.

MULTITHREADED ITERATION: An iteration scheme that follows each solution branch of a method that generates approximations to more than one root at each iteration. For example, Muller's method solves a quadratic equation for the corrections to the initial guess, but in the usual (single-threaded) scheme, the larger correction is simply discarded. Multithreaded Muller follows both roots so that one initial approximation becomes two threads at the first iteration, four approximate roots at the second level, eight branches at the third iteration, and so on.

NEWTON–ARMIJO ITERATION: Newton's method in which the correction is taken to be a constant γ times the Newton vector, where γ is chosen so as to minimize the residual for $0 < \gamma \leq 1$, often by testing the size of the residual for $\gamma = 2^{-m}$ for integer m.

NEWTON–CHORD ITERATION: Same as Newton's method except that df/dx is computed only and then used for all the remaining iterations. Also known as the *parallel chord* iteration and the *frozen Jacobian* iteration. If the derivative is recomputed occasionally, the procedure is the *Shamanski* iteration.

NEWTON-POLISHING: Refining an approximate zero by Newton's iteration after the root has been found to moderate accuracy by a different algorithm.

NEWTON'S ITERATION; NEWTON–RAPHSON ITERATION: The scheme in which $f(x)$ is approximated by its first-order Taylor series approximation about the current iterate x_c, $f(x) \approx f(x_c) + (x - x_c) df/dx(x_c)$; the root of this linear proxy is taken as the next iterate. Some purists use the term *Newton–Raphson*, but the historically less accurate but shorter label *Newton's iteration* is more common.

OPTIMALLY TRUNCATED ASYMPTOTIC SERIES: If an asymptotic series is divergent, then for a given ϵ, the error decreases as more terms are added up to some $N_{opt}(\epsilon)$, and then increases. The *optimal truncation* is to include only those terms up to and including $O(\epsilon^{N_{opt}(\epsilon)})$. An optimally truncated series is said to be a *superasymptotic* approximation.

ORACLE, ROOTFINDING: A theorem or algorithm (justified by a theorem) that provides information about zeros, such as the number of real roots on an interval.

PARAMETRIC INTERVAL POLYNOMIALIZATION (PIP): Replacing a transcendental function $f(x, \lambda)$ not merely by a polynomial approximation in x that is valid at a particular value of the parameter λ but rather by an approximation that is accurate on a finite interval in λ.

PERIODIC: A function $f(x)$ is periodic with period p if and only if

$$f(x+P) = f(x)$$

for all x.

POLYNOMIAL: A finite sum of monomials.

POLYNOMIALIZATION: The replacement of a transcendental function by a polynomial approximation that is a truncated Chebyshev series.

PSEUDORESULT: A z for the equation $F(z;\vec{a}) = 0$ whose coefficients $\vec{a}$ are known within a vector of tolerances $\vec{\epsilon}$ if z is the exact solution of $F(z,\tilde{a}) = 0$ for some set of approximate coefficients $\tilde{a}$ whose difference from $\vec{a}$ lies within the specified tolerances. The set of z for all possible coefficients $\tilde{a}$ that lie within the specified tolerances of $\vec{a}$ is the *pseudoresult set.*

PSEUDOSPECTRAL METHOD: An algorithm that uses an interpolation grid to determine the coefficients of a spectral series. Synonyms are *orthogonal collocation*, *method of selected points*, and *method of discrete ordinates*.

PUISEUX SERIES: Series, often in fractional powers, for the root of a single polynomial equation in a single unknown as a function of the parameter of the coefficients of the polynomial.

QUARTIC: A polynomial of fourth *degree.*

QUASI-NEWTON: A variant of Newton's method in which the equation $\vec{\vec{J}}(\vec{x}^{(m+1)} - \vec{x}^{(m)}) = \vec{f}(\vec{x}^{(m)})$ is solved only *approximately* at each iteration.

QUINTIC: A polynomial of fifth *degree.*

RESIDUAL FUNCTION: The result, usually denoted $R(x; a_0, a_1, \ldots, a_N)$, when an approximate solution u_N is substituted into a differential, integral, or matrix equation. The residual function is identically zero if the approximate solution is exact.

RESULTANT: The resultant of two polynomials $f(x, y)$ and $g(x, y)$ with elimination of x is a polynomial in y only, $R(f, g)(y)$, such that all roots of the system $f = g = 0$ have y values equal to roots of the resultant $R(y)$; the resultant can be computed by straightforward algebraic procedures such as Sylvester's dialytic method.

ROOT: A value of x such that the expression $f(x)$ is zero. Synonym for *zero.*

ROOTS AT INFINITY: Suppose that a polynomial system is such that the jth equation is of degree d_j. Define a modified system by omitting all terms from all equations except those terms in the jth equation which are of exactly degree d_j. The *roots at infinity* are the solution to this system of homogeneous polynomials. Zeros at infinity are improbable in theory but common in applications.

SEPARATE-SYMBOL STRATEGY: A singular perturbation method in which a series in both powers of ϵ and powers of $1/\log(\epsilon)$ is computed by defining $\nu \equiv -1/\log(\epsilon)$ and then performing a bivariate expansion as if (ϵ, ν) were completely independent.

SHAMANSKI ITERATION: A quasi-Newton method in which the Jacobian is recomputed and factored only every k iterations, where k is a user-choosable positive integer and can be varied adaptively.

SIMPLE ROOT: A point x_* such that $f(x_*) = 0$, but df/dx is not zero at this point. Also called a *single root* or a *root of multiplicity one.*

SPECTRAL: A catch-all term for all methods (including pseudospectral techniques) that expand the unknown as a series of global, infinitely differentiable expansion functions.

STENARITHM: An algorithm that is very useful only for a particular function or a narrow class of functions. From the Greek $\sigma\tau\epsilon\nu o\sigma$, "narrow." (Compare with "stenokates" in biology, which are creatures occupying narrow ecological niches.)

SURD: A fractional power such as $\sqrt{2}$ or $x^{7/8}$.

SYMMETRIC: (i) Of a matrix: $A_{ij} = A_{ji}$. The longer but more precise term *centrosymmetric* is sometimes used as a synonym.
(ii) Of a function $f(x)$: $f(x) = f(-x)$ for all x.

TENSOR PRODUCT BASIS: A multidimensional basis whose elements are the products of one-dimensional basis elements. In two dimensions,

$$\Phi_{mn}(x, y) \equiv \phi_m(x)\,\phi_n(y).$$

TENSOR PRODUCT GRID: A multidimensional grid whose MN points are chosen from the corresponding one-dimensional grids:

$$\underline{x}_{ij} \equiv (x_i, y_j), \qquad i = 1, \ldots, M, \text{ and } j = 1, \ldots, N.$$

Some prefer *Cartesian product grid* or *direct product grid.*

TRANSCRITICAL BIFURCATION POINT: A point where two branches of a solution $x(\lambda)$ cross. Synonyms are *crossing point* and *branch-crossing bifurcation.* For both $\lambda > \lambda_{\text{bif}}$ and $\lambda < \lambda_{\text{bif}}$, two different solutions exist that meet at the bifurcation point. Newton's method fails at a transcritical bifurcation point, but it is possible to "shoot the bifurcation point" or switch branches (Chapter 8). Sometimes loosely called simply a *bifurcation point* without the adjective "transcritical." This is a codimension-one bifurcation.

TRIPLE POINT: A codimension-one bifurcation in which $dx/d\lambda = \infty$ but there are branches on both sides of the bifurcation, unlike at a fold point. Also called a *hysteresis point.*

TRUNCATION ERROR: The error made by neglecting all coefficients a_n in the spectral series such that $n > N$ for some *truncation* N.

UNDERRELAXATION: A modification to an algorithm, such as Newton's iteration, where the correction computed by the algorithm is multiplied by a constant γ where $\gamma < 1$.

Bibliography

[1] M. ABRAMOWITZ AND I. A. STEGUN, *Handbook of Mathematical Functions*, Dover, New York, 1965. (Cited on pp. 130, 200, 273, 274, 325, 335)

[2] T. M. ACHO, *Derivation of eigenrelations for the Sturm-Liouville boundary value problems with interior singularities*, Quart. Appl. Math., 65 (2007), pp. 375–383. (Cited on p. 107)

[3] F. ACTON, *Numerical Methods As They Should Be*, Harper and Row, New York, 1970. (Cited on p. 3)

[4] J. A. ADAM, *Critical layer singularities and complex eigenvalues in some differential equations of mathematical physics*, Phys. Rep., 142 (1986), pp. 263–356. (Cited on p. 107)

[5] G. ADOMIAN, *A review of the decomposition method and some recent results for nonlinear equations*, Math. Comput. Modelling, 13 (1992), pp. 17–43. (Cited on p. 275)

[6] ——, *Solving Frontier Problems of Physics: The Decomposition Method*, Kluwer, Norwell, MA, 1994. (Cited on p. 275)

[7] J. C. ALEXANDER AND J. A. YORKE, *The homotopy continuation method: Numerically implementable topological procedures*, Trans. Amer. Math. Soc., 242 (1978), pp. 271–284. (Cited on p. 168)

[8] E. L. ALLGOWER AND K. GEORG, *Introduction to Numerical Continuation Methods*, no. 45 in Classics in Applied Mathematics, SIAM, Philadelphia, 2003. (Cited on pp. 151, 158, 169, 170)

[9] E. L. ALLGOWER, K. GEORG, AND R. MIRANDA, *The method of resultants for computing real solutions of polynomial systems*, SIAM J. Numer. Anal., 29 (1992), pp. 831–844. (Cited on p. 373)

[10] P. AMORE, *Solving the Helmholtz equation for Membranes of arbitrary shape: Numerical results*, J. Phys. A, 41 (2008), p. 265206. (Cited on p. 106)

[11] P. AMORE AND F. M. FERNANDEZ, *Mathematical analysis of recent analytical approximations to the collapse of an empty spherical bubble*, J. Chem. Phys., 138 (2013), p. 084511. (Cited on p. 8)

[12] E. D. ANGELOVA AND K. I. SEMERDZHIEV, *Methods for the simultaneous approximate derivation of the roots of algebraic, trigonometric and exponential equations*, U. S. S. R Comput. Maths. Math. Phys., 22 (1982), pp. 226–232. (Cited on pp. 86, 91, 92)

[13] W. S. ANGLIN, *Mathematics and history*, Math. Intell., 4 (1982), p. 10. (Cited on p. 401)

[14] V. I. ARNOLD, *Catastrophe Theory*, Springer-Verlag, Berlin, 1983. (Cited on p. 219)

[15] V. I. ARNOLD, *Polymathematics: Is mathematics a single science or a set of arts?*, in Mathematics: Frontiers and Perspectives, V. I. Arnold et al., eds., AMS, Providence, RI, 2000, pp. 403–416. (Cited on p. 149)

[16] A. P. AUSTIN, P. KRAVANJA, AND L. N. TREFETHEN, *Numerical algorithms based on analytic function values at roots of unity*, SIAM J. Numer. Anal., to appear. (Cited on p. 96)

[17] B. W. BADER, *Tensor-Krylov methods for solving large-scale systems of nonlinear equations*, SIAM J. Numer. Anal., 43 (2005), pp. 1321–1347. (Cited on p. 120)

[18] B. W. BADER AND R. B. SCHNABEL, *Curvilinear linesearch for tensor methods*, SIAM J. Sci. Comput., 25 (2003), pp. 604–622. (Cited on p. 120)

[19] ——, *On the performance of tensor methods for solving ill-conditioned problems*, SIAM J. Sci. Comput., 29 (2007), pp. 2329–2351. (Cited on p. 120)

[20] G. A. BAKER, JR., *Padé approximants*, in Advances in Theoretical Physics 1, Academic Press, New York, 1965, pp. 1–50. (Cited on pp. 256, 257)

[21] ——, *Essentials of Padé Approximants*, Academic Press, New York, 1975. (Cited on pp. 256, 257)

[22] S. BARNETT, *Companion matrix analog for orthogonal polynomials*, Linear Algebra Appl., 12 (1975), pp. 197–208. (Cited on pp. 26, 27, 91, 402, 403)

[23] ——, *Greatest common divisors from generalized Sylvester resultant matrices*, Linear Multilinear Algebra, 8 (1980), pp. 271–279. (Cited on p. 379)

[24] ——, *Congenial matrices*, Linear Algebra Appl., 41 (1981), pp. 277–298. (Cited on pp. 27, 379)

[25] ——, *Division of generalized polynomials using the comrade matrix*, Linear Algebra Appl., 60 (1984), pp. 159–175. (Cited on pp. 27, 186)

[26] S. BARNETT, *Multiplication of generalized polynomials with applications to classical orthogonal polynomials*, SIAM J. Alg. Disc. Meth., 5 (1984), pp. 457–462. (Cited on p. 27)

[27] S. BARNETT, *Euclidean remainders for generalized polynomials*, Linear Algebra Appl., 99 (1988), pp. 111–122. (Cited on p. 27)

[28] ——, *A Bezoutian matrix for Chebyshev polynomials*, in Applications of Matrix Theory, M. J. C. Gover and S. Barnett, eds., Oxford University Press, Oxford, 1989, pp. 137–149. (Cited on pp. 27, 379)

[29] D. A. BARRY, S. J. BARRY, AND P. J. CULLIGAN-HENSLEY, *Algorithm 743: WAPR: A FORTRAN routine for calculating real values of the W-function*, ACM Trans. Math. Software, 21 (1995), pp. 172–181. (Cited on p. 245)

[30] D. A. BARRY, P. J. CULLIGAN-HENSLEY, AND S. J. BARRY, *Real values of the W-function*, ACM Trans. Math. Software, 21 (1995), pp. 161–171. (Cited on p. 245)

[31] D. A. BARRY, J. Y. PARLANGE, L. LI, H. PROMMER, C. J. CUNNINGHAM, AND E. STAGNITTI, *Analytical approximations for real values of the Lambert W-function*, Math. Comput. Simulation, 53 (2000), pp. 95–103. (Cited on pp. 143, 245)

[32] ——, Erratum to "*Analytical approximations for real values of the Lambert W-function*" [*Mathematics and Computers in Simulation* 53 (2000), 95–103] Math. Comput. Simulation, 59 (2002), p. 543. Correction. (Cited on p. 143)

[33] M. BARTON, *Solving polynomial systems using no-root elimination schemes*, Comput. Aided Design, 43 (2011), pp. 1870–1878. (Cited on p. 386)

[34] M. BARTON AND B. JUETTLER, *Computing roots of polynomials by quadratic clipping*, Comput. Aided Geom. Design, 24 (2007), pp. 125–141. (Cited on p. 383)

[35] D. J. BATES, J. D. HAUENSTEIN, AND A. J. SOMMESE, *Efficient path tracking methods*, Numer. Algorithms, 58 (2011), pp. 451–459. (Cited on p. 159)

[36] D. J. BATES, J. D. HAUENSTEIN, A. J. SOMMESE, AND C. W. WAMPLER, *Numerically Solving Polynomial Systems with Bertini*, SIAM, Philadelphia, 2013. (Cited on pp. 158, 159, 239, 371, 394)

[37] Z. BATTLES AND L. N. TREFETHEN, *An extension of MATLAB to continuous functions and operators*, SIAM J. Sci. Comput., 25 (2004), pp. 1743–1770. (Cited on pp. 6, 22, 25, 52, 55, 57, 407)

[38] C. M. BENDER, K. A. MILTON, S. S. PINSKY, AND L. M. SIMMONS, JR., *A new perturbative approach to nonlinear problems*, J. Math. Phys., 30 (1989), pp. 1447–1455. (Cited on pp. 164, 263, 264)

[39] C. M. BENDER AND S. A. ORSZAG, *Advanced Mathematical Methods for Scientists and Engineers*, McGraw-Hill, New York, 1978. (Cited on pp. 42, 256, 257, 306, 425)

[40] S. N. BERNSTEIN, *Quelques remarques sur l'interpolation*, J. Mathematische Annalen, 79 (1918), pp. 1–12. (Cited on p. 22)

[41] J. BERRUT, M. S. FLOATER, AND G. KLEIN, *Convergence rates of derivatives of a family of barycentric rational interpolants*, Appl. Numer. Math., 61 (2011), pp. 989–1000. (Cited on p. 161)

[42] M. V. BERRY, *Asymptotics, superasymptotics, hyperasymptotics*, in Asymptotics Beyond All Orders, H. Segur, S. Tanveer, and H. Levine, eds., Plenum, Amsterdam, 1991, pp. 1–14. (Cited on p. 285)

[43] M. V. BERRY AND C. J. HOWLS, *Hyperasymptotics*, Proc. R. Soc. Lond. Ser. A, 430 (1990), pp. 653–668. (Cited on p. 285)

[44] M. BERZINS, *Adaptive polynomial interpolation on evenly spaced meshes*, SIAM Rev., 49 (2007), pp. 604–627. (Cited on p. 67)

[45] D. A. BINI, L. GEMIGNANI, AND J. R. WINKLER, *Structured matrix methods for CAGD: An application to computing the resultant of polynomials in the Bernstein basis*, Numer. Linear Algebra Appl., 12 (2005), pp. 685–698. (Cited on p. 394)

[46] T. BJØNTEGAARD, E. M. RØNQUIST, AND Ø. TRÅSDAHL, *Spectral approximation of partial differential equations in highly distorted domains*, J. Sci. Comput., 52 (2012), p. 6. (Cited on p. 9)

[47] A. D. BOOTH, *Numerical Methods*, Butterworths Scientific Publications, London, 1955. (Cited on p. 367)

[48] L. C. BOTTEN, M. S. CRAIG, AND R. C. MCPHEDRAN, *Complex zeros of analytic functions*, Comput. Phys. Comm., 29 (1983), pp. 245–259. (Cited on pp. 96, 100)

[49] J. P. BOYD, *A Chebyshev polynomial method for computing analytic solutions to eigenvalue problems with application to the anharmonic oscillator*, J. Math. Phys., 19 (1978), pp. 1445–1456. (Cited on p. 261)

[50] J. P. BOYD, *The nonlinear equatorial Kelvin wave*, J. Phys. Oceanography, 10 (1980), pp. 1–11. (Cited on p. 7)

[51] ——, *The rate of convergence of Hermite function series*, Math. Comput., 35 (1980), pp. 1309–1316. (Cited on p. 7)

[52] ——, *A Sturm-Liouville eigenproblem with an interior pole*, J. Math. Phys., 22 (1981), pp. 1575–1590. (Cited on p. 107)

[53] ——, *Analytical approximations to the modon dispersion relation*, Dynam. Atmos. Oceans, 6 (1981), pp. 97–101. (Cited on pp. 7, 253, 258, 293)

[54] ——, *The effects of meridional shear on planetary waves, Part II: Critical latitudes*, J. Atmos. Sci., 39 (1982), pp. 770–790. (Cited on pp. 8, 107)

[55] ——, *The optimization of convergence for Chebyshev polynomial methods in an unbounded domain*, J. Comput. Phys., 45 (1982), pp. 43–79. (Cited on p. 7)

[56] ——, *Theta functions, Gaussian series, and spatially periodic solutions of the Korteweg-de Vries equation*, J. Math. Phys., 23 (1982), pp. 375–387. (Cited on p. 424)

[57] ——, *Equatorial solitary waves, Part II: Envelope solitons*, J. Phys. Oceanography, 13 (1983), pp. 428–449. (Cited on p. 142)

[58] ——, *The asymptotic coefficients of Hermite series*, J. Comput. Phys., 54 (1984), pp. 382–410. (Cited on p. 7)

[59] ——, *The double cnoidal wave of the Korteweg–de Vries equation: An overview*, J. Math. Phys., 25 (1984), pp. 3390–3401. (Cited on p. 424)

[60] ——, *Complex coordinate methods for hydrodynamic instabilities and Sturm-Liouville problems with an interior singularity*, J. Comput. Phys., 57 (1985), pp. 454–471. (Cited on pp. 8, 20)

[61] ——, *Polynomial series versus sinc expansions for functions with corner or endpoint singularities*, J. Comput. Phys., 64 (1986), pp. 266–269. (Cited on pp. 34, 107)

[62] ——, *Spectral methods using rational basis functions on an infinite interval*, J. Comput. Phys., 69 (1987), pp. 112–142. (Cited on pp. 7, 35, 36)

[63] ——, *Exponentially convergent Fourier/Chebyshev quadrature schemes on bounded and infinite intervals*, J. Sci. Comput., 2 (1987), pp. 99–109. (Cited on pp. 318, 417)

[64] ——, *Orthogonal rational functions on a semi-infinite interval*, J. Comput. Phys., 70 (1987), pp. 63–88. (Cited on pp. 7, 35, 36)

[65] ——, *Chebyshev domain truncation is inferior to Fourier domain truncation for solving problems on an infinite interval*, J. Sci. Comput., 3 (1988), pp. 109–120. (Cited on p. 7)

[66] ——, *New directions in solitons and nonlinear periodic waves: Polycnoidal waves, imbricated solitons, weakly non-local solitary waves and numerical boundary value algorithms*, in Advances in Applied Mechanics, T.-Y. Wu and J. W. Hutchinson, eds., vol. 27 in Advances in Applied Mechanics, Academic Press, New York, 1989, pp. 1–82. (Cited on p. 424)

[67] ——, *Periodic solutions generated by superposition of solitary waves for the quarticly nonlinear Korteweg-de Vries equation*, Z. Angew. Math. Phys., 40 (1989), pp. 940–944. (Cited on p. 424)

[68] ——, *The envelope of the error for Chebyshev and Fourier interpolation*, J. Sci. Comput., 5 (1990), pp. 311–363. (Cited on p. 4)

[69] J. P. BOYD, *The energy spectrum of fronts: The time evolution of shocks in Burgers' equation*, J. Atmos. Sci., 49 (1992), pp. 128–139. (Cited on p. 7)

[70] ——, *The rate of convergence of Fourier coefficients for entire functions of infinite order with application to the Weideman-Cloot sinh-mapping for pseudospectral computations on an infinite interval*, J. Comput. Phys., 110 (1994), pp. 360–372. (Cited on p. 7)

[71] ——, *A Chebyshev polynomial interval-searching method ("Lanczos economization") for solving a nonlinear equation with application to the nonlinear eigenvalue problem*, J. Comput. Phys., 118 (1995), pp. 1–8. (Cited on pp. 26, 27, 28, 30, 55, 66, 399)

[72] ——, *Construction of Lighthill's unitary functions: The imbricate series of unity*, Appl. Math. Comput., 86 (1997), pp. 1–10. (Cited on p. 424)

[73] ——, *Global approximations to the principal real-valued branch of the Lambert W-function*, Appl. Math. Lett., 11 (1998), pp. 27–31. Erratum: in Eq. (4), 11/36 should be $\sqrt{2}(11/36)$. (Cited on p. 143)

[74] ——, *Weakly Nonlocal Solitary Waves and Beyond-All-Orders Asymptotics: Generalized Solitons and Hyperasymptotic Perturbation Theory*, vol. 442 of Mathematics and Its Applications, Kluwer, Amsterdam, 1998. (Cited on pp. 7, 42, 124, 277, 285)

[75] ——, *The devil's invention: Asymptotic superasymptotics and hyperasymptotics*, Acta Appl. Math., 56 (1999), pp. 1–98. (Cited on pp. 7, 42, 277, 285)

[76] ——, *Chebyshev and Fourier Spectral Methods*, Dover, New York, 2001. (Cited on pp. 6, 18, 21, 22, 23, 24, 25, 34, 35, 36, 41, 57, 68, 71, 73, 97, 169, 261, 315, 318, 359, 405, 414)

[77] ——, *Chebyshev polynomial expansions for simultaneous approximation of two branches of a function with application to the one-dimensional Bratu equation*, Appl. Math. Comput., 143 (2002), pp. 189–200. (Cited on p. 261)

[78] ——, *Computing zeros on a real interval through Chebyshev expansion and polynomial rootfinding*, SIAM J. Numer. Anal., 40 (2002), pp. 1666–1682. (Cited on pp. 55, 407)

[79] ——, *Hyperasymptotics and the linear boundary layer problem: Why asymptotic series diverge*, SIAM Rev., 47 (2005), pp. 553–575. (Cited on p. 42)

[80] ——, *Asymptotic Fourier coefficients for a C^∞ bell (Smoothed-"Top-Hat" Function) and the Fourier Extension Problem*, J. Sci. Comput., 29 (2006), pp. 1–24. (Cited on p. 7)

[81] ——, *Computing real roots of a polynomial in Chebyshev series form through subdivision*, Appl. Numer. Math., 56 (2006), pp. 1077–1091. (Cited on pp. 27, 30)

[82] ——, *Computing real roots of a polynomial in Chebyshev series form through subdivision with linear testing and cubic solves*, Appl. Math. Comput., 174 (2006), pp. 1642–1648. (Cited on pp. 27, 30)

[83] ——, *Computing the zeros, maxima and inflection points of Chebyshev, Legendre and Fourier series: Solving transcendental equations by spectral interpolation and polynomial rootfinding*, J. Engrg. Math., 56 (2006), pp. 203–219. (Cited on pp. 27, 30, 36, 40, 55, 77, 79, 92, 408)

[84] ——, *Computing the zeros of a Fourier series or a Chebyshev series or general orthogonal polynomial series with parity symmetries*, Comput. Math. Appl., 54 (2007), pp. 336–349. (Cited on pp. 39, 47, 92)

[85] ——, *Polynomialization of Kepler's equation through Chebyshev polynomial expansion of the sine: A non-iterative rootfinder*, Appl. Numer. Math., 57 (2007), pp. 12–18. (Cited on pp. 71, 261)

[86] J. P. BOYD, *The Blasius function: Computations before computers, the value of tricks, undergraduate projects, and open research problems*, SIAM Rev., 50 (2008), pp. 791–804. (Cited on p. 324)

[87] ——, *Multiscale numerical algorithms for weather forecasting and climate modeling: Challenges and controversies*, SIAM News, 41(9) (2008). (Cited on p. 397)

[88] ——, *Chebyshev expansion on intervals with branch points with application to the root of Kepler's equation: A Chebyshev-Hermite-Padé method*, J. Comput. Appl. Math., 223 (2009), pp. 693–702. (Cited on p. 261)

[89] ——, *Large-degree asymptotics and exponential asymptotics for Fourier coefficients and transforms, Chebyshev and other spectral coefficients*, J. Engrg. Math., 63 (2009), pp. 355–399. (Cited on pp. 7, 22, 24, 398)

[90] ——, *Chebyshev spectral methods and the Lane-Emden problem*, Numer. Methods: Theory Appl., 4 (2010), pp. 142–157. (Cited on p. 34)

[91] ——, *The near-equivalence of five species of spectrally-accurate radial basis functions (RBFs): Asymptotic approximations to the RBF cardinal functions on a uniform, unbounded grid*, J. Comput. Phys., 230 (2011), pp. 1304–1318. (Cited on p. 40)

[92] ——, *New series for the cosine lemniscate function and the polynomialization of the lemniscate integral*, J. Comput. Appl. Math., 235 (2011), pp. 1941–1955. (Cited on p. 106)

[93] ——, *Numerical, perturbative and Chebyshev inversion of the incomplete elliptic integral of the second kind*, Appl. Math. Comput., 218 (2012), pp. 7005–7013. (Cited on pp. 140, 141, 326)

[94] ——, *Bivariate rootfinding: Spectrally-enhanced Weyl exclusionary subdivision with Marching Triangles, Newton-polishing and duplicate-pruning*, J. Comput. Phys., submitted. (Cited on p. 392)

[95] ——, *A comparison of companion matrix methods to find roots of a trigonometric polynomial*, J. Comput. Phys., 246 (2013). (Cited on pp. 47, 78, 80, 92)

[96] ——, *Finding the zeros of a univariate equation: Proxy rootfinders, Chebyshev interpolation, and the companion matrix*, SIAM Rev., 55 (2013), pp. 375–396. (Cited on pp. xiv, 17, 19, 23, 37, 318)

[97] ——, *A Fourier companion matrix (multiplication matrix) with real-valued elements: Finding the roots of a trigonometric polynomial by matrix eigensolving*, Numer. Math.: Theory, Methods Appl., 6 (2013), pp. 586–599. (Cited on pp. 47, 92)

[98] ——, *Rational Chebyshev series for the Thomas-Fermi function: Endpoint singularities and spectral methods*, J. Comput. Appl. Math., 231 (2013), pp. 90–101. (Cited on p. 35)

[99] ——, *Polynomializing a transcendental equation: Four ways to compute the inverse of the complete elliptic integral of the first kind*, Comput. Math. Appl., submitted. (Cited on pp. 140, 330)

[100] J. P. BOYD AND M. BUSCH, *Chebyshevizing algebraic geometry: Computing resultants and implementing polynomial division using Chebyshev polynomials*, Appl. Math. Comput., to appear. (Cited on pp. 26, 379)

[101] J. P. BOYD AND Z. D. CHRISTIDIS, *Instability on the equatorial beta-plane*, in Hydrodynamics of the Equatorial Ocean, J. Nihoul, ed., Elsevier, Amsterdam, 1983, pp. 339–351. (Cited on p. 107)

[102] J. P. BOYD AND D. H. GALLY, *Numerical experiments on the accuracy of the Chebyshev-Frobenius companion matrix method for finding the zeros of a truncated series of Chebyshev polynomials*, J. Comput. Appl. Math., 205 (2007), pp. 281–295. (Cited on pp. 27, 30)

[103] J. P. BOYD AND S. E. HAUPT, *Polycnoidal waves: Spatially periodic generalizations of multiple solitary waves*, in Nonlinear Topics of Ocean Physics: Fermi Summer School, Course LIX, A. R. Osborne, ed., North-Holland, Amsterdam, 1991, pp. 827–856. (Cited on p. 424)

[104] J. P. BOYD AND A. NATAROV, *A Sturm-Liouville eigenproblem of the Fourth Kind: A critical latitude with equatorial trapping*, Stud. Appl. Math., 101 (1998), pp. 433–455. (Cited on p. 107)

[105] J. P. BOYD AND J. R. ONG, *Exponentially-convergent strategies for defeating the Runge Phenomenon for the approximation of non-periodic functions, Part II: Multi-interval schemes*, Appl. Numer. Math., 61 (2011), pp. 460–472. (Cited on pp. 28, 67)

[106] J. P. BOYD, C. RANGAN, AND P. H. BUCKSBAUM, *Pseudospectral methods on a semi-infinite interval with application to the hydrogen atom: A comparison of the mapped Fourier-sine method with Laguerre series and rational Chebyshev expansion*, J. Comput. Phys., 188 (2003), pp. 56–74. (Cited on pp. 36, 404)

[107] J. P. BOYD AND B. A. SADIQ, *Computing the real roots of a Fourier series-plus-linear-polynomial: A Chebyshev companion matrix approach*, Appl. Math. Comput., 219 (2012), pp. 819–826. (Cited on pp. 47, 70, 71, 92)

[108] J. P. BOYD AND R. VISSER, *Rootfinding through global Newton iteration and Chebyshev polynomials for the gain of a balanced oscillator*, Math. Comput., 182 (2006), pp. 166–174. (Cited on pp. 141, 142, 145, 146, 261, 324)

[109] J. P. BOYD AND L. WANG, *Asymptotic coefficients for Gaussian radial basis function interpolants*, Appl. Math. Comput., 216 (2010), pp. 2394–2407. (Cited on p. 424)

[110] J. P. BOYD AND Y. C. XIU, *Parity symmetry with respect to both $x = 0$ and $x = L$ requires periodicity with period $4L$: Connections between computer graphics, group theory and spectral methods for solving partial differential equations*, Appl. Math. Comput., 219 (2012), pp. 3461–3469. (Cited on p. 82)

[111] J. P. BOYD AND F. YU, *Comparing six spectral methods for interpolation and the Poisson equation in a disk: Radial basis functions, Logan-Shepp ridge polynomials, Fourier-Bessel, Fourier-Chebyshev, Zernike polynomials, and double Chebyshev series*, J. Comput. Phys., 230 (2011), pp. 1408–1438. (Cited on p. 106)

[112] L. E. BRANSCOME, *The Charney baroclinic instability: Approximate solutions and modal structures*, J. Atmos. Sci., 40 (1983), pp. 1393–1409. (Cited on p. 301)

[113] W. L. BRIGGS AND V. E. HENSON, *The DFT: An Owner's Manual for the Discrete Fourier Transform*, SIAM, Philadelphia, 1995. (Cited on p. 97)

[114] J. D. BROWN, M. T. CHU, D. C. ELLISON, AND R. J. PLEMMONS, eds., *Proceedings of the Cornelius Lanczos International Centenary Conference*, SIAM, Philadelphia, 1994. (Cited on p. 21)

[115] K. M. BROWN, *A quadratically convergent Newton-like method based upon Gaussian elimination*, SIAM J. Numer. Anal., 6 (1969), pp. 560–569. (Cited on p. 9)

[116] ——, *Computer oriented algorithms for solving systems of simultaneous nonlinear algebraic equations*, in Numerical Solution of Systems of Nonlinear Algebraic Equations, G. D. Byrne and C. A. Hall, eds., Academic Press, New York, 1973, pp. 281–348. (Cited on p. 9)

[117] K. M. BROWN AND S. D. CONTE, *The solution of simultaneous nonlinear equations*, in Proceedings of the 22nd National Conference of the ACM, ACM, New York, 1967, pp. 111–114. (Cited on p. 9)

[118] E. E. BURNISTON AND C. E. SIEWERT, *The use of Riemann problems in solving a class of transcendental equations*, Proc. Cambridge Philos. Soc., 73 (1973), pp. 111–118. (Cited on p. 247)

[119] ——, *On the solution of certain algebraic equations*, J. Comput. Appl. Math., 4 (1978), pp. 37–39. (Cited on p. 247)

[120] R. CAMASSA, D. D. HOLM, AND J. M. HYMAN, *A new integrable shallow water equation*, in Advances in Applied Mechanics, volume 31, T.-Y. Wu and J. W. Hutchinson, eds., Academic Press, New York, 1994, pp. 1–33. (Cited on p. 332)

[121] C. CARSTENSEN, *A note on simultaneous rootfinding for algebraic, exponential, and trigonometric polynomials*, Comput. Math. Appl., 27 (1994), pp. 7–14. (Cited on pp. 92, 93)

[122] C. CARSTENSEN AND M. S. PETKOVI'C, *On some interval methods for algebraic, exponential and trigonometric polynomials*, Computing, 51 (1993), pp. 313–326. (Cited on pp. 92, 93)

[123] C. CARSTENSEN AND M. REINDERS, *On a class of higher order methods for simultaneous rootfinding of generalized polynomials*, Numer. Math., 64 (1993), pp. 69–84. (Cited on pp. 92, 93)

[124] J. CHABERT, ed., *A History of Algorithms: From the Pebble to the Microchip*, Springer-Verlag, New York, 1999. (Cited on p. 127)

[125] T. F. CHAN, *Newton-like pseudo-arclength methods for computing simple turning points*, SIAM J. Sci. Stat. Comput., 5 (1984), pp. 135–148. (Cited on p. 151)

[126] E. D. CHARLES AND J. B. TATUM, *The convergence of Newton-Raphson iteration with Kepler's equation*, Celestial Mechanics and Dynamical Astronomy, 69 (1998), pp. 357–372. (Cited on pp. 129, 140)

[127] R. C. Y. CHIN, *On the matched asymptotic solution of the Troesch problem*, J. Comput. Appl. Math., 7 (1981), pp. 181–186. (Cited on p. 302)

[128] K. W. CHOW AND C. F. WU, *The superposition of algebraic solitons for the modified Korteweg-de Vries equation*, Commun. Nonlinear Sci. Numer. Simul., 19 (2014), pp. 49–52. (Cited on p. 424)

[129] C. W. CLENSHAW AND A. R. CURTIS, *A method for numerical integration on an automatic computer*, Numer. Math., 2 (1960), pp. 197–205. (Cited on pp. 25, 49, 51, 52, 55, 416, 417)

[130] P. COLWELL, *Solving Kepler's Equation over Three Centuries*, William-Bell, Richmond, VA, 1993. (Cited on pp. 6, 119, 127)

[131] L. COMTET, *Advanced Combinatorics: The Art of Finite and Infinite Expansions*, Reidel, Amsterdam, 1974. (Cited on p. 289)

[132] J. W. COOLEY AND J. W. TUKEY, *An algorithm for the machine calculation of complex Fourier series*, Math. Comput., 19 (1965), pp. 297–301. (Cited on p. 22)

[133] R. M. CORLESS, *Generalized companion matrices in the Lagrange basis*, in Proceedings EACA, L. Gonzalez-Vega and T. Recio, eds., 2004, pp. 317–322. (Cited on p. 403)

[134] ——, *On a generalized companion matrix pencil for matrix polynomials expressed in the Lagrange basis*, in Symbolic-Numeric Computation, D. Wang and L.-H. Zhi, eds., Trends in Mathematics, Birkhäuser, Basel, 2007, pp. 1–15. (Cited on p. 403)

[135] R. M. CORLESS, G. H. GONNET, D. E. G. HARE, D. J. JEFFREY, AND D. E. KNUTH, *On the Lambert W function*, Adv. Comput. Math., 5 (1996), pp. 329–359. (Cited on p. 245)

[136] R. M. CORLESS AND G. LITT, *Generalized companion matrices for polynomials not expressed in monomial bases*, unpublished, University of Western Ontario, 2000. Available on Corless's website. (Cited on p. 403)

[137] D. COX, J. LITTLE, AND D. O'SHEA, *Ideals, Varieties and Algorithms: An Introduction to Computational Algebraic Geometry and Commutative Algebra*, Undergraduate Texts in Mathematics, Springer-Verlag, New York, 1992. (Cited on pp. 371, 425)

[138] J. H. CURTISS, *Faber polynomials and the Faber series*, Amer. Math. Monthly, 78 (1971), pp. 577–596. (Cited on p. 39)

[139] H. DANKOWICZ AND F. SCHILDER, *Recipes for Continuation*, SIAM, Philadelphia, 2013. 584 pp. (Cited on p. 158)

[140] I. DAUBECHIES, *Ten Lectures on Wavelets*, SIAM, Philadelphia, 1992. (Cited on p. 424)

[141] D. DAVIDENKO, *On a new method of numerically integrating a system of nonlinear equations*, Dokl. Akad. Nauk, 88 (1953), pp. 601–604. (Cited on pp. 150, 162)

[142] ———, *On the approximate solution of a system of nonlinear equations*, Ukraine Mat. Z., 5 (1953), pp. 196–206. (Cited on pp. 150, 162)

[143] H. T. DAVIS, *Introduction to Nonlinear Differential and Integral Equations*, Dover, Mineola, NY, 1962. (Cited on p. 190)

[144] P. J. DAVIS, *Interpolation and Approximation*, Dover Publications, New York, 1975. (Cited on p. 90)

[145] W. R. DAVIS (GENERAL EDITOR), M. T. CHU, P. DOLAN, J. R. MCCONNELL, L. K. NORRIS, E. ORTIZ, R. J. PLEMMONS, D. RIDGEWAY, B. K. P. SCAIFE, W. J. STEWART, J. W. YORKE, JR., W. O. DOGGETT (ASSOCIATE EDITOR), B. M. GELLAI (ASSOCIATE EDITOR), A. A. GSPONER (ASSOCIATE EDITOR), AND C. A. PRIOLI (CONSULTING EDITOR), eds., *Cornelius Lanczos: Collected Published Papers with Commentaries*, North Carolina State University, Raleigh, NC, 1998. (Cited on p. 21)

[146] D. DAY AND L. ROMERO, *Roots of polynomials expressed in terms of orthogonal polynomials*, SIAM J. Numer. Anal., 43 (2005), pp. 1969–1987. (Cited on pp. 26, 27, 30, 91, 402, 403)

[147] B. H. DAYTON, T.-Y. LI, AND Z. ZENG, *Multiple zeros of nonlinear systems*, Math. Comp., 80 (2011), pp. 2143–2168. (Cited on p. 182)

[148] N. G. DE BRUIJN, *Asymptotic Methods in Analysis*, 3rd edition, Dover, New York, 1981. (Cited on p. 289)

[149] F. DE CASTILHOS CORAZZA AND J. V. DE OLIVEIRA MARCOS LUCIO CORAZZA, *Application of a subdivision algorithm for solving nonlinear algebraic systems*, Acta Sci. Tech., 30 (2008), pp. 27–38. (Cited on p. 393)

[150] D. W. DECKER AND H. B. KELLER, *Path following near bifurcation*, Comm. Pure Appl. Math., 34 (1980), pp. 149–175. (Cited on p. 151)

[151] M. DELLNITZ, O. SCHÜTZE, AND Q. ZHENG, *Locating all the zeros of an analytic function in one complex variable*, J. Comput. Appl. Math., 138 (2002), pp. 325–333. (Cited on pp. 96, 99, 101)

[152] L. M. DELVES AND J. N. LYNESS, *A numerical method for locating the zeros of an analytic function*, Math. Comput., 21 (1967), pp. 543–560. (Cited on pp. 95, 96, 106)

[153] L. M. DELVES AND J. N. LYNESS, *On numerical integration round a closed contour*, Math. Comput., 21 (1967), pp. 561–577. (Cited on pp. 95, 96, 99, 100)

[154] J. E. DENNIS, JR., AND R. B. SCHNABEL, *Numerical Methods for Nonlinear Equations and Unconstrained Optimization*, vol. 16 of Classics in Applied Mathematics, SIAM, Philadelphia, 1996. (Cited on pp. 119, 124)

[155] A. DHOOGE, W. GOVAERTS, Y. A. KUZNETSOV, H. G. E. MEIJER, AND B. SAUTOIS, *New features of the software MatCont for bifurcation analysis of dynamical systems*, Math. Comput. Modelling Dyn. Sys., 14 (2008), pp. 147–175. (Cited on pp. 170, 371)

[156] R. B. DINGLE, *Asymptotic Expansions: Their Derivation and Interpretation*, Academic, New York, 1973. Out of print. Available online at michaelberryphysics.wordpress.com/rb-dingles-book-on-asymptotics (Cited on p. 285)

[157] E. DOEDEL, H. B. KELLER, AND J. P. KERNÉVEZ, *Numerical analysis and control of bifurcation problems. II. Bifurcation in infinite dimensions*, Internat. J. Bifur. Chaos Appl. Sci. Engrg., 1 (1991), pp. 745–772. (Cited on p. 170)

[158] T. A. DRISCOLL, *Automatic spectral collocation for integral, integro-differential, and integrally reformulated differential equations*, J. Comput. Phys., 229 (2010), pp. 5980–5998. (Cited on p. 22)

[159] T. A. DRISCOLL AND L. N. TREFETHEN, *Schwarz-Christoffel Mapping*, vol. 8 of Cambridge Monographs on Applied and Computational Mathematics, Cambridge University Press, Cambridge, U.K., 2002. (Cited on p. 106)

[160] A. E. DUBINOV AND I. D. DUBINOVA, *How one can exactly solve some problems in plasma physics*, J. Plasma Phys., 71 (2005), pp. 715–728. (Cited on p. 244)

[161] J. ÉCALLE, *Les fonctions résurgentes*, Université de Paris-Sud, Paris, 1981. (Cited on p. 285)

[162] A. ERDELYI, ed., *Higher Transcendental Functions: The Bateman Manuscript Project*, McGraw-Hill, New York, 1953. Three volumes plus two additional volumes published under the different title *Integral Transforms*. (Cited on pp. 324, 325)

[163] H. A. EVESHAM, *Origins and development of nomography*, Ann. Hist. Comput., 8 (1986), pp. 324–333. (Cited on p. 222)

[164] ———, *The History and Development of Nomography*, CreateSpace Independent Publishing Platform, Amazon.com, 2010. (Cited on p. 222)

[165] G. FARMELO, *The Strangest Man: The Hidden Life of Paul Dirac*, Faber and Faber, London, 2009. (Cited on p. 197)

[166] R. T. FAROUKI AND T. N. T. GOODMAN, *On the optimal stability of the Bernstein basis*, Math. Comput., 65 (1996), pp. 1553–1566. (Cited on p. 28)

[167] R. T. FAROUKI AND V. T. RAJAN, *On the numerical condition of polynomials in Bernstein form*, Comput. Aided Geom. Design, 4 (1987), pp. 191–216. (Cited on p. 64)

[168] W. M. FAUCETE, *A geometric interpretation of the general quartic polynomial*, Amer. Math. Monthly, 103 (1996), pp. 51–57. (Cited on p. 232)

[169] D. FENG AND T. H. PULLIAM, *Tensor-GMRES method for large systems of nonlinear equations*, SIAM J. Optim., 7 (1997), pp. 757–779. (Cited on p. 120)

[170] D. A. FIELD AND A. P. MORGAN, *A quick method for determining whether a second-degree polynomial has solutions in a given box*, IEEE Comput. Graphics Appl., 2 (1982), pp. 62–65. (Cited on p. 355)

[171] R. FLETCHER, *Practical Methods of Optimization*, 2nd edition, John Wiley and Sons Ltd., Chichester, U.K., 1987. (Cited on p. 315)

[172] G. R. FLIERL, *Simple applications of McWilliams' "A note on a consistent quasigeostrophic model in a multiply connected domain,"* Dyn. Atmos. Oceans, 1 (1977), pp. 443–453. (Cited on p. 72)

[173] M. S. FLOATER AND K. HORMANN, *Barycentric rational interpolation with no poles and high rates of approximation*, Numer. Math., 107 (2007), pp. 315–331. (Cited on p. 161)

[174] G. E. FORSYTHE, *Generation and use of orthogonal polynomials for data-fitting with a digital computer*, J. SIAM, 5 (1957), pp. 74–88. (Cited on p. 182)

[175] L. FOX AND I. B. PARKER, *Chebyshev Polynomials in Numerical Analysis*, Oxford University Press, London, 1968. (Cited on p. 71)

[176] ———, *Chebyshev Polynomials in Numerical Analysis*, 2nd edition, Oxford University Press, London, 1968. (Cited on pp. 412, 414)

[177] W. FRASER AND M. W. WILSON, *Remarks on the Clenshaw-Curtis quadrature scheme*, SIAM Rev., 8 (1966), pp. 322–327. (Cited on pp. 49, 55, 417)

[178] F. N. FRITSCH, R. E. SHAFER, AND W. P. CROWLEY, *Algorithm 443: Solution of the transcendental equation $w\exp(w) = x$*, Commun. ACM, 16 (1973), pp. 123–124. (Cited on p. 245)

[179] A. FROMMER, *A unified approach to methods for the simultaneous computation of all zeros of generalized polynomials*, Numer. Math., 54 (1988), pp. 105–116. (Cited on pp. 91, 92)

[180] P. A. FROST AND E. Y. HARPER, *An extended Padé procedure for constructing global approximations from asymptotic expansions: An explication with examples*, SIAM Rev., 18 (1976), pp. 62–91. (Cited on p. 257)

[181] T. FUKUSHIMA, *Precise and fast computation of Lambert W-functions without transcendental function evaluations*, Comput. Appl. Math., 244 (2013), pp. 77–89. (Cited on pp. 143, 245)

[182] W. GANDER AND D. GRUNTZ, *The billiard problem*, Maple Tech. J., 7 (1992), pp. 44–49. (Cited on p. 239)

[183] J. GARLOFF AND A. P. SMITH, *Investigation of a subdivision based algorithm for solving systems of polynomial equations*, Nonlinear Anal., 47 (2001), pp. 167–178. (Cited on p. 393)

[184] W. GAUTSCHI, *On the condition number of algebraic equations*, Numer. Math., 21 (1973), pp. 405–424. (Cited on pp. 64, 180)

[185] ———, *The condition of polynomials in power form*, Math. Comput., 33 (1979), pp. 343–352. (Cited on pp. 26, 183, 379)

[186] ———, *The Lambert W-functions and some of their integrals: A case study of high-precision computation*, Numer. Algorithms, 57 (2011), pp. 27–34. (Cited on p. 143)

[187] K. O. GEDDES, *Chebyshev nodes for interpolation on a class of ellipses*, in Theory of Approximations with Applications, A. G. Law and B. N. Sahney, eds., Academic Press, New York, 1976. (Cited on pp. 39, 113)

[188] K. O. GEDDES, *Block structure in the Chebyshev–Padé table*, SIAM J. Numer. Anal., 18 (1981), pp. 844–861. (Cited on p. 45)

[189] K. O. GEDDES AND J. C. MASON, *Polynomial approximation by projections on the unit circle*, SIAM J. Numer. Anal., 12 (1975), pp. 111–120. (Cited on pp. 39, 96)

[190] R. GELARO, *Alexander-Yorke continuation: Numerically finding all the stationary solutions in a spectral model*, in Nonlinear Hydrodynamic Modelling: A Mathematical Introduction, H. N. Shirer, ed., no. 271 in Lecture Notes in Physics, Springer-Verlag, New York, 1987, pp. 225–263. (Cited on p. 168)

[191] W. M. GENTLEMAN, *Implementing Clenshaw-Curtis quadrature*, Commun. ACM, 15 (1972), pp. 353–355. (Cited on pp. 416, 417)

[192] ——, *Implementing Clenshaw-Curtis quadrature: I. Methodology and experience*, Commun. ACM, 15 (1972), pp. 337–342. (Cited on pp. 416, 417)

[193] ——, *Implementing Clenshaw-Curtis quadrature: II. Computing the cosine transformation*, Commun. ACM, 15 (1972), pp. 343–346. (Cited on p. 417)

[194] T. C. GERMANN AND S. KAIS, *Large order dimensional perturbation theory for complex energy eigenvalues*, J. Chem. Phys., 99 (1993), pp. 7739–7747. (Cited on p. 258)

[195] A. GIL, J. SEGURA, AND N. TEMME, *Numerical Methods for Special Functions*, SIAM, Philadelphia, 2007. (Cited on pp. 8, 244, 256, 323, 326)

[196] R. GLOWINSKI, H. B. KELLER, AND L. REINHART, *Continuation-conjugate gradient methods for the least squares solution of nonlinear boundary value problems*, SIAM J. Sci. Stat. Comput., 6 (1985), pp. 793–832. (Cited on p. 151)

[197] S. GOEDECKER, *Remarks on algorithms to find roots of polynomials*, SIAM J. Sci. Comput., 15 (1994), pp. 1059–1063. (Cited on p. 175)

[198] E. M. GOL'BERG AND V. N. MALOZEMOV, *Estimates for the zeros of certain polynomials*, Vestmol Leningrad Univ. Math., 6 (1979), pp. 127–135. Translated from *Vestnik Leningrad Univ. Mat., Mekh., Astr.*, no. 7 (1973), 18–24. (Cited on pp. 26, 91, 403)

[199] P. GONNET, S. GÜTTEL, AND L. N. TREFETHEN, *Robust Padé approximation via SVD*, SIAM Rev., 55 (2013), pp. 101–117. (Cited on pp. 46, 257, 258)

[200] I. J. GOOD, *The colleague matrix, a Chebyshev analogue of the companion matrix*, Quart. J. Math., 12 (1961), pp. 61–68. (Cited on pp. 26, 403)

[201] J. A. GRANT AND A. GHIATIS, *Determination of the zeros of a linear combination of Chebyshev polynomials*, IMA J. Numer. Anal., 3 (1983), pp. 193–206. (Cited on p. 27)

[202] J. A. GRANT AND A. A. RAHMAN, *Determination of the zeros of a linear combination of generalized polynomials*, J. Comput. Appl. Math., 42 (1992), pp. 269–278. (Cited on p. 27)

[203] I. GRATTAN-GUINNESS, *The Norton History of the Mathematical Sciences: The Rainbow of Mathematics*, W. W. Norton, New York, 1997. (Cited on pp. 222, 223)

[204] J. GULLBERG, *Mathematics from the Birth of Numbers*, W. W. Norton, New York, 1997. 1000 pp. (Cited on p. 237)

[205] T. L. HANKINS, *Blood, dirt and nomograms: A particular history of graphs*, Isis, 90 (1999), pp. 50–80. (Cited on p. 222)

[206] J. HE, *Homotopy perturbation method: A new nonlinear analytical technique*, Appl. Math. Comput., 135 (2003), pp. 73–79. (Cited on p. 263)

[207] M. E. HENDERSON AND H. B. KELLER, *Complex bifurcation from real paths*, SIAM J. Appl. Math., 50 (1990), pp. 460–482. (Cited on p. 169)

[208] R. D. HENDERSON, *Dynamic refinement algorithms for spectral element methods*, Comput. Methods Appl. Mech. Engrg., 175 (1999), pp. 395–412. (Cited on p. 57)

[209] N. J. HIGHAM, *The numerical stability of barycentric Lagrange interpolation*, IMA J. Numer. Anal., 24 (2004), pp. 547–556. (Cited on p. 75)

[210] B. J. HOENDERS AND C. H. SLUMP, *On the determination of the number and multiplicity of zeros of a function*, Computing, 47 (1992), pp. 323–336. (Cited on p. 348)

[211] ——, *On the exact number of zeros of a set of equations*, Computing, 30 (1992), pp. 137–147. (Cited on p. 348)

[212] M. H. HOLMES, *Introduction to Perturbation Methods*, Springer-Verlag, New York, 1995. (Cited on p. 285)

[213] A. S. HOUSEHOLDER, *The Numerical Treatment of a Single Nonlinear Equation*, McGraw-Hill, New York, 1970. (Cited on pp. 20, 345)

[214] I. ICHIM AND I. MOLNAR, *A Bairstow's type method for trigonometric polynomials*, Numer. Math., 67 (1994), pp. 251–259. (Cited on pp. 92, 93)

[215] S. J. JACOBS, *A variable order pseudospectral method for two-point boundary value problems*, J. Comput. Phys., 88 (1990), pp. 169–182. (Cited on pp. 25, 52, 55, 57)

[216] M. A. JENKINS AND J. F. TRAUB, *A three-stage variable-shift iteration for polynomial zeros and its relation to generalized Rayleigh iteration*, Numer. Math., 14 (1970), pp. 252–263. (Cited on pp. 4, 176)

[217] D. S. JONES, *Introduction to Asymptotics: A Treatment Using Nonstandard Analysis*, World Scientific, Singapore, 1997. (Cited on p. 285)

[218] C. JORDAN, *Traité des substitutions et des équations algébriques*, Gauthiers-Villars, Paris, 1870. (Cited on pp. 176, 188)

[219] D. J. KAVVADIAS, F. S. MAKRI, AND M. N. VRAHATIS, *Locating and computing arbitrarily distributed zeros*, SIAM J. Sci. Comput., 21 (1999), pp. 954–969. (Cited on pp. 343, 348, 383)

[220] D. J. KAVVADIAS AND M. N. VRAHATIS, *Locating and computing all the simple roots and extrema of a function*, SIAM J. Sci. Comput., 17 (1996), pp. 1232–1248. (Cited on pp. 348, 383)

[221] D. KAYA AND S. M. EL-SAYED, *Adomian's decomposition method applied to systems of nonlinear algebraic equations*, Appl. Math. Comput., 154 (2004), pp. 487–493. (Cited on p. 275)

[222] H. B. KELLER, *Numerical solution of bifurcation and nonlinear eigenvalue problems*, in Applications of Bifurcation Theory, P. Rabinowitz, ed., Academic Press, New York, 1977, pp. 359–384. (Cited on p. 151)

[223] ——, *Numerical Methods for Two-Point Boundary-Value Problems*, Dover, New York, 1992. (Cited on pp. 151, 169, 170, 172)

[224] C. T. KELLEY, *Solving Nonlinear Equations with Newton's Method*, vol. 1 of Fundamentals of Algorithms, SIAM, Philadelphia, 2003. (Cited on pp. 119, 123, 149)

[225] R. B. KING, *Beyond the Quartic Equation*, Birkhäuser, Boston, 1996. (Cited on pp. 190, 191)

[226] R. B. KING AND E. R. CANFIELD, *An algebraic algorithm for calculating the roots of a general quintic equation from its coefficients*, J. Math. Phys., 32 (1991), pp. 823—825. (Cited on p. 190)

[227] ——, *Icosahedral symmetry and the quintic equation*, Comput. Math. Appl., 24 (1992), pp. 13—28. (Cited on p. 190)

[228] R. KNOX, G. MARLIN, AND J. SWAN, *The Quotable Knox*, Ignatius Press, Chicago, 1996. (Cited on p. 193)

[229] V. KOWALENKO, M. L. GLASSER, T. TAUCHER, AND N. E. FRANKEL, *Generalised Euler-Jacobi Inversion Formula and Asymptotics beyond All Orders*, vol. 214 of London Mathematical Society Lecture Note Series, Cambridge University Press, Cambridge, U.K., 1995. (Cited on p. 285)

[230] E. E. KRAMER, *The Nature and Growth of Modern Mathematics*, Princeton University Press, Princeton, NJ, 1981. (Cited on p. 195)

[231] P. KRAVANJA AND M. VAN BAREL, *A derivative-free algorithm for computing zeros of analytic functions*, Computing, 63 (2000), pp. 69–91. (Cited on p. 96)

[232] ——, *Computing the Zeros of Analytic Functions*, vol. 1727 of Lecture Notes in Mathematics, Springer, Heidelberg, 2000. 112 pp. (Cited on p. 96)

[233] P. KRAVANJA, M. VAN BAREL, O. RAGOS, M. N. VRAHATIS, AND F. A. ZAFIROPOULOS, *ZEAL: A mathematical software package for computing zeros of analytic functions*, Comput. Phys. Commun., 124 (2000), pp. 212–232. (Cited on pp. 95, 96, 101)

[234] J. H. LACASCE AND J. PEDLOSKY, *Baroclinic Rossby waves in irregular basins*, J. Phys. Oceanography, 12 (2002), pp. 2828–2847. (Cited on p. 112)

[235] C. LANCZOS, *Trigonometric interpolation of empirical and analytical functions*, J. Math. Phys., 17 (1938), pp. 123–199. Reprinted in *Cornelius Lanczos: Collected Papers with Commentaries*, W. R. Davis, eds., North Carolina State University, Raleigh, NC (1997), vol. 3, pp. 3-221–3-297. (Cited on p. 21)

[236] C. LANCZOS, *Applied Analysis*, Prentice-Hall, Englewood Cliffs, NJ, 1956. (Cited on p. 311)

[237] F. M. LARKIN, *A combined graphical and iterative approach to the problem of finding zeros of functions in the complex plane*, Comput. J., 7 (1964), pp. 212–219. (Cited on p. 112)

[238] P. W. LAWRENCE, R. M. CORLESS, AND D. J. JEFFREY, *Algorithm 917: Complex double-precision evaluation of the Wright omega function*, ACM Trans. Math. Software, 38 (2012), article no. 20. (Cited on p. 245)

[239] J. C. LE GUILLOU AND J. ZINN-JUSTIN, eds., *Large-Order Behaviour of Perturbation Theory*, North-Holland, Amsterdam, 1990. (Cited on p. 285)

[240] R. A. LEATHERS AND N. J. MCCORMICK, *Closed-form solutions for transcendental equations of heat transfer*, ASME J. Heat Transfer, 118 (1996), pp. 970–973. (Cited on p. 247)

[241] V. I. LEBEDEV, *On formulae for roots of cubic equations*, Soviet J. Numer. Anal. Math. Modelling, 6 (1991), pp. 315–324. (Cited on p. 203)

[242] L. LEWIN, *Dilogarithms and Associated Functions*, Macdonald, London, 1958. (Cited on p. 236)

[243] T.-Y. LI, *Solving polynomial systems*, Math. Intelligencer, 9 (1987), pp. 33–39. (Cited on pp. 151, 168)

[244] E. LOMBARDI, *Oscillatory Integrals and Phenomena beyond All Algebraic Orders*, vol. 1741 of Lecture Notes in Mathematics, Springer-Verlag, New York, 2000. (Cited on p. 285)

[245] R. LUCK AND J. W. STEVENS, *Explicit solutions for transcendental equations*, SIAM Rev., 44 (2002), pp. 227–233. (Cited on p. 111)

[246] Y. L. LUKE, W. FAIR, G. COOMBS, AND R. MORAN, *On a constant in the theory of trigonometric series*, Math. Comput., 19 (1965), pp. 501–502. (Cited on pp. 70, 87)

[247] I. V. MAKRELOV AND H. I. SEMERDZHIEV, *Methods for the simultaneous determination of all zeros of algebraic, trigonometric and exponential equations*, U. S. S. R. Comput. Maths. Math. Phys., 24 (1984), pp. 99–105. (Cited on pp. 91, 92)

[248] I. V. MAKRELOV AND K. I. SEMERDZHIEV, *On the convergence of two methods for the simultaneous find of all roots of exponential equations*, IMA J. Numer. Anal., 5 (1985), pp. 191–200. (Cited on pp. 91, 92)

[249] D. MANOCHA AND J. DEMMEL, *Algorithms for intersecting parametric and algebraic curves, I: Simple intersections*, ACM Trans. Graph., 13 (1994), pp. 73–100. (Cited on p. 375)

[250] ——, *Algorithms for intersecting parametric and algebraic curves, II: Multiple intersections*, Graph. Models Image Process., 57 (1996), pp. 81–100. (Cited on p. 375)

[251] J. MAROULAS AND S. BARNETT, *Further results on the qualitative theory of generalized polynomials*, IMA J. Appl Math., 23 (1978), pp. 33–42. (Cited on p. 27)

[252] ——, *Polynomials with respect to a general basis. I. Theory*, J. Math. Anal. Appl., 72 (1978), pp. 177–194. (Cited on pp. 27, 375)

[253] ——, *Some new results on the qualitative theory of generalized polynomials*, IMA J. Appl. Math., 22 (1978), pp. 53–70. (Cited on p. 27)

[254] ——, *Greatest common divisor of generalized polynomials and polynomial matrices*, Linear Algebra Appl., 22 (1979), pp. 195–210. (Cited on p. 379)

[255] ——, *Polynomials with respect to a general basis. II. Applications*, J. Math. Anal. Appl., 72 (1979), pp. 599–614. (Cited on p. 27)

[256] S. L. MARSHALL, *On the analytical summation of Fourier series and its relation to the asymptotic behaviour of Fourier transforms*, J. Phys. A, 31 (1998), pp. 9957–9973. (Cited on p. 424)

[257] V. P. MASLOV, *The Complex WKB Method for Nonlinear Equations I: Linear Theory*, Birkhäuser, Boston, 1994. (Cited on p. 285)

[258] J. C. MASON AND D. C. HANDSCOMB, *Chebyshev Polynomials*, Chapman and Hall/CRC Press, Boca Raton, FL, 2003. (Cited on pp. 55, 81)

[259] C. MAVRIPLIS, *Adaptive mesh strategies for the spectral element method*, in Analysis, Algorithms and Applications of Spectral and High Order Methods for Partial Differential Equations, C. Bernardi and Y. Maday, eds., Selected Papers from the International Conference on Spectral and High Order Methods (ICOSAHOM '92), Le Corum, Montpellier, France, 22–26 June 1992, North-Holland, Amsterdam, 1994, pp. 77–86. Also in Comput. Meths. Appl. Mech. Engrg., vol. 116. (Cited on pp. 55, 57)

[260] J. MAYNARD, *Resonant ultrasound spectroscopy*, Physics Today, 49 (1996), pp. 26–31. (Cited on p. 325)

[261] J. H. MCCABE, *A formal extension of the Padé table to include two point Padé quotients*, J. Inst. Math. Appl., 15 (1975), pp. 363–372. (Cited on p. 257)

[262] J. H. MCCABE AND J. A. MURPHY, *Continued fraction expansions which correspond to power series expansions at two points*, J. Inst. Math. Appl., 17 (1976), pp. 233–247. (Cited on p. 257)

[263] J. E. MCCUNE, *Exact inversion of dispersion relations*, Phys. Fluids, 9 (1966), pp. 2082–2084. (Cited on p. 96)

[264] R. J. Y. McLeod and M. L. Baart, *Geometry and Interpolation of Curves and Surfaces*, Cambridge University Press, New York, 1998. (Cited on pp. 221, 319)

[265] J. M. McNamee, *A bibliography on roots of polynomials*, J. Comput. Appl. Math., 47 (1993), pp. 391–394. (Cited on p. 176)

[266] ——, *A supplementary bibliography on roots of polynomials*, J. Comput. Appl. Math., 78 (1993), p. 1. (Cited on p. 176)

[267] ——, *An updated supplementary bibliography on roots of polynomials*, J. Comput. Appl. Math., 110 (1999), pp. 305–306. (Cited on p. 176)

[268] ——, *A 2002 update of the supplementary bibliography on roots of polynomials*, J. Comput. Appl. Math., 142 (2002), pp. 433–434. (Cited on p. 176)

[269] ——, *Numerical Methods for Roots of Polynomials—Part* I, Elsevier, New York, 2007. (Cited on pp. 87, 176)

[270] J. M. McNamee and V. Y. Pan, *Efficient polynomial root-refiners: A survey and new record efficiency estimates*, Comput. Math. Appl., 63 (2012), pp. 239–254. (Cited on pp. 96, 112, 176)

[271] ——, *Numerical Methods for Roots of Polynomials—Part* II, Elsevier, New York, 2013. (Cited on pp. 87, 176, 203)

[272] H. W. Menard, *The Ocean of Truth*, Princeton University Press, Princeton, NJ, 1986. (Cited on p. 365)

[273] P. D. Miller, *Applied Asymptotic Analysis*, American Mathematical Society, Providence, RI, 2006. (Cited on pp. 42, 277, 306)

[274] B. Montagnes, *The Story of Father Marie-Joseph Lagrange*, Paulist Press, New York, 2006. (Cited on p. 411)

[275] R. E. Moore, R. B. Kearfott, and M. J. Cloud, *Introduction to Interval Analysis*, SIAM, Philadelphia, 2009. (Cited on p. 383)

[276] A. Morgan, *Computing all solutions to polynomial systems using homotopy continuation*, Appl. Math. Comput., 24 (1987), pp. 115–138. (Cited on pp. 169, 394)

[277] A. P. Morgan, *A method for computing all solutions to systems of polynomial equations*, ACM Trans. Math. Software, 9 (1983), pp. 1–17. (Cited on p. 169)

[278] ——, *Solving Polynomial Systems Using Continuation for Scientific and Engineering Problems*, Prentice-Hall, Englewood Cliffs, NJ, 1987. (Cited on pp. 80, 168, 239, 394)

[279] A. Morgan, *Solving Polynomial Systems Using Continuation for Scientific and Engineering Problems*, vol. 57 of Classics in Applied Mathematics, SIAM, Philadelphia, 2009. (Cited on pp. 158, 169, 371)

[280] F. R. Moulton, *An Introduction to Celestial Mechanics*, 2nd edition Macmillan, New York, 1958. (Cited on p. 127)

[281] B. Mourrain and J. P. Pavone, *Subdivision methods for solving polynomial equations*, J. Symbolic Comput., 44 (2009), pp. 292–306. (Cited on p. 393)

[282] N. I. Muskhelishvili, *Singular Integral Equations*, Noordhoff, Groningen, 1953. (Cited on p. 247)

[283] Y. Nakatsukasa, V. Noferini, and A. Townsend, *Computing the common zeros of two bivariate functions via Bezout resultants*, Numer. Math., to appear. (Cited on pp. 379, 380, 394)

[284] F. W. J. OLVER, D. W. LOZIER, R. F. BOISVERT, AND C. W. CLARK, eds., *NIST Handbook of Mathematical Functions*, Cambridge University Press, New York, 2010. (Cited on pp. 7, 40, 130, 140, 274, 324, 325)

[285] R. PACHÓN, R. B. PLATTE, AND L. N. TREFETHEN, *Piecewise-smooth Chebfuns*, IMA J. Numer. Anal., 30 (2010), pp. 898–916. (Cited on p. 22)

[286] V. Y. PAN, *Solving a polynomial equation: Some history and recent progress*, SIAM Rev., 39 (1997), pp. 187–220. (Cited on pp. 96, 112)

[287] ——, *Approximating complex polynomial zeros: Modified Weyl's quadtree construction and improved Newton's iteration*, J. Complexity, 16 (2000), pp. 213–264. (Cited on pp. 96, 112)

[288] R. B. PARIS AND D. KAMINSKI, *Asymptotics and Mellin-Barnes Integrals*, Cambridge University Press, Cambridge, U.K., 2001. (Cited on p. 285)

[289] R. B. PARIS AND A. D. WOOD, *Stokes phenomenon demystified*, IMA Bulletin, 31 (1995), pp. 21–28. (Cited on p. 285)

[290] W. S. PATERSON, *On preferring iteration in a transformed variable to the method of successive substitutions*, Chem. Engrg. Sci., 41 (1986), pp. 601–602. (Cited on pp. 262, 264)

[291] H. PEITGEN AND P. RICHTER, *The Beauty of Fractals*, Springer-Verlag, Berlin, 1986. (Cited on p. 132)

[292] H.-O. PEITGEN, M. PRÜFER, AND K. SCHMITT, *Global aspects of the continuous and discrete Newton method: A case study*, Acta Appl. Math., 13 (1988), pp. 123–202. (Cited on p. 132)

[293] G. PETERS AND J. H. WILKINSON, *The $Ax = \lambda Bx$ and generalized eigenproblem*, SIAM J. Numer. Anal., 7 (1970), pp. 479–492. (Cited on pp. 374, 375)

[294] W. PFEIFFER, *A graphical method for finding complex roots and its application to plasma physics problems*, J. Comput. Phys., 33 (1979), pp. 397–404. (Cited on p. 112)

[295] G. PICKETT AND Y. MILLEV, *On the analytic inversion of functions, solution of transcendental equations and infinite self-mappings*, J. Phys. A, 35 (2002), pp. 4485–4494. (Cited on pp. 244, 247)

[296] G. W. PLATZMAN, *An exact integral of complete spectral equations for unsteady one-dimensional flow*, Tellus, 16 (1964), pp. 422–431. Corrigendum, 17, 150 (1965). (Cited on p. 7)

[297] W. H. PRESS, B. H. FLANNERY, S. A. TEUKOLSKY, AND W. T. VETTERLING, *Numerical Recipes: The Art of Scientific Computing*, Cambridge University Press, New York, 1986. (Cited on pp. 176, 200)

[298] W. C. RHEINBOLDT, *Numerical Analysis of Parameterized Nonlinear Equations*, vol. 7 of University of Arkansas Lecture Notes in the Mathematical Sciences, John Wiley, New York, 1986. (Cited on pp. 149, 170)

[299] ——, *Numerical continuation methods: A perspective*, J. Comput. Appl. Math., 124 (2000), pp. 229–244. (Cited on p. 170)

[300] A. RUHE, *Algorithms for the nonlinear eigenvalue problem*, SIAM J. Numer. Anal., 10 (1973), pp. 674–689. (Cited on pp. 8, 61, 374)

[301] S. M. RUMP, *Verification methods: Rigorous results using floating-point arithmetic*, Acta Numer., 19 (2010), pp. 237–449. (Cited on p. 383)

[302] D. G. SAARI AND J. B. URENKO, *Newton's method, circle maps and chaotic motion*, Amer. Math. Monthly, 91 (1984), pp. 3–17. (Cited on p. 132)

[303] K. SABBAGH, *The Riemann Hypothesis*, Farrar, Straus, and Girouk, New York, 2003. (Cited on p. 343)

[304] T. SAKURAI, P. KRAVANJA, H. SUGIURA, AND M. VAN BAREL, *An error analysis of two related quadrature methods for computing zeros of analytic functions*, J. Comput. Appl. Math., 152 (2003), pp. 467–480. (Cited on pp. 96, 101)

[305] H. E. SALZER, *Complex zeros of the error function*, J. Franklin Inst., 260 (1955), pp. 209–211. (Cited on p. 273)

[306] ——, *Lagrangian interpolation at the Chebyshev points* $x_{n,\nu} \equiv \cos(\nu\pi/n), \nu = o(1)n$*; some unnoted advantages*, Comput. J., 15 (1972), pp. 156–159. (Cited on p. 75)

[307] P. T. SAUNDERS, *An Introduction to Catastrophe Theory*, Cambridge University Press, Cambridge, U.K., 1980. (Cited on p. 219)

[308] H. A. SCARTON, *Method of eigen-valleys*, J. Comput. Phys., 11 (1973), pp. 1–11. (Cited on p. 112)

[309] R. B. SCHNABEL AND P. D. FRANK, *Tensor methods for nonlinear equations*, SIAM J. Numer. Anal., 21 (1984), pp. 815–843. (Cited on p. 120)

[310] C. SCHULTZ, *Bezier clipping is quadratically convergent*, Comput. Aided Geom. Design, 26 (2009), pp. 61–74. (Cited on p. 383)

[311] A. SCHWEIKARD, *Trigonometric polynomials with simple roots*, Inform. Process. Lett., 39 (1991), pp. 231–236. (Cited on pp. 92, 240)

[312] A. SCHWEIKARD, *Real zero isolation for trigonometric polynomials*, ACM Trans. Math. Software, 18 (1992), pp. 350–359. (Cited on pp. 92, 240)

[313] T. W. SEDERBERG, *Algorithm for curve intersection*, Comput. Aided Design, 21 (1989), pp. 547–554. (Cited on p. 372)

[314] T. W. SEDERBERG AND T. NISHITA, *Curve intersection using Bezier clipping*, Comput. Aided Design, 22 (1990), pp. 538–549. (Cited on p. 383)

[315] H. SEGUR, S. TANVEER, AND H. LEVINE, eds., *Asymptotics beyond All Orders*, Plenum, New York, 1991. (Cited on p. 285)

[316] A. V. SERGEEV, *Summation of the eigenvalue perturbation series by multivalued Padé approximants—application to resonance problems and double wells*, J. Phys. A.: Math. Gen., 28 (1995), pp. 4157–4162. (Cited on p. 258)

[317] A. V. SERGEEV AND D. Z. GOODSON, *Summation of asymptotic expansions of multiple-valued functions using algebraic approximants: Application to anharmonic oscillators*, J. Phys. A: Math. Gen., 31 (1998), pp. 4301–4317. (Cited on p. 258)

[318] R. SEYDEL, *Practical Bifurcation and Stability Analysis: From Equilibrium to Chaos*, 2nd edition, Springer-Verlag, Heidelberg, 1994. (Cited on p. 170)

[319] ——, *From Equilibrium to Chaos: Practical Bifurcation and Stability Analysis*, 3rd edition, Elsevier, Amsterdam, 2010. (Cited on pp. 169, 172)

[320] R. E. SHAFER, *On quadratic approximation*, SIAM J. Numer. Anal., 11 (1974), pp. 447–460. (Cited on pp. 46, 259, 318)

[321] J. SHEN AND L. L. WANG, *Sparse spectral approximations of high-dimensional problems based on hyperbolic cross*, SIAM J. Numer. Anal., 48 (2010), pp. 1087–1109. (Cited on pp. 13, 398)

[322] E. C. SHERBROOKE AND N. M. PATRIKALAKIS, *Computation of the solutions of nonlinear polynomial systems*, Comput. Aided Geom. Design, 10 (1993), pp. 379–405. (Cited on p. 393)

[323] C. E. SIEWERT, *An exact analytical solution of an elementary critical condition*, Nuclear Sci. Engrg., 51 (1973), p. 78. (Cited on p. 247)

[324] ——, *On computing eigenvalues in radiative transfer*, J. Math. Phys., 21 (1980), pp. 2468–2470. (Cited on p. 247)

[325] ——, *An exact expression for the Wien displacement constant*, J. Quant. Spectroscopy Radiative Transfer, 26 (1981), p. 467. (Cited on p. 247)

[326] C. E. SIEWERT AND E. E. BURNISTON, *An exact analytical solution of Kepler's equation*, Celestial Mech., 6 (1972), pp. 294–304. (Cited on p. 247)

[327] ——, *Exact analytical solutions of* $z\exp(z)=\alpha$, J. Math. Anal. Appl., 43 (1973), pp. 626–632. (Cited on pp. 245, 247)

[328] ——, *Solutions of the equation* $z\exp(z)=a(z+b)$, J. Math. Anal. Appl., 46 (1974), pp. 329–337. (Cited on p. 247)

[329] ——, *An exact analytical solution of* $x\coth x=\alpha x^2+1$, J. Comput. Appl Math., 2 (1976), pp. 19–21. (Cited on p. 114)

[330] C. E. SIEWERT AND C. J. ESSIG, *An exact solution of a molecular field equation in the theory of ferromagnetism*, Z. Angew. Math. Phys., 24 (1973), pp. 281–286. (Cited on p. 247)

[331] C. E. SIEWERT AND J. S. PHELPS, III, *On the solution of* $a\tan(\xi-k\pi)+\tanh\xi=0$, J. Comput. Appl. Math., 5 (1979), pp. 99–101. (Cited on p. 247)

[332] ——, *On solutions of a transcendental equation basic to the theory of vibrating plates*, SIAM J. Math. Anal., 10 (1979), pp. 105–111. (Cited on pp. 247, 248)

[333] I. H. SLOAN AND W. E. SMITH, *Product integration with the Clenshaw-Curtis quadrature scheme*, Numer. Math., 34 (1980), pp. 387–401. (Cited on pp. 55, 57)

[334] M. W. SMILEY AND C. CHUN, *An algorithm for finding all solutions of a nonlinear system*, J. Comput. Appl. Math., 137 (2001), pp. 293–315. (Cited on pp. 383, 393)

[335] W. K. SMITH, *Inverse Functions*, Macmillan, London, 1966. (Cited on p. 329)

[336] M. A. SNYDER, *Chebyshev Methods in Numerical Approximation*, Prentice-Hall, Englewood Cliffs, NJ, 1966. (Cited on p. 81)

[337] A. J. SOMMESE AND C. W. WAMPLER, II, *The Numerical Solution of Systems of Polynomials Arising in Engineering and Science*, World Scientific, Singapore, 2005. (Cited on pp. 80, 158, 239, 371, 394)

[338] W. SPECHT, *Die Lage der Nullstellen eines Polynoms* III, Math. Nach., 16 (1957), pp. 369–389. (Cited on pp. 26, 91, 403)

[339] ——, *Die Lage der Nullstellen eines Polynoms* IV, Math. Nach., 21 (1960), pp. 201–22. (Cited on pp. 26, 91, 403)

[340] F. STENGER, *Summary of sinc numerical methods*, J. Comput. Appl. Math., 121 (2000), pp. 379–420. (Cited on p. 34)

[341] F. STENGER, *Handbook of Sinc Numerical Methods*, Chapman & Hall/CRC, Boca Raton, FL, 2010. 482 pp. (Cited on p. 34)

[342] B. Y. STERNIN AND V. E. SHATALOV, *Borel-Laplace Transform and Asymptotic Theory: Introduction to Resurgent Analysis*, CRC Press, New York, 1996. (Cited on p. 285)

[343] H. J. STETTER, *Condition analysis of overdetermined algebraic problems*, in Computer Algebra in Scientific Computing (CASC 2000), V. G. Ganzha, E. W. Mayr, and E. V. Vorozhtsov, eds., Springer-Verlag, New York, 2000, pp. 345–356. (Cited on p. 179)

[344] ——, *Numerical Polynomial Algebra*, SIAM, Philadelphia, 2004. (Cited on pp. 26, 91, 403)

[345] B. STURMFELS, *Polynomial equations and convex polytopes*, Amer. Math. Monthly, 105 (1998), pp. 907–922. (Cited on pp. 185, 189)

[346] ——, *Solving algebraic equations in terms of $\mathcal{A}$-hypergeometric series*, Discrete Math., 210 (2000), pp. 171–181. (Cited on pp. 188, 189, 251)

[347] ——, *Solving Systems of Polynomial Equations*, vol. 97 of CBMS Regional Conference Series, American Mathematical Society, Providence, RI, 2002. (Cited on pp. 185, 188, 189, 251)

[348] A. A. SUVERNEV AND D. Z. GOODSON, *Perturbation theory for coupled anharmonic oscillators*, J. Chem. Phys., 106 (1997), pp. 2681–2684. (Cited on p. 258)

[349] J. TODD, *Special polynomials in numerical analysis*, in Numerical Approximation, R. E. Langer, ed., University of Wisconsin Press, Madison, WI, 1959, pp. 423–446. (Cited on p. 351)

[350] A. TOWNSEND, Y. NOFERINI, AND Y. NAKATSUKASA, *Vector spaces of linearizations for matrix polynomials: A bivariate polynomial approach*, manuscript. (Cited on p. 379)

[351] A. TOWNSEND AND L. N. TREFETHEN, *An extension of Chebfun to two dimensions*, SIAM J. Sci. Comput., 35 (2013), pp. C495–C518. (Cited on pp. 380, 394)

[352] L. N. TREFETHEN, *Spectral Methods in MATLAB*, SIAM, Philadelphia, 2000. (Cited on pp. 22, 24)

[353] ——, *Approximation Theory and Approximation Practice*, SIAM, Philadelphia, 2012. (Cited on pp. 5, 6)

[354] L. N. TREFETHEN ET AL., *Chebfun Version* 4.0. `http://www.maths.ox.ac.uk/chebfun/`, 2011. The Chebfun Development Team. (Cited on pp. 6, 20, 22)

[355] L. N. TREFETHEN AND R. S. SCHREIBER, *Average-case stability of Gaussian elimination*, SIAM J. Matrix Anal. Appl., 11 (1990), pp. 335–360. (Cited on p. 61)

[356] H. UMEMURA, *Resolution of algebraic equations by theta constants*, in Tata Lectures on Theta, II, D. Mumford, ed., Birkhäuser, Boston, 1984, pp. 3.261–3.272. (Cited on pp. 176, 188)

[357] M. VAN DYKE, *Perturbation Methods in Fluid Mechanics*, 2nd edition, Parabolic Press, Stanford, CA, 1975. (Cited on pp. 283, 284)

[358] V. M. VAĬNBERG, V. D. MUR, V. S. POPOV, AND A. V. SERGEEV, *Strong-field Stark effect*, JETP Lett., 44 (1986), pp. 9–13. (Cited on p. 258)

[359] D. VEBERIC, *Lambert W function for applications in physics*, Comput. Phys. Commun., 183 (2012), pp. 2622–2628. (Cited on p. 244)

[360] J. VIGNES, *New methods for evaluating mathematical computations*, Math. Comput. Simul., 20 (1999), pp. 227–249. (Cited on pp. 199, 203)

[361] C. WAMPLER AND A. SOMMESE, *Numerical algebraic geometry and algebraic kinematics*, Acta Numer., 20 (2011), pp. 469–567. (Cited on p. 239)

[362] M. J. WARD, W. D. HENSHAW, AND J. B. KELLER, *Summing logarithmic expansions for singularly perturbed eigenvalue problems*, SIAM J. Appl. Math., 53 (1993), pp. 799–828. (Cited on p. 297)

[363] E. WASSERSTROM, *Root finding of polynomials as an initial value problem*, J. Comput. Phys., 8 (1971), pp. 304–308. (Cited on p. 163)

[364] ——, *Numerical solutions by the continuation method*, SIAM Rev., 15 (1973), pp. 89–119. (Cited on pp. 163, 169)

[365] J. A. C. WEIDEMAN AND S. C. REDDY, *A MATLAB differentiation matrix suite*, ACM Trans. Math. Software, 26 (2000), pp. 465–519. (Cited on p. 326)

[366] P. WEIDNER, *The Durand-Kerner method for trigonometric and exponential polynomials*, Comput., 40 (1988), pp. 175–179. (Cited on pp. 91, 92, 407)

[367] E. W. WEISSTEIN, *The CRC Concise Encyclopedia of Mathematics*, 2nd edition, CRC Press, Boca Raton, FL, 2002. (Cited on p. 189)

[368] H. WEYL, *Randbernergungen zu Hauptproblemen der Mathematik, II, Fundamentals atz der Algebra and Grundlagen der Mathematik*, Math. Z., 20 (1924), pp. 131–151. (Cited on pp. 96, 112)

[369] E. WHITTAKER AND G. ROBINSON, *The Calculus of Observations*, 4th edition, Dover, Mineola, NY, 1967. (Cited on pp. 186, 427, 429)

[370] E. T. WHITTAKER, *A formula for the solution of algebraic and transcendental equations*, Proc. Edin. Math. Soc., 36 (1918), pp. 103–106. (Cited on p. 186)

[371] D. C. WILCOX, *Perturbation Methods in the Computer Age*, DCW Industries, La Cañada, CA, 1995. (Cited on p. 249)

[372] J. H. WILKINSON, *Rounding Errors in Algebraic Processes*, Prentice-Hall, Englewood Cliffs, NJ, 1963. (Cited on p. 63)

[373] K. G. WILSON, *Critical phenomena in* 3.99 *dimensions*, Physica, 73 (1974), pp. 119–128. (Cited on p. 263)

[374] K. G. WILSON AND M. E. FISHER, *Critical exponents in* 3.99 *dimensions*, Phys. Rev. Lett., 28 (1972), pp. 240–243. (Cited on p. 263)

[375] J. R. WINKLER, *Condition numbers of a root*, Appl. Numer. Math., 38 (2001), pp. 275–285. (Cited on p. 180)

[376] ——, *A companion matrix resultant for Bernstein polynomials*, Linear Algebra Appl., 362 (2003), pp. 153–175. (Cited on p. 403)

[377] ——, *The transformation of the companion matrix resultant between the power and Bernstein polynomial bases*, Appl. Numer. Math., 48 (2004), pp. 113–126. (Cited on p. 403)

[378] ——, *A unified approach to resultant matrices for Bernstein basis polynomials*, Comput. Aided Geom. Design, 25 (2008), pp. 529–541. (Cited on p. 403)

[379] E. WITTEN, *Quarks, atoms, and the* $1/n$ *expansion*, Phys. Today, 33 (1980), pp. 38–43. (Cited on p. 263)

[380] E. M. WRIGHT, *Solution of the equation* $z \exp(z) = \alpha$, Bull. Am. Math. Soc., 65 (1959), pp. 89–93. (Cited on p. 244)

[381] J.-J. XU, *Interfacial Wave Theory of Pattern Formation: Selection of Dendritic Growth and Viscous Fingering in Hele-Shaw Flow*, Springer-Verlag, New York, 1998. (Cited on p. 285)

[382] V. L. ZAGUSKIN, *Solution of Algebraic and Transcendental Equations*, Pergamon Press, New York, 1961. Translated from the Russian by G. O. Harding. (Cited on p. 4)

[383] W. I. ZANGWILL AND C. B. GARCIA, *Pathways to Solutions, Fixed Points, and Equilibria*, Computational Mathematics, Prentice-Hall, Englewood Cliffs, NJ, 1981. (Cited on pp. 162, 166)

[384] Z. ZENG, *A numerical elimination method for polynomial computations*, Theoret. Comput. Sci., 409 (2008), pp. 318–331. (Cited on pp. 379, 394)

[385] J. ZHANG, *Symbolic computation on complex polynomial solution of differential equations*, J. Symbolic Comput., 22 (1996), pp. 345–354. (Cited on p. 39)

Index